Praise for the first edition of The Rough Guide to Weather

"Brisk and authoritative for all weather consumers, whether sun seeker or tornado chaser."

New Scientist

"May in Greece should be warm, rather than hot, with temperatures in the mid-60s and just a slim chance of rain. How do I know? My bible on the shelf above my desk is *The Rough Guide to Weather*."

David Wickers, Good Housekeeping

" A terrific book ... lovely photographs and amazingly complete world weather pages."

Davis Instruments Weather Club

"Far outshines similar books ... I strongly recommend this book."

The Weather Doctor (website)

"It's packed – no, crammed – with everything from cloud types to rainbow formation to weather summaries for cities around the world."

Dayton News

"A book to dip into and enjoy, again and again."

Good Book Guide

"Excellent."

Adventure Travel

Credits

The Rough Guide to Weather

Editing: Peter Buckley
Layout: Peter Buckley & Andrew Clare
Picture research: Robert Henson
Proofreading: Amanda Jones
Production: Aimee Hampson & Katherine Owers
Cartography: Maxine Repath, Katie Lloyd-Jones & Ed Wright

Rough Guides Reference

Series editor: Mark Ellingham
Editors: Peter Buckley, Duncan Clark, Matthew Milton, Ruth Tidball, Tracy Hopkins, Joe Staines, Sean Mahoney
Director: Andrew Lockett

Cover images

Front cover: courtesy of Alamy – lightning strikes in Arizona
Inside front cover: © Kenneth D Langford – mammatus descending from a thunderstorm
Back cover: © Terry Robinson/NCAR – strong vertical contrasts in wind and air mass shaping a Kelvin-Helmholtz pattern

Publishing Information

This second edition published April 2007 by
Rough Guides Ltd, 80 Strand, London WC2R 0RL
345 Hudson St, 4th Floor, New York 10014, USA
Email: mail@roughguides.com

Distributed by the Penguin Group
Penguin Books Ltd, 80 Strand, London WC2R 0RL
Penguin Putnam, Inc., 375 Hudson Street, NY 10014, USA
Penguin Group (Australia), 250 Camberwell Road, Camberwell, Victoria 3124, Australia
Penguin Books Canada Ltd, 90 Eglinton Avenue East, Toronto, Ontario, Canada M4P 2YE
Penguin Group (New Zealand), 67 Apollo Drive, Mairongi Bay, Auckland 1310, New Zealand

Printed in Italy by LegoPrint S.p.A

Typeset in DIN, Myriad and Minion to an original design by Duncan Clark

432 pages; includes index

A catalogue record for this book is available from the British Library

ISBN 13: 9-781-84353-712-0
ISBN 10: 1-84353-712-5

1 3 5 7 9 8 6 4 2

THE ROUGH GUIDE to

Weather

by

Robert Henson

ROUGH GUIDES

www.roughguides.com

Acknowledgements

My deep appreciation goes to the many scientists who gave of their time and expertise to review sections of this book, and to all of the researchers who took time to meet with me on visits to the UK Met. Office, the World Meteorological Organization (WMO), Australia's Commonwealth Scientific & Industrial Research Organisation (CSIRO) and the Australian Bureau of Meteorology Research Centre (BMRC), as well as to the people who arranged these visits: Andy Yeatman (UK Met. Office), Haleh Kootval (WMO), Paul Holper (CSIRO) and Paul Della-Marta and Mark Jenkin (BMRC). Reviewers (generally listed below with their 2002 affiliations) included: **BMRC**: Andrew Watkins, Blair Trewin, Diana Greenslade. **National Center for Atmospheric Research (NCAR)**: Caspar Ammann, Ben Bernstein, Gordon Bonan, Paul Charbonneau, John Firor, David Johnson, Sue Schauffler, Kevin Trenberth. **US National Oceanic and Atmospheric Administration (NOAA)**: Sim Aberson, Hugh Cobb, John Cortinas, Ben Felzer, Michael Fortune, JoAnn Joselyn, Robert Jubach, Vern Kousky, Fanthune Moeng, Tom Schlatter, David Schultz, Mel Shapiro, Matt Wolf. **University Corporation for Atmospheric Research (UCAR)**: Norma Beasant, David Hosansky, Sergey Sokolovskiy.

Also: Howard Bluestein (University of Oklahoma), Philippe Bougeault (Météo-France), Tim Brown (Las Cumbres Observatory), Vilma Castro (University of Costa Rica), Natalia Chakina (Hydrometeorological Centre of Russia), Stanley Changnon (Illinois State Water Survey), Cecilia Conde (National Autonomous University of Mexico), Walt Dabberdt (Vaisala, Inc.), Anthony Del Genio (NASA), James Fleming (Colby College), Marco Antonio Salas Flores (Mexican Institute of Water Technology), Rob Folkert (National Institute of Public Health and the Environment, The Netherlands), Agusti Jansa (National Institute of Meteorology, Spain), Greg Jenkins (Pennsylvania State University), Mark Jury (University of Zululand), Laurence Kalkstein (University of Delaware), George Kallos (University of Athens), Marc Killinger (Boulder, Colorado), Paul Kocin (The Weather Channel), Hiroki Kondo (Meteorological Research Institute, Japan), Raymond L. Lee, Jr, (US Naval Academy), Malcolm Lee (UK Met. Office), Vincenzo Levizzani (National Research Council, Italy), Gary Lezak (KSHB-TV, Kansas City, Missouri), Diana Liverman (University of Arizona), János Mika (Hungarian Meteorological Service), Patrick McCarthy (Environment Canada), Lars Olsson (Swedish Meteorological and Hydrological Institute), Ales Poredos (Meteorological Office, Ljubljana, Slovenia), Mohan Ramamurthy (University of Illinois at Urbana-Champaign), Marilyn Raphael (University of California, Los Angeles), Kevin Rae (South African Weather Service), Linda Shapiro (Longmont Clinic, Colorado), Pete Stamus (Colorado Research Associates), John Teather (BBC), Donald Wilhite (National Drought Mitigation Center, US), Ding Yihui (National Climate Centre, China), Fuqing Zhang (Texas A&M University).

I couldn't have finished this book without the feedback and support of my friends and family, particularly Joe Barsugli, Brad Bradford, Zhenya Gallon, Chuck Henson and Ann Thurlow. Lucy Warner and Jack Fellows graciously allowed me a part-time sabbatical from UCAR. The Mount Washington Observatory gave permission to adapt material I'd written for its radio show, *The Weather Notebook*. Nita Razo (UCAR Communications), UCAR's Co-operative Program for Operational Meteorology, Education and Training (COMET), and *Weatherwise* magazine (through its annual photo contest) went beyond the call of duty in making imagery available. Karyn Sawyer (UCAR) and Peggy LeMone, Linda Mearns and Diane Rabson (NCAR) furnished useful tips and leads. WMO, NOAA's National Climatic Data Center, Columbia University's International Research Institute for Climate and Society, and the University of East Anglia's Climatic Research Unit provided crucial help in assembling data for the city climate charts. I'm also very grateful for assistance from libraries at NCAR, NOAA and the University of Colorado. John Wilkens generously provided input on weather in the cinema.

Throughout this project I've been fortunate to have the discerning judgement and good humour of Robert Shepard, a truly matchless friend and agent. The dedicated staff at Rough Guides – in particular, Orla Duane and Jonathan Buckley for the first edition, and Duncan Clark, Peter Buckley and Andrew Lockett for the second edition – made the entire process of putting the book together not only manageable but enjoyable. I am also indebted to Maxine Repath and her cartography team, Edward Wright and Katie Lloyd-Jones, at Rough Guides for skilfully crafting the city-by-city charts.

Special thanks go to my über-reviewers, Emily CoBabe, Roger Heape, Matt Kelsch, Amy Marks, Stu Naegele and Stephan Sylvan; to Rob Wood for sharing his insight on rainbows; and to Rene Munoz for providing a non-stop flow of juicy weather clippings and tidbits. This book is dedicated to Barbara Henson and Jane Kelsch, two much-loved sisters.

Contents

Introduction

Savour it, ignore it, obsess over it – however you approach weather, it's a significant part of our daily lives. Since the earliest times when humans scurried into their caves to avoid a storm, people have taken the weather into account. No doubt those same primitive people stood in front of their dwellings a few hours later to admire a rainbow or gaze at a spectacular cloud bank on the horizon.

What's changed over the centuries is how people interpret the dance of meteorological elements. Most of us no longer worry, as our ancestors did, about currying favour with those spirits who shuffle the atmospheric cards to punish or reward us. Today, when the weather doesn't go our way, we tend to look for a more scientific explanation. It's the fault of the jet stream, a low-pressure centre, or trendy scapegoats like El Niño and the North Atlantic Oscillation – or even global warming.

There is still a lot we don't understand about weather. We can now predict some aspects of climate as much as a year in advance, but nobody can tell you if a cold front will arrive on your doorstep a fortnight from now, or if a thunderstorm will strike at precisely 4pm tomorrow. We understand the basic physical laws that drive our atmosphere, but we're still hampered by our inability to observe every nuance of the present weather and by the finite (though rapidly increasing) speed of the computers that project weather into the future. There's also another factor restraining our knowledge: even as we learn more about different parts of our atmosphere, the interplay between them can produce something more than the sum of its parts – a crescendo that atmospheric scientists call "non-linear behaviour".

Where does that leave you, the consumer of weather information? Your cable TV service probably offers weather details on at least one, if not several, of its channels. You can find hundreds of maps and forecasts on the Internet and can dig up enough raw data to satisfy even the most hardcore of atmospheric appetites. But as the gurus of the information revolution keep reminding us, data isn't the same thing as information. You might be searching for a very specific forecast, say to plan a wedding or to take a boat trip. Maybe you're heading into the backcountry for a few days and want to know the weather signs that could spell trouble. Or perhaps you're travelling to a city halfway across the world and you need a sense of the typical weather at your destination, including the worst as well as the best that you might expect.

The *Rough Guide to Weather* aims to help you get the weather knowledge you're seeking. We've collected descriptions and statistics of the weather in dozens of countries and over two hundred destinations around the world. We also take you behind the scenes of the government forecasting centres, the TV studios and other places where your daily dose of weather information is crafted. For all the gains that forecasting has made with the help of computer guidance, humans have not yet been rendered obsolete. Particularly when the weather turns threatening, skilled forecasters can go a step beyond computer guidance and save lives in the process. You'll learn how the experts decide what to tell you about the upcoming weather and what they may choose to withhold due to limits of time and space, their own uncertainty, politics and other factors. In this second edition of the book, we've added key weather events of recent years, such as the 2003 European heat wave and 2005's disastrous Hurricane Katrina. You'll find coverage on several new topics, such as the weather on other planets (see p.372). We've also enhanced the section on global climate change – a more pressing topic than ever – and updated the city-by-city charts so that they reflect conditions through 2005.

In the end, weather is what people choose to make of it. Every maze of red and blue fronts on the TV screen, every weather warning that crackles across the radio, passes through the filter of our own likes, dislikes, hopes and fears. It's hoped that this book makes the weather you experience as enjoyable, understandable and as memorable as possible.

Robert Henson
Boulder, Colorado

Chapter one
The ingredients

The ingredients

As perfectly sculpted as a thunderstorm may look, or as glass-smooth as a blue sky may seem, the endless procession we call weather doesn't arrive at our doorstep in a static package. The atmosphere around us is crafting weather every second. The detritus from one batch of clouds gets recycled to form another. Air currents loop from one cyclone to the next in endless succession. To understand what drives this pageant, you have to pull back the curtain and take a look at the invisible processes going on backstage. Each player is important – even a molecule of water clinging to a mote of dust, or a proton sailing toward us from the Sun, has a significant role. In its eclectic quest to understand how tiny events add up to weather, meteorology draws from chemistry, physics, mathematics and other disciplines. You don't have to know the science of weather to understand tomorrow's forecast, but if you get to know the ingredients, you'll be drawn to delve deeper into the complex, fascinating, unpredictable world that is weather.

What's weather anyway?

Sunlight peering through overcast skies. A burst of rain. Lightning that splits the heavens in two. A great gust of wind. They're all ingredients that make up weather. Yet while the existence of weather may be a constant, the weather itself is constantly changing. The same is true of our ideas about weather. People have always marvelled at the atmosphere, but our concepts of weather have changed enormously throughout the centuries.

Magic and superstition ruled many people's experience of the weather for millennia. As recently as the eighteenth century, scientists had no idea that local weather was controlled by great swirls of air – at greater or lesser pressure – moving from place to place. Even by 1910, cold fronts and warm fronts had yet to be recognized and named as we now know them. Two great twentieth-century observation tools, **satellite** and **radar**, have shown us how weather looks on a large scale. Computers now tell us where important weather features should be three, five, or even ten days in advance. Even though these projections aren't always correct, we now acknowledge that weather is a sub-set of the known physical world, something that can be understood and predicted – in principle, if not always in practice.

Weather BC

The **earliest humans** – foragers, hunters and eventually farmers – were utterly vulnerable to weather and climate. Some of the earliest aspects of civilization, such as clothing and shelter, were adaptations to the weather. As cultures slowly coalesced, they began to accumulate rules and sayings about the weather's behaviour. Such weather lore persists to this day, especially among indigenous people. Weather stones called "mourners" in Gaelic folk culture grew darker as a result of temperature and moisture, revealing the state of the current atmosphere and, it was believed, the weather to follow. When seasonal rains failed in eastern Australia, the Yarralin people – noting that rain there usually moved from west to east – believed that other Aboriginals living to the west had somehow blocked the rains.

On a larger scale, climate and its long-term cycles affected people's choice of habitat, and this had much to do with how ancient cultures evolved. Some groups gathered where the climate was pleasing, while others may have been forced by competition into less favourable climes, where they learned to adapt. In his survey of civilization's growth, *Guns, Germs, and Steel*, Jared Diamond asserts that the spread of people and plants occurred far more readily across east-west than north-south belts. Weather conditions aren't always identical across a given latitude, but the length of day and the intensity of sunlight are.

NCAR / UCAR / NSF

Across Earth's lush rainforests, temperatures often change more from day to night than from summer to winter.

During the last **glaciation** (ending some 10,000 years ago), Earth's sea level was more than 100m/330ft lower than today. Great Britain was joined to France by a land bridge that traversed what is now the English Channel. This gave England a climate more like that of modern-day Germany: hotter

Weather vs climate

Science-fiction master **Robert Heinlein** once pointed out, "Climate is what you expect; weather is what you get." That, in a nutshell, is the difference between the two. **Weather** refers to the day-to-day vagaries of the atmosphere, the conditions that change from hour to hour and from day to day. **Climate** deals with the average of weather over time. Climate statistics can tell you much about the weather, like a brief description of a person. But once you meet that weather face to face, you may be surprised to discover certain important details that didn't show up in the introduction. For instance, a given day's high temperature may be far different from the climatological average for that day. Desert locales can get a year's worth of rain in 24 hours and hardly any other moisture in the remaining 364 days. Climate itself is more variable than we once believed. Some regions that were covered by a kilometre-thick ice sheet 20,000 years ago are now temperate – but they might become virtually sub-tropical in another century or two.

in the summer, colder in the winter. Similar land bridges joined east Asia to Alaska and Australia to Tasmania and New Guinea. In the window between the retreat of the ice and the inundation of the Bering Strait bridge, people migrated across the Strait, spreading themselves (by roughly 11,000 BC) over present-day North America.

As the climate warmed and the glaciers melted, humans occupied more and more parts of the globe, although the going wasn't always smooth. For instance, a break in the warm-up (from about 6200 to 5600 BC) caused the level of the Black Sea to fall more than 100m/330ft. People had settled around its shores during this 600-year interval, but they suffered a rude awakening when the post-glacial warm-up resumed. Geologic records show that the rising Mediterranean Sea poured into the Black Sea in a cataclysmic torrent that may have been the key source of the Bible's flood myth.

With the rise of organized religions came a more systematized view of how weather worked: in most cases, the gods were assumed to be in charge. Rainmaking rituals were part of **Egyptian culture** by 3500 BC, if not earlier. Weather elements often personified the deities of the time, including Zeus, the Greek god of lightning; Daz Bog, the Slavic creator of lightning and the Sun; and Indra, the Indian god of thunderstorms. Priests were the chief scientists of the day, deciding how and when to supplicate the gods for particular weather. Even so, people observed that weather patterns followed a certain prevailing logic when the gods were otherwise occupied. The annual cycle of heat and cold was recognized in **Chinese** and **Greek calendars** and often tied to astronomical events such as the appearance of constellations. One of the Bible's oldest books, Job, notes that "Fair weather cometh out of the north" – still a decent rule of thumb, at least for the Northern Hemisphere, since north winds often herald the arrival of high pressure and calm weather.

In his *Meteorologica* (written around 340 BC) **Aristotle** described hail, thunder, winds and other weather features in far more detail than any of his

Life in the Ice Age

It wasn't exactly a bowl of cherries for the protohumans who endured the last Ice Age. When Neanderthals ranged through Europe, present-day France and Germany were barren tundra and most of the British Isles and Scandinavia were under a sheet of ice more than 1.6km/1 mile thick. Similarly, massive ice sheets ranged as far south as the current US states of Iowa and Illinois. Debris left in the Midwest by departing glaciers now makes for some of the richest agricultural soil on Earth. Thanks to sea levels that were several hundred feet lower, the Aleutian Islands of Alaska formed a land bridge that allowed people to stream across and populate North America. The La Brea tar pits of Los Angeles teem with the skeletons of mammoths, sabre-toothed cats and other creatures that thrived in the sub-tropics. Although Earth probably averaged 3–6°C/5–10°F colder than today during the last Ice Age, the tropics didn't cool down much at all. This means the pole-to-tropic temperature contrasts were heightened, and gigantic storms were probably a regular occurrence close to the margins of the ice.

predecessors. Many of his theories miss the mark in the light of contemporary knowledge: for instance, he believed that the Sun calmed winds by drying up "exhalations" from the ground. Nonetheless, for centuries, *Meteorologica* was considered *the* book on weather. A few other voices emerged in the centuries that followed. For example, in about 1000 AD, **Ibn Al-Haitham** used the length of twilight (the period between sunset and full darkness) to deduce the height of the atmosphere. Throughout the Middle Ages, however, most scholars simply interpreted Aristotle, or expanded his findings. Mariners and explorers devised weather "signs" that helped to explain local conditions, sometimes successfully. Many others simply watched the sky and relied on the legends and sayings of their ancestors.

The age of observing

It was technology – in particular, three instruments created in the seventeenth century – that led to our modern understanding of weather. **Galileo** experimented with a device that used the expansion of a liquid (water, initially) to show the temperature of the air. Other inventors later sealed mercury in glass and attached a scale to create a bona fide **thermometer**. Following fast on the thermometer's heels in the late-seventeenth century were the **barometer** (for measuring air pressure) and the **hygrometer** (for relative humidity).

Widespread by the eighteenth century, these tools led to a worldwide boom in weather observing. Several United States founders were part of the craze. **Thomas Jefferson** duly noted weather conditions at the signing of the Declaration of Independence on July 4, 1776. (For the record, it was 22.5°C/72.5°F in Philadelphia at 1pm.) Jefferson kept a continuous weather journal from that year until shortly before his death a half-century later. He also mused on how settlement patterns in the emerging States might have been altering climate and suggested a co-ordinated, twice-daily network of weather observers that could help decipher such a development.

The multitalented **Benjamin Franklin** made his own contributions. On October 21, 1743, Franklin was thwarted from seeing a lunar eclipse by clouds that rolled into Philadelphia. Later, corresponding with his Boston brother – who'd had a clear view of the eclipse before the sky became overcast hours later – Franklin realized that the weather system obscuring his own view must have shifted from Philadelphia northeast to Boston, even though the clouds he saw in Philadelphia had moved in *from* the northeast. Franklin thus hit on the fact that midlatitude weather systems – regardless of the local wind direction – generally move from west to east (which is true in both Northern and Southern Hemispheres, as we'll see later). From 1732 to 1757, Franklin published *Poor Richard's Almanac,* a mix of weather lore and dubious long-range forecasts, based on astrological signals, similar to the almanacs that had swept Europe two centuries before.

True weather forecasting had to wait until there was a quick way to pool observations. The **telegraph** adequately filled this need. By the 1850s, networks of observers scanned the skies and began sending reports to Washington DC, London, Paris, St Petersburg and other cities. Before long, it was possible for government agencies to combine these reports, see where weather systems were going, warn the public of storms, and issue "indications" that appeared in the rapidly growing daily press.

Announcer Noelle Middleton and meteorologist T H Clifton collaborated on an early BBC weathercast in 1954.

High-tech as this was, the new forecasts had their limitations. Weather maps tracked high and low pressure centres, and forecasters learned how to connect these to certain kinds of weather, but there was no unifying concept of how the highs and lows related to each other and what made them develop and subside. Forecasters were also unable to tell what was happening above ground level, where so much weather is shaped.

The twentieth century

As it turned out, we needed to see the atmosphere for what it was: a giant pool of fluid sloshing across the surface of the planet. A small group of pioneering meteorologists refined this vision around the time of World War I. Based in Bergen, Norway, the **Bergen School**, led by the father-and-son team of **Vilhelm** and **Jacob Bjerknes**, began to explore ideas based on fluid dynamics, and thereby discovered the importance of the boundaries where cold and warm air masses bumped into each other. They labelled these boundaries "fronts", using the wartime analogy of armies facing off. Suddenly a world of weather features came into focus, and the daily weather map was transformed.

Bjerknes family

In 1919, Jacob Bjerkness and colleagues were at the helm of a meteorological revolution.

But what made **fronts** move? The Bergen School believed that winds at upper levels pushed the surface fronts forward, and they developed a three-dimensional theory of fronts to strengthen their case. Their concepts were bolstered by a growing set of upper-level observations gathered from kites, balloons and high-altitude weather stations. Starting in the 1930s, weather services around the world began launching weather instruments by balloon each day. The packages radioed data back to Earth from more than 20km/12 miles high. Finally, forecasters could track the upper-level features that choreographed the weather below on a day-to-day basis.

Revelatory in the 1960s, full-disc satellite photos of Earth's clouds are now routine.

How to deal with all this information also needed to be addressed. **Digital computers** afforded the answer. A team of scientists in Princeton, New Jersey, carried out the first computerized weather forecast in 1950. It was useless in itself – the 24-hour outlook took about six days to complete! – but invaluable as a sign of what would follow. Things became far more sophisticated in the succeeding years. By 1960, computers around the globe were predicting major weather features a day or two in advance. Human experts filled in the local blanks. Scientists were also putting another new instrument, **radar**, to work in locating water droplets and ice crystals, thus allowing the location of rains and snows to be tracked from a distance every few minutes.

One of the classic photographic icons of the 1960s was the appearance of Earth from space as a lustrous, cloud-laced, blue-green marble. Eye-popping at first, such photos quickly became routine tools in weather forecasting. In pointing out the raw isolation of our planet, satellite photos helped make us far more aware of our environment. Chemists and meteorologists teamed up to study the local effects that triggered and trapped air pollution. In big cities, ozone alerts became part of the daily weather outlooks.

Today's weather is more effectively predicted and more closely scrutinized than ever. Three-day forecasts in the US are now more accurate than the two-day outlooks of the 1980s. A torrent of data is available to all on the Internet. Colourful, information-packed TV weathercasts have left their dowdy predecessors in the dust. However, we haven't yet reached meteorological perfection. Small-scale weather features such as thunderstorms cannot be nailed down much more than a few minutes ahead of time. The chaotic nature of weather may well prohibit local forecasts of more than about two weeks in advance.

On top of all of this, woven through the ups and downs of daily weather, is the ominous counterpoint of **climate change**. Although the **greenhouse effect** was identified a century ago, it gained relatively little attention until the late 1980s, when a spike in global temperatures coincided with ominous projections from **global climate models**. The 1990s were likely the warmest decade of the last millennium, and the first decade of the twenty-first century may end up even warmer. Aside from a very few outspoken dissidents, the

Your great-great-great-grandparents' cold world

Regardless of what our industrial society is doing to Earth's climate, nature can produce surprises of its own accord. Sketchy observations and recent inferences show that, from about 1550 to 1850, a good part of the planet was as much as 1.0°C/1.8°F cooler than today. That may not sound like much, but it was enough to profoundly affect society in Europe and North America, where the temperature drop from medieval times was felt most strongly. Famines dogged Europe, particularly in 1594–97 and in 1693, when perhaps one out of every three Finlanders died. Expanding glaciers swallowed up entire Swiss villages, and paintings of the era show people skating on canals that virtually never freeze over today. Sunspot records show the Sun was fairly inactive during the heart of the Little Ice Age, with incoming solar energy perhaps dipping by a quarter of a percent or so. It's possible, but difficult to prove, that this lowered solar energy was at least partly to blame for the cooldown. One factor that's well known is the sun-screening action of volcanoes, which were especially active in the 1700s and early 1800s. The epochal eruption of Indonesia's Tambora on April 5, 1815, is the most likely culprit for 1816's "year without a summer". June snows and freezes plagued New England, and Europe was caught in relentless drear. Holed up with fellow writers in a summer house near Lake Geneva, teenager Mary Shelley penned a dark novel that reflected her mood: *Frankenstein*.

world's scientists now agree that emissions of greenhouse gases (primarily carbon dioxide) bear a substantial share of the blame for the global temperature rise.

In the light of climate change, the quicksilver shifts of our daily weather have taken on a whole new cast. For very different reasons than before, science has once again put us in the role of our superstitious ancestors: we suspect that how we behave affects the clouds and wind and the warmth we experience. As writer Jay Rosen observed in 1989, just as the greenhouse effect was entering popular awareness, "The happy atmosphere of the weather report will be difficult to maintain, for the weather can no longer serve as a haven from history".

Welcome to our atmosphere

An old pop song tells us that "love is like oxygen". A politician concerned about the environment is accused of being "out in the ozone". Computer programs that exist only in a promoter's press releases are dubbed "vapourware".

Oxygen, ozone and **water vapour** are three of the most familiar ingredients in the life-sustaining soup we call the **atmosphere**. It's a complicated brew: some of its elements weren't even discovered until well into the twentieth century. You'll find only a tiny bit of molecular difference when you compare the air that hangs over an industrial city, simmers in a sauna or lies atop ice at the South Pole. However, just as a pinch of cayenne pepper can transform a recipe, several of the most scant ingredients of the atmosphere – which can show up at one part in a billion or even less – make a huge difference to how the atmosphere behaves and how it feels to us.

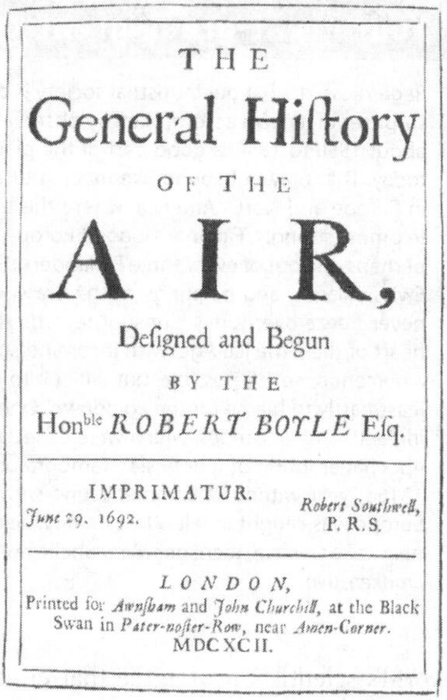
THE

General Hiſtory

OF THE

AIR,

Deſigned and Begun

BY THE

Honble ROBERT BOYLE Eſq.

IMPRIMATUR. Robert Southwell, P. R. S.

June 29. 1692.

LONDON,

Printed for Awnſham and John Churchill, at the Black Swan in Pater-noſter-Row, near Amen-Corner.

MDCXCII.

NOAA Photo Library

Fiery and foul

If there's more to the atmosphere than we once thought, there's also less. Aristotle believed in four elements – earth, air, fire and water – but he also thought the Sun's heat interacted with water and earth to produce "vapours" that caused rain and snow and "exhalations" that fuelled the wind. Over a thousand years later, Shakespeare's Hamlet bemoaned a "foul and pestilent congregation of vapours", and his King John fretted about an "exhalation in the sky".

With the arrival of the seventeenth century and the scientific renaissance, people acquired the tools to discover just what constituted air. One of the earliest ideas came from **Leonardo da Vinci** and several followers. They pushed for the notion of "fire air", which they thought helped to fuel blazes and support life, and "foul air", which they believed inhibited the same processes. In the 1770s, **oxygen** was discovered and identified as "fire air". Before he literally lost his head in the French Revolution, pioneering chemist **Antoine-Laurent Lavoisier** demonstrated that living creatures took in oxygen and released carbon dioxide, just like a slow-burning fire. (Plants do the opposite, an important aspect of the climate-change debate.)

Although it's probably the main component that springs to mind when we think of air, oxygen constitutes only about 21 percent of the atmosphere. A far greater portion, around 78 percent, is made up of **nitrogen**, identified (also in the 1770s) as "foul air". Nitrogen is actually a rather benign element of our atmospheric blend; it can't support life, however. Nitrogen is such a stable molecule that only a few natural processes can remove it from the air, including the intense heat of a lightning bolt. Once atmospheric nitrogen returns to the soil, it helps nourish plant life. At the same time, bacteria in the soil and ocean can both remove nitrogen from the air and help put it back. This is just one of many important cycles that link the atmosphere to what scientists call the **biosphere** – the world of plants, animals and other organisms.

Oxygen and nitrogen are the biggies of the atmosphere. Together, they constitute about 99 percent of it. **Water vapour** accounts for much less than a percent overall, although in humid regions it can occupy as much as 3 percent of the ground-level air. The rest of the atmosphere is divided into dozens of players. Some of these are amazingly stable. The so-called noble gases – argon, neon, helium, krypton and xenon – can stay in the air for many millions of years.

The air also includes **solid particles,** or **aerosols,** ranging from the microscopic to the visible. Salt escapes from the ocean in sea spray, dust is scoured from the ground by wind, and a variety of particles are ejected from volcanoes. Most of these take only brief trips of a few hours to a few days through the atmosphere before they settle to earth or sea, although volcanic debris can stay airborne for several years if it's shot high enough. Also, meteorites deposit a few million tonnes of dust each year as they whisk into our atmosphere and disintegrate.

There's also a group of sparse, but highly reactive, molecules that, like rabble-rousers at a gathering, tend to influence others. For instance, the **hydroxyl radical (OH)** makes up less than a trillionth of the atmosphere, yet OH is critical to clearing the atmosphere of hydrocarbons and other pollutants. Molecules like these have far more of an influence on the air than their numbers might suggest, and it's their enhancement – or destruction – by human activity that adds a question mark to our atmosphere's future.

How the atmosphere evolved

In the billion or so years after the cataclysmic **Big Bang**, as Earth coalesced out of cosmic debris, the lighter elements found in the Sun (**hydrogen** and **helium**) gradually escaped. Volcanoes spewed out much of the stuff now found in our atmosphere, including methane, nitrogen, the noble gases and carbon dioxide, as well as the water vapour that condensed to form the world's oceans. In fact, carbon dioxide probably dominated our atmosphere for billions of years, much as it does on Mars and Venus, where it makes up more than 90 percent of the air.

Oxygen wasn't a major player until life came along in the form of primitive bacteria. Like the plants that would follow them eons later, these bacteria carried out **photosynthesis**: using sunlight, they took in carbon dioxide and emitted oxygen. Once some of the oxygen was converted to **ozone**, it formed a shield for ultraviolet light and allowed more forms of life to develop. A host of chemical readjustments followed. Gradually, the atmosphere took on its present shape – or, at least, the shape it had for roughly a billion years before industrial society arrived. Oxygen and nitrogen still take up the same 99 percent of dry air they did before industry arrived, but the rest of the picture has changed notably.

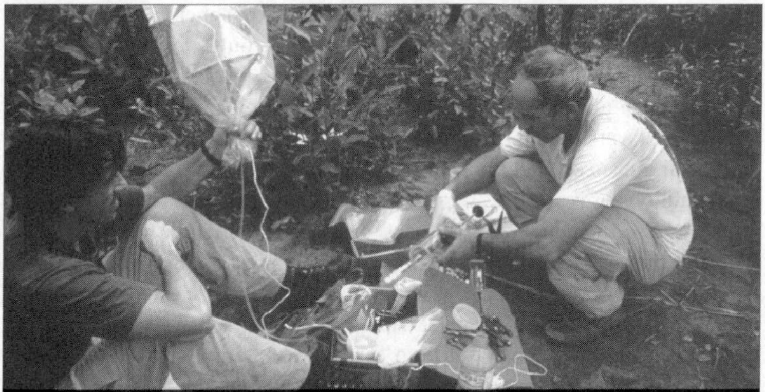

Scientists trek to the African rainforest and other remote locales to sample air chemistry at work.

Humans are a large part of this changing picture. Of course, we've burned wood, coal and other substances for millennia, but from the late-eighteenth century onward, industrial culture spread widely enough to affect Earth's air quality on a regional and global scale. Before the most obvious pollutants from coal were addressed, they triggered horrific smog episodes in Europe and North America that killed thousands in the first half of the twentieth century. The term "smog" was coined in Britain to describe a combination of smoke, mainly from coal burning, and fog.

Another source of burning, the **internal combustion engine**, has made possible a different kind of nasty brew. Compounds emitted by cars interact with sunlight to form ozone and other troublesome chemicals. Although ozone in the stratosphere helps protect us from ultraviolet radiation, it causes health problems when it's at ground level. This type of pollution has become endemic to sunny, lower- and mid-latitude cities from Los Angeles and Mexico City to Rome and Bombay.

Many kinds of pollution, including the soot produced from coal and heavy industry, wash out of the air in only a few days (although acidic rainfall has produced its own set of ill effects on forests and lakes). Many cities and regions have been able to make substantial improvements in their air quality simply by finding cleaner ways to burn. However, the key byproduct of internal combustion remains carbon dioxide, and it's quite tough to eliminate. A very stable molecule, carbon dioxide can stay in the atmosphere for over a hundred years. About 25 percent of the carbon dioxide produced by humans is taken up by the oceans. Plants remove some of the rest, but they aren't doing so quickly enough to keep up with the expansion of human industry. Together with other greenhouse gases, increasing levels of carbon dioxide have taken the concept of pollution to a new level by raising the spectre of human-triggered global warming (see p.150).

The lighter side of humidity

Step off a plane in Miami or Calcutta and you may feel as if you've been hit by a tonne of hot, moist bricks. When air is extremely muggy, it feels thicker and heavier than usual. Actually, though, tropical air is less dense than its cold, dry counterpart when both are at the same air pressure. One reason is the way that temperature works. The molecules in warmer air are moving more energetically, so they tend to take up more space at a given pressure. What's more, a molecule of water vapour only weighs about half as much as a two-atom molecule of either oxygen or nitrogen – the main components of dry air. If you packed some water vapour into a closed container of dry air, you'd be adding weight to the contents, but in real life air moves and mixes. This means that water vapour ends up muscling out some of the dry-air molecules and occupying their space. The result is less atmospheric mass and lighter air. The only thing that's become heavier is the perspiration-drenched shirt on your back.

Layers of the atmosphere

The sky is the very definition of vastness. When we take a boat to sea or find ourselves on a treeless plain, we're reminded just how all-encompassing the sky can feel. It's easy to believe the atmosphere must go on forever, stretching outward and upward.

Try this perspective, though: Earth as seen from the US space shuttle, over 160km/100 miles above sea level. From this vantage point, the atmosphere looks paper-thin, a sheer layer that barely covers the globe, like cellophane stretched around a beach ball. In fact, the atmosphere is just about that skimpy relative to Earth. More than 90 percent of our atmosphere lies within 32km/20 miles of the planet's surface, a mere sliver compared to Earth's diameter of nearly 13,000km/8000 miles.

This filmy coating is all it takes to produce the enormous variety of weather on the surface where we live. The "weather layer" of the atmosphere, in fact, is even thinner: only a few miles deep. Above that are several other atmospheric layers. Each one has its own temperature profile, cloud types, chemical make-up and impacts – some minor, some major – on the inhabitants of Earth.

The ground floor

The **troposphere** is where most of our weather action takes place. It's the bottom of the atmosphere, extending from ground and sea level up to around 10–16km/6–10 miles high, depending on the season and location. It's tallest wherever and whenever the air is warmest. Mount Everest stretches more than halfway up through the troposphere, and many other mountains are tall enough to affect how the air in this layer moves. And move it does. The Latin root of "troposphere" means "to turn", and the air in this swirling realm is indeed turning and overturning all the time. In a turbulent thun-

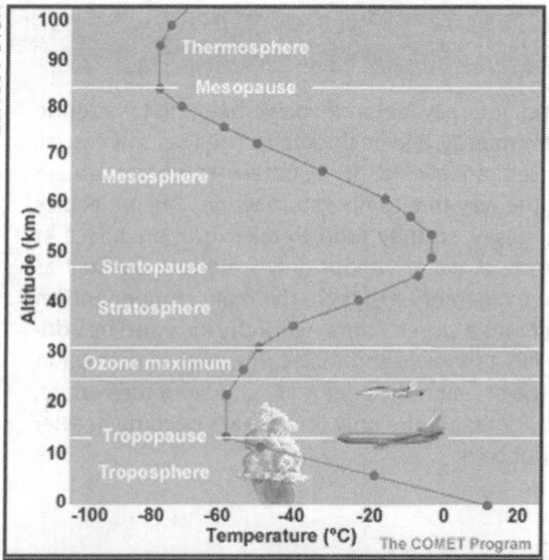

UCAR / COMET

Although the atmosphere has several layers, most weather is focused in the troposphere.

derstorm it may rise at speeds of over 160km/100 miles per hr. Usually, the upward and downward motions are far more gentle, typically less than 1.6kph/1mph. The average horizontal motions we experience as wind are quite a bit stronger. A typical breeze is 16–24kph/10– 15mph.

What keeps the troposphere moving are contrasts between warm and cold, moist and dry. The play of sunlight across Earth varies with the season, the time of day, and where it's cloudy and clear. The result is a patchwork of air masses that have either been heated by sunlight or cooled by its absence. These contrasts are greatest near the ground, where soil and ocean have a chance to add their stored heat to the Sun's. Overall, the heating differences are most obvious between the poles and the tropics, a contrast that forms the source of the fronts and cyclones driving our daily weather (see "Where does the wind go?", p.23).

Crickets and other temperature scales

Even after the **thermometer** was invented, it wasn't obvious how to set its readings. Scientists bickered over which points on the scale should be tied to something in the natural world. Some favoured "blood heat", or body temperature. In northern Europe, where thermometers were refined in the eighteenth century, people didn't expect air temperature to soar beyond blood heat. That made it a natural point on which **Daniel Gabriel Fahrenheit** fixed the high end of his scale, which he calculated as 100°F. Zero Fahrenheit was measured using a set mixture of ice, water and salt. In between was the freezing point of fresh water, set at 32°F. Although the Fahrenheit scale, published in 1724, nicely enclosed the range of typical European weather, **Anders Celsius** adopted a different approach. He used the freezing and boiling points of water – 0°C and 100°C respectively. This made each Celsius degree 1.8 times larger than a Fahrenheit degree. Although the Celsius scale was coarser, its logic fitted nicely with the metric system that emerged later. Celsius is now the international standard, although Fahrenheit is still what most people in the United States use. Physicists prefer degrees **Kelvin** – identical to a reading in Celsius plus 273.15°. The baseline of this scale, zero Kelvin, is thus equal to –273.15°C (roughly –459°F) and is also known as **absolute zero**. Scientists have cooled atoms to within a few billionths of a degree of this ultimate chill, while outer space averages a nippy 3°K. For a biologically based temperature scale, you can't beat **crickets**. The metabolism of these cold-blooded creatures speeds up with temperature, so you can estimate how warm it is (in degrees F) by counting the number of times a cricket chirps within a 14-second period and then adding 40 to the count. For degrees C, count the chirps within 7 seconds and add 5.

Air that's warmed becomes less dense than its surroundings and rises as a result. With height, however, the air becomes thinner and thus chillier. Cool air goes through the opposite process, typically settling, compressing and warming with time (except during calm nights, when the air can chill down as it pools in darkness and Earth radiates heat to space). The patchwork of the troposphere, then, is in a state of perpetual upheaval, seeking to equalize conditions through a never-ending series of highs, lows, fronts and storms.

As any mountain climber knows, air gets colder the higher you go. The drop-off rate averages 6.0°C for every kilometre you climb (or 3.3°F for every 1000ft). At certain times and places – when moist layers are present, for example, or when there's a cold pool trapped near the ground – this relationship doesn't hold, but it's a good rule of thumb. In the upper troposphere, where most jets fly (around 10–13km/6–8 miles at mid-latitudes), it's always well below freezing.

Where the ozone is

Compared to our weather layer, the **stratosphere** – the layer above the troposphere, around 16–50km/10–30 miles – is a topsy-turvy place. Temperatures plummet to as low as –80°C (–112°F) in the tropopause (the troposphere–stratosphere boundary), but then they actually rise as you ascend through the stratosphere. Near the top, you're again close to the freezing point.

This temperature flip-flop occurs mainly because of the stratosphere's role as a shield for harmful ultraviolet sunlight. The **ozone layer**, centred in the lower stratosphere, is the prime absorber of ultraviolet light. The top of the ozone layer takes the biggest hit from the incoming rays, which allows it to heat up

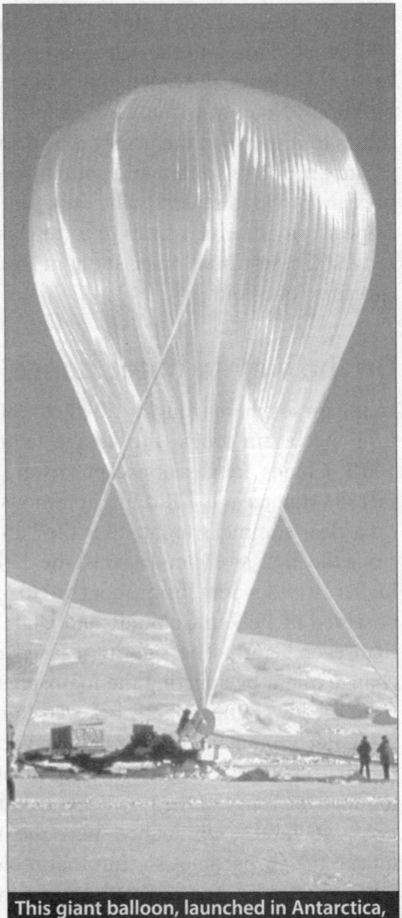

NSF

This giant balloon, launched in Antarctica, takes a research payload as heavy as a family car to heights of 38km/24 miles.

the most. The sunlight itself actually helps generate ozone by splitting oxygen molecules. Together with the mixing of ozone downward into the troposphere, a natural cycle has kept the ozone layer in balance for millennia.

In recent decades, humans have disturbed the stratosphere from below. Industrial chemicals have wafted upward and jeopardized our ultraviolet defence shield. Working in tandem with sunlight and stratospheric clouds, these chemicals (chlorofluorocarbons, or CFCs) have depleted the ozone layer by several percent worldwide. For a few weeks each spring above parts of the Antarctic, the loss is closer to 100 percent (see p.163).

Only a small number of aircraft – mainly research planes – fly at stratospheric heights. There's long been concern, however, that more extensive flying at these altitudes, such as by a fleet of passenger planes, may cause additional damage to the ozone layer. Major studies, such as a 1998 review by the US National Academy of Sciences, point to a number of unresolved questions. Nobody knows, for example, exactly how long it takes for aircraft fumes to disperse in the poorly mixed stratosphere.

Since the air is so thin and cold in the stratosphere, very little moisture is present for cloud making. The only type found is the **nacreous**, or mother-of-pearl, cloud, which tends to form close to the poles and above mountain ranges. These clouds filter sunlight in a way that gives them a luminous spectrum of colour.

On the way to outer space

In the **mesosphere** (roughly 50–80km/30–50 miles high), the air's temperature once again drops with height, falling to readings well below –73°C/–100°F. Clouds are absent except for rare **noctilucent** clouds, first described in 1884 (the term is Latin for "luminous at night"). These can be seen only on a clear summer's night, after the Sun has set, but while its rays can still reach the high-level cloud against the dark sky. Widely reported in the past at higher latitudes only (from about 50° to 65°N and S), noctilucent clouds were observed in 1999 in Colorado and Utah, much further away from the poles than ever before. The sighting caused much speculation on how climate change might be affecting the mesosphere. One hypothesis is that methane from the lower atmosphere, believed to rise and react with sunlight to form the water vapour that makes up noctilucent clouds, is now being joined by carbon dioxide, which helps to cool the upper atmosphere – perhaps enough to move the range of the noctilucent clouds closer to the equator.

As we head even higher, into the **thermosphere** (80–650km/50–400 miles), the air becomes so thin that molecules have difficulty keeping each other in check. One molecule might travel a distance the width of the English Channel before colliding with another. Because the air thins out so drastically with height, the air "temperature" can climb dramatically to up to

The weight of it all

Living in it as we do our whole lives, it's easy to forget that air is heavy stuff. All of the air that rests on top of an adult's head (an area of about 500 square centimetres or one-half square foot) weighs roughly 400 kg, or about half a tonne! Thankfully, our bodies are up to this challenge. There's enough outward force from our bodies and the air inside us to keep us from imploding. We sense the true force of air only when it's blowing against us in a strong wind, or when we move upward quickly enough. As we ascend in an elevator or airplane, the air pressure around us drops quickly, and the air inside us tries to escape, causing discomfort in our ears or our stomachs. Many people claim to be able to predict weather by feeling it "in their bones". What's likely going on is that a sharp pressure drop (which usually signals a dramatic weather change) is enough to cause the air within sensitive tissues – such as those inflamed by arthritis – to expand and trigger pain.

1800°C/3200°F. However, with so few molecules buzzing around, the heat isn't that intense; in a pressurized suit, a space-shuttle astronaut would do just fine.

Like many other distinctions, the one that separates our atmosphere from outer space is rather fuzzy. Above a height of approximately 95km/60 miles, the air sorts itself out by weight, and the lightest molecules (hydrogen and helium) gradually rise to the top. The atmosphere doesn't so much end, as dribble down to an infinitesimal thinness through a zone a few hundred miles high. From about 650km/400 miles outward, the atmosphere is constantly being drained off. A few hundred tonnes of stray hydrogen molecules escape into space each year. The vast bulk of the atmosphere stays in place, though, held fast by gravity.

The thinness of the upper atmosphere allows forces other than brute mechanics to take over. The intensity of the unfiltered sunlight at these heights breaks up molecules and leaves them electrically charged. When an especially potent batch of magnetized plasma arrives via the solar wind, a **geomagnetic storm** may result. The strongest of these can last for days, often fouling up radio signals and satellite orbits. Such storms are now predicted and tracked by earthbound sensors and by satellites deployed as far as a million miles from home. (Geomagnetic storms also produce the ethereal beauty of the aurora borealis and australis. See "The Sun and Earth", below.)

The Sun and Earth

Consider the **total solar eclipse** – a thing of undeniable beauty. When the Moon passes before the Sun, all that remains visible for a few minutes is the shimmering corona. Perhaps even more remarkable than the sheer beauty of an eclipse is its sheer improbability. The Sun is far larger than the Moon, but it's much farther away, so the two bodies – as we see them in the sky – are within one percent of each other's size. If one or the other were just a bit closer or a bit larger, there could be no eclipse as we know it. The coincidence is truly cosmic.

The more routine pas de deux of the **Sun** and **Earth** is an anchor of daily life. We watch the Sun travel across the sky each day, its path edging north or south with the **seasons**. Yet it's the wobblings of Earth – spinning on its axis and circling the Sun – that not only produce our seasons, but play a huge role in natural climate change (as opposed to the kind in which humans are implicated).

NCAR / UCAR / NSF

In just a few minutes, a solar eclipse can yield valuable data on the Sun's workings.

The source of it all

Without the **Sun**, there wouldn't be any weather, not to mention anyone to notice its absence. The Sun derives its power from **nuclear fusion** within its superheated core, whose centre simmers at roughly 15 million degrees C/27 million degrees F. As the nuclei of two hydrogen molecules fuse, they produce helium and release energy that radiates upward to the Sun's outer layer. That zone of overturning gases and intense magnetism channels the energy in complex ways, forming sunspots, flares and other solar features. It's this varied structure and the constant changes within it that make the Sun far more than a homogeneous blob.

The atmosphere surrounding the Sun isn't nearly as blistering as the solar core, but at more than a million degrees C, it's still so hot that the atoms within it can't hold onto their electrons. The result is a stew of charged particles (**plasma**) that expands outward into the coolness of deep space as the **solar wind**, moving at speeds in the range of 320km/200 miles a second. Meanwhile, the plasma churning inside the Sun emerges at the surface in the form of active regions – a crude equivalent of disturbed weather on Earth. **Sunspots** are the best-known type of active-region disturbance. Although sunspots appear darker than the rest of the Sun, they're bursting with magnetic energy, so their overall effect is to increase the amount of plasma streaming out from the Sun and help fuel "gusts" in the solar wind. From the Sun's perspective, Earth is like a pinpoint in space, so only a few of these plasma-laden gusts manage to hit the tiny target of our upper atmosphere, but once there, they can wreak magnetic and electrical havoc (see opposite).

Solar energy on Earth

Let's assume the front of your body spans about nine square feet, or roughly a square metre. If you went to the outer edge of our atmosphere and faced the Sun head on, you'd intercept the energy equivalent of about 14 one-hundred-watt light bulbs. The precise number – 1368 watts per square metre – has traditionally been called the **solar constant**. This term isn't quite accurate,

Forces behind the Northern and Southern Lights

The beautiful sky show known as **aurora borealis** (or **australis**, in the Southern Hemisphere) is, in fact, a manifestation of molecular violence. Aurorae are produced by the solar wind, a stream of charged particles from the Sun. Because Earth's magnetic field is a very good plasma shield, the solar wind usually flows around it much like a stream flows around a rock. At certain times the magnetism in the solar wind is especially "geoeffective" – it's especially intense and oriented so that it clashes with Earth's magnetic field. When a batch of geoeffective plasma arrives, electrons trapped on the dark side of Earth are energized. They travel downward toward the pole along magnetic field lines and strike atoms and molecules at heights between 80–1000km/50–600 miles. Aurorae develop as each particle struck emits a characteristic hue by element: red and yellow-green from oxygen, blue from nitrogen, etc. Aurorae can occur anytime, but they're especially frequent in a one- or two-year span close to the eleven-year peaks in the solar magnetic cycle; this millennium's first peak occurred in 2001. The most intense aurorae can be seen as far away from the poles as 30°N or S (in 1909, one aurora was spotted near the equator).

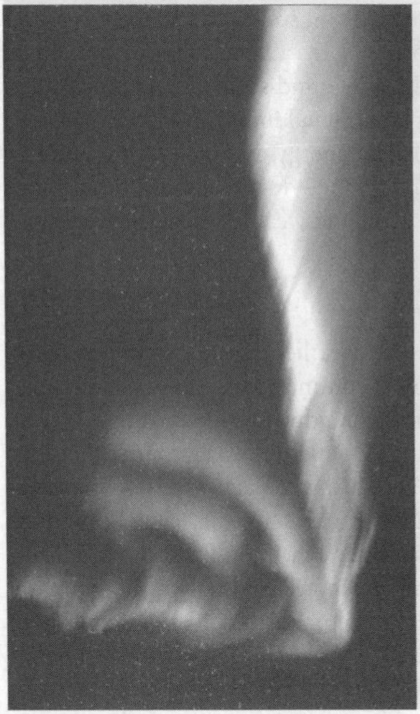

Aurorae are just one consequence of the high-altitude magnetic and electrical paroxyms known as **geomagnetic storms**. Most of these are caused by solar storms – giant ejections of as much as a billion tons of magnetized gas from the Sun. Occasionally, a geomagnetic storm is induced by a strong solar wind emerging from a quiet region on the Sun called a coronal hole. Despite the beauty of their auroral calling cards, geomagnetic storms can scramble radio waves and corrupt power grids on Earth. Six million people in the Montreal area lost power for up to nine hours following a geomagnetic storm on March 13, 1989. By heating the upper atmosphere, geomagnetism also exerts a drag on low-orbiting satellites, pulling them downward until a reboost can be accomplished. Space-weather experts rely on satellites to monitor the Sun's action; they can now provide a day or two of advance warning to help industry and government prepare for the worst storms, and aurora lovers to get into position.

however. Thanks to ever more precise observations from satellites, we now know the Sun's output rises and falls by as much as 0.1 percent over short periods, as well as during the eleven-year solar cycle. Larger variations occur over centuries and millennia, helping to trigger such climatic events as the Little Ice Age that put post-Renaissance Europe in the deep freeze (see box, p.9). And ultraviolet energy can rise and fall much more than the Sun's total output, varying by as much as a factor of 10 across the solar cycle.

Scientists are still grappling with the "**faint sun paradox**". From three to four billion years ago, when life began on Earth, the Sun was only 75 percent as radiant as it is now. Yet the best evidence shows that Earth's temperature was comparable to today's. Greenhouse gases were probably more prevalent than today, which may explain why Earth was relatively toasty despite the weaker Sun.

The basic engine of our atmosphere's behaviour is the contrast between sunny and not-so-sunny places. Every spot on the planet gets an equal amount of darkness and daylight, but how it's distributed varies by **latitude**. On the equator, every day of the year is twelve hours long, while the North and South Poles get six months of daylight and six months of darkness each year. Mid-latitudes get the familiar pattern of shorter days in the winter and longer ones in the summer.

However, the sunlight that hits Siberia can't hold a candle to that of Sumatra. Both places get the same amount of daylight each year, but closer to the poles, the Sun sits lower on the horizon, so its rays strike the planet at a sharper angle (see diagram, opposite). The same amount of sunlight is spread over a much larger area near the poles, and so there's less solar heating per patch of land. The net result is a perpetual imbalance: the tropics are always getting more than their share of solar energy, and the poles are forever shortchanged. The atmosphere responds by moving heat from the equator toward the poles, and voila – we have wind and weather (see Chapter 2).

A little seasoning

What makes the **seasons**? Perhaps no other weather question is so misunderstood and incorrectly answered. The most common knee-jerk answer is that Earth is closer to the Sun during summer than during winter. If this were all there was to it, then the whole Earth would experience summer at the same time, but we know that's not the case. The **tilting of Earth's axis** is what gives us seasonality. If Earth stood straight up within the circle it makes around the Sun, there would be no such thing as summer or winter. Instead, Earth leans at an angle of about 23.5° (see diagram) from celestial north, the top of an imaginary line cutting through our solar system. Earth's tilt allows the Sun to face the North Pole in June and the South Pole in December. As each hemisphere turns to face the Sun, the land and air warm up and summer arrives.

A close look at the calendar reveals that the seasons are not created equal. People in the Northern Hemisphere get something of a break: their winter is only about 89 days long, while their summer lasts roughly 94 days. This is tied to the fact that the Sun doesn't sit squarely in the middle of Earth's orbit. Instead, it's displaced about 3 percent of the way toward one side. Because of the laws of angular momentum, Earth has to hustle as it passes closer to the Sun, just as the water in a drain rotates more quickly as it nears the centre. Earth's closest approach to the Sun each year happens to fall on January 4, so the seasons near that date (northern autumn and winter, southern spring and summer) are shorter than those on the other side of the calendar (northern spring and summer, southern autumn and winter). The 28-day brevity of February is an adjustment to this astronomical fact.

An oblique and eccentric story

People change, haircuts change – why not **orbits**? Earth has spun around the Sun for a long time, but the process hasn't been as routine as you might imagine. Over millions of years, each one of the three major **orbital parameters** has gone through variations, and those variations have played a gigantic role in the climate on Earth. None of them are short-fuse enough to worry about in our own lifetimes, but they do shed light on how a small change in our climate system can have a huge impact.

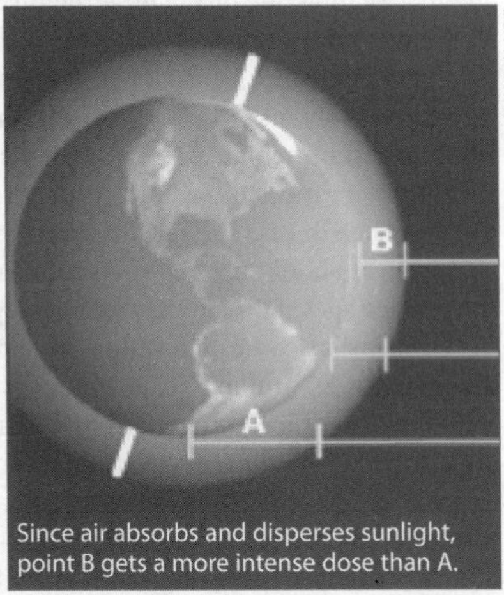

UCAR / COMET

Since air absorbs and disperses sunlight, point B gets a more intense dose than A.

- The tilt of Earth – its **obliquity** – varies from about 21.8° to 24.4° and back over a cycle that lasts about 41,000 years. A greater tilt means warmer summers and colder winters across the globe. When the tilt decreases, it can help bring on an ice age (all else being equal), because the progressively cooler summers can't melt the past winter's snow.
- The offset of the Sun from the centre of Earth's orbit, discussed above – in other words, the eccentricity of the orbit – is now on a slow decrease. Currently Earth is about 3 percent closer to the Sun, and receives about 7 percent more solar energy, in early January than in early July. The orbit was about twice as eccentric when the most recent Ice Age began, a little more than 100,000 years ago. Higher eccentricity means more contrast in solar heating between the nearest and farthest annual approaches of Earth to the Sun (perihelion and aphelion, respectively).

- A slight wobble of Earth's rotational axis, and a slight wobble of Earth's orbit around the Sun, combine to produce **precession** – a slow advance in the date of perihelion. Right now, it occurs on January 4, but each year it moves ahead about 25 minutes. By the year 11,750, it'll fall in June, and people in the Northern Hemisphere will endure a winter about five days longer than it is now. When a hemisphere is closer to the Sun during summer than during winter – as is now the case in the Southern Hemisphere – the seasonal contrast in heating is heightened (although the vast southern oceans more than compensate for this by smoothing out seasonal temperature swings).

What on earth does all this mean for climate? James Croll, a janitor in a Glasgow university, speculated in the 1860s that orbital variations might be behind the onset and departure of ice ages. It took until the 1940s for the Serbian scientist **Milutin Milankovitch** to complete the precise work that pointed in the same direction. By blending the three orbital parameters above, Milankovitch found that certain combinations could reduce the energy reaching one or the other hemisphere by a few percent. Interestingly, it's the mellower combinations – those that spread the radiation most evenly throughout the year – that apparently lead to ice ages. The idea, in a nutshell, is that relatively mild winters allow for heavier snow to fall at high latitudes, while cool summers help keep the higher-latitude snow from melting.

Computer modelling and other tools have largely supported the Milankovitch thesis. The various orbital parameters coincide for the Northern Hemisphere about every 100,000 years, and that's the rough frequency of the last few ice ages, as deduced from geologic records. However, the book isn't closed on what causes glaciation. When the most recent round of ice ages began about 2.7 million years ago, they occurred about every 41,000 years, corresponding nicely to variations in Earth's obliquity (see above). About a million years ago, the ice-age frequency dropped to once every 100,000 years or so, with short periods between each glaciation. It's not quite clear what is setting the current ice-age frequency – and as best we can tell, there have been stretches of up to a billion years without any major glaciation at all. It's possible that the Milankovitch cycles are able to turn glaciation on and off simply because Earth's continents are now located where they shelter the Arctic Ocean and facilitate its freezing.

Experts disagree strongly on when the next ice age is due. Many believe that orbital cycles won't conspire to bring it on for thousands of years – perhaps as many as 50,000. But William Ruddiman (University of Virginia) argues that human-driven changes in land use have postponed the arrival of an ice age that should already have begun. Perhaps the biggest wild card is global warming. If the more dire scenarios come to pass, ice sheets in Greenland and western Antarctica could disappear in the next few hundred years, which would in turn affect the timing of any subsequent glacial expansion.

Where does the wind go?

Wind is the flywheel of the Sun's heat engine, the motion produced by raw solar energy as it reaches Earth. It sculpts our weather, as it forms clouds, fronts and countless other weather features out of thin air. Yet for a surprisingly long time, people didn't quite know what to make of the wind. Classical and medieval scientists wrestled with the concept of wind as something distinct from the air itself. As the make-up of the atmosphere became clearer, physicists began to plot and explain the wind's typical course over Earth. It wasn't until the twentieth century that such wind-driven features as cold and warm fronts were conceptualized. We're still learning a great deal about wind. The strangely erratic streaks of damage produced within hurricanes and tornadoes, for instance, are only now being unravelled with high-resolution radar and fine-scale analysis.

Our windblown planet

Every spot on Earth has its characteristic pattern of breezes, gusts, gales and calm spells. Why does the wind favour certain directions? In ancient Greece, the winds were given names connected to deities – northerlies were *Boreas*, southerlies were *Notos*, and so forth. (Remember that when meteorologists say "northerlies", they mean winds blowing *from* the north.) Aristotle believed there were more varieties of northerly than of southerly wind. He attributed this to rain and snow that fell at higher latitudes, stimulating bursts of wind that emerged from Earth and blew toward the equator. Wacky as this might seem, it was the most widely accepted explanation for winds until **Edmund Halley** – he of the famous comet – came along. In the 1680s, the British astronomer took a close look at records of wind direction from around the globe (at least as far as European explorers had ventured) and connected these reports to early readings from the newly-invented barometer.

This Hawaiian tree serves as a freakish testimony to the power of ever-present trade winds.

Halley discovered a key part of our global circulation: warm air rises near the equator, flows out toward the poles at high levels, and is replaced below by a steady stream of surface winds flowing toward the equator. These are called the **trade winds**, possibly an allusion to their usefulness in carrying out trade via ship. (To "blow trade" once meant to follow a regular course.)

Trades blow from the northeast in the Northern Hemisphere and from the southeast in the Southern Hemisphere, converging on the equator. Halley thought the trade winds had this east-to-west component because they followed the noonday Sun as it traversed the globe. In fact, the easterly component of the trades, and the westerly component of the prevailing winds at mid-latitudes, are both due to the Coriolis force – a byproduct of our rotating planet. Imagine a bundle of air at the North Pole that starts flowing southwards. The closer it gets to the equator, the faster Earth spins beneath it. Without the wind itself taking a turn, the southward path is soon angling toward the west, and thus this north wind has become a northeast wind. A mirror-image effect happens in the Southern Hemisphere, so trades here blow from the southeast.

Since the trade winds only extend to the sub-tropics, Halley's simple model needed more detail. In 1735 **George Hadley** added a second loop, this one in the mid-latitudes. He correctly figured out that the air flowing up and out at high altitudes from the tropics would eventually subside – as it does around 30°N and S – and some of it would flow poleward as it descended, with a component from west to east. These are called the **prevailing westerlies**, and they are the prime weather makers across much of Europe, Asia and North America, as well as South Africa and parts of South America.

The zones of subsiding air near 30°N and S have brought misery to many across the planet. Sinking air tends to dry out and warm up, and some of the world's bleakest **deserts** (the US Mojave, Africa's Sahara, the Great Indian Desert and Australia's Nullabor) are clustered near these latitudes. Sailors feared the listless winds of the

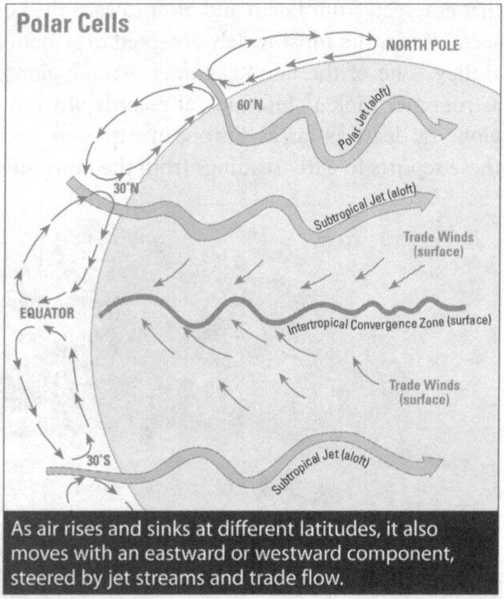

As air rises and sinks at different latitudes, it also moves with an eastward or westward component, steered by jet streams and trade flow.

30° belts. As legend has it, these were dubbed the "horse latitudes" because large ships had to toss horses and other heavy cargo overboard to avoid being becalmed.

The last major leg of global circulation was discovered by American scientist **William Ferrel**, who in 1856 came up with a three-cell model of hemispheric winds. Ferrel balanced the subsident zones at 30°N and S with belts of rising air at 60°N and S and subsidence at the poles. The result (see diagram, opposite) is a set of side-by-side circuits that pull air upward, carry it north or south, and deposit it at another latitude. On average, the surface winds blow with either an eastward or westward angle based on the Coriolis force. While it's been tweaked a little since 1856, Ferrel's theory of global circulation still gets the important details right.

A frontal assault

Any portrait of average global wind will fail to capture the nuances of even one day's weather. It would be unthinkable for the wind to blow from the southwest at every point across the US, Europe and northern Asia, as it does in the ultra-simplified Ferrel model. Real winds surge, fade and clash across a crazy quilt of directions and strengths.

Meteorologists knew this a century ago, but they didn't have a solid theory to explain it. Weather maps showed centres of high and low pressure circled by **isobars** – lines showing equal pressure. Wind arrows spun around these highs and lows on the map. Yet nowhere did the arrows collide to form what we now think of as a **front**. The concept simply wasn't part of the weather vocabulary of 1900. Instead, forecasters relied on a byzantine set of guidelines to explain and predict local weather. These rules of thumb left the door open for colossal blunders.

All this changed with the advent of **frontal theory**, an intellectual battering ram forged in the cool northern air of Bergen, Norway. At the so-called Bergen School, **Vilhelm Bjerknes** assembled a feisty group of scientists – including his son, Jacob – who turned the world of weather forecasting upside down in the 1920s. By thinking of Earth's atmosphere as a fluid, they took the notion of discontinuity (think of the line separating oil and vinegar in a jar of salad dressing) and applied it to weather. They named these sharp boundaries of pressure and temperature "fronts", and labelled them as warm or cold, based on whichever air mass was advancing.

To understand fronts, it's helpful to first look at **highs** and **lows** (see diagram overleaf). An "H" on a surface weather map denotes a centre of high pressure at ground level. Basically, there is more air sitting on top of this imaginary H than there is around it, so the high resembles a mound of air, with winds cascading downward and outward from it. An "L", or low-pressure centre, is the opposite beast. There's more air around the low than at

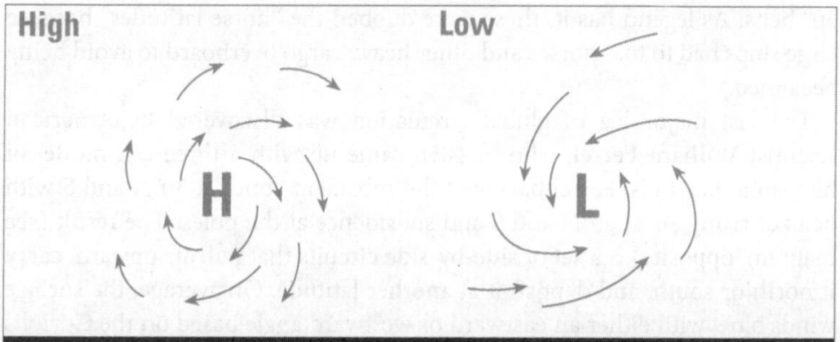

Air flows outward from a high and inward toward a low. In the Northern Hemisphere (shown here), the flow also angles clockwise around a high and counterclockwise around a low.

its centre, so winds feed into the low and ascend through it. Thanks to the Coriolis force, winds spin clockwise as they depart a high and counter-clockwise as they approach a low in the Northern Hemisphere (the reverse is true south of the equator). The highs and lows that frequent the mid-latitudes form and decay with the help of weather features aloft.

Fronts are bands of lowered pressure that arc outward from low-pressure centres. Each front serves as a dividing line between highs of different atmospheric qualities. For example, highs often form over regions where the land (or sea) are fairly uniform, so this gives the air within each one a fairly consistent texture. If a high from the Arctic strikes in January, you can expect cold weather. The Bermuda high in summer helps deliver hot, humid conditions to eastern North America. While air masses do modify as they move over a different part of the globe whence they formed – Arctic air gradually loses its bite as it moves over warm land – they generally retain enough identity so that most frontal passages produce a distinct change in the weather. Incidentally, fronts don't have to stretch across a continent; even a sea breeze is a micro-scale front.

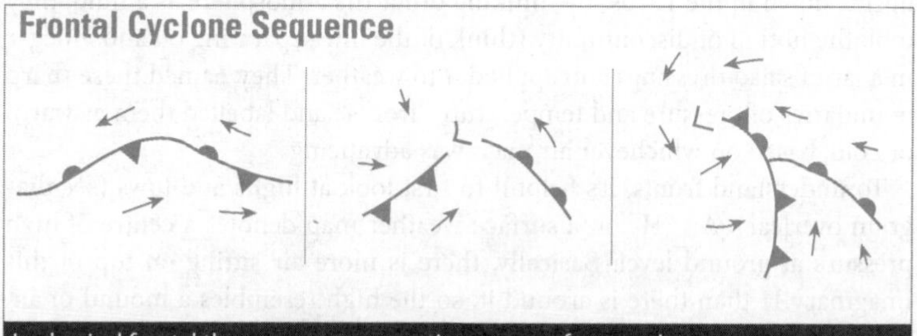

In classical frontal theory, a zone separating warmer from cooler air buckles as a core of low pressure intensifies along it.

A concept worth flushing

Everyone knows that the water in a toilet bowl flows in the opposite direction if you cross the equator ... or does it? Charlatans in equatorial countries have been using this notion for years to make a fast buck and dazzle visitors. With a tour group in tow, they carry a bowl with a drain across the equator and demonstrate how the water flow changes. They may be good salespeople, but they're contradicted by the **Coriolis force.** This effect of the rotating Earth is what makes low-pressure systems spin counter-clockwise in the Northern Hemisphere and clockwise in the Southern Hemisphere. But the Coriolis force is most evident on large-scale circulations like hurricanes and winter storms. The smaller a system, the less time and distance there is for the flow to change direction. Dust devils – typically only a few metres across – rotate in either direction in both hemispheres, and even some tornadoes rotate contrary to what you'd expect within a given hemisphere. Unless the water in a toilet bowl has days to drain, Coriolis won't make it spin any differently. The flow after a flush is far more dependent on the shape of the bowl, how the water enters, and other such factors. Moreover, the Coriolis force is strongest at the poles and weakest at the equator; hurricanes never form at 0° because they can't rotate there.

The thread that runs through the jumble of highs, lows and fronts in the Bergen School scheme is the **polar front**. This is the semi-permanent boundary between the temperate air of mid-latitudes and colder air closer to the poles. Bjerknes and his group linked the rising band of air in the Ferrel model to a polar front at the surface. Although the polar front straddles the globe at an average latitude of about 50 to 60°N, it sinks toward the equator in winter and retreats toward the poles in summer. On top of this seasonal shift is a whole series of northward and southward thrusts (see diagram below-opposite) that represent the individual cold and warm fronts we're used to seeing on the weather map.

Weather at altitude

Fronts aren't merely lines on the ground. Cold air tends to sink, so a **cold front** slides forward like a mass of syrup flowing over a waffle, sloping backward with height. The uplift of an approaching front often helps to trigger intense but relatively brief showers and thunderstorms. A **warm front**, on the other hand, is tilted forward, with the cold air retreating and the warm air overspreading it. This typically produces a longer period of cloudiness with the possibility of steady rain or snow. Not all parts of the world get regular cold and warm fronts in line with the Bergen model. Classical warm fronts do not exist in Australia, for instance. Cool fronts often sweep in from the Southern Ocean, but the continental air that succeeds them is so dry that clouds and rain seldom form during the warm-up.

The Bergen School came up with an influential multi-step sequence for the evolution of a low along a front (see opposite). But what makes a low or

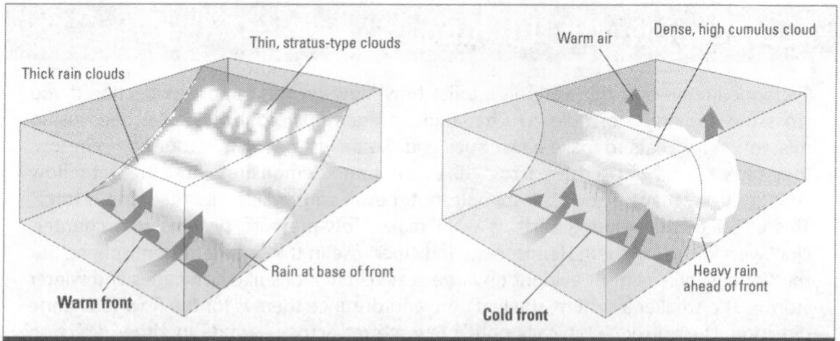

A cold front (right) often pushes up intense showers and thunderstorms along its leading edge as it ploughs forward. Along a warm front, warm air flows above cool air, often leading to lighter but more extensive rain or snow.

high form in the first place? As observations of the upper air became more frequent, starting in the 1930s, it became clear that the upper atmosphere held the key.

Roughly 5.5km/3.4 miles above sea level lies the mid-point of the atmosphere – the 500-millibar (or 500-hectopascal) pressure level. About half of the atmosphere's mass sits above this level and half below. Warm air in the lower levels tends to push this surface upward, while cold air depresses it. On a 500-millibar (500mb) map, therefore, "highs" and "lows" signify heights rather than pressure values; they represent how far up you'd need to go to reach the 500mb pressure surface above a certain point.

Although TV weather people usually refer to the **jet stream** in the singular, there is more than one of them. The best-known is the band of enhanced upper-level winds called the **polar jet**. Typically some 350–500km/200–300 miles wide and about 5–10km/3–6 miles above sea level, it circles the globe

Lynne Host / Mt Washington Observatory

Ice-encrusted much of the year, this weather station sits over 1.6km/1 mile high on top of Mount Washington, New Hampshire.

from west to east through the mid-latitudes of the Northern Hemisphere. Another polar jet runs from west to east in the Southern Hemisphere. Each of these rivers of air can separate into branches that rejoin hundreds of miles downstream. Within the polar jet are small pockets of intensified wind called **jet streaks** that can blow at speeds of 320kph/200mph or more. Meanwhile, two **sub-tropical jets** separate tropical from mid-latitude air as they flow around the globe close to 30° N and S – again from west to east. More-localized **low-level jets** can funnel air northward or southward over a few hundred miles. These usually develop in the vicinity of intense storm systems and last for a day or two, although some – like the **Somali jet** that swings from east Africa to India, feeding the summer monsoon – can persist for months. Situated only a mile or so above ground, low-level jets can blow at speeds topping 130kph/80mph.

Highs and lows on the ground are usually pushed anywhere from 160–800km/100–500 miles ahead of their upper-level counterparts. In the Northern Hemisphere, a southward dip (a trough of low pressure) in the jet stream induces upward motion on its east side. This helps to generate a surface low along the nearest front. As the trough moves east, it urges the low along. The low may intensify if the right conditions are present, such as a strong contrast between air masses below, or upper-level winds that diverge markedly.

Some surface lows don't rely on upper-level support. "Heat lows" form in the summer desert due to the rising of sun-heated air each day. Heat lows don't cause much weather in themselves, although they can help strengthen a monsoon-type pattern by pulling in moist air from afar. Tropical cyclones, including hurricanes, rely on upper winds to be relatively weak in order to build the chimney-like circulation that allows them to grow.

Face to face with the jet stream

Among its myriad of influences, World War II took aviation to new heights. Fighter planes regularly soared above 8km/5 miles. There they found a strange, strong wind that blew at unheard-of speeds. One Nazi spy-plane reported 170-knot winds (310kph/195mph) before it crashed in the eastern Mediterranean. In November 1944, two US squadrons approaching Tokyo from the southwest were suddenly swept forward by unexpected tailwinds that made bombing impractical. Meanwhile, Japan was putting the jet stream to work by launching some 10,000 "balloon bombs". A few hundred of these small, hydrogen-inflated craft traversed the Pacific and made it to US shores, although their arrival was hushed up by wartime censors. Most of them landed in the Pacific Northwest, where frequent winter rains quenched the risk of explosion or fire – although not completely. Seven members of an Oregon family came upon a balloon bomb while picnicking in May 1945. Six of them were killed – they were the only civilian casualties of World War II in the 48 contiguous US states. By war's end, the jet stream had made a name for itself, but it was years before upper-air measurements brought its ever-changing structure to light.

Taking the weather's pulse

According to rock legend Bob Dylan, you don't need a weatherman to know which way the wind is blowing. He's got a point. Wet your finger, put it into the breeze, and you'll sense the wind's direction right away. You could do the same thing with a basic wind vane or, for that matter, with a Doppler weather radar that costs millions of US dollars.

Humans made do with the most cursory weather instruments for millennia. Rainfall was measured numerically as early as the fourth century BC in India, and Egyptians have measured the height of the life-giving Nile River for 5000 years using gauges dubbed nilometers. Exceptions like these aside, most weather was measured qualitatively: strong or light wind, cloudy or fair skies, dry or humid air. Numbers didn't enter the picture in a big way until the seventeenth century, when a set of Renaissance scientists figured out how to tease information out of the weight of the air, the heat within it, the strength of the wind, and the water vapour floating through it all. These new instruments transformed the way we see weather. Before long, people across the world were using a common scientific language to record the state of the atmosphere.

As instruments took flight aboard balloons, rockets and airplanes, we began to get our first truly 3-D grasp of the way weather worked. Yet nearly all instruments before World War II did their work *in situ*. **Radar** was the first widespread device that could sense the weather from afar and report back on it. Then came the weather **satellite**, a Cold War spin-off that blossomed in the 1960s and 1970s. Now, **digital sensors** and processors have reinvented classic weather instruments. Despite all the show-stopping imagery from satellites and radar, our current global weather-observing network still hinges on the lowly *in situ* instruments – thermometers, barometers and the like.

NOAA

A Frenchman of the 1600s experiments with an early barometer.

Blowin' in the wind

Perhaps because you didn't need a weatherman to measure it, **wind direction** was one of the first elements of the air that

© Bob Henson

The Shattuck Windmill Museum in Oklahoma is helping to preserve a legacy that goes back centuries.

came under regular scrutiny. Ancients in China, Babylon, Egypt, Greece and elsewhere developed **wind vanes** before the time of Christ. The most famous of these is the Tower of the Winds, built by astronomer Andonikos in the Roman agora of Athens around 50 BC. The concrete tower included a frieze with carvings of the deities associated with each of the eight wind directions (N, NE, E, SE, etc). Above this was a bronze Triton that spun on an axis, a rod in his right hand always pointing into the wind. Triton is long gone, but the 12m/40ft tower remains, along with sundials and the remnants of a water clock. England's Royal Meteorological Society uses the Tower of the Winds as its symbol.

Wind vanes survived the Dark Ages and made something of a comeback in the pre-Renaissance era, when they crowned many European churches. Even without knowing the exact speed of the wind, people already saw it as a useful energy source. Over 10,000 windmills were spinning across England by 1400, long before the devices became synonymous with the Netherlands. Meanwhile, the art of weather observing was making slow progress. At Oxford, **Walter Merle** kept one of the earliest daily weather journals known, from 1337 to 1344. A hundred years later, German cardinal **Nicholas Cusanus** came up with a scheme for sensing humidity via the amount of moisture absorbed by a bundle of wool.

A burst of scientific invention spread across Europe in the seventeenth century with some of the world's most renowned minds turning their attention to the techniques of weather observing. The atmosphere – always at hand – was one of the prime subjects of study.

▶ **Temperature** Although the ancient Greek physician Galen explored the idea of degrees of hot and cold, these weren't quantified as we now know them until centuries later. Galileo experimented with the heat-induced expansion of liquid in a tube, but the sealed, liquid-in-glass thermometer was actually invented in 1660 by a member of royalty – Ferdinand II, the Grand Duke of Tuscany and one of the famed Medici clan. He and other early makers of thermometers experimented with water and wine; mercury later became the standard. The two best-known temperature scales, Celsius and Fahrenheit, were both developed in the early 1700s (see box, p.14). By 1800, special maximum and minimum thermometers could indicate each day's high and low temperature through a small in-glass element.

- **Pressure** Measuring the weight of the air had to wait, ironically, until people believed in its absence. In the 1640s, most scientists didn't think a vacuum could occur in nature. Gasparo Berti proved them wrong with a bold experiment. He built a glass tube several storeys high alongside his house in Rome, filled the tube with water, closed the top and opened the bottom into a water-filled pail. The water in the tube dropped to a certain point, then stopped, leaving a space that Berti correctly concluded was a vacuum. Another Roman scientist, Evangelista Torricelli, built the first true barometer between 1643 and 1644 using mercury, about fourteen times denser than water. The mercury stood about 0.76m/30in in Torricelli's tube, as opposed to the water height of roughly 9m/30ft in Berti's. Today, we still refer to pressure in units of length (eg 29.92in of mercury) even though actual mercury barometers are rare. Most household barometers are now aneroid: they measure the expansion and contraction of a sensitive bellows held open by an internal spring.
- **Humidity** Centuries after Cusanus thought of weighing the water in wool to measure moisture in the air, people were building hygrometers using strings, boards and other stretchable devices that expanded or contracted as the air become more moist or dry. This bizarre line-up included an ox's intestine, a rat's bladder and strips of whale bone; even wild oats were hooked up to dials that recorded the twisting and untwisting of their sprouts. Human hair became the era's most popular tool for measuring humidity. A single lock of degreased human hair stretches about 2.5 percent in length as the relative humidity (see

© Bob Henson

Roof-mounted mobile stations became a fixture of US weather research in the 1990s.

box, below) goes from near 0 to 100 percent. By the twentieth century, many weather observers had moved from hygrometers to sling psychrometers made of side-by-side thermometers, one kept dry and the other wrapped in muslin and dampened at measurement time. When the instrument is slung (carefully) in a circle, the muslin-coated thermometer cools to a reading known as the wet-bulb temperature. Standard tables then relate the wet-bulb reading to relative humidity.

▶ **Windspeed** The first anemometer, built around 1450 by Leon Battista Alberti, featured a tiny board hanging from a wind vane. As the wind speed increased, the board swung upward with its tip following a curved scale. Much fancier versions, using plates, spheres and other objects, were created into the eighteenth century, and similar designs are found today in schoolchildren's weather kits. Britain's Follett Osler created the first practical recording anemometer by measuring the pressure on a plate that turned to face the wind. Osler cleverly combined this with a rain gauge so that a movable pen traced the readings from both. One of the most common mechanical designs, a cup anemometer, uses a set of three angled cups that catch the wind as they rotate around a spindle; the wind speed is inferred from the rotation rate.

▶ **Cloud cover and visibility** Traditionally, human observers have classified clouds by their height and appearance (see box, p.415). With the advent of aviation, clouds and visibility began to take on a new importance. Nephoscopes measure the speed and direction of cloud motion, while ceilometers use beams of light reflected off clouds to calculate ceiling height.

Humidity: it's all relative

What does "90 percent humidity" really mean? Try thinking of water vapour as something that cohabitates with dry air; water vapour becomes more prevalent as winds bring it in or lakes and oceans evaporate more of it. The warmer the dry air is, the more water vapour can coexist with it, which is where the "relative" in relative humidity comes in. Many describe this process as the air "holding" moisture, although sticklers disapprove of the linguistic convenience. If the air is cooled past a certain point, some of the vapour in it must condense – like the droplets that cling to an icy glass on a hot day. Air only needs to warm up by about 11°C/20°F in order to hold roughly twice as much water vapour. In other words, air at 21°C/70°F can hold nearly double the moisture of air at 10°C/50°F. A relative humidity of 90 percent means that you're getting 90 percent of all the water vapour that can exist at the current air temperature. Of course, this level of humidity feels much worse on a hot day than when it's frigid. That's why meteorologists prefer to use something called the **dew point**, a temperature-like value that serves as an absolute measure of how much water vapour is in the air. If you cool air down to its dew point, the relative humidity is, by definition, 100 percent. A dew point of 15°C/59°F feels only mildly humid, while most people would consider a dew point of 25°C/79°F extremely muggy, regardless of the relative humidity. Once dry air warms to above 32°C/90°F, it can coexist with a vast amount of water, so much that it never gets saturated under everyday conditions. You may hear somebody tell you that they sweated through "95 degrees [F] and 95 percent humidity". Such an extreme would not only be intolerable, it's virtually impossible. Some of the world's most oppressively humid regimes would be hard-pressed even to reach 32°C/90°F while maintaining 90 percent RH.

Satellites are another key source of cloud cover data. Visibility at ground level can be judged by whether certain objects at a known distance, such as buildings or mountains, are obscured or not.

Along with these core instruments, a wealth of others have evolved, measuring such things as the duration of sunshine. By the early twentieth century, technology allowed the continuous recording of many weather variables. Microprocessors now allow almost every type of atmospheric quality to be measured digitally with incredible precision. Some high-tech instruments can deduce temperature, moisture, wind and other variables by examining how the character of an electromagnetic wave (wavelength, speed, etc) changes as it travels through air across a short distance from transmitter to receiver.

Up, up and away

These days it's a tranquil pastime of the middle classes, but **hot-air ballooning** was once a heated scientific endeavour. The French brothers **Joseph-Michel** and **Jacques-Elienne Montgolfier** pioneered this particular mode of travel in 1783 and not long afterwards researchers around the world were taking barometers and thermometers with them on air journeys. An English mathematician and his co-pilot suffered frostbite and asphyxia in 1865 while observing the weather from a balloon nearly 10km/6 miles high. By the 1920s, automated instruments were climbing by kite, balloon and aircraft and recording whatever they found. However, many readings were lost as instruments crashed to the ground far from their launch sites.

Radio saved the day in the 1930s by allowing weather data to be transmitted as it was collected. Within a few years, countries

NOAA

Radiosondes were a vital tool in the 1940s – and still are.

around the world were launching hundreds of the small instrument-and-balloon packages known as **radiosondes** each day. Using compact sensors that have grown ever smaller and more sophisticated with time, radiosondes send back temperature, pressure and humidity data. Wind speed and direction were traditionally inferred by tracking the path of the radiosonde; now, the motion is calculated using **Global Positioning System (GPS)** sensors. The data gathered daily from hundreds of radiosonde journeys continues to serve as the framework of our global weather picture, supplemented by other observations and by computer models.

What about beaming a picture of Earth itself from space? In his 1865 book *From the Earth to the Moon,* futurist/novelist **Jules Verne** wrote of lunarnauts viewing "rings of clouds placed concentrically round the terrestrial globe". Fellow dreamer and novelist **Arthur C Clarke**, in the 1940s, was among the first to imagine satellites stationed in space. Some of the first cloud photos from space were taken from 1946 to 1952 by Nazi-manufactured A-4 rockets that had been confiscated by the US. Still, the meteorological Holy Grail remained a device that could take image after image without plummeting to Earth.

The launch of the Soviet Union's Sputnik satellite in 1957 goosed the US into stepping up its own satellite development. In April 1960, an experimental US **weather satellite** went into orbit. Five years later, a successor had delivered enough images to assemble the first mosaic portrait of the entire planet, and in 1966 routine coverage began. Today, roughly a dozen weather satellites from the US, China, Japan, India, Russia and a European consortium keep a continuous watch on Earth, and another dozen or so experimental weather satellites are also in the sky. Most satellites are either **geosynchronous** (rotating with the planet above a fixed point at heights of about 36,200km/22,500 miles) or **polar-orbiting** (circling the planet from pole to pole about every 100 minutes at altitudes near 800km/500 miles).

Typically, satellites can detect energy from more than one wavelength, which expands their range of products far beyond a simple photograph.

- **Visible** imagery provides the crisp black-and-white satellite photos often shown on the evening weathercast and on weather websites. These devices rely on sunlight reflected from clouds and Earth.
- **Infrared** sensors detect the heat, rather than visible light, rising from clouds and from water vapour. Since they don't rely on sunlight, they can furnish photos 24 hours a day. The infrared photos on TV weathercasts are often colour-enhanced, with the colours denoting cloud temperature. The most vivid hues are normally keyed to the highest (and thus coldest) cloud tops near the centres of thunderstorms, hurricanes and other big weather systems.

Another revolutionary weather sensor from the twentieth century was **radar**, developed in the 1930s and refined in World War II as a technique for

How to read a weather map

There's method in the madness of your local TV weathercaster when he/she is rambling on about curved, studded lines hanging from a little red "L". The fronts on a basic weather map were devised by Bergen School meteorologists early in the twentieth century (see p.25). It took years for them to be accepted by scientists in much of Europe and the US, and they didn't become part of pop culture until TV and newspapers made them a fixture in the 1950s. Here's a meteorological phrasebook to help you interpret surface weather maps.

- "L" denotes low pressure, "H" high.
- The thin lines sometimes shown as wrapping around Hs and Ls are **isobars** – lines that connect points of equal surface pressure. By and large, the surface wind blows along the isobars, but it tilts slightly inward toward low pressure. Because winds are driven by pressure contrast, they're almost always strongest where the isobars are most tightly packed. Quickly intensifying lows or strengthening highs may induce the wind to cross isobars at a sharper than usual angle from high toward low.
- Lows, highs and isobars are drawn on a map only after the pressure at each station is adjusted to a sea-level equivalent. The idea is to keep the lower pressure that's always present at high-altitude stations from skewing the entire map (and producing a misleading "L" across the mountains day after day).
- **Cold fronts** are usually drawn in blue with triangular barbs; **warm fronts**, in red with semicircles. (The barbs and semicircles allow fronts to be recognized in black and white.) **Occluded fronts**, less common than the first two, occur where a cold front overtakes a warm front; they're traditionally drawn in purple with rectangles. Troughs of low pressure that have little temperature change across them, such as dry lines in the southern US plains, are sometimes depicted as dashed lines or as fronts spiked with brown rectangles.

tracking aircraft. The word is an acronym for *ra*dio *d*etection *a*nd *r*anging, which is exactly what radar does. Microwaves travel outward from a rotating transmitter at the speed of light in brief pulses, about a thousand each second. Each pulse is intense – carrying about a megawatt of energy – but lasts only around a millionth of a second. Thus, when averaged over time, a typical weather radar emits about 1000 watts of energy, roughly the same as a microwave oven.

A tiny fraction of each radar pulse bounces back from objects it encounters along the way. The time that elapses from departure to return indicates how far away the object is, and the strength of the returning signal is a clue to the object's size and/or density (large, dense objects usually reflect a bigger fraction of the energy that hits them). By converting these signals into displays, one can locate objects and track their motion. Weather was an irritant to wartime radar operators: the blobs of energy that came back from rain and snow hindered their ability to watch for aircraft. After the war, however, it became obvious that this wartime bane was a peacetime boon.

Since the 1980s, **Doppler radar** has come into vogue among weather watchers working for government and in the television industry, especially in the United States. A Doppler radar does everything its older counterpart can do and more. It measures the frequency of a returning signal as well as its strength and timing. Just as your ear can sense if an ambulance is approaching or leaving by the change in pitch of its siren, a Doppler radar monitors the difference in frequency between the signals it transmits and receives, in order to tell which direction rain, snow and other particles in the air are moving. It can spot potential tornadoes by looking for a 180° change in the direction of strong winds across a distance of less than 3.2km/2 miles, often coinciding with a hook-shaped region on one side of the storm. Not every signal means a twister is on the ground, but the Doppler network installed across the US in the 1990s has more than doubled the warning lead time, which now averages over ten minutes. Portable Doppler radars also came onto the scene in the 1990s, allowing scientists to profile some of the world's wildest weather up close.

Radar is the most famous member of a whole family of sensors, each of which send out pulses of energy and analyses the returns. **Profilers** are upward-pointing radars that track the winds well above ground level. **Lidars** are radar-like devices that use laser beams rather than microwave energy; this enables them to watch the small-scale motions of tiny air particles.

Solid observations are crucial not only to daily weather forecasting but also to tracking climate change. Global standards and long-term station records help ensure that the data show actual trends rather than variations in measurement technique. The biggest gap in measurements continues to be across the oceans, which cover almost three-quarters of the globe. Ships often take measurements along their paths, and fixed buoys can be the closest thing to a land-based weather station. Still, the observational net remains thin over the oceans. A new technique that may help fill in the oceanic gaps involves measuring a tiny atmosphere-induced delay in GPS signals to infer the state of the atmosphere across large distances.

Sun, sky and colour: the optics of weather

It's a lazy summer evening by the lake. The Sun melts from its usual yellow into a blazing red as you watch it set over the water. As the bottom of the Sun noses below the horizon, the whole sphere seems to enlarge and flatten, as if somebody were stretching it against the land. Finally, the uppermost sliver of the Sun starts to disappear, and a twinkle of green light appears at the last second.

You've just seen the **optics** of the atmosphere at their most dramatic – and illusory. The Sun is actually neither yellow, nor red, nor green. It doesn't really expand as it sets; instead, the atmosphere bends and spreads sunlight to make it appear so. And by the time you saw the base of the Sun hit the horizon, the entire sphere was already below your line of sight; the atmosphere's bending of solar rays allowed you to watch the Sun even after it set. Like the reflections in a carnival fun house, the colour and light we see in the sky aren't always what they seem.

© Bob Henson

Sunset brings out the best in atmospheric optics.

Blue skies, sort of

Several things happen to **rays of light** as they complete their eight-minute journey from the Sun and pass through our molecule-laden, particle-cluttered atmosphere. First and foremost, they get scattered. One way of visualizing a ray of light is as a series of waves, with peaks that are separated by a constant distance. The colour of the light depends on the distance between those peaks – a number known as the **wavelength**. Visible light includes wavelengths as low as 400 nanometres (violet) to as high as 700nm (red). A nanometer is one billionth of a metre, or about forty billionths of an inch. In between these extremes are the other familiar colours of the rainbow: blue, green, yellow and orange, in order. Flanking the visible spectrum on either side are the invisible wavelengths of ultraviolet (less than 400nm) and infrared (greater than 700nm). With a little help from the atmosphere, this distribution of light explains why the sky is blue. If a ray of incoming sunlight hits a particle whose diameter is much smaller than the light's wavelength, the light gets scattered in all directions, including toward Earth, and the scattering is greatest for the shortest wavelengths. This is the case for **nitrogen** and **oxygen molecules**, which make up the bulk of the atmosphere. They're only a few nanometers across, a tiny fraction of the wavelengths of visible light. Thus, the bulk of the scattered light falls at the lower end of the visible spectrum. Violet has the shortest wavelength, but sunlight delivers much more energy in blue – the next colour up the spectrum – so blue is the overall colour we see scattered toward us when we look up on a clear day.

Of course, the rich indigo above a mountain meadow isn't the same as the copper haze above a polluted city. When the air is hazy or dirty, it's full of many more large particles than you find in clean, clear air. This means that more of the sunlight is scattered – not just the smaller violet and blue wavelengths, but other colours as well. This pushes the blend of skylight away from a deep blue and toward paler, lighter shades. When soot and other visible particles are present, they add their own colour to the mix.

All about rainbows

Since water and ice are ideally transparent, they allow light to do a whole host of interesting things. The mechanism behind the beauty of the **rainbow** was explained hundreds of years ago in a tag-team effort by **Rene Descartes** and **Isaac Newton**. Descartes, the French philosopher-scientist, theorized in 1637 that the rainbow was produced by sunlight reflected toward a viewer by raindrops. Descartes was on target as far as he went, but he couldn't correctly explain why the rainbow had colours. Newton, through his innovative experiments with glass prisms in 1666, showed how raindrops separate light into the familiar colours of the rainbow. The full spectrum of visible light is usually interpreted as seven bands, corresponding to seven wavelength regions – red, orange, yellow, green, blue, indigo and violet. Newton added indigo to the list largely for philosophical rather than perceptual reasons. Actually, rainbows have an indefinite number of hues. Through a cognitive process called metamerism, it seems we tend to clump the colours across this continuous spectrum into the rainbow bands we perceive.

It's easy to know where to look for a rainbow: opposite the Sun, against a shower or thunderstorm. In the mid-latitudes, most rainbows occur in the late afternoon and evening, when the Sun is toward the west and the depart-

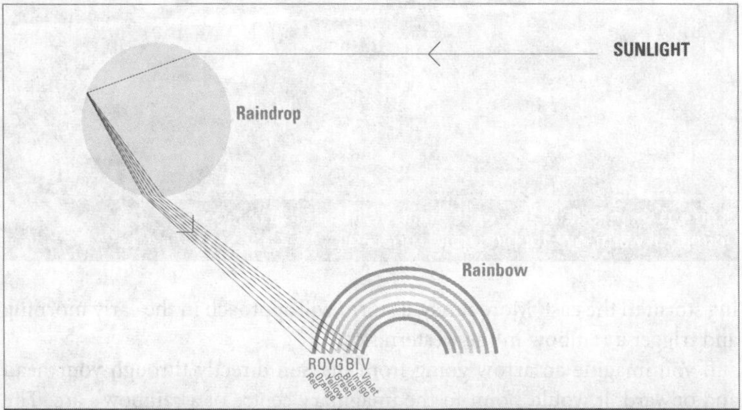

The sunlight enters a raindrop, tiny variations in the entrance angle translate into a separation of hues as the light reflects and exits the drop. If conditions are right, the colours show up as a rainbow across the sky from the sun.

Castles in the air

We're used to thinking of mirages as oases in the desert, surrounded by palm trees. In real life, our most likely encounter with a **mirage** is on the highway, where water may seem to shimmer in the distance. Mirages rely on a property of the atmosphere called **refraction** – the bending and spreading of light rays as they pass through air layers of different densities. On a bright, sunny day, heating at the surface – especially above a quick-to-heat surface like pavement or sand – lowers the density of the air nearest the ground, where usually it's the most compressed. The light rays bend upward as they hit this less dense layer near the ground, so that the light from something in the distance (like the sky) may actually be coming from below the horizon when it hits our eyes. A bright patch may thus appear on our visual image of the highway. Tiny variations in air temperature and density give the mirage its shimmering, waterlike texture. If the vertical profile is reversed (warm air on top of cool), a surface-based feature may appear above ground. Such is the case with the **fata morgana**. Sometimes used as a synonym for any mirage, it is in fact a particularly rare one usually seen at sea. The fata morgana is named after Morgan le Fay, the shape-changing half-sister of legend's King Arthur. This enchantress lived in an undersea castle and lured sailors to their death by reflecting her castle's image into the sky. Likewise, the fata morgana stretches ground-based features upward so that the sea surface becomes a row of mountains or a bank of turreted castles. In 1906, when **Robert Peary** thought he saw a mountainous land just beyond reach in the Arctic Ocean, it was almost certainly the fata morgana that fooled him. It took seven more years for another explorer, **Donald MacMillan**, to reach the site and discover that the so-called Crocker Land was only an illusion. **Henry Wadsworth Longfellow** saluted the famed fata morgana in his poem of the same name: "I approach and ye vanish away, I grasp you, and ye are gone/But ever by night and by day, The melody soundeth on."

Weatherwise Magazine / Seji Miyauchi

Icebergs seem to float above the Antarctic, thanks to a mirage formed in warm, overtopping frigid surface air.

ing storm to the east. More rarely, a storm will approach in the early morning and trigger a rainbow in the western sky.

If you imagine an arrow going from the Sun directly through your head and onward, it would point to the imaginary centre of a rainbow's arc. The angle from that line to the arc itself is always 42°. To explain that number, you have to look inside a **raindrop** (see diagram, p.39). When sunlight enters a raindrop, some of it bounces off the back of the drop and re-emerges on the same side it entered, but at an angle. This angle is a little different for each entrance point, but a good deal of the existing light is clustered around an angle of 42° from the entering light. The wavelengths that produce the colours of the main or primary rainbow each emerge at slightly different angles close to 42°, thus yielding the familiar bands, with red always at the top.

A careful rainbow-watcher will notice several quirks that can be easily explained. Rainbows tend to be brighter along their legs, near the ground, than at the top of their arcs; this is a result of the shape of raindrops. As we'll see in the Rain section (p.52), most large raindrops don't actually resemble teardrops. Since these squashed drops are flatter than they are tall, the reflections in the vertical dimension (which produce the top of the rainbow) are less focused than those in the horizontal (which produce the rainbow's feet). The brightness and crispness of the entire rainbow depends on the density and character of the raindrops, as well as the clarity of the atmosphere in between rainbow and viewer.

Sometimes a rainbow has a twin, a fainter arc well above the main bow. This outer or **secondary rainbow** is produced by light that's reflected not just once, but twice, within different drops than the ones that produce the main rainbow. The secondary rainbow is always angled at 51° from the imaginary centre point discussed above. Because of the extra reflection that turns each ray of rainbow light more than 180°, the secondary rainbow's colours are reversed: red appears at the bottom, violet at the top.

Much white light gets reflected toward the violet side of a rainbow. Thus, you may notice that the sky beneath a bright single rainbow is noticeably lighter than the sky above it. Likewise, the sky above a secondary rainbow is slightly brighter than the sky between the bows. The in-between region is called **Alexander's dark band**, after the sharp-eyed Alexander of Aphrodisias.

In theory, at least, the rainbow family extends beyond twins. Light that's reflected three times within a drop should produce a third rainbow – but it would appear across the sky from its siblings, around the Sun itself. Since this third rainbow would be even fainter than the first two to begin with, the chances of seeing it amid sunlight are basically nil. Only a handful of sightings have ever been reported.

Haloes and glories

Rainbow-like features paint the sky in several other ways. When the Sun or Moon is behind a thin sheet of cirrus cloud, it may be encircled by a **halo**, typically reddish-yellow at its inner edge and whitish beyond. Like rainbows, haloes are produced by light passing through airborne water, but in this case the water is frozen. The ice clouds that produce most haloes are made up of

© Gary Herbert

Sun dogs appear on either side of this dramatic halo, with a tangent arc at top.

six-sided plates and pencil-shaped crystals. Light that enters these crystals departs at a variety of angles, and the crystals themselves may be oriented randomly. Still, there's a concentration of light close to a 22° angle from the Sun or moon. As with rainbows, haloes can occur at other angles, too.

The plate-shaped crystals aren't always mixed up grab-bag style; larger ones fall nearly flat. In this case, the visible reflections occur mostly along a horizontal line passing through the Sun. These are **sun dogs** – small, bright patches, often in rainbow colours, that flank the Sun on either side at a 22° or greater angle. Since they're formed by a more organized array of crystals than

The mysterious green flash

Jules Verne's 1882 novel **Le Rayon vert** (**The Green Ray**) focused on a rarely seen optical phenomenon. In the last moment of sunset at sea, the remaining glimmer of sun may turn "a green which no artist could ever obtain on his palette ... If there is a green in Paradise, it cannot be but of this shade". Skeptics long thought the **green flash** might have been a visual after-image produced by staring at the setting sun, but it can also occur at sunrise. Others considered it a result of sunlight passing through waves on the water, yet it's also been reported over the desert. The flash can be explained by thinking of the atmosphere as a prism that separates the disk of the setting Sun into a rainbow spectrum: violet on top, red on the bottom. Violet would normally be the last visible colour, but since the atmosphere tends to scatter blue and violet light out of the direct beam, what's left in it is green. Photographers struggle with the green flash's brevity (a second or less in most places) and its tiny dimensions (too small for anything but enormous telephoto lenses to capture). At high latitudes, the Sun can hover just at, or below, the horizon for hours in the summer, making for a prolonged show of greenery. On October 16, 1929, an Antarctic exploration team led by **Richard Byrd** watched the green flash come and go over a 35-minute span.

those that produce haloes, sun dogs can feature more well-defined colours, although they don't normally approach the brilliance of rainbows.

Some very different rings of light – **the glory** – may be one inspiration for what artists have placed around the heads of spiritual icons for centuries. When a plane flies above a cloud bank on a sunny day, a small, roughly circular ring – usually in very pastel shades of red – may form around the plane's shadow. It's caused by **diffraction**: the interactions of sunlight at various wavelengths and angles hitting the faintly obscured area around the plane's distinct shadow. Even though the shadow reduces the total amount of light, cloud droplets can return some of the remaining light and, ironically, make this region brighter and more colourful than the areas in direct sun. The width of a glory depends on the size of the water droplets within the cloud; some glories are concentric. Halo-painting artists may also have been inspired by the more common **heiligenschein**, a bright light seen immediately around the shadow of your sunlit head on a variety of wet and dry surfaces.

Long before airplanes existed, glories were created by people perched on mountaintops. If you stand on a peak with the Sun at your back and a cloud bank below, you might see a huge-looking, elongated shadow of your body on the cloud, with a ring of glory surrounding your head. It's easy to believe that myth-makers drew on this spectacular effect in depicting the holiness of Christian angels, Greek gods and Indian icons.

Climate zones

Nobody really experiences climate per se. You can experience weather any time just by stepping outside, but climate is the average of sun, rain, snow, wind and other elements playing out over the long haul. This makes it an abstraction, built from the fragments of weather that caress or assault us. Even so, climate is a useful way to think about the sum total of weather and the ways in which it can differ so dramatically from place to place.

It wasn't until the late 1800s – by which time Europeans had touched most corners of the world – that scientists could truly begin to assemble the jigsaw pieces of world weather. The first part of the twentieth century became the age of **climate zones**. Climatologists drew lines to separate one weather regime from another, just as natural scientists had been doing with plants and animals. The difficulty is, climate has few of the obvious break points available to naturalists. It's easy to distinguish a rose from a cantaloupe, but what amount of rainfall ought to separate a semi-arid from an arid region? Added to this, weather doesn't sit still: it flows across artificial climate boundaries with ease.

Because weather is such a fluid quantity, climatologists lean on averages to carry out their work, a process fraught with peril. San Francisco, California, has roughly the same annual precipitation as St Petersburg, Russia, but the two climates are vastly different. When does most of the rain or snow fall? How much of it falls as rain and how much as snow? Does it tend to fall in occasional bursts or in more frequent, gentle rounds? How dry are the driest years, and how wet are the wettest? These are the kinds of questions climatologists address as they try to build a coherent picture from our world of weather.

From pole to tropic

Schoolchildren quickly learn that it's cold at the North and South Poles and hot at the equator. The world's north–south temperature gradient was the first step in climate mapping. Even the ancient philosophers of Greece – who theorized that Earth was round centuries before anyone could prove it – had a sense of this temperature distribution. Working with very little raw data, **Parmenides** (c. 500 BC) broke Earth's surface into a torrid tropical zone, two temperate zones in the mid-latitudes, and two frigid zones beyond that. The Greek astronomer **Hipparchus** further subdivided these realms into belts based on the amount of sunlight each latitude gets at the summer solstice. He called these belts *klimata* – the root of the English word climate.

The ancient Greeks had a decent fix on things, as it turns out. The **tropics** are indeed a separate climatic realm. Their strict definition is the same as it was to Parmenides: the land between the **Tropics of Cancer** and **Capricorn** (23.5°N and S). Because the tilt of Earth's axis is 23.5°, these two latitudes mark the poleward limit of full sun, the points where the Sun is directly overhead at noon on the summer solstice.

A more subtle definition of the tropics involves **temperature**. You know you're in tropical territory when a typical day's low and high span more degrees than the difference between the average temperature on a summer and on a winter

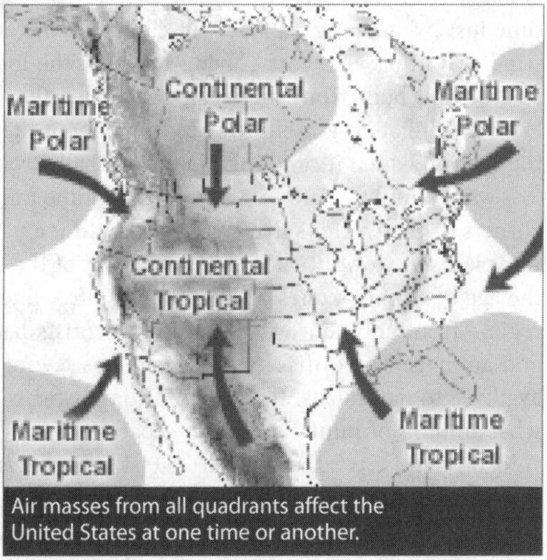

Air masses from all quadrants affect the United States at one time or another.

UCAR / COMET

day. Indeed, nighttime is often called "the winter of the tropics". This view of climate would be unthinkable in a place like Beijing, where the summer–winter difference is in the order of 30°C/54°F, but in the tropics the seasons usually vary by less than 8°C/14°F. Here, rainfall variations are far more dramatic than annual temperature swings, and locals typically refer to wet and dry seasons rather than warm and cold seasons. There's an added twist in Latin America, where the astronomical winter, which is usually quite dry, is referred to as "summer" and the wet astronomical summer is called "winter". It's a vestige of the region's cultural roots in Spain, where winter happens to be the wettest time of the year.

Many of the world's **deserts** are clustered in the **sub-tropics**, close to 30°N and S, where air arrives from other latitudes at high levels and then descends (see "Where does the wind go?", p.23). Closer to the poles, the restless westerlies prevail, and the day-to-day variety of weather is the most vivid. The label "temperate" isn't quite accurate here, since it can be bitterly cold or broilingly hot, so climatologists now favour the label "mid-latitudes" for the zones from about 30° to 60°N and S.

Frigidity is an obvious trait of the **polar zones** north and south of the **Arctic** and **Antarctic Circles** (66.5°N and S). What's less obvious is how little it snows in this frozen realm. The air is so cold at all levels that it carries only scarce amounts of water vapour. In terms of moisture, much of the polar zone could be classified as a desert. The poles' famous round-the-clock summer sunlight is another misleading notion. True, the Sun may be above the horizon for most of the summer, but the solar angle is so low that it feels more like early morning or late evening around the clock. Temperatures seldom stray far above freezing, and the melting ice helps form frequent fogs across much of the Arctic. Winter, on the other hand, is pretty much what one would expect – dark and brutally cold.

More to the picture

Take a closer look at the globe (see diagram overleaf), and you'll easily find exceptions to the north–south temperature rule. Along the 60°N latitude belt, January brings average temperatures that are well below –30°C/–22°F in parts of Siberia and as mild as 5°C/41°F off the coast of Scotland. At latitude 30°N, the average readings in July soar above 35°C/95°F across northern Saudi Arabia, while they stay below 25°C/77°F to the west along the ocean-cooled northwest coast of Africa. Clearly, when it comes to temperature, latitude isn't destiny.

Virtually all of the deviations in the north-to-south temperature regime can be explained by **continents** – where they're located and what's on them. Most of the planet's land area (about 70 percent of it) is located in the **Northern Hemisphere**. Since land heats and cools more readily than water, the north

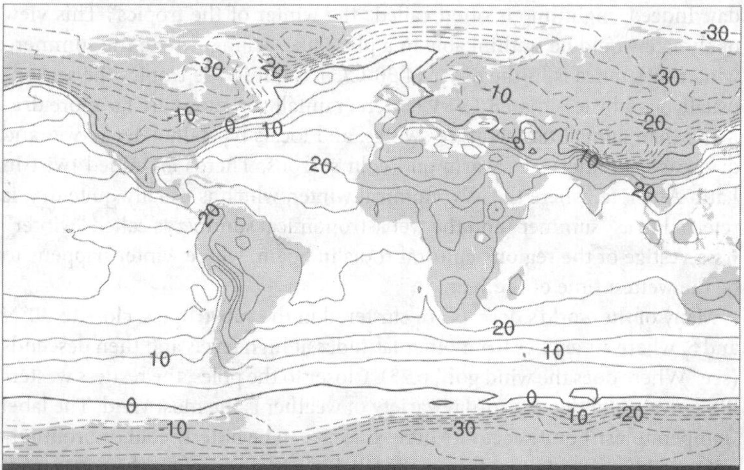

NCAR / ECMWF

This twenty-year analysis shows the average temperature for January across the globe.

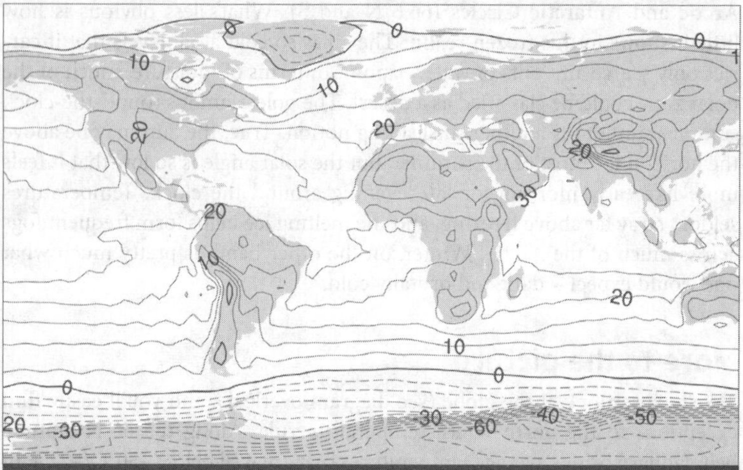

NCAR / ECMWF

Same as above, except for July.

half of the globe is where the biggest temperature swings occur on both a day-to-day and seasonal basis. Notice how North America, Europe and Asia are cooler than the surrounding oceans in winter (January) but warmer than the oceans in the summer (July). This phenomenon is less striking across the **Southern Hemisphere** because neither Africa, South America nor Australia cover much territory south of 30°S.

The pattern of **ocean currents** further sculpts the climate (see diagram, opposite). Driven by prevailing winds, the largest-scale ocean currents are

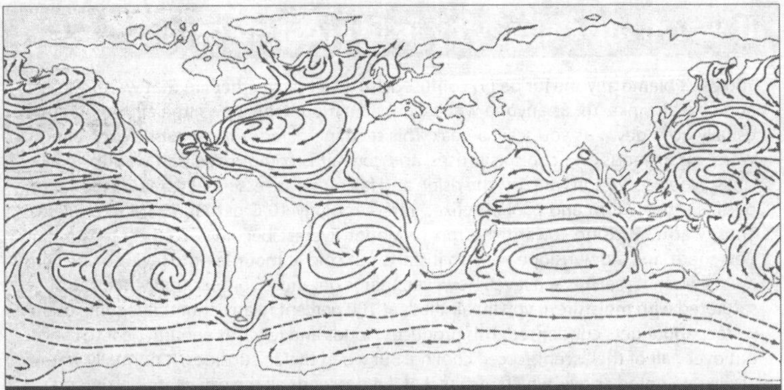

Sea-surface circulation patterns mirror those of the atmosphere, flowing westward across the tropics and generally eastward at higher latitudes.

clockwise loops in the Northern Hemisphere and counter-clockwise loops in the Southern Hemisphere. The poleward flow of warm water helps to moderate the climate across western Europe and western Canada, while the return flow of cooler water leads to milder summers than latitude would predict along the coasts of California, Spain and Morocco, Ecuador and Chile, and Namibia.

Mountains are the next-biggest climate player. Their sheer height makes it possible to experience snow near the equator. Mountain ranges that intercept moisture-laden winds produce some of the world's wettest climates, often paired with an arid zone on the other side of the slopes.

Another key climate element is the presence of **large inland bodies of water**, like the Black Sea or North America's Great Lakes. These help keep temperatures more moderate, and they tend to enhance rainfall or snowfall near their coastlines. Bands of converging wind result from differences in surface wind speed and direction, which are in turn induced by differences in friction over land and water. These bands can produce huge amounts of rain or snow immediately near coastlines. During its ferocious winter of 1976–77, Buffalo, New York, notched 53 consecutive days of snowfall en route to a seasonal total of 507cm/199in.

Climate and the land

Step from a cool forest into a warm meadow on a sunny day, and you've had a brush with the relationship between **plants** and **climate**. Everything on the face of the globe absorbs and reflects sunlight based on its colour and composition. In turn, the global patterns of rainfall and sunlight carve out distinct zones of vegetation. If a region gets enough rain or snow year round to compensate for what's lost through evaporation, **thick forests** can thrive.

If warm air rises, why is Everest so cold?

You can't blame anyone for being confused. It's a fact of weather life that warm air rises and cold air sinks. Yet as anyone who's climbed a mountain knows, the air temperature usually goes down as you scale a peak. This seeming paradox has a subtle explanation. Warm air expands and cools as it rises, and cool air is compressed and warms up as it sinks. When conditions are dry, the rising and falling air are essentially trading identities, because they warm and cool at a fixed rate – roughly 10°C per km (5.5°F per 1000ft). Thus, warm air rising toward the top of Mount Everest becomes cold by the time it gets there, and a downslope wind that starts out cold at mountaintop level can be very toasty when it reaches sea level. However, things get more complicated when the air is saturated with moisture. If you lift air that's at 100 percent relative humidity, some of the water vapour gets condensed. This produces clouds and releases enough heat to cancel out over half of the ascent-forced cooling. But when the air subsides, it normally warms at the rate noted above, which means that if it descends all the way to its starting point, it ends up warmer than it was originally. Because of this asymmetry, the atmosphere is constantly shifting to adjust to temperature variations, both horizontally and vertically.

Forests cover about a third of Earth's land area, including high-latitude areas where precipitation is light but evaporation is low as well. **Grasses** dominate about one-fifth of Earth's surface, where the rains are lighter or more concentrated in the summer months, when evaporation is at its strongest. **Prairies** segue from tall grass to short grass toward the more arid end of this spectrum. Grasses may also take hold where forests have been cut or burned for agriculture, as is the case across vast tracts of tropical rainforest. As much as 20 percent of Earth's land surface is occupied by **desert**, where cacti and shrubs endure by storing water (a year's worth may fall in a single day) and drawing on below-ground aquifers. The rest of the planet is a mix of ice, mountains and ground-hugging tundra vegetation – the only plant life possible in the chilly, damp reaches of northernmost North America and Eurasia.

Plants formed the basis for the first major advance in global climate classification beyond the Greeks. A number of scientists, many in Germany, spent the 1800s relating plants to the characteristics of weather and climate. **Wladimir Köppen** synthesized and built on this work and, in 1900, began what would become more than a half-century of mapping global climate. Köppen built a base of five climate types – dry, tropical rainy, warm temperate forest, snowy forest and polar – and broke these into sub-groups with a set of temperature-based thresholds. A "snowy forest" climate, for instance, has an average temperature below freezing in its coldest month and above 10°C/50°F in its warmest month. Examples include the northeast US, eastern Canada and most of Russia. Köppen breaks off a sub-zone where fewer than four months in the year average above 10°C/50°F; here, deciduous trees struggle and evergreens dominate.

Although it still commonly appears in textbooks, the Köppen classification scheme isn't the only one. By focusing on plants, temperature and precipitation amounts, Köppen downplayed a key aspect of climate: the balance between **precipitation** and **evaporation**. Our planet has an essentially fixed amount of water, so in the long term, what goes up must come down. This doesn't happen equally, however. Much of Australia's rainfall gets sucked up quickly, leaving the continent with few major rivers. Equatorial Brazil, on the other hand, tends to get much more rain than it can evaporate, so a lot of the excess flows to the sea via the Amazon. In the 1930s, US climatologist **Charles Thornthwaite** – who cut his research teeth in Oklahoma studying the US Dust Bowl – created a system based on the ratio of precipitation to evaporation, regardless of the plants that populate an area.

The art of classifying climate has been eclipsed in recent years by the study of how climate evolves on the geologic scale of millennia as well as on the more tractable periods within our cultural record. **Radiometric dating** and other tools of the geologist and archeologist have shed much light on how our atmosphere has arrived at its present-day state. These techniques should prove invaluable as the task of assessing tomorrow's climate becomes ever more critical (see "A primer on climate change", p.150).

Chapter two
The wild stuff

The wild stuff

Some of the most intense and terrifying weather features are among the hardest to decipher. Scientists are still working to understand exactly how a thunderstorm's energy becomes focused enough to form a tornado, or just how much rain a raging torrent may dump. We do know more than ever about when such events are possible and how to protect ourselves against them. If our fascination with atmospheric trouble has a plus side, it's the way in which safety messages now make their way into news reports, saving lives on a routine basis. Wild weather has a role to play in basic science, too. When researchers take a closer look at the worst that weather can dish out, they can learn much about how the atmosphere as a whole operates.

Rain

It drizzles, it pours, it pelts. **Rain** falls on the just and the unjust, winners and losers alike. In some ways, rain is the planet's great equalizer, yet rain falls anything but equally around the globe. You could spend a hundred years in the middle of the Sahara and collect less rain than an Hawaiian villager might see on a single wet day.

The life of a raindrop

Rain is only one feature on the vast, endless loop scientists call the **hydrologic cycle**. A good place to hop aboard this cycle is over the **ocean**. Each day, trillions of gallons of water escape from the sea surface to join the atmosphere as **water vapour**. This takes energy. As each molecule of H_2O evaporates, it absorbs a tiny bit of heat. The warmer the water, the more evaporation occurs. Lakes and other bodies of water also add moisture to the air, but more than 80 percent of the planet's water vapour comes from the sea.

It takes a little over a week for the average water-vapour molecule to fall back to Earth as rain. During this short spell of flight, it may get swept thousands of miles. In fact, much of the rain that falls on continents is comprised of moisture from far-off oceans. But before it can fall as rain, our wayfaring molecule has to join thousands of others to make a raindrop. This happens inside **clouds**, where the relative humidity is at, or just over, 100 percent. This air is saturated – it's been cooled (by lifting, usually) to the point where some of its water vapour must condense onto the nearest particle of salt, soot or other material. The end product is a **cloud droplet**, but one that's initially too small to fall as rain by itself.

There are two main avenues for raindrop production. The simpler of the two is the **warm rain process**. It predominates when most of a cloud is positioned lower than the freezing level, as is often the case in the **tropics**. A growing cumulus cloud might have one or two hundred droplets scattered through an area the size of your fingertip. Each of these droplets is approximately 0.001cm/0.0004in wide, less than half the width of a strand of human hair. As these droplets fall at varying speeds, based on their sizes, and bump into each other, some of them grow larger than others. The bigger they get, the faster they fall, and along the way, they coalesce with smaller droplets they collide with. This allows them to fall even faster and to absorb still more of their peers. In as little as half an hour, a **raindrop** is born, perhaps a hundred times wider than the many thousands of cloud droplets it incorporated.

When ice is present in a cloud, a different process – **cold rain** – takes the lead. **Ice crystals** are able to form only on a few special kinds of nuclei (primarily dust), so they're far outnumbered by water droplets. However, the crystals grow more quickly than the droplets, since water vapour tends to be drawn more to ice than to liquid water when in the presence of both. Some of these fast-growing crystals are soon big enough to qualify as **snowflakes**. Through mechanisms that aren't yet fully understood, the crystals throw off tiny shards that attract water vapour of their own. The result is **snow**, forming far above ground, that turns into rain on its descent through warmer air.

How fast and how much?

Only a fraction of the moisture in a typical thunderstorm becomes rain. Even the raindrops that exit a storm may not survive their trek to earth. When a rainshower forms above extremely dry air, the rain sometimes evaporates before it reaches the earth. The streaky curtain that hangs from the cloud base is called **virga**. It's a common sight across arid and semi-arid regions, especially in the summer. (For definitions of other cloud types, see p.415).

Shallow, low-slung clouds – the kind you might see hugging a coastline or banked against a mountain range – don't have the vertical extent necessary to produce heavy rain. These make **drizzle**, the misty droplets that may be only a tenth of the diameter of a bona fide raindrop. Drizzle-producing clouds can be less than 300m/1000ft thick. The sun might be shining on top of a coastal mountain even as people on the beach are hauling out their raingear.

© Bob Henson

From puffy cumulus below to gossamer cirrus above, a thunderstorm can stretch 16km/10 miles high.

When air is being forced strongly upward – whether by winds blowing up-slope, the dynamic force of an approaching upper-level low, the buoyancy of a hot and humid air mass, or all of the above – a much deeper cloud may form. **Thunderstorms** can stretch more than 16km/10 miles from top to bottom, and the temperatures inside may range from 15°C/ 59°F at cloud base to –60°C/–76°F within the icy **cirrus cloud** at its crown. With this much of a cooldown for air parcels that rise to the top, a lot of moisture gets wrung out, and the droplets and ice crystals that are left behind have more room to collide and coalesce on their way down. All this leads to the potential for seriously heavy rain. Especially when thunderstorms are embedded in a **tropical cyclone**, the amounts they can dump are astonishing.

Even a less imposing garden-variety shower can rain with surprising gusto. Anyone who's visited the wet tropics knows how a seemingly innocuous batch of cloud can produce torrents in a few minutes, then nonchalantly move on. What's helping these modest showers to pour is the sheer amount of water inside them. A nice, juicy **tropical cumulus** might have 50gm/2oz of water in a space the size of your living room. That doesn't sound too impressive until you realize just how large even a tiny cloud is. A cumulus that's only about 1.6km high and wide (1 mile in each direction) could contain over 3.8 million litres/1 million gallons of water. It doesn't take much time for the droplets in this dense a cloud to reach raindrop size.

Rain shadows and seasons

If all the rain and snow that fell throughout the course of a year were melted and spread evenly over the planet, it would form a layer roughly 1000mm/39in deep. Among the regions that average close to this global norm are western Europe and the prairies of the US and Canada. However, many places get substantially more, or much less, than this. The most dramatic rainfall contrasts within small areas are tied to **mountainsides**. The prevailing global wind patterns tend to dampen the west sides of mid-latitude mountains and the eastern sides of tropical peaks. These are some of the most dependably rainy places on earth, although a serious drought can alter things enough to dry out even the most sodden slopes. On the opposite sides of these mountains, you'll often find **rain shadows**. These regions get far less precipitation than their neighbours just over the ridge. Even a shift of a few miles or kilometres can cut average rainfall by half or more. Across the **Big Island of Hawaii**, rainfall drops from a yearly average of around 5100mm/200in at Kahuna Falls (which faces the trade winds), to less than 250mm/10in on the downwind Kohala Coast.

Most places tend to get more rain at one time of year than another. Usually the highest amounts are in the summer, if only because the air tends to be warmer and/or more humid. Mediterranean-style climates – like those of Italy or coastal California – are the exception that proves the rule. Here, persistent high pressure keeps the land drier in the summer, while winter storms bring the bulk of the rain. **Monsoons** – including the proto-typical ones across southern and eastern Asia – bring the most spectacular seasonal shifts in rainfall. After a bone-dry winter and a brutally hot spring, **India** becomes a land of greenery and relief when the summer monsoon hits. A few months later, the circulations shift and the heavy rains move south toward Indonesia and northern Australia, only to return the following summer.

NCAR / UCAR / NSF

For better or worse, the perennial torrents of the summer monsoon are a fact of life in India.

Drought (see p.110) can strike almost anywhere, but there are some parts of the globe where it simply doesn't rain much at all. These tend to be clustered around latitudes 30°N and 30°S, where the overall global circulation tends to produce sinking air and the most extensive **deserts** on earth (see "Climate zones", p.43).

All in all, it seems to be raining a bit more, and a bit harder, than it used to. Data spanning the entire Earth's surface is hard to come by, but the best estimates indicate that average annual global precipitation over land rose by perhaps one percent in the twentieth century. In many locations, the moisture that does fall is now focused in more intense bursts as opposed to longer, steadier rains. The most likely culprit is a **global increase in water vapour**,

The real shape of rain

While songwriters have long thrown teardrops and raindrops into the same lyric for emotional effect, artists often use the former to illustrate the latter. Imagine the shock of a youngster if you told them that raindrops are actually shaped more like hamburger buns than teardrops. The tiniest cloud droplets are almost perfectly spherical, but as they increase in size, there's more water inside relative to the surface area of the droplet. This means there's less surface tension to keep the sphere tight, so the droplet slowly evolves into a more oblong shape as it descends. A full-sized raindrop falls with its flat face downward. The pressure of the air below can form a dimple at the drop's centre (think of a doughnut whose hole isn't fully punched through). When we watch raindrops fall, our eyes may be tricking us into seeing a teardrop – the speed of the drop can produce a persistence effect in our visual field, much as the blur of a speeding car makes it seem longer than it is.

as suggested by data from balloon-borne instruments and satellites over the past several decades. Warmer oceans tend to evaporate more of their water, so the apparent water-vapour increase is one of the many hints that **global warming** is making its presence felt (see "A primer on climate change", p.150). On the other hand, warmer temperatures also help rainfall to evaporate more quickly, so the global risk of drought is not necessarily going down.

Snow

Nothing can lure the fun-loving child out of a grown adult more quickly than a snowstorm. Particularly in places where it doesn't fall very often, **snow** is a holiday-maker, an enforced interruption to our routine. Both kids and kids at heart fling snowballs; the more meditative among us stroll through a

© Carlye Calvin

Snowbound in Colorado.

panorama of softened sound and muted colour. There's shovelling, to be sure, but on the whole it seems a fair exchange for a day or two of magnificent transformation.

For those who live on the northern flanks of North America, Europe and Asia, snow is integral to everyday life. Transportation, housing, food – any, or all, of these may depend on the quality and timing of the winter snowpack. Airplanes and snowmobiles have introduced some flexibility to life in the Arctic, but they haven't altered its essence. Every winter, trucks rattle along a Canadian ice highway that runs over 560km/350 miles from the town of Yellowknife to diamond mines on the other side of the Arctic Circle. When the snow and ice melt, the highway is gone – until the next winter.

Shapes in the sky

The first step towards understanding snow is to ditch the stereotypical image of a monotonous field of white. There are endless variations on snow, starting with its **hue**. Specialized algae that live within snow can tint it blue, green or pink. Pollen falling from cypress and pine trees can even produce yellow snow. Snow's variety starts in the air. Every schoolchild learns that no two snowflakes are alike, but that assumption was proven incorrect in 1988. Two Colorado researchers examining snowflakes preserved from a research flight found conjoined twins: two perfectly matched columns attached to each other. Given the uncounted numbers of flakes that form each year across the world, it's highly likely that there are other duplications out there. Whether or not each one is original, the shapes snow can take are stunning.

Every snowflake begins as a single, microscopic **ice crystal** that usually forms on a mote of dust in the middle of a cloud. Among all the particles that float through the air – including salt, soot and other debris – only a select few have the right molecular structure to allow ice to form on them. **Windblown dust** is the most common of these.

Raindrops grow mainly by bumping into each other, while ice crystals grow into snowflakes by attracting **water vapour**. In a process called **deposition**, the vapour goes directly onto the ice without ever passing through a liquid phase. As the crystal grows, **temperature** is the main factor that controls what shape it will take. If the air in its cloudy womb is hovering close to freezing, the snowflake is likely to turn into an unadorned **needle** or **plate**. The classically pretty, six-branched flakes known as **dendrites** tend to develop at a cooler temperature, ranging as low as about –18°C/0°F. **Hexagonal plates** (less frilly than dendrites) form when it's as cold as –28°C/–18°F. Any colder and the result is usually a **six-sided column**.

As they descend, some ice crystals bump into **supercooled** water droplets. These have remained liquid below freezing because they lack nuclei that would help ice form. The result of these collisions, called **riming**, is less flaky,

Snow photographer Wilson Bentley captured over 5000 flakes, including these classics that reveal the favoured six-side structure.

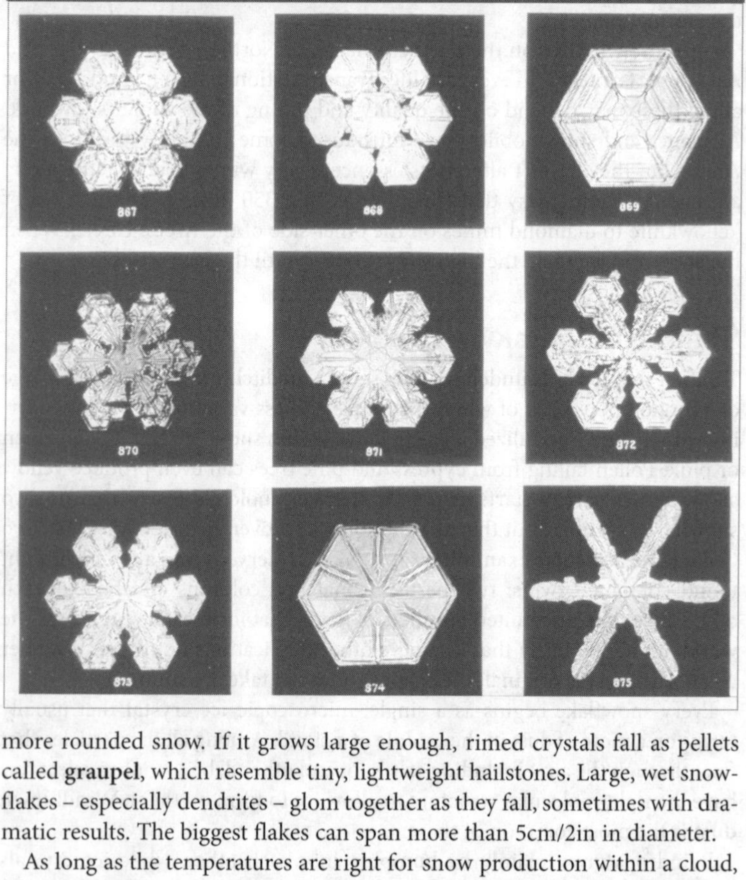

more rounded snow. If it grows large enough, rimed crystals fall as pellets called **graupel**, which resemble tiny, lightweight hailstones. Large, wet snowflakes – especially dendrites – glom together as they fall, sometimes with dramatic results. The biggest flakes can span more than 5cm/2in in diameter.

As long as the temperatures are right for snow production within a cloud, it's not all that relevant how cold it is at ground level. In fact, it's usually snowing in the topmost layers of a summertime thunderstorm, even as temperatures sizzle near the ground. Snow can accumulate on grass and other surfaces that are well above freezing, as long as it falls quickly enough to compensate for melting on the warm ground. By the same token, it's never really too cold to snow. The peak amount of moisture in the air does go down with air temperature, so if it's extremely cold at all levels it's unlikely to snow very hard. However, if a batch of near-freezing air moves on top of a shallow layer of much colder air, then heavy snow can fall into the frigid surface layer, even if temperatures near the ground are below –18°C/0°F. Indeed, some of the worst snowstorms on record have occurred in temperatures far below freezing (see box on New York's blizzard of 1888, opposite).

The geography of snow

Low-pressure centres that sweep across mid-latitudes in the winter (see "Where does the wind go?", p.23) bring most of the northern continents their snow. Many such storms have a **zone** where the temperature and moisture profiles are just right to yield the best ice-crystal production. This zone can range from about 80–160km/50–100 miles wide, and is typically located about that far to the left of the surface low as it moves along. The classic result is a swath of heavier snow bracketed by lighter amounts on either side. Radar often shows a **bright band** marking the transition aloft from rain to snow production. Forecasting where the rain/snow line will occur at the surface is a difficult task, however. A span of less than 50km/30 miles can separate drenching rain from paralysing snow. (See "Reading between the lines", p.136.)

The **Southern Hemisphere** isn't renowned for its snow, although Antarctica certainly has plenty of it, and hefty amounts can fall each year on various low-latitude mountains and on the highlands of Chile, Argentina, South Africa, Australia and New Zealand. **The Northern Hemisphere** hogs most of the world's snowfall, mainly because it has more land area on which to generate cold air masses. By January, the majority of Canada, the extreme northern US, northeast Europe, Russia, Mongolia and northeast China – along with many Northern Hemisphere mountain ranges – are snow-covered and remain so until spring. More sporadic snow-cover comes and goes across the bulk of the US, Europe and northern Asia. Further south, snowfalls are infrequent, though not necessarily minor. Close to the sub-tropics, a freak storm in February 1895 gave Houston 51cm/20in of snowfall and New

New York's blizzard of the ages

For North Americans, the winter of 1887–88 was an equal opportunity offender. The Midwest's worst snowstorm on record struck between January 12 and 14. Drifts topped 460cm/15ft and temperatures dropped as low as –47°C/–52°F. Some farmers in the Dakotas literally froze in their fields, and dozens of rural schoolchildren perished while walking home. Urbane New Yorkers looked with pity to the west, but their own crisis would come only a few weeks later. A rainy, spring-like morning on Sunday, March 11, turned into a three-day onslaught of snow, gale-force winds and bitter cold. At the height of the storm, Central Park stayed below –12°C/10°F for over 24 hours. More than 100cm/39in of snow blanketed the land from New Haven, Connecticut, to Albany, New York. Drifts piled as high as second-storey windows. Manhattan seemed to fall in on itself as snow-packed awnings and telegraph poles crashed to the ground. More than 400 people died, including some trapped in snowdrifts for days. One New York newspaper screamed, "Now We Know What a Dakota Blizzard Is". On the city's Third Avenue elevated train line, a storm-triggered accident injured more than twenty people. It was the last straw for many who'd been lobbying for underground lines to relieve congestion and noise. Thus was born the silver lining of the Blizzard of '88: New York's subway system.

How to make an avalanche

The long-suffering towns of **the Alps** know the power of snow too well. For centuries, before control techniques were perfected, they were plagued by periodic **avalanches**. More than 2000 people died in 1618 when tons of snow slid into Plurs, Switzerland, and the Chiavenna Valley of Italy. Avalanches can strike anywhere there's enough snow: disastrous ones struck sites in Peru, China and India last century, and in 1836 eight people were killed at the British town of Lewes. North America's deadliest avalanche occurred on March 1, 1910, when a train stranded for five days near Wellington, Washington, was finally hit by a wall of snow, killing 96 passengers.

Conditions are ripe for an avalanche when a light coating of early-season snow warms and refreezes over a long period (days to weeks) of relative dryness. Water vapour from the still warm ground seeps up through the snowpack, and the interlocking snowflakes morph into rounded crystals that can't keep the snowpack as firmly bonded. From this point on, the base can disintegrate in a loose-snow avalanche, which typically starts small and – like a snowball – gains momentum. For a slab avalanche, you need a wet, heavy snowpack on top of this loosened base. The slightest trigger – perhaps a skier, or (increasingly, these days) a high-powered snowmobile – can set a 3m/10ft-deep slab of snow sliding downhill at speeds that can top 160kph/100mph. Slopes of 30°–45° are the most vulnerable, as steeper slopes tend to shed their snow in more manageable bits.

When fully buried by an avalanche, less than half of all victims survive for more than half an hour, as the snow melted by their breath gradually freezes into an icy mask that cuts off oxygen supply. Shrewd outdoorspeople swim toward the surface of an avalanche, as if riding a wave; if burial seems certain, they quickly form a breathing space with their arms. They also carry digital beacons that send GPS signals to help searchers find them if they're trapped in a snowbank. Finally, if they must cross a vulnerable snowfield, they do so singly rather than in a pack.

Orleans 20cm/8in. At roughly the same latitude, Jerusalem was blanketed by some 60cm/23in of snow in January 2000. Even Miami Beach saw snowflakes in January 1977 – a chilling affair that hasn't been duplicated since.

Mountains are practically synonymous with snow. Their high altitude means they experience cold air more consistently, and the peaks serve as a barrier to wring the moisture out of ascending winds. Any side of a mountain can get substantial snows. Some parts of the Tibetan plateau see snow more often in the monsoon-dampened summer than in the winter. The **west-facing slopes** of mountains at mid-latitudes encounter the prevailing **westerlies**, so they see the most prodigious amounts of snow on earth (as long as they're cold enough). Japan's **Mount Ibuki** holds the world record for snow on the ground with a staggering 1182cm/465in. In the US state of Washington, **Mount Baker** broke the seasonal world record held by neighbouring Mount Rainier when it notched 2898cm/1140in in the winter of 1998–99. Cities next to mountain ranges can also rack up huge snowfalls if cold, moist winds blow up-slope for long periods. The **Denver, Colorado**, area has seen over 100cm/39in fall in a single storm.

Even the **tropics** can experience snowfall with the help of altitude. When a mid-latitude winter storm drifts toward the equator, it can bring once-in-a-lifetime snows to higher terrain. In December 1997, **Guadalajara, Mexico**, suffered its first snowfall since 1881. Nor was it a mere dusting: parts of town saw 41cm/16in.

When snowfall exceeds melting in one spot for many years, a **glacier** forms. **Antarctica's glaciers** are legendary with good reason: they hold some 70 percent of all the fresh water in the world. Permanent glaciers dot many other parts of the world, including several **Andean peaks** as well as the African volcanoes **Kilimanjaro** and **Kirinyaga**. Low-latitude glaciers are sensitive barometers of global warming. The famed snow fields of Kilimanjaro shrank 80 percent during the twentieth century, and at least one study predicts them to vanish by 2020. However, this widely publicized decline may be the result of sparse snowfall more than a local temperature increase. A better example of warming at work would be the extensive glaciers of the Peruvian Andes, which are receding dramatically. Warmer global temperatures don't necessarily mean that all glaciers will shrink right away. In parts of Antarctica, warmer temperatures could bring increased snowfall; as temperatures there inch closer to freezing, the air tends to carry more snow-producing water vapour.

Freezing rain and sleet

For the most part, winter ice doesn't do much more than inconvenience people for a few hours. Worldwide, it's far less common than snowfall, making it an unusual enough phenomenon to grab people's attention. The shimmering beauty of an **ice storm**, for example, offers a gratifying encounter with the work of Mother Nature while packing a truly destructive force. The worst ice storms (see p.63) can uproot trees and power lines, bringing civilization to a standstill for days. In some regions, ice-coated pavements are the leading cause of broken bones and other winter-weather injuries. Vehicles, too, can slip and slide just as easily as people do, resulting in many deaths on the road. Ice build-ups on aircraft wings have led to many an aviation disaster.

Pelted by pellets

Winter's subdued answer to summertime hail is **ice pellets**, also known as soft hail in the UK and sleet in the US (in the UK, sleet refers to a mix of rain and snow). A hailstone forms within a strong thunderstorm updraught (see p.70) and falls into ground-level temperatures that may be quite hot. Ice pellets, on the other hand, form in flatter, stratiform clouds. Since the upward motions are much weaker, ice pellets never attains the gargantuan sizes of the biggest hailstones.

Unlike hail, ice pellets verge on becoming rain and then pull back. The key is a layer of above-freezing air sandwiched within the heart of a winter storm. An ice pellet begins as a snowflake, then falls into the warmer middle layer and melts either partially or completely. Before reaching the surface, though, it hits a ground-hugging layer of air where the temperature is again below 0°C/32°F. This freezes the melted snowflake into a rounded, solid mass.

Because ice pellets are sometimes associated with big winter storms, they tend to occur in the parts of the world where intense cyclones have access to frigid surface air as well as warm maritime air flowing on top of it. Ice pellets typically occur in eastern North America, western and central Europe, and Japan – some of the same areas prone to the most serious bouts of freezing rain (see below). Even in the most favoured locales, you're unlikely to experience ice pellets for more than a few days in a given winter.

Unlike all other forms of winter precipitation, ice pellets are most likely to occur around midday. One theory exists that the relative warmth of the daytime helps to nourish the warm, middle layer of a storm, thereby giving snowflakes a chance to melt and refreeze with ease.

Freezing rain vs freezing drizzle

Only a slight tweak to the conditions for ice pellets can yield one of winter's most treacherous visitors, **freezing rain**. If the sandwiched-in warm layer is sufficiently thick, a snowflake may completely melt yet not have time to refreeze as an ice pellet before it strikes the ground. In this case, it refreezes

Twigs stand little chance against the power of frozen water.

on whatever surface it hits – but only if the surface is adequately chilled. Ideally, the air at ground level has been well below freezing for at least a few hours, if not for several days. Sometimes only the most exposed surfaces, such as highway bridges and overpasses, are cold enough for rain to freeze on them. That's why a motorist driving on wet roads at temperatures near freezing can run into an unexpected ice patch across a bridge.

When a cloud lacks the nuclei to create large snowflakes that melt and form full-sized raindrops, **drizzle** may result (see Rain, p.52). **Freezing drizzle** begins as supercooled water droplets that remain liquid while in air that's below freezing. Like freezing rain, these droplets may pass through a layer above 0°C/32°F and then turn to ice on contact once they hit the cold ground. But since freezing drizzle starts out in liquid form, there's nothing to melt, so the warm layer isn't really necessary. Some recent studies have found that a large amount of freezing drizzle – at least in the United States, and perhaps elsewhere – forms without the presence of any above-freezing layer at all.

Ground zero

Freezing rain and freezing drizzle have their favourite haunts, and those don't always coincide. The most likely place for freezing drizzle in North America is a swath through the centre of the continent, from the high plains of Texas up to Hudson's Bay. Another familiar zone is along the northern and eastern fringes of the continent, in the cold maritime air often prevalent from Alaska across Canada and south to New England. This air tends to be relatively clean, with few ice nuclei – a good environment for freezing drizzle to form. The peak zones see as many as 30 or 40 hours of freezing drizzle a year, spread out over a few days: St Johns, Newfoundland, gets over 100 hours per year. Most areas experience only 10 or 20 hours a year, if that much, although by and large freezing drizzle tends to last longer than freezing rain.

The numbers are roughly similar across western and central Europe: here, Atlantic winds can easily surmount Arctic air near the ground. The prime spots are slightly inland: they include central England, as well as the prairies of the Ukraine and a belt just northeast of the Alps from southern Germany to Romania. The mountains and adjacent lowlands of central China are another designated region at risk, as moist air from the Indian Ocean overrides persistent Siberian air at lower levels. Some Chinese mountains have reported weeks of continuous glazing, with astounding ice build-ups of more than 30cm/10in.

Ice storms – the result of heavy freezing rain – are most common in the vicinity of big winter storms with large temperature contrasts. Some of the most famous icings in American history have occurred from the Deep South northeast to New England. The states of Oklahoma, Arkansas and Texas

The ice storm that hit the south-central US in Christmas 2000 put this tree and millions of others in jeopardy.

were hit by an ice storm on Christmas 2000 that left millions of trees damaged or destroyed and some half a million people without electricity for days. Thousands of people across rural Arkansas were without power, water, heat and telephone for as long as a week. A similar disaster struck the more urban setting of Montréal in January 1998 (see box, below).

Ice storms do their dirty work by coating everything in sight – tree limbs, power lines, highways – with a glaze whose thickness can exceed 5cm/2in. Even a thinner coating of ice can be surprisingly heavy, since it carries roughly ten times the weight of a similar depth of wet snow. Although normally

Montréal's icy onslaught

The sounds were inescapable: tree limbs cracking, power lines sparking, all of them crashing to the ground in various directions. From Monday to Friday, January 5–9, 1998, the zone from northern New England to the St Lawrence Valley was locked in a strange and deadly weather pattern. While sub-freezing air held firm near the ground, wave after wave of Atlantic moisture got pulled north on top of it by an energetic branch of the jet stream. This produced more and more rain, followed by more and more freezing. Across the region, over four million people lost electricity, some for as long as a month. The most concentrated weather war zone was **Montréal**, where over 400 shelters housed more than 15,000 refugees. The storm brought down more than 129,000km/80,000 miles of power lines across Quebec. Fears of a firestorm percolated as the city, on two occasions, came close to a complete loss of water pressure.

Oddly enough, this weather disaster – the worst in Canada's history – may be related to a warming of the tropical Pacific. Some computer models indicated that the record 1997–98 El Niño (see "El Niño and La Niña", p.114) helped create the unusual jet-stream configuration that led to the colossal ice storm.

flexible in high winds, power lines and transmission towers (not to mention trees) were not designed to bear huge amounts of weight. The cold northwest winds that often follow an ice storm make matters worse as they add to the stress on ice-coated trees and power lines.

Both freezing rain and freezing drizzle can seriously threaten aircraft. Radar and satellite displays can pinpoint the whereabouts of cloud or rain, but neither offers a foolproof method of locating exactly where the air is below freezing (though new types of radar may soon help). The deicing devices attached to most planes are designed to prevent cloud-sized water droplets from coating the aircraft. They're not meant to protect against the larger droplets that can freeze onto vulnerable parts of the plane, with adverse effect. Deicing fluid, applied before take-off to prevent ice build-up, helps, but it only works for a limited time after application. Each year an average of about 25 plane crashes in the US are blamed on in-flight icing. Most of these, however, involve aircraft smaller than the typical passenger plane.

Fog

If you're reasonably warm and comfortable, then a thick cloak of **fog** can be one of the most sublime pleasures weather has to offer. Every hard edge in the landscape softens; a moist cocoon of cloud envelops both your physical and mental world, and everything around seems to shrink to a more manageable size.

Fog is literally a cloud on the ground. Like clouds in the sky, it comes in myriad shapes and forms. Fog can span a single valley or a thousand miles of ocean. Lowlands, mountains and the sea each have their own preferred types. Natives of some tropical islands have never seen fog, while the people in sections of coastal California or the Scottish highlands see it day after day, a backdrop to their everyday lives.

Nature's fog machine

As with any other cloud, the trick to **making fog** is to either cool the air so much that some water vapour is effectively squeezed out, or to add so much water vapour that some of it is forced to condense into cloud. Either way, you're left with more moisture than can remain in vapour form at a given temperature. Some of it then gets deposited onto salt, dust, soot or whatever else is lying around. Each of the resulting water-coated particles measures in the region of 10–20 micrometers/400–800 millionths of an inch.

It doesn't take much moisture to make fog. If a batch of fog materialized in an ordinary living room, it might represent only about 3.1 grams of water, or about one-tenth of an ounce. That's barely enough to coat the bottom of a

Putting fog to work

There aren't many deserts where the air feels soft and moist, as it does along the western fringes of **Peru** and **Chile**. Cold water that upwells along the coast keeps the air ultra-stable most of the time. Towns like **Chungungo**, Peru, may get less than 5cm/2in of rain a year, even as the chill ocean helps produce fog and light drizzle day after day. Despite local subsidies, families may spend as much as 10 percent of their income on their water supply. As part of a demonstration project, Chungungo took advantage of the perennial fog by harvesting it. A set of 75 collectors – each about the size of a billboard – were installed by a Canadian research centre on a ridge above the town. Intercepting the moist sea breeze, the collectors gleaned roughly 15,000 litres/3900 gallons of water, which were piped downhill to Chungungo. Although the system was abandoned in 2003 for a variety of reasons, some of them political, similar techniques are being used in other fog-shrouded coastal deserts from Namibia to Oman.

drinking glass. An identical amount of water can produce a thin fog or a pea-souper, depending on whether it's distributed across a small or large number of particles. Back in Victorian England, when unrestricted coal-burning was the rule, there was so much airborne soot for moisture to cling to that the era's "stinking fogs" were far thicker than they are now (see box, p.69).

Fog close to the **ocean** tends to form on **airborne salt**, a process that can happen even when the relative humidity (see box, p.33) is as low as 70 percent. The thin obscuration that results – **haze** – can lower visibility to a few miles or kilometres. In cities, the tons of pollution spewed out by cars and factories leads to the urban equivalent of haze: **smog**, a word that conjoins smoke and fog. (Hawaii even has something called vog, a product of water adhering to the ash and other debris emitted by the island's volcanic peaks.)

The lines between haze, smog and fog are rather fuzzy. Water is more attracted to some particles (like salt) than to others. If lots of these water-attracting (hygroscopic) particles are at hand, then fog can form at humidities on the order of 95 percent rather than the fully saturated 100 percent. In major urban areas, pollutants alone can restrict views below the fog criterion without the help of moisture. Some dank days are informally tagged as foggy or hazy when they're actually just dirty.

Where and when

The **season of fog** depends on what kind of **landscape** you inhabit. On some mountaintops, it's year round. Air that blows toward mountains cools down as it's forced upward, so that cloud banks commonly envelop higher elevations, especially when they're close to a moisture source such as a large lake or the sea. In an average year, fog shrouds the summit of New England's Mount Washington six out of every seven days, and Kenya's Mount Kirinyaga gets at least some fog almost every day.

The tentacles of a river valley stand in relief as fog settles into its lower elevation.

The waning sunlight and shortened days of autumn and winter set up ideal conditions for **radiation fog** to form across large stretches of the mid-latitudes, such as central Europe and central China. Usually this happens under clear skies after dark, as Earth's surface radiates heat to space and cools down. Moisture left behind by rains earlier in the day can help. Before long the ground-level air approaches 100 percent relative humidity. If there's a slight breeze, some moisture might be deposited on the ground as dew, but when conditions are calm the moisture is more likely to cling to airborne particles. At first this happens above those land features that conduct the least heat, such as snowbanks and river valleys. In time these foggy patches spread to cover the entire ground, although perhaps to a depth of only 3–6m/10–20ft at first. The top of the fog bank now serves as the radiative equivalent of the Earth's surface. As it transfers heat to space, it cools further, thus helping the fog to grow upward. By morning a deck of radiation fog might be over 30m/100ft deep.

Another kind of inland fog occurs more often in the spring. When a warm, moist air mass spreads over colder ground, it can be cooled to saturation and **advection fog** may develop (in weather jargon, advection refers to the wind moving some atmospheric element, such as moisture). Advection fog can occur any time of day; it tends to form en masse and sweep across the landscape, sometimes in a stiff breeze, rather than following the patchy formation sequence of radiation fog.

The **ocean** can produce fog in two ways: by warm air passing over cold water, or cold air passing over warm water. Cold, dry air masses that cruise southeast off Canada's ice sheets may pick up moisture that congeals into fog as they flow onto the northwest Atlantic off **Newfoundland**. The same is true for bitter Siberian highs that move over the **Sea of Japan**. The wispy, dramatic fog that results is called "sea smoke" – it's not unlike the clouds of steam you might see radiating from hot pavements after a summer shower. Off the west coasts of the **Americas** and **southern Africa**, the chilly ocean currents generate regular fogs, especially during the spring and summer. Local air circulations that change by the hour help determine whether the fog rolls in or stays out to sea. The **San Francisco Bay** is a perfect haven for fog. During the warm season, each day's sea breeze pushes Pacific fog into the bay through the Golden Gate, the break in the coastline that gives the famous bridge its name.

A clear demise

Breaking up fog can be a difficult process. Fog particles actually sink to the ground at about 1cm/0.4in per second, so an ordinary fog would settle to the ground in about an hour or so if cooling at the top didn't keep it replenished. Morning **sun** can be an effective way to disperse radiation fog. Sunlight filtering through the fog warms the ground, and the near-ground air may warm enough to dissipate the lowest fog layer by allowing the cloud droplets to evaporate into water vapour. Tiny eddies of air then help the evaporation and clearing to work its way upward. We often say such a fog is lifting, although to be more precise, it's eroding from below. Sunlight also warms the air within the fog bank itself, helping evaporate and dissipate it from within – a process often called burning off fog.

Another, quite different way to dissipate fog is to spread a **bank of clouds** well above it. Whether it's day or night, overhead clouds restrict the heat loss

Weatherwise Magazine / Pat Timm

Once it lifts off the ground, fog is no more disruptive than any other cloud.

London's killer smog

For some 300 years, coal-burning in London transformed the city's frequent fogs into sickening smogs. The Victorian taste for dark wallpaper was in part a clever way to disguise soot, and Dickens wrote of snowflake-sized soot particles. Still, nobody anticipated the disaster of December 4–8, 1952, when a smog of epic proportions settled in beneath a stagnant dome of high pressure. By Friday, the smog's second day, it was hard to see across the street through the yellowish muck, and within the next two days the visibility had dropped to as low as 0.3m/1ft. Since this wasn't precisely a weather disaster, there were no warnings *per se* on radio or TV; many Londoners made things worse by going home and stoking up their stoves. London's daily fatality rate jumped from the usual 300 to nearly 1000, mostly due to respiratory failures. In total, the unnamed onslaught either caused or hastened some 4000 deaths.

A similar smog attack in Donora, Pennsylvania, four years earlier had killed 20 people and sickened hundreds. London cleaned up its act in a hurry with bans on black smoke and a shift toward smokeless fuels; the last major London smog was in 1962. The city's present-day air isn't exactly pristine, but central London now gets as much as 50 percent more winter sunshine than it did when coal was king.

from the top of the fog, which puts a damper on the fog replenishment process. Brute force will also help. **Strong winds** ripping across the top of a fog bank can stir up turbulence and disperse the moisture.

Sometimes none of these natural processes does the trick. Winter fog can last for weeks in sheltered valleys like those of the American West or northern Italy. Some airports have gone so far as to inject fog with sprays of salt, ice or other particles that attract moisture away from the fog droplets. One technique applied in Utah has turned fog banks into light snow. For sheer drama, it's hard to outdo the strategy of Britain's Royal Air Force during World War II. Along the runways of key air bases, trenches full of fuel were set ablaze in foggy conditions. The fire-driven updrafts cleared the way for the Allies.

Thunderstorms

It's with very good reason that **lightning** evolved as a symbol of divinity in both the Bible and the Koran, as well as in Greek mythology and countless other religious and secular traditions. Few natural events are as familiar yet as awe-inspiring as a **thunderstorm**. The best estimate is that some 2000 thunderstorms are raging across the planet's surface at any one moment. That's a lot of soaked picnics and sleepless nights.

The world couldn't do without thunderstorms, of course. They furnish some areas with more than half of their annual rainfall. Lightning even converts nitrogen into a form that fertilizes the soil. Even so, thunderstorms exact a heavy toll for the good they achieve. Hundreds of people around the world are killed each year by lightning. Many hundreds more fall victim to tornadoes, flash flooding and hailstones.

What makes a thunderstorm?

Few places on Earth are free of thunderstorms. Wherever there is enough heat and moisture, they are likely to form. **Mountains** help trigger them by forcing air upward. The only spots that are virtually free of thunder are those where it hardly ever rains, chiefly the parched **Sahara** and the bitter reaches of the **Arctic** and **Antarctic**. Even here, however, surprises are possible. The northernmost settlement in the US – **Barrow, Alaska**, perched on the Arctic Ocean – got its first thunderstorm in living memory on June 19, 2000. It terrified animals and children who didn't know what to make of the flashing light and deafening sound.

In order to produce **thunder** and **lightning**, you need a cloud tall enough to pull moisture up into the sub-freezing realm over 5km/3 miles high. These clouds are **cumulus**, the bubbly, cauliflower-like formations that typify summer. Cumulus can form any time of year as long as there's warmth below and coolness above, the same set-up – meteorologists call it **convection** – by which a hot burner makes water boil and bubble. Data from a weather balloon can tell forecasters how much fuel there is for thunderstorms by showing how much contrast, or instability, exists between the layers of air. Once cumulus clouds reach a height where **ice crystals** form, they become **cumulonimbus**. Here the upward motion gradually slows. With the help of stronger upper-level winds, the cloud spreads out in a sheet-like formation called the **anvil** (for its resemblance to the top of a blacksmith's anvil).

© Bob Henson

Towering cumulus clouds boil up into a fast-growing thunderstorm that could spew hail and high winds within the hour.

As ice crystals form, they collide with each other and with still-unfrozen water droplets. Current theory has it that each bump and jostle results in a tiny bit of **electrical charge** being generated, just as scraping your shoes against carpet on a dry day gives your body a static charge. All this takes place in a larger, storm-wide **electric field** that's enhanced by the collisions. If the field becomes intense enough, lightning soon results.

The many faces of lightning

Experts refer to **lightning** by pairs of initials. **IC** means **intracloud**; it's the kind of lightning that stays aloft, dancing through and around the thunderhead itself. Each IC flash lasts, on average, about a quarter of a second. **CGs** are **cloud-to-ground flashes**, produced when a tentative **leader** of electrons descends from a storm and is met by a **return stroke** of positive charge from the ground. Although common sense tells us that lightning comes down from above, it's actually the return stroke heading upward that produces the visible flash. Each leader pauses for a tiny fraction of a second a few times along its downward path – sometimes splitting along the way – as it hunts for the most efficient route. The resulting twists and turns give lightning its crooked, forked appearance. Each CG includes a series of anywhere from one to two dozen pulses of electricity (the average is four). These produce the characteristic flicker of a CG, which can last up to several seconds.

About 10 percent of CGs have leaders that bring **positive** instead of negative charge to the ground. The long, continuous currents from these positive CGs trigger more than their share of fires, as many of them strike well away from the rainy part of a thunderstorm. Overall, of course, lightning is a beneficial part of Earth's ecosystem. Some species, such as Jack pine, rely on fire in order to reproduce. In the days before campers and cigarettes, only nature could provide the needed spark. The human desire to limit forest fires has added to the potential damage from lightning. In some places, a century's worth of dead and living vegetation now waits to burn, as it did across the western US in the disastrous, smoky summer of 2002.

The intense heat of lightning, estimated at 30,000°C/54,000°F, can cause instant havoc for people as well. Entire football teams have been felled by a single flash. Only a minority of people struck by lightning are killed, but the injuries – physical, neurological and psychological – can be severe and long-lasting.

To help **track lightning**, ground-based **sensors** have been used since the 1980s by utility companies, fire control agencies and others in North America and more recently in Europe, Asia and South America. These sensors calculate the locations of CGs and plot them on up-to-the-minute maps. **Satellites** have collected long-term data on all lightning (ICs and CGs alike)

"Spider lightning" courses through the sky as a large thunderstorm complex approaches.

© Kenneth D Langford

since the mid-1990s. They've shown that the global average is about 30 to 40 flashes per second, less than half of what we once thought it was.

As the heat expands air in and near a lightning flash, the world's most familiar shock wave – **thunder** – radiates outward. Thunder can resemble a sharp crack or a low, sustained roll, depending on the density of the air, the orientation of the bolt, and the bolt's distance from the observer. Thunder travels about 1km every 3 seconds (or 1 mile every 5 seconds), so by counting the seconds between flash and rumble, you can estimate how far lightning is from you. A CG can strike as far as 32km/20 miles away from a storm's core, far enough so that haze or obstacles may prevent any visual awareness of a storm before the bolt from the blue arrives. Indeed, most lightning injuries and deaths occur at a storm's outset, before the hazard is obvious. Guidelines now recommend taking shelter if thunder can be heard within 30 seconds of a lightning flash. (See "Safety tips", p.398).

The quintessential bolt from the blue reaches out from the top of a Florida thunderstorm on a moonlit night.

Weatherwise Magazine / Nathan Gates

What makes a storm green?

Many cloud-watchers have noticed the peculiar green tint that sometimes appears in the heart of a strong thunderstorm. A popular explanation had it that hail or ice in the storm was refracting and absorbing the light to favour green wavelengths. In the mid-1990s, a pair of scientists took a spectrophotometer to Oklahoma and Florida, hunting for green storms. They found fifteen cases where various shades of green appeared. Some possible causes were ruled out; for instance, the green wasn't being reflected from vegetation. However, some of the hail-producing storms failed to produce green cloud at all. The results hinted that more than one factor must be at work behind this distinctive hue.

Perhaps the biggest weather find of the late 1990s was a whole new class of lightning that surges upward from storm tops. Pilots and amateur observers had reported these strange, colourful flashes for decades, but their reports were largely dismissed. Low-light video taken from ground and space finally proved them right. The fancifully-named discoveries include **sprites** (red pillars that extend up to 90km/55 miles), **elves** (broad disks of light, often on top of sprites) and **blue jets** (beams that shoot out from storm tops). Their impacts on aircraft, air chemistry and radio waves are only beginning to be explored.

Thunderstorm types

Most **thunderstorms** spit out only a few lightning flashes and a modest amount of rain before dissipating. These so-called **single-cell storms** (meaning a single updraught) pop up and die down so quickly on a summer afternoon that meteorologists refer to them, a tad scornfully, as "popcorn convection". They're doomed from the start by lackadaisical upper winds that keep them becalmed. Once rain begins, it cools the air below and cuts off the storm's energy in less than an hour. Sometimes new single-cell storms can form along the cool air outflowing from old ones, or several cells may coalesce to form a **multicell**, possibly dropping small hail and somewhat heavier rain. When a strong cold front is marching through, a **squall line** may form. This band of connected cells moves through quickly with strong wind, heavy rain and perhaps hail or even a brief tornado.

The biggest movers and shakers of the storm world are **supercells**. These behemoths only form when instability is quite strong and, typically, when upper-level winds strengthen with height. This keeps the storm moving and keeps its top ventilated, so that warm, moist air is pulled in from below as in a chimney. Many of the jaw-dropping elements of the strongest thunderstorms – torrential rain, large hail, hurricane-force wind and violent tornadoes – occur with supercells. Because supercells are so well differentiated, they can cause a variety of trouble at different points beneath them during their

Lightning strikes again – and again

If lightning never struck more than once in the same spot, the makers of lightning rods would have a tough time of it. Since that day in June 1752 when **Benjamin Franklin** (inventor of the lightning rod) tapped into storm electricity with a kite and a key, we've known why tall objects tend to attract cloud-to-ground strikes. The Empire State Building in New York is hit about two dozen times a year on average. A lightning rod takes advantage of lightning's nature by urging it to the ground along a safe channel. Church steeples often attract lightning but fail to dispel it safely, which must have puzzled those who gathered there for safety during storms. Europeans once rang church bells to "scatter the thunderstorms" and inhibit strikes. Some church towers bore the inscription *Fulgura frango* ("I break up lightning"). Hundreds of bellringers were apparently struck and killed prior to 1800. A few lucky (or unlucky?) souls have been hit more than once and survived. One US park ranger was reportedly struck seven times in 35 years. He lost his hair, eyebrows and one big toenail, but lightning never managed to take his life.

lifespan of six hours or more. Like the tornadoes they spawn, supercells are most common in the mid-latitudes, especially the central and eastern US; they occur less often in parts of northern and western Europe, eastern and southern Asia, Argentina, South Africa and Australia.

The largest thunderstorm groups on earth have a suitably big name: **mesoscale convective systems (MCSs)**. Think of an MCS as a storm on steroids. Actually a collection of storms – typically organized as a cluster or a squall line – it can span 160–320km/100–200 miles and last for more than twelve hours. Partly due to their sheer size, MCSs can generate buckets of rain (over 250mm/10in can fall) and vast amounts of lightning (upwards of 10,000 strikes in an hour, or about three per second). They are known for batches of spidery lightning that stretch through the cloud almost from horizon to horizon. Record-breaking floods of 1993 along the US Mississippi and Missouri rivers were fed by more than 70 MCSs and smaller, related systems that traversed the plains through the summer.

MCSs favour the moist heat of the warm season across the mid-latitudes and tropics, from the Americas to Africa and Asia. In many places, they peak during the overnight hours, as smaller storms merge and low-level jet streams intensify. Once rolling, they often prowl well into the next day. If an MCS forms or moves over the ocean, it can serve as the nucleus of a **hurricane** or **typhoon**.

Hail

Hail isn't the most common weather hazard, but it can be one of nature's most efficient demolition tools. Anyone who's been caught in a hailstorm can tell you that the assault can really hurt. Hail kills an untold number of animals across the world each year and destroys many tonnes of crops. Over 200 people were injured by a 1995 megastorm that struck an outdoor festival in Fort Worth, **Texas**, with little warning and few places to take cover. In other hail-prone areas, such as **China** and **India**, hailstorms have been known to kill more than 100 people. Deaths in the US are rare, but in 1979 a three-month-old Colorado infant died after being struck by an orange-sized hailstone, and in 2000 a Fort Worth man was killed by hail when he tried to move his car out of harm's way.

A hailstone's trip to earth

We have thunderstorms to thank, or blame, for hail's existence. It takes the strong updraughts and the cyclonic spin of a severe storm to make big hail. That's why it tends to occur in the same parts of the world that are most prone to severe thunderstorms. Surveys from France, South Africa, Canada and India show that the lion's share of the hailstones they get are less than 2.5cm/1in wide, but all of these countries have seen hailstones more than three times as large. Smaller hail is far more common, occurring just about anywhere it thunders. Even non-thundery winter storms can be vigorous enough to produce tiny, soft hailstones called **graupel**.

The world's most infamous **hail zones**, such as **northeast Colorado** and the **pampas of Argentina**, are high-altitude plains in the lee of mountains. When a storm forms over elevated terrain, a bigger percentage of it extends above the freezing layer, giving hailstones a chance to grow. Hail, in fact, served as one of the most important clues to ancient weather observers like Aristotle that the atmosphere gets colder with height.

A **hailstone** usually begins as a tiny ice crystal and grows by encountering a special kind of moisture – **supercooled water droplets**. In strong storms, the air is rising so quickly that some of the inflowing water vapour

NCAR / UCAR / NSF

Hail research often involves members of the public who comb their neighbourhood for the largest stones, as did this rural couple.

The great hail war

If you're stuck in a car during a hailstorm, the racket may be enough to make you think you're in a war zone. Indeed, humans have treated hail as if it were a living, breathing enemy. Europeans of the 1800s fired rockets and cannons at hailstorms, hoping to either deflect the storm or break its hail into less damaging fragments. A more scientific approach was adopted shortly after World War II, when cloud seeding showed promise as a way to reduce hail damage. The idea was to drop huge quantities of silver iodide from aircraft into the parts of a storm laden with supercooled water. Ideally, the water would freeze on the silver iodide particles instead of on hailstones, thus limiting the hail's size. This sounded good in theory, but the actual effect on hail proved nearly impossible to measure. Hailstorms are so variable to begin with that nobody could say for sure how a given storm would have evolved had it not been seeded. Today, the rainclouds above several US states are still being seeded to reduce hail, despite the lack of published evidence to support this technique.

doesn't have a chance to freeze onto a solid nucleus, like a bit of dust or salt. These supercooled droplets can remain liquid even at temperatures well below 0°C/32°F. Suspended in the frigid air, all they're waiting for is a surface on which to freeze – and a growing hailstone does the job nicely.

Many hailstones take only a single trip through a storm. Growing as they ascend, they're buoyed on winds that can blow upward at over 160kph/100mph. Eventually, the hailstone is either flung out of the updraft or becomes heavy enough to fall through it. Updraughts in a hail-producing storm tend to be tilted, so most falling hailstones quickly exit the updraught and descend to earth. However, if you slice open a **giant hailstone**, you're likely to find evidence of multiple journeys. Most giant stones have an onion-like structure, with perhaps four or more bands of clear, solid ice separated by milky layers. The milkiness is due to tiny air bubbles trapped within and between ice crystals during the freezing process. These layers may be 50 to 90 percent less dense than solid ice.

As best we can tell, giant hailstones get their layers by being recycled. Before they fall out of a storm, they're caught in the inflow and channelled through the updraught for another round trip. It's thought that the alternating milky and smooth layers represent different parts of the storm, where the conditions allow for either slow growth of solid ice or quick, irregular growth of opaque ice. This long-standing concept has been challenged somewhat by recent work that employs computer models to follow a simulated hailstone through a cloud. Instead of several up-and-down trips, these models point more toward a single, tight helix. In this view, a hailstone circles the updraught as it rises, moving through the same quadrants of the storm with each rotation.

In its final drop to earth, a giant hailstone may glue itself to a few smaller companions, giving it a spiky, irregular shape. Such was the case for the

NOAA

A massive hailstorm left waist-high drifts of ice across roads and car parks near Cheyenne, Wyoming, in August 1985.

world's heaviest hailstone, which fell on Coffeyville, Kansas, in 1970. About twenty smaller stones congealed to form the final product. It measured roughly 14cm/5.5in in diameter and weighed 759 grams/1.69 pounds. An even larger hailstone fell near Aurora, Nebraska, in 2003: it spanned 17.8cm/7in in diameter, although its weight fell just shy of the Coffeyville record-holder.

"The bigger they are, the harder they fall" doesn't necessarily hold true for hail. Gigantic stones tend to be spun out toward the edge of a storm, where it may scarcely be raining. These grapefruit- or softball-sized missiles are few in number, and they can be separated by 30m/100ft or more. Observers have noted that some of them don't seem to be falling very fast. The layers of low-density ice inside big stones could be making a difference by reducing their downward momentum. Another factor is **spin**. Many large hailstones are shaped somewhat like hockey pucks – wider than they are thick – and they may be spinning around their centres at more than forty revolutions a second. One study found that the spin, interacting with wind transitions, could be slowing the stone and curving its path, much like a baseball or cricket ball that's given a twist by the pitcher. The analogy is apt, since hailstones can hit the ground at pitched-ball speeds of more than 144kph/90mph.

On impact

For farmers across the United States, hail is a big consideration. On average, about 2–3 percent of **agricultural output** across the US is destroyed by hail each year; roughly a quarter of farm output is covered by hail insurance. It's a must in places like the "Hail Alley" of Colorado, Nebraska and Wyoming.

A 1985 hailstorm in Cheyenne (Wyoming's capital and the continent's hail epicentre) left behind icy drifts as high as car windows (see photo, p.77).

A single misbegotten hailstone can kill a person or destroy a car's windshield, but it takes a lot more hail to ruin a crop. Perhaps surprisingly, it's not gigantic hail that farmers fear most, but the denser, smaller variety. This type tends to fall within heavy rain and can easily blow sideways in a strong wind. A thick batch of pea- to walnut-sized hail can virtually destroy a field of wheat or corn in minutes. Some crops are even more vulnerable. Hailstones only 0.5cm/0.2in in diameter can damage the delicate leaves of soybean and tobacco plants. Since heavy hail is produced in a particular part of a storm, it tends to be localized, often falling in streaks as narrow as 0.5km/0.3 miles and only a few miles or kilometres long. The biggest and meanest storms produce **hail swaths**, multiple streaks that can stretch for over 400km/250 miles.

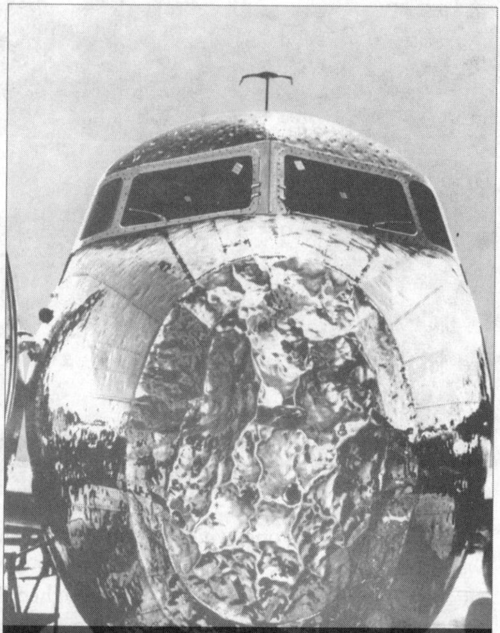

Cars aren't the only vehicles that hail can ruin, as evidenced by this plastered plane.

NCAR / UCAR / NSF

Sydney's day of ice

It doesn't snow in Sydney to speak of, but Australia's biggest city makes up for it with occasional blizzards of hail. New Year's Day 1947 brought one such storm, with hailstones the size of oranges. A more recent pummelling came on the evening of April 14, 1999 – quite late in the autumn for hail in Sydney. A garden-variety thunderstorm paralleling the New South Wales coast drew little notice from forecasters at first. Abruptly, it moved inland and grew into a monster over Sydney's eastern suburbs. The largest hailstones spanned more than 8cm/3in. Over more than five hours, some 40,000 vehicles and 20,000 structures were damaged by the storm. In parts of Kensington and Eastlakes, every home experienced some type of damage. With a total bill in the region of AU$1.5 billion, the hailfall ranks as Australia's costliest natural disaster.

When hailstorms hit cities, our houses and vehicles (aside from gardens) tend to suffer the most. A round of hail and strong wind pounded **Munich, Germany**, on July 12, 1984, causing more than $1 billion in damage to some 70,000 homes and 200,000 cars. Similar events destroyed many thousands of vehicles across **Denver** in 1990 and **Sydney** in 1999 (see box opposite).

Tornadoes

Terrifying and mesmerizing at the same time, **tornadoes** have earned every inch of their fearsome reputation. They produce the strongest winds on the planet – close to 480kph/300mph. If that weren't enough, tornadoes can be astoundingly flaky. A tornado – often informally referred to as a twister – might last only seconds or churn for upwards of an hour. It can be as narrow as a house or as wide as a town. Its path might be a straight line, an arc, or even, on rare occasions, a circle.

The US is pockmarked by about 1000 twisters a year, far more than the number reported throughout the rest of the world combined. Each year a dozen or so American tornadoes are classified as "violent", meaning their winds have topped 320kph/200mph. Yet the twister risk in many other places is underrated. The United Kingdom averages 33 tornadoes a year, though the majority of them are really quite weak. Per unit land area, that's not too different from the rate in parts of the north-

NCAR / UCAR / NSF

Even if a tornado looks wispy, its winds can still howl above hurricane force.

east United States. A few other places see enough tornadoes to take them seriously, including southern Canada, western Russia, France and the Low Countries, Australia, South Africa, the river valleys of eastern China and the pampas of Argentina. Many other countries will get a stray twister on occasion. The only spot that rivals the US for deadly tornadoes is the populous Ganges Valley of northeast India and Bangladesh, where hundreds of residents have been killed in the last couple of decades.

Spin control

Tornadoes are masters at taking the **spin in the air** around them and rendering it lethal. Like a pirouetting ballerina who circles ever faster as she brings in her arms, tornadoes gain speed by tightening and focusing rotation in and near a thunderstorm.

Unlike other atmospheric swirls, such as dust devils (see "Quick spins", p.99), a tornado can only develop out of a **parent thunderstorm**. That's why the US is so twister-prone. The Great Plains, Midwest and South are prime breeding grounds for the violently rotating thunderstorms called **supercells** (see "Thunderstorms", p.69). They form as air arrives from several different directions at different heights. Sultry breezes flowing north from the Gulf of Mexico are overlaid by cooler, stronger west winds imported from the Pacific Ocean and wrung dry by the Rocky Mountains. Sometimes a warm, dry layer off the desert Southwest flows in between. These contrasts of temperature and moisture make for highly unstable conditions, and the change in winds with height helps induce in the storm a slow cyclonic turning (counter-clockwise in the Northern Hemisphere). No other part of the world has quite the geographic blend to produce intense, rotating storms as frequently as North America does. Since tornadoes rely on thunderstorms, they're most common

Tornado-proof?

Countless folk tales have proclaimed that a region is protected from twisters by certain geographic features, such as a hill or river. The people of Topeka, Kansas, assumed they were being kept safe by a rise called Burnet Mound – that is, until the US city was ravaged by a killer twister on June 8, 1966. If tornadoes can indeed go wherever they like, they appear to prefer some places more than others. America's **Tornado Alley**, stretching from north Texas to Minnesota, sees more action than any other place on the planet. Even on the local scale, variations can be dramatic. Dallas County, Texas, has seen about three times more tornadoes of note than adjacent Tarrant County, the home of Fort Worth. Just north of Geneva, Switzerland, lies a mini-Tornado Alley of sorts. A violent twister travelled some 60km/38 miles through the Jura Mountains from near Oyonnax, France, to Croy, Switzerland, on August 19, 1890. History repeated itself on August 26, 1971, when a tornado of similar strength covered a near-identical path. Destructive winds reported in the same valley in 1624, 1768 and 1842 may also have been tornadic.

during the afternoon and evening in the spring and early summer – May is the peak US month – but they have been observed at all times of the day and throughout the year. European twisters have little seasonal preference.

In the 1970s and 1980s, dozens of **storm chasers** began prowling the Great Plains, many of them working for universities and federal labs. They quickly learned to recognize the visual cues that preceded tornadoes. Often a storm's rotation is visible in the billowing walls of cumulus on its south end, where tornadoes are most likely. As the updraught intensifies, a broad, flat, rotating mass called the **wall cloud** may descend from the main storm. Hailstones bigger than walnuts or even grapefruit can fall, products of the strong updraught. Tapered lines of cloud extend outward, spoke-like, from the wall cloud.

With the help of visual reports like these, along with radar returns highlighting the areas at risk, specific warnings are now issued in the US twenty minutes or more before many twisters develop, although some tornadoes still aren't caught by radar. More general areas of alert (called tornado watches), usually about the size of a Midwestern state, go into effect hours in advance. Yet the final ingredient that causes a tornado to emerge from a rotating storm still eludes researchers. Several large studies have canvassed the southern Great Plains since 1994. Using wind-probing Doppler radars mounted on trucks and tiny weather stations on top of sedans, scientists have probed tornadic storms in detail. They're now sifting through the results. Rain-driven downdraughts that descend next to a thunderstorm's main updraught appear to be a key piece of the puzzle.

Many tornadoes, especially outside the US, live and die in ways that depart from the standard picture. Conditions in a storm may only come together briefly to produce a short-lived, weak twister, and even in the States these far outnumber their more violent cousins. Although a weak tornado may develop winds of 160kph/100mph or more, its damage is typically limited by a width that averages less than 60m/200ft and a path that usually extends less than 16km/10 miles. By comparison, the most damaging US twister of recent years – passing through the south suburbs of **Oklahoma City** on May 3, 1999 – spanned 1.2km/0.75 miles at its peak and traversed 61km/38 miles. Its peak winds topped 320kph/200mph most of the way.

Tornadoes that emerge from supercells tend to rotate cyclonically, like their parent storms. This appears to be true in the Southern as well as Northern Hemisphere, although statistics are hard to come by. Tornadoes themselves are too small and fleeting to be ruled by the Coriolis force that makes large low-pressure centres rotate cyclonically, so a small percentage of twisters are anti-cyclonic.

The Oklahoma City tornado of 1999 wrecked many thousands of homes.

What a tornado does

The brute force of the wind within a tornado can cause a great deal of damage, some of it weird and unexpected. Trains have been thrown off track, houses stripped to their foundations, and entire forests levelled. Cancelled checks and other bits of paper have travelled over 160km/100 miles after being lifted into the air by twisters. Even a humble piece of straw can penetrate a plank of wood when driven by tornadic winds. Although most creatures don't survive intact, some people have taken wild rides in tornadoes – via bathtubs, mattresses or other impromptu vehicles – and lived to tell their tales.

Tornadoes can seem capricious, destroying one house while leaving the one next door almost untouched. Careful analysis has shown that large tornadoes often have multiple vortices – mini-tornadoes that spin around the parent funnel. These tiny whirls may leave narrow streaks of intensified damage. Building construction can vary as well: a brick house may be able to withstand the same force that levels its wooden neighbour. In the US, mobile homes are notorious for flying apart in twisters. Their occupants account for more than 40 percent of all US tornado fatalities between 1985 and 2005, far out of proportion to their share of the total populace. Cars and trucks are also quite vulnerable. The best shelter is a small interior room on the lowest floor of a sound building.

Tornado censors of 1900

NOAA

For over fifty years, the word "tornado" was taboo in US weather forecasts – ironically, because of early detection success. Military meteorologist **J. P. Finley** took it upon himself to issue the world's first large-scale tornado alerts in the 1880s. Using teletype reports from his own network of hundreds of observers scattered across the country, Finley marked off the regions where he expected tornadoes might form in the next few hours. His accuracy was impressive but, because of byzantine federal politics, Finley's work was halted in 1887. Long after the political storm had passed, scientific insecurity and fear of public panic continued to suppress tornado warnings of any sort, even after radio came on the scene. Finally, in 1948, a bizarre coincidence changed things. A tornado on March 20 wrecked US$10 million worth of aircraft at **Tinker Air Force Base** (pictured), near Oklahoma City. Two meteorologists on base were ordered to develop a scheme for predicting tornadoes. They set to work, and five days later, they gamely issued a tornado warning. The base was again hit by a tornado, but this time, because most planes had been secured, the damage was far less severe. The scheme developed at Tinker became the nucleus for the US watch/warning system put in place in the 1950s and still used today.

Officials once urged people to open windows if a tornado approached. The idea was to equalize the air pressure and keep the house from exploding. Engineers in the 1970s put this notion to rest. The air pressure inside even the strongest tornado is only about 10 percent less than normal. It's believed that tornadic winds lift up a building's roof by tucking under its eaves, after which the walls collapse (which can give the appearance of an explosion). Opening windows may actually increase the pressure on the roof and hasten destruction – not to mention putting the occupant in danger from wind-blown debris.

For more on tornado precautions, see "Safety tips", p.398. Tornado strength is assessed by several US and UK intensity scales; see p.404.

Hurricanes and tropical cyclones

The weather was probably quite tranquil on the Caribbean island of Barbados on the night of October 9, 1780: the calm before the storm. We know what it was like the next evening, thanks to Sir George Rodney. He and his fellow British colonials cowered in the government house, watching walls a metre thick crumble in a wind worse than any they'd ever seen. The next day, Rodney wrote: "Had I not been an eyewitness, nothing could have induced me to have believed it. More than six thousand persons perished, and all the inhabitants are entirely ruined."

Such is the power of the atmosphere's trump card. Hurricanes – also called **typhoons** in the northwestern Pacific and **cyclones** in the Indian Ocean and South Pacific – have induced more abject misery more swiftly than any other weather feature on earth. The 1780 hurricane killed a total of 20,000 people on several islands. India and Bangladesh have fared even worse. Since 1700, more than a million people have perished in the face of tropical cyclones arriving from the north Indian Ocean.

The horror of hurricanes is intermingled with the romance of the tropics. One nightclub in New Orleans created a drink in honour of hurricanes, and hurricane parties were once a staple of coastal culture. Most people now know enough to base any such bashes well inland. In the US, hurricane damage continues to grow along with coastal development, but improved warnings and public awareness kept death tolls well under 100 per year from the 1970s into the start of the twenty-first century. That changed in a horrific way with 2005's Hurricane Katrina, whose vicious winds and floods teamed

NASA

Hurricane Mitch was one of the Caribbean's most powerful hurricanes on record, yet the worst damage occured in floods after Mitch had peaked.

Katrina, Rita, Wilma – and more

The degree to which climate change may be intensifying tropical cyclones across the world is a topic of spirited scientific discussion, but there's no debating the impacts seen in recent years. Five hurricanes struck or sideswiped the United States in 2004, with four of those slapping woebegone Florida. Meanwhile, a record 10 typhoons raked Japan. Things got even worse in 2005, at least for North America. Based on measurements of their lowest pressure, the year brought three of the six most powerful Atlantic hurricanes ever observed: **Wilma** (the strongest), **Rita** (fourth strongest) and **Katrina** (sixth strongest). All three made devastating landfalls. Wilma slammed Cancún and south Florida, Rita pounded the Texas/Louisiana border, and Katrina swept through Miami before bringing mind-boggling misery and over 1300 deaths to the New Orleans area and the Mississippi coast. All told, there were 27 names assigned to Atlantic hurricanes and tropical storms in 2006 – so many that the National Hurricane Centre reverted to its backup naming list (the Greek alphabet) for the first time ever.

Other parts of the world haven't been spared. Japan endured an unprecedented string of ten landfalling typhoons in 2004 (the previous record was four). Australia was hammered in early 2006 by a trio of cyclones, each packing winds of over 250 kph (155 mph). It's the first time the region has seen three storms of that calibre in a single season. The central Pacific saw its strongest-ever cyclone, Ioke, in 2006. And Supertyphoon Saomai ripped into the east coast of China that same year – the most intense typhoon to strike the nation in more than half a century.

The damage and misery wrought by 2005's Hurricane Katrina topped anything produced by a single US storm for many decades. The storm's toll in property damage cost tens of billions of dollars, and that doesn't include the cost of restoring the New Orleans levee system to its previous condition – much less the billions more needed to make the system able to withstand a direct hit from a Category 5 hurricane. Some observers linked the Katrina debacle to climate change from the outset, while others denied any connection. Certainly a storm like Katrina doesn't require global warming in order to flex its muscle. Though they're quite rare, hurricanes on a par with Katrina have developed in the Atlantic since records began. However, the gradual warming of tropical waters over the last several decades has made it easier for storms like Katrina to intensify. As for Katrina's landfall in New Orleans, that was simply the luck (albeit the bad luck) of the meteorological draw. Several near-misses had grazed the city since 1965, when Hurricane Betsy produced flooding that killed dozens. Betsy prompted the US Corps of Engineers to build the city-girdling levee system that was in place when Katrina struck. That system was built to withstand only a Category 3 storm, a fateful move sure to be analyzed by critics for years to come – especially since Katrina itself had weakened to Category 3 strength by the time it made landfall.

up with government ineptitude and an ill-prepared populace to kill more than 1700 residents of Louisiana and Mississippi (see box).

Ironically, the worst disasters often happen as hurricanes dissipate, when slow-moving remnants can dump prodigious rains on mountainous areas. A dying **Hurricane Mitch** (see image, above) produced floods that killed some 10,000 in Honduras and Nicaragua in the autumn of 1998. In 1974 **Typhoon Nina** stalled just west of Shanghai and dropped more than 1.6m/63in of rain over three days. When two huge reservoirs failed, almost 100,000 Chinese died.

Ingredients of a hurricane

At heart, a **hurricane** is a heat engine that sucks energy from **tropical oceans**, often carting it northward or dispersing it along the way in the form of wind, rain and muggy air. Hurricanes always form over warm ocean waters. Since it's so much denser than air, water can hold a tremendous amount of heat. Hurricanes feed on the warmth of the sea surface, as well as on the extra heat released into the air when a molecule of water vapour from the ocean condenses as rain. Experience shows that ocean temperatures usually need to be above 26°C/79°F for a hurricane to thrive. Since it takes a couple of months for ocean temperatures to peak after the summer solstice, hurricanes are most likely in late summer and early autumn. Late August and September bring the most Atlantic activity.

There's another requirement for hurricane growth: the **upper winds** mustn't be too strong, or the embryonic circulation will get pulled apart. The key incubation areas (see map, opposite) all lie between about 10° and 30° latitude north and south – in and near the tropics, but far enough from the equator so that Earth's turning can impart a cyclonic spin to the storm.

Hurricanes can grow out of a variety of low-pressure centres across the tropical oceans, but the usual seeds are clusters of showers and thunderstorms called **tropical disturbances** or **tropical waves**. These pockets of disturbed weather slide like pearls along and near a necklace of converging **easterly** (west-flowing) **trade winds** that encircle the tropics. This band, called the **intertropical convergence zone (ITCZ)**, shifts north and south to one side of the equator or the other every six months. Hundreds of waves form close to the ITCZ each year, but only some of them evolve into closed centres of low pressure. The closed low acts as a nucleus around which thun-

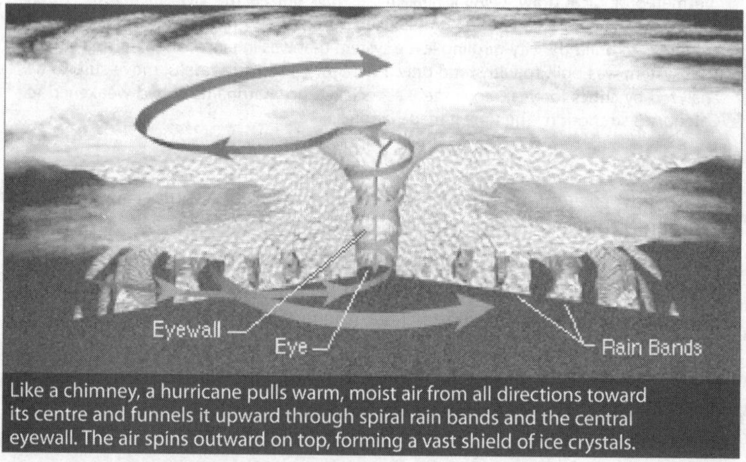

UCAR / COMET

Like a chimney, a hurricane pulls warm, moist air from all directions toward its centre and funnels it upward through spiral rain bands and the central eyewall. The air spins outward on top, forming a vast shield of ice crystals.

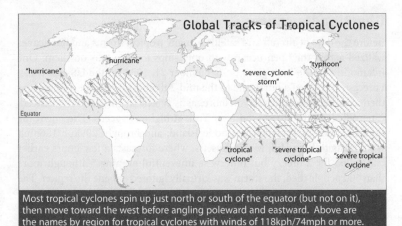

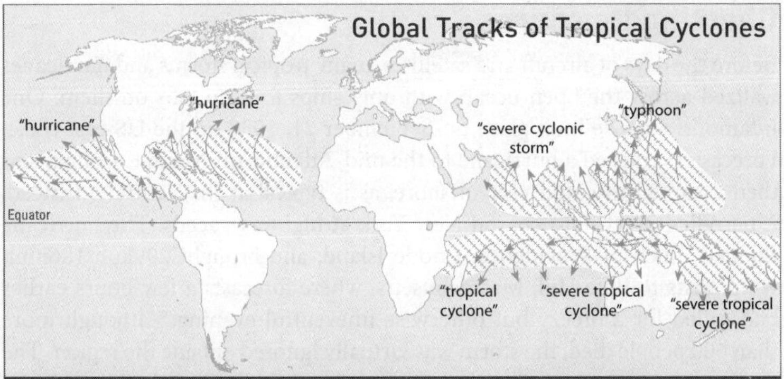

Most tropical cyclones spin up just north or south of the equator (but not on it), then move toward the west before angling poleward and eastward. Above are the names by region for tropical cyclones with winds of 118kph/74mph or more.

derstorms can further focus; it may be a chicken-and-egg question whether the thunderstorms induce the closed low or vice versa. Either way, the rising motion in the storms helps to draw in surrounding air, which rushes in to fill the vacuum that would otherwise result. The stronger these winds are, the more warmth and moisture they can absorb from the ocean below. Predicting when this process will unfold successfully and when it will fail is still an uncertain art.

After a few hours or a few days, the sustained winds in a closed tropical low, or tropical depression, may reach 62kph/39mph. Once forecasters verify this trend – by sending a plane in to take measurements, or by watching cloud motion on satellite – they give the depression, now called a **tropical storm** or cyclonic storm, a name (see box, p.89). Once winds reach 118kph/74mph, the storm is described as a **hurricane**, **typhoon** or **tropical cyclone**, depending on its location. There are about 80–90 such systems each year around the globe. This average is remarkably reliable in most years, even when some parts of the world are hyperactive and others stagnant.

It's possible that greenhouse-warmed oceans are adding extra punch to the hurricanes that do form. Several studies in 2005–06 lent credence to this idea, including a global survey that revealed nearly twice as many systems packing winds above 209kph/130mph than was the case in the 1970s. Critics pointed to errors inherent in hurricane analyses of past decades, but they also found themselves confronting the Atlantic's astounding 2005 season (see box, p.85). Over the past century, hurricane production in the Atlantic has oscillated on an apparent multi-decadal cycle. It was active from the 1920s to the 1960s, then quiet until the mid-1990s, when activity again picked up. It's expected to begin waning again by the 2010s or 2020s, assuming that global change doesn't skew the cycle.

Charting its course

Before the days of aircraft and satellites, many tropical storms and hurricanes waltzed across the open ocean with only ships to keep tabs on them. One infamous surprise took place on September 21, 1938 on the US east coast. Forecasters tracked a hurricane to the mid-Atlantic, off North Carolina, and then assumed it would veer offshore, as is typical at that latitude. Instead, it barrelled north into coastal New York at highway speeds. The hurricane flooded much of Providence, Rhode Island, and brought 299kph/186mph wind gusts to Blue Hill, Massachusetts, where forecasts a few hours earlier had called for a breezy but otherwise uneventful evening. Although more than 600 people died, the storm was virtually ignored outside the region. The front page was stolen by British prime minister Neville Chamberlain's visit to Adolf Hitler and the Munich Pact finalized a few days later.

Aircraft began probing hurricanes and collecting weather data during World War II, and the practice became standard shortly afterwards. Radar in the 1940s, and satellites in the 1960s, gave humans their first look at the complete spectacle of a full-blown hurricane. Spiral bands of thunderstorms pinwheel around the centre. In the strongest storms, the **central eye** – a zone of clear skies and calm winds – appears as a picture-perfect circle.

With the help of current tools – including instrument packages that drop from airplanes, gathering and transmitting data as they parachute to sea

South America's stray hurricane

People in Brazil aren't used to hurricanes, to put it mildly. In fact, no hurricane had ever been reported there until 28 March 2004, when a mysterious system packing winds up to 137 km/hr (85 mph) swept onto the coast of Santa Catarina province. Forecasters from Brazil and from the US National Hurricane Centre had been tracking the storm by satellite. Although no tropical cyclones had ever been officially recorded in the South Atlantic, the swirl of clouds had the clear eye and other telltale features of a bona fide hurricane. Despite the absence of a precedent, Brazil's meteorological service issued warnings and got people out of harm's way. The storm ended up destroying many hundreds of coastal homes, but only one death was reported.

Christened Hurricane Catarina for its landfall location, the storm continued to stir things up even after it died. Some meteorologists questioned whether it was truly a hurricane. Many global-change activists pointed to Catarina as an illustration of a planet gone awry. In fact, it turns out that Catarina had some of the classic earmarks of a hurricane, but not all of them. For instance, the showers and thunderstorms circling its centre were more shallow than usual. And Catarina can't easily be pinned on climate change, because the waters over which it formed were actually slightly cooler than average for that time of year. The storm was clearly unprecedented in its landfall impacts, but analysts poring over satellite records found at least one similar (if weaker) system in 1994 that stayed well out to sea. In short, Catarina remains a cipher – a bizarre weather event that pushed the buttons of a world increasingly jittery about climate change.

What's in a hurricane name?

Most weather events come and go like nameless strangers, but each hurricane gets the human touch of a moniker all its own. Centuries ago, the worst hurricanes in Spanish-speaking areas were named after the saint's day on which they occurred. A more elaborate naming tradition started with **Clement Wragge**, a forecaster in Australia during the 1880s. Wragge dubbed cyclones after Greek gods, Polynesian maidens and politicians with whom he disagreed. The idea caught on in America through George Stewart's 1941 novel *Storm*, in which a Pacific weather forecaster named storms after women. After the US armed forces began naming Pacific typhoons in 1945, forecasters applied military lingo to Atlantic hurricanes starting in 1950 (Able, Baker, Charlie, etc), then switched to women's names in 1953 (starting with Alice, Barbara and Carol). The practice made it far easier for the public to follow a specific storm over a few days, but the sexist overtones of using only female names soon became impossible to ignore. Again, Australia set the pace by switching in 1975 to a scheme that alternated men's and women's names. The US, and much of the rest of the world, followed suit in 1979; that year, David and Frederic were the two most destructive Atlantic storms.

Atlantic and Pacific hurricanes are now named from lists that rotate every few years. In 1999 a group of 14 nations affected by typhoons in the northwest Pacific took on a mix of indigenous names that includes roses, singing birds, favourite foods and at least one transportation hub: Kai-tak, Hong Kong's old airport. Cyclones in the North Indian Ocean went nameless until 2004, when a new naming system was created by the 16 countries bordering the region.

– forecasters can now track hurricanes with ease. **Computer guidance** helps plot their course, with fair accuracy in most cases, up to five days in advance. When the upper-level winds that steer hurricanes are especially weak or poorly diagnosed, the track forecasts are the toughest. Fortunately, if a hurricane sits in one place for too long, it often churns up enough cool water from below to quell its strength. Strong upper-level winds can also weaken or destory a hurricane by tilting the circulation, blowing off the tops of its thunderstorms, or otherwise jeopardizing its symmetry and integrity.

Although forecasters can usually tell when a hurricane is doomed, it's much harder to predict when strengthening will occur. Even if all the ingredients seem to be on hand, not all storms will thrive as expected. The worst storms often go through a phase of **rapid intensification** that can transform them from minimal hurricanes to monsters in less than 24 hours. Hurricane Wilma shocked onlookers in 2005 as it vaulted from tropical-storm status to become the Atlantic's strongest hurricane on record (based on barometric pressure) in less than a day's time. **Satellite-based radar** can now scan some parts of the ocean for patches of warm water that are several times deeper than usual, extending as far down as 150m/500ft compared to the usual 30–45m/100–150ft. This extra fuel seems to be one element that hastens rapid strengthening.

Rosenberg Library / Galveston, texas

The deadliest natural disaster in US history was the hurricane that struck Galveston, Texas, on September 8, 1900. It killed perhaps a third of the city's inhabitants – an estimated 9000 people.

Landfall and beyond

The classical motion of a Northern Hemisphere hurricane is westward through the tropics, with a gradual veering toward the north, and then a sharp "recurving" to the north and east at about 30°N, where the hurricane often encounters the westerly (east-flowing) winds that predominate at mid-latitudes. In the Southern Hemisphere, the tracks tend to be a mirror image: southwesterly, then southeasterly. The curve in these favoured paths tends to put the east coasts and the equator-facing sides of continents at the greatest risk of a hurricane strike (see map, p.87). Tropical waters are at their warmest toward the western Pacific and western Atlantic, which adds to the east-coast risk, although the northwest Australian coast – which adjoins the warm Indian Ocean – is a particular hot spot. The northwest Pacific is the world capital of tropical cyclones, producing more than a third of the global total in a typical year. Some two dozen each year thread their way through otherwise tranquil islands; many go on to threaten Vietnam, China or Japan.

The air tends to sink on the periphery of a hurricane or typhoon to compensate for all the rising air within the showers and storms. This often produces strangely calm weather in the hours before a hurricane arrives. **Winds** and **clouds** then increase, and periods of squally weather gradually intensify, punctuated by calmer spells. The **eyewall** – a ring of the most intense squalls – produces most of the wind damage. After **Hurricane Andrew** struck Florida in 1992, surveyors found evidence for mini-bursts of wind that caused severe damage across areas of less than a city block in no more than five seconds. A direct hurricane strike brings the eye across, producing a respite of a few minutes to an hour or more, inevitably followed by the other side of the eyewall, a trick of nature that has lured more than a few people to their deaths.

Size matters...or does it?

Like the people they're named after, hurricanes vary tremendously. There's actually little correlation between the size of a tropical cyclone and its strength. Some of the most destructive and powerful hurricanes are relatively small, and there have been numerous big, lumbering systems that packed little punch. The core of **Hurricane Andrew**, which ravaged south Florida in 1992, could fit snugly within the large eye of Hugo, another multi-billion dollar storm that struck South Carolina in 1989. The most intense US hurricane at landfall was the Labor Day storm of 1935, which moved across the Florida Keys and killed over 400 people. As strong as it was – its winds probably exceeded 250kph/155mph – the Labor Day storm was a tiny affair. So was **Tracy**, which virtually destroyed the Australian city of Darwin on Christmas Day 1974. The true Godzilla of tropical cyclones was 1979's **Typhoon Tip**, which was gigantic as well as powerful. At Tip's heart was the lowest barometric pressure ever measured at sea level (870 millibars, or about 25.70 inches) with fierce winds to match. Tip produced tropical-storm gales across an enormous width of 1100km/680 miles. For all its strength and size, Tip moved fairly harmlessly across the western Pacific, brushing near Japan while in a far weakened state. **Hurricane Gilbert**, which churned across the Gulf of Mexico in 1988, was another behemoth, with the lowest pressure ever observed in the Atlantic basin and a giant circulation to boot. Gilbert weakened before its two landfalls, lessening its damage to the Yucatan peninsula and the northeast Mexican coast.

Wind is only one part of a hurricane's bag of destructive tricks. The **sea surface** bulges up a few inches in the low-pressure eye, but a far greater rise, or storm surge, is produced as the storm's waters are funnelled into the coast. If the waters offshore are especially shallow, if the coastline is concave, or if a hill or mountain blocks the coast – or if any combination of the above occurs – the surge can be 6m/20ft or more. (Should a powerful hurricane ever approach New York City from the right direction – an unlikely but conceivable event – the city's harbour would help induce a storm surge that could inundate a good part of Manhattan Island.) As a hurricane moves inland and loses its oceanic fuel, friction helps weaken the winds, but it also helps set up a wind structure conducive to **tornadoes**. As many as sixty twisters have been reported in landfalling hurricanes. Also, as happened with Mitch and Nina (see p.84), heavy rains can trigger devastating floods.

Leaving the immediate coast is the best way to avoid the risk of hurricane. However, evacuation can be a delicate matter. If too many people leave from safer spots inland, they can clog highways and hinder the escape of those truly at risk along the immediate coast. Some US cities have emergency plans in place to house people in large structures such as stadiums when evacuation times are short, although Katrina showed how such a plan can quickly devolve into chaos (see p.85). Well-built homes on high ground can be equipped to withstand many hurricanes with the help of clips that hold roofs in place and shutters that block debris from shattering windows.

For more on hurricane precautions, see "Safety tips", p.398. Hurricane strength is assessed most often with the Saffir-Simpson intensity scale; see p.403.

NOAA

The roots left exposed at the base of this tree reveal coastal erosion, a common byproduct of coastal storms.

Coastal storms

Hurricanes are the most familiar of the storms that ravage coastlines, but they aren't necessarily the sturdiest. For all its immense power, a hurricane can be quite finicky. Should the sea surface be a touch too cool, or should a stiff wind tilt the central chimney of rising air a bit too much, the balance of forces is destroyed and the hurricane becomes history.

There's another kind of oceanic storm, one that doesn't wilt in the presence of cold air and isn't easily torn to shreds. You've seen such a storm if you've ever tried to walk the shores of Cape Cod during one of the vicious storms known as **nor'easters**, or stood on the Normandy coast in the teeth of a wet January gale. These systems – great swirls of low pressure that pull cold air south and warm air north – don't rate a name as they approach. Some of them are named afterwards, however, once they've blasted miles of beachfront, smashed windows, unroofed homes, and robbed people of their lives. Years later, they're remembered by location or date: the Ash Wednesday storm, the Columbus Day storm, the Great Storm of 1703, the Storm of the Century. Less renowned than hurricanes or tornadoes, **coastal storms** can still pack hurricane-force winds and cause tremendous damage to buildings and ecosystems along the shore.

A product of contrasts

Coastal storms are part of the huge family of **low-pressure centres** that wend their way from west to east around the surface of the globe at mid-latitudes. Most places, whether inland or on the coast, see dozens of these lows come and go every year. Each low might bring rain, thunderstorms, snow, ice or perhaps just a shift in the wind. Most of these lows form ahead of equatorward dips in the upper-level jet streams that encircle the Northern and Southern hemispheres (see "Where does the wind go?", p.23). Such dips, or troughs, help nourish the circulation of a surface low and tend to keep it moving eastward, especially in the presence of jet streaks. These bundles of wind, blowing at speeds that can top 320kph/200mph, descend from the upper atmosphere and help enhance upward and downward motion at lower levels.

Once a low is generated, it needs **moist air** in order to make rain or snow. The ocean needn't be the only source, since humid winds can travel hundreds of miles inland, or a large lake can provide moisture. However, lows that form or move over the sea have little trouble staying moist. Like a hurricane, a coastal low feeds off the warmth of the ocean and the water that evaporates from it. Unlike a hurricane, a coastal low can ingest warm, moist air from the ocean even as it pulls in cold, continental air on its other side. Coastal lows are structured so that this asymmetry in temperature doesn't hinder them. In fact, the contrast between warm and cold can actually energize their circulations. Some of the greatest coastal storms have occurred when a powerful low moves close to a mass of cold air located near the coast. As warm air rises on top of the blob of cold air, heavy rain and snow can be wrung out, and the tightening temperature and pressure contrasts often produce gale-force winds. Waves can build for days across thousands of miles of ocean before they batter the shore.

Along the US east coast, cold air often lodges against the Appalachians as far south as the Carolinas for days at a time. This "cold air damming" shifts the effects of the mountain range closer to the coast and heightens the likelihood of a coastal storm producing snow or ice rather than rain.

In general, east coasts around the world are usually favoured for the worst coastal storms, particularly snowstorms. That's because of the huge anti-cyclonic oceanic gyres that tend to bring warm water toward the poles along the east coasts of continents, just at the same place where cold air may be moving off the continent. In **Australia**, the intense cyclones that spin up in the Tasman Sea are dubbed East Coast lows. The action is a bit further offshore in **Asia**, where cold air is typically entrenched across the continent in mid-winter, so lows in this region usually pull east of Japan toward the Gulf of Alaska.

Western Europe has also experienced its fair share of coastal storms, thanks to the **Gulf Stream**. This mild ocean current flows up the US east coast; an extension continues all the way northeast toward Norway. The warm waters encourage some coastal storms to spin up near the British Isles or the continent, while others manage to migrate across the North Atlantic. Similarly, coastal lows can surf the Pacific's **Kuroshia current** and its extensions and intensify all along the way to the Gulf of Alaska. Many end up slam-

Associated Press

Calm after the storm: Londoners survey damage after the gale of 1987.

ming into the mountains of the southeast Alaskan coast, where the town of Ketchikan is drenched with an average of close to 4000mm/157in of rain each year.

Europe's big ones

Between December 26 and 28, 1999, two intense coastal lows swung inland across **France**, killing over 100 Europeans. The first low – the more northerly and stronger of the two – brought wind gusts of up to 219kph/136mph and knocked out much of France's electric power for days. The strongest winds in half a century littered Paris with debris. Rains from the second storm forced the Seine to overflow its banks. Some 10,000 trees in the grounds of Versailles palace were destroyed, including 80 percent of the palace's historic trees, such as the pine from Corsica planted during Napoleon's reign.

The French nightmare reminded many British people of their own hurricane-like experience, the **Great Storm of 1987**, whose intensity caught many by surprise. Winds and rain had been forecast for the night of 15–16 October, but the emphasis had been on heavy rain. As the coastal low approached, BBC weatherman Michael Fish announced there would be no hurricane, in response to a viewer's phoned-in query about a separate system far out in the western Atlantic. Although technically accurate, his reassurance may have been lost on Gorleston, Norfolk, where winds gusted to 196kph/122mph. Six of the namesake trees at Sevenoaks were felled, along with 15 million others across southeast England. London's power was knocked out for the first time since the Blitz, and the total damages came close to one billion UK pounds.

The 1987 storm was dubbed by some as the nation's worst in three hundred years, calling to mind the Great Storm of 1703. Two weeks of bad weather that November had driven hundreds of ships to the southeast coast. On the night of November 26–27, winds topped hurricane strength as a low crossed England from southwest to northeast. Coupled with the work of a seasonally high tide, the winds piled some 700 ships into a bend of the Thames. The storm knocked over some 800 houses and killed about 125 people on land and as many as 8000 at sea. The most famous account came from **Daniel Defoe**, who lived in London at the time and surveyed the damage in town and at sea. Defoe edited a set of journalistic accounts of "the particular dreadful effects of this tempest", and he penned a poetical essay full of political and spiritual portent: "And every time the raging element/Shook London's lofty towers, at every rent/The falling timbers gave, they cry'd REPENT."

Although it didn't do its worst in London, the deadliest modern-day coastal storm to strike Europe brought massive destruction to England's east coast and the low countries on February 1, 1953. A rapidly deepening low pulled hellacious gales onshore from Norfolk to Kent. Seawalls were breached en masse, and coastal flooding was so severe that some residents reportedly escaped only by punching holes in the ceilings of their flats. Just over three hundred died in England, but the picture was far worse in The Netherlands, where the storm's main surge coincided with high tide. A massive dike failure flooded over 2000 square km/770 sq miles and killed some 1800 people. The dual catastrophe led to decades-long flood control projects in both countries, including the Thames Barrier that now protects London.

Was it really the Perfect Storm?

Plucked from obscurity by a little-known writer, it gained fame on the best-seller lists. The coastal storm of October 31, 1991, is now part of American pop culture as "The Perfect Storm", made famous by Sebastian Junger's 1997 book and the 2000 film of the same name. But did the event really live up to its image?

Other coastal storms have gained more intensity, inflicted more damage, killed more people, or followed the classic life cycle of a coastal storm with more fidelity. What made the 1991 storm so exceptional was a rare, prolonged confluence of factors that led to exorbitant wave heights, measured at 21m/70ft and estimated at 30m/100ft – as high as any in the twentieth century in the north Atlantic. The storm's embryo was a moderately strong, but fairly typical, coastal low. As its upper-level support slowed with time, the surface centre slowed to a crawl off the Canadian Maritimes; it eventually moved back west toward New England, providing several days for wave heights to build. As cold air fed into the low from the northwest, the remnants of Hurricane Grace moved in from the southeast, adding a shot of tropical energy. The low developed an unusually hybrid structure, and satellite images briefly showed an eye. A US hurricane-hunter plane on training duty took a peek and found all the earmarks of a minimal hurricane. However, given the marginal conditions for growth and the already-high public awareness, forecasters opted to leave the storm unnamed. It brought 8m/26ft waves to the New England shore, destroying about 200 homes and damaging thousands, including the seaside retreat of the senior President Bush at Kennebunkport, Maine.

High points of US coastal lows

Record-setting storms happened to strike both sides of North America during 1962. The **Ash Wednesday** coastal storm takes the prize as the nation's most relentlessly damaging of the twentieth century, due largely to its timing and longevity. The low moved slowly up the coast over a four-day period (March 6–10) that coincided with the year's highest tides. An unusually strong mass of high pressure centred over maritime Canada helped feed cold easterly winds that battered the US coast from Massachusetts to North Carolina. The easterlies went on to help dump heavy snow as they ascended the Appalachians. Later in the year, on Columbus Day (October 12), the Pacific Northwest was mauled by a fast-moving system that cruised up the Oregon and Washington coasts – much closer to land than usual – in a matter of hours. With the remnants of a typhoon tucked in its core, the low brought wind gusts to 201kph/131mph at Oregon's Mount Hebo and damaged over 50,000 homes across the state.

One of the most powerful and most accurately-predicted lows in American history was dubbed the **Storm of the Century** even before it struck on March 12–14, 1993. A jet stream plunging well into the Gulf of Mexico gave this low a kick-start unusually far south. Feeding off warmer than normal Gulf waters, the low intensified and barrelled into the northwest coast of Florida with a 3.7m/12ft storm surge. A ferocious line of thunderstorms extended south all the way to Cuba, causing widespread damage with wind gusts estimated to be 220kph/137mph. Staying just inside the East Coast as it moved north, this low lived up to the dire forecasts spun by computer models. It produced windblown snows from north Florida to Maine, as deep as 127cm/50in in places. The storm dropped the barometric pressure to unprecedented lows as it became the first weather event ever to close all the major East Coast airports from Washington, DC, northward. Some 270 people died – many of them the victims of heart attacks due to shovelling heavy, wet snow.

Other windstorms

"They call the wind Maria", or so a classic tune from a 1950s musical informs us. A few years earlier, the forecaster-protagonist of the 1941 novel *Storm* did just that (and perhaps inspired the song) by dubbing a particularly mean cyclone Maria. The fixedness of our earth's landscape helps to shape the wind into familiar patterns that come and go. Across the world, people can't seem to resist naming the wind *something* – especially the types of wind that regularly visit the same place.

Blowing hot and cold

Some of our most recognizable winds come into existence as a result of **mountains** and **hills** that channel air streams. As wind blows toward and up a mountain range, the space between ground level and the top of the atmosphere narrows, so the air accelerates in response, like water rushing through a squeezed garden hose. Upsloping winds can produce lots of rain and snow, but after they pass a crest and head downslope, they take on a wholly different cast. Downsloping wind picks up momentum as it descends, which helps it to blow even more vigorously.

The most famous **downslope winds** are either markedly hot or cold, depending on which of two factors governs. Descending air is normally compressed as it sinks, which helps to warm it up. This is the mechanism that dominates in the hot, dry downslope winds epitomized by the **Santa Ana** of the Los Angeles Basin or the **brickfielders** of Sydney. The equivalent in the Alps is the **föhn**, a term often used as a general label for warming downslope winds. In winter, a föhn can bring delightfully mild temperatures, but at the cost of howling gusts that can set teeth on edge and wreak havoc. In Boulder, Colorado, a 1982 **chinook** – the term means "snow-eater" – peaked at 220kph/137mph.

When a strong cold front reaches a mountain range, the cool, dense air may flow down the slope with vigour. Even though it warms through compression, it's still colder than the air it's replacing. These biting winds are known as the **bora** along the Adriatic Sea, **canyon flow** in the Salt Lake Valley of Utah and the **mistral** in France.

Both warm and cold downsloping winds typically last only a few hours at their strongest, although a persistent large-scale regime might bring several bouts of wind over a couple of days. Downslope winds do tend to occur in conjunction with plenty of sunshine, thanks to the evaporation of whatever cloud droplets may be in the air as it expands and descends. A thick, dramatic mass of cloud called a **föhn wall** might be seen banked against the ridge line. Such a cloud takes shape just over the ridge, where the air's ascending, and can hug the mountains for hours. If a cold bora-type wind is strong enough, a few drops of rain or snow can be flung over areas a few kilometres downstream, even where the sky directly overhead may be clear. When moisture is especially scant, the only cloud visible above a downsloping wind might be the smooth, graceful arc of a lenticular or cap cloud above the highest peaks.

Downbursts and derechoes

Another kind of windstorm – a bane to farmers and aviators alike – is a **downburst**, marked by cool, moist air descending from a thunderstorm. It

Tough weather to dress for

Where mid-latitude mountains meet the plains, downslope winds can trigger spectacular temperature shifts. One of the most dramatic took place in Spearfish, a tiny US town in the Black Hills of South Dakota. On the morning of January 22, 1943, a frigid Arctic air mass was muscled aside by a balmy downslope wind. The temperature in Spearfish at 7.30 am rocketed from –4°F to 45°F (– 20°C to 7°C) in a mere two minutes – the most rapid temperature shift ever measured anywhere. The mercury continued to gyrate through the morning and finally dropped back to its frigid starting point a couple of hours later.

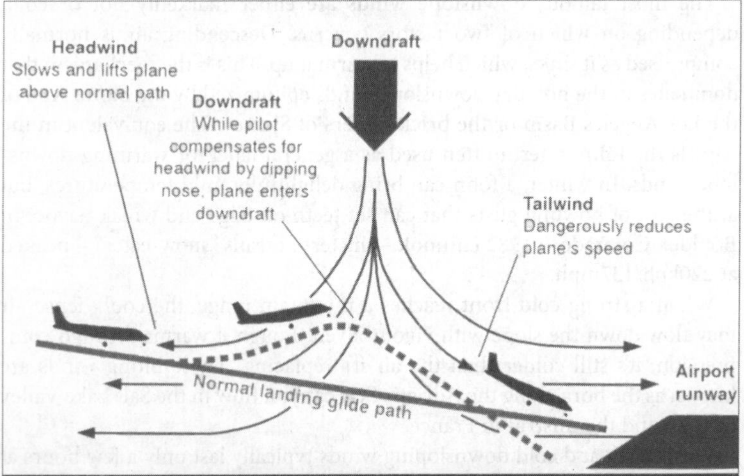

The outflowing winds of a downburst pose a special hazard to low-flying aircraft.

was long recognized that thunderstorms could cause severe damage on the ground without a tornado being evident, yet nobody could explain what was going on. Meanwhile, as air travel grew in the 1960s and 1970s, more and more planes were crashing to the ground in and near storms. Tornado guru **Theodore Fujita** made a connection as he flew over damaged cornfields in the US Midwest following the world's biggest outbreak of twisters on April 3, 1974. Fujita noticed that some of the downed cornstalks weren't arranged in tornado-induced spirals, but were instead flayed outward in a starburst pattern. It reminded Fujita of bombing damage he'd surveyed in World War II. As it turns out, this kind of meteorological bomb was not only ravaging crops, but causing aircraft mishaps such as a 1975 crash at New York's Kennedy Airport that killed 113 passengers.

Fujita and colleagues later verified that the tons of rain and hail in a cloud can trigger brief, yet destructive, rain-cooled downbursts of air blowing at up to 240kph/150mph. The smallest of these – less than 4km/2.5 miles across – are labelled **microbursts**. Even a weak storm positioned over dry surface air can produce a surprisingly strong downburst as rain evaporates and cools the air below. A plane entering a microburst first encounters a headwind that slows its descent and raises its nose. The pilot's logical response – tipping the nose downward – can send the plane hurtling to the ground when it reaches the tailwind on the other side of the microburst.

A blend of surface instruments, radar and software now tracks this dangerous type of wind shear (wind difference across space) near airports in the US and many other storm-prone countries. Downbursts are a hazard away from airports, too. Nine New York schoolchildren were killed in November 1989 when their school's cafeteria wall collapsed. The event, New York's deadliest-ever windstorm, was first tagged as a tornado, but Fujita and other experts later concluded it was most likely a downburst.

The king of downbursts is the **derecho**, a long-lasting windstorm that can leave a swath of damage a few miles or kilometres wide and more than 800km/500 miles long. It's especially common from the US Midwest into central Canada, where severe thunderstorms can take on a particular temperature and wind structure that lends itself to long-lived jets of descending wind. Once called prairie hurricanes, and sometimes referred to in Canada as blowdowns, derechoes may pack winds of well over 160kph/100mph that blow for upwards of an hour at any given spot. Derechoes shred crops and can even deroof buildings – damage that's often mistaken for tornadic mayhem until specialists take a closer look.

Quick spins

There are plenty of small-scale atmospheric vortices that keep our air spinning. Most of the world's warmer oceans and lakes, including the Mediterranean – and even some colder ones, like the Great Lakes of North

NOAA

Cyclonic winds sculpt the ocean surface in this classic photo of a waterspout off the Florida Keys.

America – see **waterspouts**, also known in Italy as *trombe marine* (marine trumpets). Like other tornadoes, these too emerge from showers and thunderstorms, but a waterspout is considered a different animal unless and until it moves over land.

A typical waterspout forms along the boundary that separates standard marine air from the cooler air mass beneath a rain (or snow) shower. It's a 30- to 60-minute-long process that begins with the appearance of a **dark spot** on the water surface. Some of these spots develop **spiral bands** that reflect inflowing winds. Next comes a **ring of spray** just above the water surface. A few minutes later, a **slender funnel** may descend from the cloud base; sometimes the visible funnel stops short of the sea and leaves a transparent section in between. The typical waterspout spans no more than 75m/250ft (often as little as 9m/30ft) and lasts 20–30 minutes before it pulls rain-cooled air into its updraught and dissipates.

Graceful as they may look, waterspouts can be deadly. Most develop winds between gale and hurricane force, and a few produce gusts above 240kph/150mph. A number of boaters have been killed in such US haunts as North Carolina's Outer Banks and the Florida Keys. If the conditions are right, a pack of three or more waterspouts may appear along a line of convergent winds (see the photo on p.99). Even the heat from island volcanoes can induce swarms of waterspouts, such as the dozens that formed downwind of Surtsey in 1963. Over land, some weaker tornadoes are classified as **landspouts** because they form through a similar process as waterspouts. One

Weatherwise Magazine / Mark Hamilton

Oil fires triggered in Kuwait during the Persian Gulf War in 1991 spawned this dramatic fire whirl.

You say diablo, I say sirocco

A hot wind by any other name is just as enervating. The people of Los Angeles and San Francisco put up with occasional Santa Ana and diablo winds, respectively, that funnel down the canyons flanking these cities. Southern Europeans are familiar with a torrid desert wind that blows in from the Sahara or Middle East; it's called the *leveche* in Spain, the *sirocco* in Italy, and the *levanto* in the Canary Islands. Since this wind arrives at low levels, it's not downsloping – just furiously hot. When desert air from interior Australia blows into the southeast coast, Sydneysiders call it a *brickfielder*, after the dusty industrial areas scoured by the same wind in an earlier era. Wintry North American air masses sometimes careen into and over the mountains of Central America; if they reach the Pacific coast, they produce a *papagayo* in Nicaragua and Guatemala and, to the northwest, a *Tehuantepecer* in the Gulf of Tehuantepec. Hurricanes often skirt the same Pacific coast in the summer, although they seldom make landfall. The southerly wind they produce has been called a *cordonazo*, or the lash of St Francis (after the Feast of St Francis, which occurs on October 4, close to the peak of hurricane season).

Texan captured a row of six simultaneous landspouts on camera. Like their marine cousins, landspouts often resemble transparent tubes with invisible mid-sections.

Anyone who's driven across parched fields on a hot summer's day knows about **dust devils**. Unlike the other spin-ups already discussed, dust devils aren't a product of rain clouds. Indeed, they thrive on clear skies, which allow the sun to produce intense heat near the ground and induce bubbles of rising air called **thermals**. All that's needed then is a slight breeze or a small variation in surface heating (as between a ploughed and unploughed field) to help create eddies that might serve as dust-devil nuclei. Because they're too small and short-lived to be affected by the turning of the Earth, dust devils have no preferred direction of rotation. They occur most commonly around noon, when solar heating is at its strongest, and just afterward. American researchers have counted as many as a hundred dust devils on a single day. Land devils or hay devils have been observed in Great Britain and other areas.

Most dust devils are puny by tornadic standards. In one Arizona survey, more than 80 percent of them spanned less than 15m/50ft, although they can extend upward for more than 0.8km/0.5 miles; their winds rarely exceed 80kph/50mph. Dust devils move with the surrounding wind, and they tend to be larger when that wind is stronger, as long as it's not too strong. Some dust devils can last for hours if buildings, hills or even vehicles happen to be arranged to channel the wind in a favourable way. Power lines have been downed and trailers knocked over by fierce dust devils in Arizona.

When a raging fire produces ground-based heat, **fire whirls** (pictured opposite) can result. These can rival weak tornadoes in their strength and size, and they can help advance a forest fire with violent intensity. Spin-ups have been seen near the heat vents from active volcanoes. Even fires in man-made settings, such as the horrific firestorm caused by the World War II bombing of Dresden, Germany, can produce vortices.

Heat waves and cold waves

They aren't the flashiest of villains, but hot and cold spells – when they're sufficiently intense and prolonged – are among the deadliest of all weather events. Everyone knows that it's risky for us warm-blooded humans to spend much time in temperatures far above or below our margins of comfort. Yet only in the past several decades have we fully realized the risks that heat and cold can pose to even the most developed cultures. An estimated 50,000-plus Europeans fell victim to the brutal heat wave of 2003 (see box, p.104). And many thousands of vulnerable people die each winter when bouts of cold sap their immune systems, leaving them more vulnerable to influenza and other cold-weather ills.

Part of the reason we haven't given heat and cold waves the respect they deserve is because they make for less-than-compelling video. In an image-driven world, it's much easier to grab people's attention with scenes of a tornado or hurricane. Hot and cold air do their dirty work far more stealthily, away from the klieg lights. Often the victims are elderly people living alone on tight budgets, unable to keep their homes temperate enough for mere survival, let alone comfort. Sociologist Eric Klinenberg refers to heat waves – in a comment that could apply to cold waves as well – as "invisible disasters that kill largely invisible people".

Although it's perpetually warm in the tropics and chilly near the poles, the systems we think of as heat waves and cold waves tend to strike the mid-latitudes, especially North America, Europe and Asia. Much of the time, the polar jet stream keeps weather systems marching across these continents at a fairly brisk clip. Temperatures rise and fall, but they rarely stray far from the seasonal norm for long. At other times, the jet stream arcs around a massive dome of high pressure extending far up into the atmosphere. Like a stone in a river, this high-pressure ridge forces the usual stream of fronts and cyclones to flow around it, sometimes for a week or longer. Somewhat paradoxically, a blocking pattern like this can produce extreme heat in the summer and vicious cold in the winter. In the summer, the clear skies above high pressure allow for maximum sunlight to reach the surface. In the wintertime, the arcing jet can pull bitter surface air from the Arctic around the ridge and southward on the east side, resulting in a cold wave that can extend well southeast of the ridge itself (or to its northeast in the Southern Hemisphere).

No matter what the time of year, forecasters keep a wary eye out for these blocking patterns. Sometimes an entire season may be prone to several

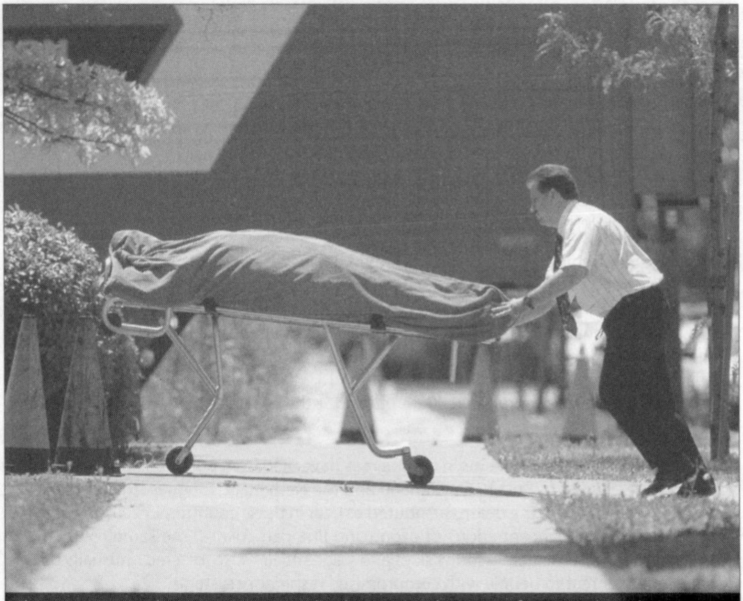

Chicago, 1995: a heat-wave victim's body is moved into a refrigerated truck.

occurrences of the same block, often due to El Niño or other ocean-atmosphere cycles (see p.118).

Too darn hot

In some parts of the world, hot spells are imported. Mediterranean-style climates often have coastal cities that stay naturally air-conditioned, with intense heat lurking just inland. A small shift in circulation can push the torrid air seaward, producing a spell of sizzling heat in areas that may not be used to it. The typical summer day in San Francisco falls short of 20°C/68°F, but on occasion the city has soared above 38°C/100°F. Likewise, Melbourne gets many summer afternoons that top out close to a pleasant 25°C/77°F, but every so often the city can roast in readings that touch 40°C/104°F. Thankfully, such incursions of heat tend to be short-lived, usually lasting only a couple of days.

A truly dangerous heat wave encompasses far more than a toasty afternoon or two. When it comes to the toll exacted by a heat wave on people, animals and plants, the duration and the warmth of the nighttime lows are the real killers. A weeklong stretch of severe heat can be far more deadly when nights stay unusually warm, erasing the few hours of relief one might otherwise get. Uncomfortably warm nights are most likely when and where the humidity

Europe feels the heat

Records melted like ice cubes across Europe during the astonishing summer of 2003. Pulses of heat plagued much of the continent in June and July, but by far the worst occurred in the first two weeks of August. Many areas topped 35°C/95°F day after day, with nights often staying above 20°C/68°F. England saw its first day of 38°C/100°F heat in nearly 300 years of record keeping; in London, tube stops were like ovens. Three other countries (Germany, Switzerland and Portugal) also notched all-time records, with the town of Amareleja, Portugal, soaring to 47.3°C/117.1°F.

The unusual strength and persistence of the heat wave produced a catastrophic death toll that went almost unnoticed at first. In France, the peak of the heat coincided with the annual vacations that close down much of Paris. Left behind were thousands of elderly residents unable to cope with the unprecedented heat. Short-staffed hospitals and morgues found themselves overwhelmed. More than 20,000 deaths were reported by September across Europe, including more than 13,000 in France alone. That toll steadily mounted, eventually topping 50,000 – making it the greatest heat-related disaster in worldwide annals.

Thousands of heat-wave victims in Europe may have died from pollution rather than the heat per se. Studies in Britain, Switzerland and the Netherlands suggest that anywhere from 10–40 percent of the deaths attributed to heat in those countries in 2003 may have been due to high concentrations of ozone and fine particulates. The same conditions that led to the intense heat – sunny skies and stagnant air – also fostered unusually dirty air, a serious threat to people with compromised respiratory systems.

is unusually high. In highly urbanized settings, buildings and pavements release the heat at night that they've absorbed by day (part of the urban heat island effect). Many of the worst heat-wave disasters of recent years have occurred in older midlatitude cities that feature dense urban layouts and large numbers of residents who lack home air conditioning.

Even when dangerous heat is obviously at hand, it's not always clear when a disaster is unfolding. Many people die indirectly in heat waves – from pollution, for instance, or from pre-existing health conditions exacerbated by the heat. It can take days or even months before the full scope of the death toll is evident. And since many of the victims are elderly, it's often assumed that the heat simply claimed people who were about to die anyway. If that were the case, then mortality ought to dip below average in the months following a heat wave. Such dips do occur, but studies show that they account for only about 20–30 percent of the excess deaths observed in a typical heat wave.

Heat waves don't correspond to latitude in the way one might think. In the subtropics, it's got to be really hot to qualify as a heat wave. Many residents of cities like Phoenix, Arizona, and Riyadh, Saudi Arabia, hardly bat an eyelash until the mercury shoots above 45°C/113°F. Stretches of truly unusual heat appear to be less deadly in climates like these – where people are already adapted to torrid conditions. Poverty can change this equation, though. Just before the summer monsoon sets in across India, temperatures can soar well above 40°C/104°F across wide areas. Several pre-monsoonal heat waves in recent years have each killed more than a thousand people, many of them landless workers forced by circumstance to toil in the elements.

Social scientists have looked closely at recent heat disasters in the midlatitudes and learned much about how to save lives. The impetus in the United States was a 1995 heat wave that was especially intense in Chicago: daytime highs reached 41°C/106°F and two nights stayed above 27°C/81°F. More than 700 Chicagoans died. Yet some city officials downplayed the event, even blaming the victims themselves for not taking precautions. After the disaster in Chicago, Laurence Kalkstein of the University of Delaware teamed up with the US National Weather Service to launch a new watch-warning system for extreme heat. The scheme, which included assigning people to check on neighbours, saved an estimated 100 lives that decade in the city of Philadelphia alone. The system was exported to Italy just before its 2003 heat wave: Rome declared 18 heat-emergency days that summer. Yet even with the warning system in place, the sheer intensity of that summer's heat produced hundreds more deaths in Rome than Kalkstein and colleagues had anticipated.

The 2003 heat wave (see box opposite) may have provided a sneak preview of what North America and Eurasia can expect in the century to come. As it raises the average worldwide temperature, global warming is also expected to boost the number of days that climb above local heat-wave thresholds. One study by British scientists found that the typical European summer of the 2040s could be warmer than the blistering summer of 2003 was.

Cold's dangerous grip

While it usually takes several days for a large-scale heat wave to set in, cold waves often arrive with stinging abruptness. The more quickly a frigid air mass can sweep in from polar latitudes, the colder the weather it's likely to produce, since there's less time for warm ground to temper the air mass. Some of the world's sharpest temperature transitions have occurred in the US Great Plains, where blue northers can send readings plummeting by as much as 25°C/40°F in less than two hours. One of the strongest such fronts on record occurred on 11 November 1911. That afternoon, just before a cold front arrived, Oklahoma City set a record high of 28°C/83°F. Before midnight, the city had plunged to a record low of –8°C/17°F.

When snow accompanies the quick onset of cold, the impact can be more brutal still. On 12–13 January 1888, more than a hundred US children, many of them rushing home from school on foot, perished in a frigid snowstorm that swept across the Dakotas with blinding speed, killing several hundred adults as well. The great "children's blizzard" struck just two months before New York experienced its own epic snowstorm (see p.59).

As with heat waves, the true measure of a cold wave is its persistence. Britain endured its coldest winter of modern times in 1962–63, when snow cover lay across most of the UK for over two months. The bulk of the nation hovered below freezing through most of January and much of February, with several spikes of cold that were even more extreme (temperatures in Yorkshire sank below –20°C/–4°F at one point).

Some of the world's coldest settled areas are in Siberia. The region's largest city, Yakutsk, manages a paltry –37°C/–35°F for its average high in January, and stretches of cold, clear weather can send the mercury even lower. Yet these hardy Russians are seldom caught unprepared, since the cold is so frequent. That's not always the case further south, where temperatures well above freezing can prove fatal to people without adequate clothing and shelter. More than 600 people died in Bangladesh when a cold wave in December 2002 and January 2003 sent temperatures below 5°C/41°F. It was one of several cold snaps that each have killed hundreds across South Asia over the past few years.

Even in relatively well-off Britain, it's believed that as many as two million people, largely elderly, suffer from "fuel poverty" – typically defined as when heating costs take up 10 percent or more of household income. Unable to keep their poorly insulated houses warm, many Britons endure the winter in perpetual chill, which increases their vulnerability to pre-existing health problems and to communicable diseases such as the flu. Each winter brings an estimated 20,000-plus deaths in Britain due to health effects stemming from too-cold housing. In a 2001 initiative, the British government set out to eradicate fuel poverty by 2016, with national and local governments issuing grants to needy homeowners to help them pay utility bills and stay warm.

Floods

Most weather events descend upon us. **Floods** rise up to meet us from below – sometimes quietly, sometimes ferociously. They're one of the few weather-related hazards that can be made far worse by the way we choose to live: where we pave the land, build houses, fell trees and dam rivers. And floods may be the most underrated natural disaster known to man. Aside from heat and cold, flooding is the biggest weather-connected killer in the world.

Unless you've seen a trickle of a creek become a torrent, you may have no idea of the sheer, brute power that mere water can exert. Experts divide flooding into **river floods** and **flash floods**, two very different creatures. Flash flooding – the more mercurial and lethal kind – can send cars tumbling down the engorged canal of a hillside town in Italy, or leave a forlorn house marooned on a freshly carved island in the midst of a rampaging Canadian river.

The suburbs of Brisbane, Queensland, were hit hard by flooding in January 1974 during a stormy Australian summer.

Except for the rare spectacle of a dam break, floods aren't solo events. They're tied to other headline-grabbing weather features, whether it be a hurricane, a series of severe thunderstorms or a melting field of snow. Often the storm steals the spotlight while the flood is treated as an afterthought, perhaps not getting the attention it deserves until many people have suffered.

What makes it flood?

Rain is the first and last word in flooding, but not the only one, of course. Everything depends on how the rain falls and how the water behaves once it's on the ground. A large amount of rain spread out over days of gentle pitter-pattering might not cause a problem, while the same rainfall packed into an hour or two could devastate a town. The key factor is whether the landscape can absorb the rain as it falls. This depends on a host of things:

- **Urbanization** Water flows more quickly and efficiently across pavement than grass. Storm drains can easily clog.
- **Soil type** Rainfall percolates through sandy soil more easily than through brick-like clay.
- **Recent weather** All else being equal, the more it has rained in the past few days or weeks, the more likely a flood becomes. Even if the rain puts an end to drought, it might be falling onto soil that's so sun-baked it can't absorb much water at first.
- **Topography** Mountains are great floodmakers – they can wring water out of moist winds blowing into them, and they can channel rains into steep channels in virtually no time at all. Most of the world's worst flash floods have been in

mountainous areas. Even arid or semi-arid regions can experience destructive flash floods if the terrain is rugged enough (or if urbanization has paved enough of it).

A **river flood** builds slowly, often from a winter's worth of snowmelt or a summer of incessant rain over a large area, perhaps more than a metre's worth (39in). As smaller tributaries feed into larger mainstems, the power of the water can multiply to astounding levels, especially in large basins like the Mississippi, which flows south through the central US, or the Yangtze, running through the heart of eastern China. These floods can destroy huge amounts of property and disrupt vast areas for weeks, but they often play out on a long enough time scale that relatively few deaths and injuries occur. ("Relatively" is the key word along China's mountain-cradled Yangtze, which is prone to landslides and dam breaks that exacerbate river floods. More than 3500 died in the Yangtze flooding of 1998, while the floods of 1931 may have killed as many as one million people.)

As the name implies, a **flash flood** happens quickly. It's often a product of one or more slow-moving thunderstorms that park over a flood-prone area, perhaps one as small as a city neighbourhood, and dump huge amounts of rain in an hour or two. The "wall of water" that flash-flood survivors often describe can raise the water height by a metre or more in less than a minute, although flash floods can easily kill without such a dramatic visual cue. Preceding rains that saturate the soil make a flash flood more likely, and the landscape makes a huge difference as well. There's some overlap between river and flash floods: the latter can occur on a small scale in a region that's already experiencing a river flood.

NOAA

Today's motorists are just as prone to get stuck – or killed – in a flood as in 1927, when the Mississippi ran wild.

What can we do?

The age-old dream of controlling rivers has triggered some of the most ambitious engineering projects in history. China's **Three Gorges project** aims to tame the Yangtze with a series of **dams** that could end up displacing a million people and costing upwards of US$30 billion. The United States alone has more than 50,000 dams along its rivers and creeks. Without them, Las Vegas, as we know it, could never have thrived in its arid landscape. Yet damming a river can eradicate fish and tamper in many other ways with local ecosystems. In the US, the expansionist mindset of river control ran headfirst into the growing environmental movement of the 1960s and 1970s. Since then, no major new dams have been built in the US, and the Three Gorges project has been roundly criticized by environmentalists.

Of course, there have been many motives for damming: recreation, water supply and hydroelectric power, among others. Still, even in their role as flood-control devices, dams and riverside levees now have a somewhat compromised reputation. A river with levees might flood less often, but once the levees are breached, the resulting flood spreads out more quickly – and sometimes catastrophically – than it otherwise would have. Bottled-up rivers also tend to collect silt at their bases, adding to the potential flood risk years down the line. Dams and levees are now being supplemented by a more variegated approach. Instead of jailing a flood, the idea is to coax it into behaving. Trees and other natural breaks can help slow and channel flood waters. Setting aside open areas along river courses gives a flood room to grow safely. Retention ponds capture flood water for constructive use.

Anyone can spot the kind of torrential, long-lasting rains that act as fuel for flooding. However, predicting floods with precision may be the toughest job in meteorology. River floods can be well anticipated several days ahead (even weeks ahead, if snowmelt is involved), with the help of automated streamflow gauges and computer models that project the water in time and space. The specific height of a major river's flood crest is hard to pin down: it depends on the inflow from myriad smaller, harder-to-diagnose streams, as well as whatever rain or snow might fall between forecast and crest. In 1997, a difference of 1.5m/5ft between the predicted spring crest and the actual one at Grand Forks, North Dakota, made the difference between a serious flood and an epochal one that destroyed much of this city of 50,000 residents.

Short-fuse floods are a forecaster's nightmare. Few national weather services around the world even try to issue formal warnings for these rapid-fire events, which might come and go before word gets to officials, much less to the public. The most sophisticated weather radars can now estimate rainfall amounts minute by minute over wide areas, but knowing the local topography well enough to pinpoint the status of a tiny mountain creek is much more difficult. Even in flat country like Houston, Texas, predicting just how much rain will fall isn't easy. Forecasters had issued a flash-flood warning on the night of June 8, 2001, as the remnants of **Tropical Storm Allison** lingered near the coast. Computer models indicated that the system could dump as much as 13cm/5in of rain by the next day, but nobody expected 64cm/25in to fall overnight across parts of town. The deluge inflicted some US$3 billion in damage and killed more than twenty people. The potential for dam

Canyons and water: a dangerous combo

The lure of a **slot canyon** – narrow and serene, with walls that ascend far higher than the canyon's width – is hard for a serious hiker to resist. Thousands traipse into these flash-flood factories each year unaware of the risk they face. A downpour in one spot can send water pouring into a slot canyon more than 32km/20 miles away, where it might not have rained at all. This was the case when twelve hikers were trapped by a flash flood in Arizona's Lower Antelope Canyon on August 12, 1997. Only one survived, and the canyon was closed for nearly a year afterward. A communications system and heavy-duty ladders are now in place. Flash floods are also a real hazard for canyoners who skitter down gorges and waterfalls with a minimum of equipment. Near Interlaken, Switzerland, 19 canyoners in a group of 53 participants and guides were killed on July 27, 1999, as a wall of water containing trees and other debris careered into them. Whenever heading into the wilderness, take care and note the presence of storms all around, not just in your vicinity. A downpour need not be overhead to be deadly.

or levee failure, as in Hurricane Katrina, is another hard-to-assess wild card in gauging flood risk.

Seeking safety from a flood has its own risk. Many flood victims die not from drowning but from being tossed into and among debris in a raging current. Cars are a particular hazard: most flash-flood deaths in the US, and many elsewhere, occur in motor vehicles. Time and again, people try to drive through a flooded roadway that looks navigable, only to lose traction and float downstream. Surprisingly, it takes only about 50cm/20in of water to float most vehicles – even large ones. Once a car's wheels have left the pavement, the vehicle floats toward the most rapid flow, which typically is close to the deepest part of the channel. By that point, there's no hope of driving away. The only reliable way to escape a flash flood is to move to higher ground as soon as possible, even if that means leaving one's vehicle and setting out on foot. Most flash floods play out in an hour or two, a small investment of time to save one's life.

Drought

Drought is one weather extreme that's defined not by what happens but by what doesn't happen – namely, rain. It's as if nature forgot that it was supposed to provide a certain amount of moisture each year. The meteorological spigots are shut down, and nobody can say just when they'll be turned on again, although El Niño may provide a hint (see p.114).

Drought is usually pictured through its effects: withered crops, empty reservoirs, dust storms. The most poignant impacts are on people, of course. Few weather-related images are as burned into our consciousness as the worry-lined, dust-scratched faces of America's drought-stricken 1930s and the hollow-eyed Ethiopians who faced famine in 1985. If a flood can destroy people in minutes, a drought is far less merciful in accomplishing the same task, and far slower in bringing relief.

What is a drought?

As plain-as-day as a drought may seem, academics and scientists have wrangled long and hard over a precise definition. Recently they've coined a set of criteria for drought types and agreed that a given drought may only meet some of these.

Meteorological drought is the obvious definition: the rains stop, or diminish to a certain degree, for some length of time. In many places it doesn't have to be bone-dry to register as a drought, especially where it normally rains year round, such as London or Fiji. In these locales, a drought might mean it's simply raining less hard when it does rain. To assess a drought here, you might note how long you've gone without getting what you consider to be a decent day's rain. In more arid climates, as in Denver or Melbourne, you could evaluate how many days it's been since it's rained at all. This is especially true in desert climates, where a single day can produce the amount of rain you'd normally expect in a year.

However you gauge it, what matters when assessing a drought is how much rain a place has received compared to the average. A year with 500mm/20in of rain would be manna for parts of Mongolia but a disaster for Miami. Flora and fauna adapt to the long-term average rainfall in a given locale, so it's the departure from that average that takes a toll. Cities that function well in a normal rain cycle may be in big trouble when a drought strikes, especially if they've expanded during times when water was plentiful.

Since drought affects us by drying up our water supply and damaging our crops, it can also be viewed through the lens of **hydrology** and **agriculture**. How low is the water level in the reservoir that your city relies on for water? Are the lands upstream getting enough rain to help recharge the system? Crops can be affected in complex ways by drought. If a dry spell is just starting, the topsoil may be parched but the deeper subsoil still moist enough to keep an established crop going.

NOAA

Day turned to night in minutes as dust storms raged across the central US during the 1930s.

Conversely, a long-standing drought may break just in time to help germinating plants get started, even if the sub-soil remains dry.

When drought becomes destiny

There's a certain self-perpetuating aspect to drought. A persistent weather pattern that inhibits rainfall for a few weeks is enough to get a drought started. Once it's going, a more subtle **interaction** between **land** and **air** takes place. The dry ground heats up quickly and sends its heat to the atmosphere more readily. As green vegetation dies off, the brighter ground reflects even more heat than before. This may help sustain the giant, cloud-free domes of air so prevalent across a drought-stricken region. However, the relative importance of this land–air interaction in keeping a drought going isn't totally clear.

Another factor, one that's become a boon to drought prediction, is the discovery of the **El Niño/La Niña** influence on world climate. These cycles, tied to the tropical Pacific (see "El Niño and La Niña", p.114), tend to create a set of preferred weather patterns that can persist for one or two years around the world. El Niño is strongly associated with drought across **northeast Brazil** and **South Africa**, and it commonly diminishes the strength of the Asian monsoon, thus triggering drought from India to Indonesia and northern Australia. La Niña tends to cause drought in places where El Niño encourages heavy rain, such as eastern **equatorial Africa** and the **central Pacific**; it's also been tied to drought across the western and central United States. These relationships aren't iron-clad, but they do serve as powerful tools in helping farmers and policymakers manage their water resources better.

Does it follow the plough?

Places like the central Sahara, where it virtually never rains, can't really experience a true drought. However, the margins around these barren lands can go from long-term drought into a more ominous, semi-permanent state of aridity, perhaps abetted by the actions of society. **Desertification**, as this is called, jumped onto the political landscape in the 1960s and 1970s, when the **Sahel** of **western Africa** slid into a drought that lasted off and on for more than two decades. In the moist years beforehand, government policy had encouraged farmers to cultivate northward toward the desert's edge. In turn, that pushed nomadic farmers and their herds into even more marginal territory further north. When the drought hit, both grazing and farming were too intensive for their land's now fragile state. Wind and erosion took hold, and the zone of infertile land grew southward.

It wasn't the first time **land use** had been fingered as a partner to drought. The catastrophic **Dust Bowl** across the central US in the 1930s – made

NCAR / UCAR / NSF

Vegetation has a tough go of it in the Sahel region of Africa, where rainfall is sporadic and often unreliable.

famous in John Steinbeck's *The Grapes of Wrath* – followed several decades of giddiness as Europeans settled what was once called the Great American Desert. Despite the warning signs of several intense droughts during the late 1800s, the land was green and fruitful by the 1920s. But the above-average rains abruptly stopped in the early 1930s, and the overworked soil soon blew away in choking dust storms that paralysed towns and cities from Texas to the Dakotas. The drought's timing – just as the Great Depression was setting in – seemed especially cruel, but it was the activist government of the **New Deal** that ended up educating farmers on land use and creating such innovations as shelter belts of trees to break the fierce prairie wind. Another southern-plains drought was nearly as intense in the 1950s, but it caused far less damage, thanks in large part to the New Deal. Some of the lessons were transplanted to Africa; shelter belts of neem trees reduced the impact of the 1970s Sahel drought in parts of Niger.

Boosters of expansionist farming once thought that "rain follows the plough" – if a marginal land were cultivated, then water would evaporate above the greened land and help keep it moist. As US political scientist **Michael Glantz** has noted, the events of the twentieth century pointed toward a more bleak aphorism: drought follows the plough. We still can't fathom how long a long-term drought influenced by human intervention might last, even in the case of the Sahel, where more regular rains had returned by the 1990s. Agricultural practices may expose the vulnerability of a given land, but the atmospheric forces that induce drought and take it away could be the main player after all. Humans may yet help trigger more extensive and intense droughts in another way: by warming the global atmosphere. (See "A primer on climate change", p.150).

El Niño and La Niña

Moody, tempestuous, violent: it's so easy to assign human qualities to **El Niño** and its companion, **La Niña**. This pair of atmospheric cycles broke into the news in the late 1990s when a remarkably strong El Niño wreaked global havoc. The pithy name fits well with headlines, and the concept – a strange, portentous warming of a faraway ocean that triggers bad weather – captured the public's fancy, especially in the US and Australia.

Millions of people around the world knew about El Niño by 1998, but few could have told you exactly what it was. Even the best scientists in the field have yet to detail all of the important factors that make El Niño and La Niña events blossom and wither. One thing that's now clear is that the brother-and-sister duo are part of an extended family of ocean/atmosphere cycles that shape weather over large chunks of the globe (see box, p.118).

The basics

El Niño and La Niña are two sides of a coin. At the most basic level, an **El Niño** happens when the surface water becomes warmer than average through the **eastern tropical Pacific**, a strip running west from the Ecuador coast. **La Niña** is the opposite – the same waters are cooler than average.

These temperature changes aren't dramatic: only about 1–2°C/1.8–3.6°F for a weak Niño and 3–4°C/5.4–7.2°F for a quite strong one. La Niña departures are even less. What gives the cycles their immense power is their geographic scope. During the El Niño of 1997–98, the warmed waters covered an area the size of the US. Vast amounts of heat were pumped into the air above, and the result was a shift in weather patterns that cascaded across much of the globe. The exact chain of events that produces some El Niño impacts at surprising distances, such as in eastern Africa, isn't always well understood.

The family name

Because there's far more to El Niño than a change in the ocean, the term is now popularly used to refer not only to the oceanic warming but also to the weather changes that go along with it. Ditto for La Niña. However, scientists have to be more precise than this. Some argue over which patch of the Pacific ought to be used to diagnose El Niño. At least four different regions have been considered. Experts refer to the whole interwoven phenomenon – El Niño, La Niña and the atmospheric changes related to them – as the **El Niño/Southern Oscillation** (ENSO). Journalists would cringe, but one scientist proposed changing "El Niño" to a more precise but far less catchy term: "the warm phase of ENSO". The flip side of El Niño became commonly known as La Niña only in the 1980s. Peruvians had never attached a label to the cooler than normal waters, since they saw no negative effects on the local ecosystem. La Niña is actually a bit of an ungainly term, since there is no female equivalent in Spanish for the Christ child. Two alternatives – seldom used outside of the scientific world – are the anti-El Niño and El Viejo (the old man).

However, thanks to careful statistical analysis, we know the general kinds of weather that El Niño tends to produce, even if we don't know precisely how it does each trick.

El Niño and La Niña typically alternate, separated by neutral or near-normal periods that can last several years. On average, each Niño or Niña takes from one to three years to play out, although the lengths vary wildly. In the long run, the in-between periods occupy about as much time as El Niños and La Niñas combined.

What makes it happen

To get to the bottom of El Niño, you have to go to its birthplace. The **tropical Pacific** is the world's broadest stretch of ocean. A boat trip from Ecuador to Papua New Guinea takes you more than a third of the way around the globe. Easterly trade winds blow almost incessantly across the length of these tropics. They push up so much water along their path that the sea level on the west side normally runs more than 50cm/ 20in higher than on the east side.

The western end of this equatorial belt, close to Indonesia, is known as the **western Pacific warm pool**. It has some of the toastiest ocean water on the planet, averaging 30–32°C/86–90°F. Meanwhile, far to the east, much cooler seas prevail. Offshore currents allow deep water to well up, supporting a rich marine life near the coasts of Ecuador and Peru.

A similar set of contrasts normally play out in the atmosphere above. Near-constant showers and thunderstorms dump huge amounts of rain each year on the islands and oceans of the warm pool. But what goes up must come down, and the air that rises in rainstorms above the warm pool tends to flow back east at high levels and then subside over the eastern Pacific, suppressing

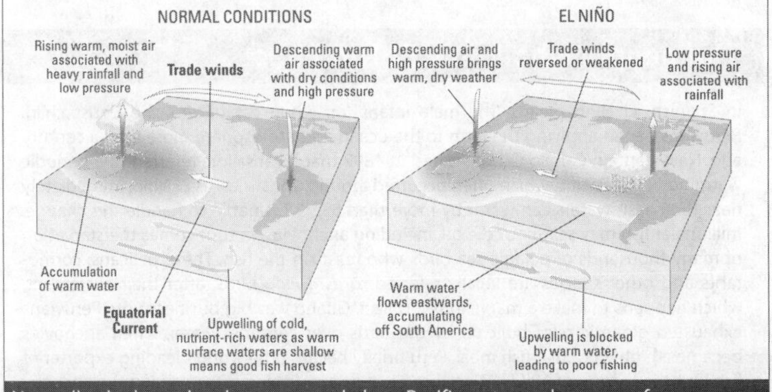

Normally, the boundary between cool, deep Pacific water and warm surface water slopes upward from west to east across the tropics (left). During El Niño, warm water builds in the east, and tropical rainfall shifts east as well.

rainfall there. Parts of the west coast of South America can go months or even years without seeing anything more than drizzle.

El Niño acts to reverse this flow of air and water. The stage is set with a gradual deepening of the warm pool waters so that they extend from the surface down as far as 200m/660ft. Every so often, a batch of especially strong and long-lived showers and thunderstorms pops up over the warm pool, coupled with a surge of westerly wind. As it presses against the ocean surface, this surge appears to help force a kink in the **thermocline** (the undersea boundary that separates warm surface water from cooler deep water). As the surge and the kink in the thermocline both push into the eastern Pacific, they help to suppress the usual cool upwelling. In a few weeks or months, warm surface water thus covers the entire tropical Pacific from shore to shore. Without the strong east-west contrast in sea–surface temperature, the trade winds weaken, sometimes even reversing direction.

Once this new air-and-water pattern is in place, it tends to sustain itself until El Niño runs out of steam. How this happens isn't totally clear, but one factor is the same thermocline kink that got things started. As the downward kink hits the east end of the Pacific, a slower-moving impulse rebounds to the west. It's eventually reflected off the shallow west end of the warm pool, and within a year or two, makes its way back across to the eastern Pacific, where it may help to shut down El Niño.

In some ways, the La Niña pattern is "even more normal than usual". Instead of weakening, the prototypical air and water circulations strengthen. Trade winds blow harder, more warm water builds up near Indonesia, and the upwelling of cool water off Ecuador and Peru intensifies. The normally heavy rains in and near the warm pool, including eastern Australia, become even more intense. In the US, the standard wintertime clashes in air mass often step up, so that brutal cold spells may alternate with unusual warmth.

The people who know El Niño best

In Spanish, El Niño means "the male infant" or, when capitalized, the Christ child. Peruvians began applying the term to the ocean and atmosphere more than a century ago. Navy captain Camilo Carrillo noted in 1892 that local sailors referred to a periodic warming of the Pacific, which often occurred around Christmas, as El Niño. Immediately near the coast, waters can warm by more than 5°C/9°F. Unable to handle the change, millions of fish may scatter or die off, including anchovies. In turn comes the starvation of many thousands of pelagic sea birds who feast on the fish. These pelicans, cormorants and other species are locally referred to as *guano* birds, after their excrement, which happens to make a marvellous fertilizer. Guano was big business until Peruvians exhausted gigantic piles built up by the birds over centuries. Later, when anchovies became a highly popular fish meal, Peru briefly became the world's leading exporter of fish. But the El Niño of 1972–73 occurred just as overfishing was ravaging the anchovy stocks, and the reverberations were felt around the world. Peru's fisheries haven't been the same since.

The factors that make La Niña come and go are even more shrouded in mystery than El Niño's. We do know there isn't any tit-for-tat: the strength and length of a Niño have little bearing on the timing or intensity of the next Niña, and vice versa.

Calling the shots

In 1982, nobody knew that a catastrophic El Niño was forming until its effects were at hand. In response, a set of instrument-studded buoys was installed across the tropical Pacific, and a new kind of weather and climate watching emerged. These bobbing platforms sent back data every day on the water temperature at and below the surface. It was the first early-warning system for El Niño and La Niña. Now, with the buoys and with satellite monitoring of the ocean, we can spot the signs of a developing Niña or Niño months before its effects reach some places.

Computers make a big difference as well. Global models specially tuned to follow El Niño and La Niña can extrapolate the data from sea and sky and help provide a glimpse of whether the coming event is likely to be a doozy or a dud. Several of the models sensed in early 1997 that an El Niño was on its way, although none predicted its colossal strength until the Niño actually began unfolding that summer. Still, that was enough time to warn of the impacts that would likely follow by year's end. Governments around the world took heed. Public awareness campaigns helped get roofs sealed, crops harvested and reservoirs drained before the impacts hit. Damages in California were far less than expected, most likely due to a combination of being prepared and a late start to the well-predicted torrential rains. (Comedians who mocked "El No-Show" in January had to eat their words in February.)

We haven't attained forecast nirvana yet. In parts of the developing world, warning systems aren't yet in place. Where the word does get out, El Niño fatigue can set in. The term became an all-purpose scapegoat across the US for what seemed like every weather event in 1997–98. Actually, El Niño or La Niña effects are seldom guaranteed. The odds may shift markedly toward wet or dry, warm or cold, but in any year the rest of the global atmosphere can counterbalance these likelihoods. For instance, rain and snow usu-

NOAA

A network of buoys stationed across the Pacific hunts for signs of El Niño and La Niña.

ally pound the US Pacific Northwest during La Niña, but conditions were unusually dry as a weak La Niña hung on during the winter of 2000–01.

Toughest of all to predict are starting and stopping times. Much goes on in the water and air of the Pacific beyond the gaze of our observing system. And the most sophisticated computer models often disagree on timing, especially during the period from April to June, when the Pacific tends to be in transition between regimes. By September, though, it's often clear who will help dictate the world's weather for the next few months: brother, sister or neither.

Cycles galore

Europeans who couldn't care less about El Niño may sit up and take notice when somebody mentions the **North Atlantic Oscillation**. Although it's less important on a global scale than ENSO, the NAO is a much bigger player in Europe's climate. Unlike El Niño, it's an atmospheric cycle that operates more or less independently of the ocean beneath it, although possible connections to faraway tropical oceans are emerging. The NAO is basically a see-saw in air pressure between the **Azores** and **Greenland**. When the NAO is positive – higher pressure over the Azores, lower pressure over Greenland – the west-to-east jet stream cuts more directly and strongly across the Atlantic. This tends to bring wet, mild winter weather to northern Europe and makes the Mediterranean winter even more invitingly sunny and dry. By contrast, the NAO's negative phase blocks the jet stream, forcing it to arc around the North Atlantic. This can bring periods of cold, dry air across northern Europe and leave southern Europe open to renegade systems that can bring lingering dampness. Since the NAO is detached from the ocean beneath it, there can be no early-warning system like the one provided by Pacific buoys for El Niño. Moreover, the NAO may swing from positive to negative or vice versa after only a few weeks in a given mode. The NAO favoured its positive phase for much of the 1980s and 1990s, giving northern Europe a string of wet and mild winters. Several computer models project a continued preference for the positive NAO in the century to come.

More widespread displacements of the polar jet stream throughout the Northern and Southern Hemisphere are dubbed the **Northern Annular Mode** and **Southern Annular Mode**. For both the NAM and SAM, a positive phase means the jet is contracted poleward in a tight ring that tends to limit incursions of polar air into midlatitudes. According to basic physics as well as computer modelling, the positive modes of the NAM and SAM should become more prevalent over time as global warming pushes the jet streams poleward. This should help feed the positive NAO trend noted above; it may also tend to push critical winter rains away from parts of southern Australia.

Then there's the **Pacific Decadal Oscillation**, a circulation pattern centreed south of Alaska that was named in 1996. The PDO is a measure of regions of warm and cool water across the North Pacific that typically takes 20 to 30 years to shift. When the PDO is positive, as was the case from the late 1970s into the 1990s, it tends to enhance El Niño's effects and diminish La Niña's; the converse is true for a negative PDO. The most evident PDO correlations to date involve fisheries: salmon and other fish in the North Pacific tend to cluster further north, toward Alaska, during the PDO's positive or warm phase. As with the PDO, the North Atlantic may also have warm and cool cycles that switch every 20 or 30 years, alternately stoking or suppressing hurricane activity. Scientists are honing the statistical tools needed to separate cycles like these from day-to-day weather as well as from long-term warming.

Chapter three
Forecasts and how to read them

Forecasts and how to read them

Professionals discovered long ago that the atmosphere is simply too complex to track without the help of computers – the more powerful, the better. Today, some of the world's peppiest machines zip through billions of calculations a second in the service of meteorology. They can't yet follow the wind through every mountain valley and around each building, which is where the human factor enters the equation. Weather presenters on television (whether or not they're meteorologists) can add local knowledge and translate the statistics churned out by computers into language we understand. The same applies to government forecasters around the world, as well as private firms that serve ever more specialized niches. Common sense and rules of thumb can get you started as an amateur forecaster. You can join other interpreters by taking your own weather readings and digging into the raw data available to all on the Internet.

Getting the word out

So you're wondering if it's going to rain tomorrow. You pick up the paper and find a weather report that's a sea of typography. The local forecast must be bobbing around somewhere in there, but you can't find it straight away. You hop onto the Internet and come up against a hundred different Web pages with tomorrow's outlook. Some of them say it'll rain; others say it won't. You turn on the TV and find yourself flying with the weathercaster through the innards of a great cloud mass somewhere off at sea. You learn there's a 50 percent chance this system will bring you rain tomorrow. Does that mean it'll rain for 12 out of 24 hours, or that half of your city will get wet – or that you could flip a coin and come away with just as much information?

As with so many other aspects of our choice-crazy culture, the challenge involved in accessing weather information isn't actually in finding it, but in deciding how much of it you can process. What used to be "the forecast" – meaning just one – has metamorphosed into unlimited weather on demand. As recently as the 1980s, the choices were far fewer and their presentation was far more drab.

Although today's smorgasbord of data and forecasts may be intimidating, it's also a spectacular success story. In the 1940s, apart from a few scattered examples, the best minds of meteorology were hard-pressed to offer useful predictions for more than a day or so in advance. The concept of cold and warm fronts was foreign to most people. Then came the arrival of **computers, radars, satellites** and **television**, all of which rose to prominence between the mid-1940s and mid-1960s. The Internet followed in the 1990s. Together, these tools and media have made a world of difference in creating useful forecasts and getting them to the public.

Despite the perennial role of the forecaster as whipping boy (if not comic relief), forecasts on the whole are better than ever. Large-scale weather shifts can often be seen a week in advance, even if the details aren't crystal-clear. And people *do* pay attention. Weather is consistently the most-watched part of the local news in the US, and the weather section is the most viewed part of the BBC's website. Part of this thirst for weather information is undoubtedly due to mobility. More and more people live and travel far from their childhood homes, so they tune in to see the kind of weather that is affecting their relatives or the town where they plan to fly to on business.

Before the advent of television, hundreds and even thousands of people were killed by big storms almost every year in North America and Europe. Now it's rare for a weather event to take more than a hundred lives in either region (Hurricane Katrina and the 2003 European heat wave notwithstanding). Sadly, this isn't the case in many developing countries, where communications links are fragile and millions of people live in meteorological harm's way.

From psychic reading to forecast

Soothsayers have always been willing to speculate about future weather (a tradition that continues in US tabloids, where the biblical end times are forever at hand). Weather proverbs have always had their grains of truth (see box, p.124). However, the art of predicting the weather based on physical law is surprisingly recent. Even after reliable techniques began to take shape in the late nineteenth century, the age of electronic mass communications had to arrive to make real the notion of the timely weather forecast available to all.

Aristotle's classical view of meteorology held sway for some 1500 years. This conventional wisdom saw weather as a phenomenon controlled by the interaction of sunlight and the planets with exhalations, vapours and other elements drawn up from the land and sea. The very word "meteorology" refers to meteors, but in a much broader sense than today – not just rocks from space, but rain, snow and everything else that blows through the atmosphere.

The earliest published weather outlooks appeared in almanacs, beginning in the 1500s. Two of the most popular by the eighteenth century were *Old Moore's Almanac* in Britain and *An Almanack for the Year 17XX* in the colonial US. (Variations can be found to this day, such as the venerable US competitors *The Farmer's Almanac* and *The Old Farmer's Almanac.)* These forecasts relied heavily on a blend of **astronomical tables** and weather signs linked to the **calendar** and the position of the **planets**. For instance, fair weather was believed to follow if the fourth night after a new moon was clear. People also looked to holidays and saints' days (see box, p.128) for guidance to the weather. The day of the week on which Christmas fell, for example, was sometimes used to predict whether the coming winter would be mild or stormy.

The invention of basic **weather instruments** and the **age of exploration** slowly changed things. Scientists seized on new data and began forging theories of storm behaviour and motion. Once weather reports from a variety of points were assembled, the connections that tied local weather to a larger scale were harder to ignore.

The Internet of its day – the **electric telegraph** – arrived in 1844, as the message "What hath God wrought!" made its way in dots and dashes from Washington to Baltimore. For the first time, people could paint a coherent picture of the weather across a vast nation on the same day that the weather occurred. The British marvelled at the world's first up-to-the-minute (or up-to-the-day, at least) weather map in 1851, when it debuted at **London's Great Exhibition** in Hyde Park. For a penny, visitors could take away a map of the British Isles that showed observations relayed to the site by train and telegraph. Later in the 1850s, you could drop by the **Smithsonian Institution** in **Washington, DC**, and check out a wall-mounted map showing the day's reports from several dozen stations from New York to New Orleans. In 1863, **France** became the first country to publish daily weather maps; other nations soon followed.

The Civil War delayed America's development of weather forecasting for a few years, while a psychological war had the same effect in Britain. **Captain Robert FitzRoy**, who had commanded *The Beagle* – on which Charles Darwin sailed around the world carrying out research in the 1830s – was named the first head of the Meteorological Department of the British Board of Trade (soon to become the **Meteorological Office**) in 1854. Although the science of weather forecasting was still in its infancy, FitzRoy was tireless in promoting it. On August 1, 1861, the Met. Office published the **world's first public forecast**: the next day's wind and weather for four UK districts. Published in the London *Times,* the outlooks were an instant hit. However,

FitzRoy was censured by scientific societies who felt the state of meteorology was then too immature for forecasts to be of any real merit. FitzRoy later spent several years fruitlessly seeking a grand theory to connect solar variations with weather (still a topic of research – see "The Sun and Earth", p.17). Discouraged and in ill health, he committed suicide on April 30, 1865. Suddenly devoid of FitzRoy's enthusiasm and with little support elsewhere, the Met. Office soon suspended its forecasts. They didn't resume until the 1870s.

Weather for the masses

Weather forecasts came to **America** in the 1870s with the establishment of the **National Weather Service**. Part of the military at first, it soon became the US Weather Bureau, a name that stuck till the 1970s. Using data from the NWS, the *New York Times* began carrying a small section that summarized the previous day's weather and gave official "indications" of the regional conditions to come. By 1900, the weather ear – a front-page box with a terse forecast – had made its way into newspapers.

Oddly enough, it took several more decades for **weather maps** to carve out their now familiar niche in US papers. The London *Times* launched the world's first daily map in 1875, and it was still going strong in the early years of the 21st century. Maps appeared in newspapers around the world as government weather services blossomed. However, in the already vast and still growing United States, it proved tougher to compile data, map it to the government's satisfaction, and get it to far-flung newspapers in time to

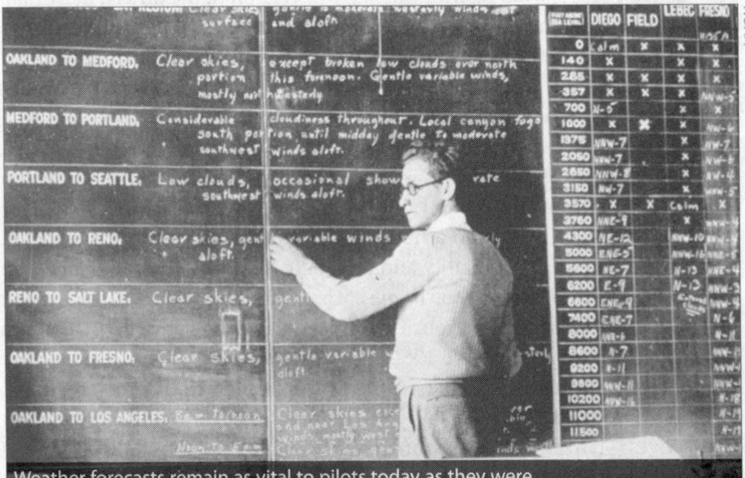

Weather forecasts remain as vital to pilots today as they were at this airport station in San Francisco, circa 1931.

NOAA

So the sky says

Folklore through the ages has linked the state of the sky to the upcoming weather. The parade of cloud that often precedes a mid-latitude storm – from high, wispy cirrus and patches of cirrocumulus to lower, rain-bearing nimbus – gave birth to the maxim, "Mackerel scales and mare's tails make lofty ships carry low sails." The presence of wind shear (wind variations with height) ahead of a storm often reveals itself in contrasting cloud motions: "When high clouds and low, in different paths go, be sure that they show it'll soon rain and blow." Clouds to the west are the source of most weather systems in mid-latitudes, as Shakespeare surmised: "The sun sets weeping in the lowly west/ Witnessing storms to come, woe, and unrest." In the book of Matthew, Jesus observes, "When it is evening, ye say, it will be fair weather; for the sky is red. And in the morning, it will be foul weather today: for the sky is red and lowering." The meteorological sentiment is expressed in a more secular format as, "Red sky at night, sailor's delight; red sky at morning, sailor take warning." What a red sky really means is that shorter wavelengths of sunlight are getting scattered, leaving a reddish cast to the sky. This effect is enhanced by dust and other particles held within a stagnant air mass. So at mid-latitudes, a red sunset implies that tranquil weather is moving our way from the west, while a red sunrise implies the air mass is on its way out. Like all weather proverbs, of course, this one offers plenty of room for error.

make deadlines. The maps were expensive for papers to reproduce and hard for some readers to interpret. Yet the government mailed millions of maps each year from local weather offices to anyone who asked for them. Things changed quickly in 1910, when a national system for routing simplified maps from the US Weather Bureau to newspapers proved wildly successful. Internal politics soon dashed the new system, however, and the weather map again disappeared from most US newspapers until 1935. That year the **Associated Press** started distributing its own version of the official daily map by facsimile to its member papers nationwide. These small, simple, black-and-white maps were the state of the art until *USA Today* debuted in 1982. That newspaper's flashy, full-page weather section was derided at first, but it proved so popular that most other US papers had to upgrade their own weather sections to compete.

In the 1920s, newspapers met with new competition in the form of **radio** when the **BBC** transmitted its first **public weather forecast** on November 14, 1922; by 1924 more than a hundred US stations offered radio weather. Some stations linked personality and credibility to create a new persona, the "weatherman", who visited your home each day and told you what to expect. On Boston's WEEI radio station, **E B Rideout** and his daily reports were a fixture from the 1920s into the 1960s. Meteorologist **Jim Fidler** launched his radio career in Muncie, Indiana, in 1934 by appealing to truckers, builders, farmers ("We plan our work by your reports", read one testimonial) and housewives ("I'm going to wash tomorrow – Jimmie said the sun will shine"). Fidler later became the first weatherman on NBC's morning TV programme *Today*.

TV gets in the act

To help keep potentially useful details out of enemy hands, **censors** put a brief, but dramatic, hold on public weather information during **World War II**. Newspapers in the US weren't allowed to publish forecasts for regions more than 250km/150 miles away. Even as the government clamped down, a new medium waited in the wings: **television**. The BBC first tested weather text on television in 1936, and an experimental New York station launched the first TV weatherperson in 1941 – **Wooly Lamb**, a cartoon character who opened each segment by singing, "It's hot, it's cold. It's rain, it's fair. It's all mixed up together."

As television exploded in the 1950s, **TV weather** became part of the daily newscast. This time, the US led the way. The BBC launched its first live TV weather segment in 1954. At about the same time, American programmers discovered they could use weather as lighthearted relief from the rest of the day's news. Meteorologist **Percy Saltzman** had already delivered weather for the Canadian Broadcasting Corporation using three puppets. Soon the US airwaves were swamped by costumes, animals, rhymesters and weather "girls". As many as a third of US weathercasters in the mid-1950s were women, many of them apparently recruited for their sex appeal. One in New York tucked herself into bed on camera each night while discussing tomorrow's forecast. Another, clad in a swimsuit, gave her reports while immersed in a huge tank of water.

Thankfully, the medium soon matured. A "seal of approval" programme was launched by the American Meteorological Society in 1960, so viewers could tell which weathercasters had the credentials to provide a serious weather report (including bona fide female meteorologists, whose numbers increased slowly but surely from the 1970s onward). Clowning didn't vanish from TV weather overnight, however. **Willard Scott** became a fixture on NBC's *Today* show in the 1980s, dressing as Carmen Miranda and Boy George in between forecasts. In Australia, Melbourne's popular **Edwin Maher** maintained a good-humoured atmosphere to his ABC reports into the 1990s, even recording a weather rap ("Your Weekend's Gonna be Ruined") with musicians Captain Jellybeard & the Scurvy Dogs. But more and more, the increasing sophistication and pizzazz of weather graphics in the 1980s and 1990s – such as colourful satellite and radar images – allowed weathercasters to stand back and let the weather itself take centre stage.

American-style goofiness never invaded the lower-key weather presentations of the BBC, although in 2000 the network launched a major graphical revamp of its weather broadcasts. This included one of the most dramatic visual innovations of recent years: "**fly-throughs**" that simulate a three-dimensional trip through a weather map or satellite photo. Sticklers gripe that these images vastly distort the vertical scale of weather features, since

Washingtoniana Division, DC Public Library

Like many of her TV contemporaries in the 1950s, Cindy Dahl (Washington, DC) drew weather maps backwards while standing behind Plexiglas.

most of the broad low-pressure centres that bring bad weather are about a hundred times wider than they are tall. Another round of brickbats flew in 2005 when the BBC added a 3-D perspective to the map itself. Southern England and Wales now loomed perpetually large compared to Scotland, a fact noted with distaste by more than a few Scots. The same graphical trick is increasingly common on US television, which gives Texas and Florida disproportionate heft compared to, say, North Dakota or New England.

America's thirst for weather information has made **The Weather Channel** a cable-TV institution, ranking with CNN and ESPN as one of the most widely available cable outlets in US households. Few gave the channel much hope for survival after its 1982 debut. Gradually, however, viewers and advertisers came around and the channel now reaches more than 70 million homes. The core programming originates from studios in Atlanta, Georgia, while local conditions and forecasts are relayed by satellite and inserted by hundreds of cable systems about every ten minutes.

Where does it come from?

Most weather information is relayed to the public by **private companies** through TV, radio, newspaper and the Internet, but it's not always apparent how much these sources rely on the government. Each nation's weather system varies in its degree of private involvement. **New Zealand** has one of the world's most privatized – its stated mission is to "grow a commercially successful business". At the other extreme, some TV weathercasts in **Italy** are delivered by military meteorologists in uniform. The **US** is something of a hybrid: it has the world's largest government weather service, operating side by side with more than one hundred private forecast providers. By and large, it is the **government agencies** that take and gather weather observations, exchange them among the rest of the world's countries, and operate most of the computer models that guide most forecasting decisions – whether they're made by a faceless government worker, or by that amiable weathercaster you watch over coffee each morning.

Thanks to its big population, geographic reach, hyperactive weather and media-saturated culture, the US employs far more people to dispense weather news than any other country. Even with the Internet, local TV weathercasters are still a key source of tomorrow's forecast – and meteorologists are probably the only scientists that people see on TV on a daily basis. There are more than 1000 on-air weathercasters in the US, while Britain has about fifty at best. Perhaps half of America's weathercasters have meteorological training, typically adding their own insight to the official forecasts and model data provided largely by the **National Weather Service.** Non-meteorologists on-air are perhaps more likely to adhere to the government forecasts, although they may or may not acknowledge the debt. Warnings for hurricanes, tornadoes and other severe weather are issued by the government and passed on more or less verbatim at virtually all stations. The US also employs hundreds of other meteorologists in the private sector, many of them working in liaison with newspapers, TV and radio stations. One of the biggest of these private companies is **Accu-weather** in State College, Pennsylvania, which claims to have the most forecasters assembled at one place (close to one hundred). Most US TV stations have arrangements with Accu-weather or some other partner from the private sector. At a minimum, these companies provide **software** that allows a weathercaster to create his or her own graphics. For those stations that want or need more input, private firms can provide ready-made displays and interpretation of the output from government models (a few private companies produce their own models as well). Accu-weather also has a stable of its own radio and TV weathercasters. For over thirty years they've furnished segments that are fed live or spliced on tape into local news bulletins around the country. Similarly, many newspapers have weather sections created wholesale each day by private firms that may be hundreds of miles away.

Britain's weather culture is dominated by the **Met. Office**. The weathercasters on the BBC's various outlets are actually Met. Office employees (mostly meteorologists) stationed at the network. Elsewhere on TV, radio and in the newspapers, a mix of Met. Office and private contractors furnishes the weather information. Much of the TV weather in Australia and Europe reflects the British rather than the American paradigm. **European countries** typically have one government TV channel and a few commercial outlets. Weathercasters tend to be meteorologists, some employed by weather agencies and others by the TV stations themselves. In **Australia**, both public and private sources provide weather graphics, but weathercasters tend to use the official forecasts from the Bureau of Meteorology.

Whatever the source of their information, TV weathercasters would be shooting themselves in the foot if they didn't avail themselves of the wealth of weather data the **Internet** offers (see Resources section p.385). The Internet has illuminated a long-standing tension between the principle of free data exchange – and the desire of governments and private weather services to make money from the weather. In order to preserve this balance, the United Nations and its World Meteorological Organization passed a resolution in 1995 allowing countries to withhold some of their less critical data from free international exchange. Still, many of the observations and computer-model results produced around the world can be found on the Web, a practice that originated with meteorology departments at a handful of US universities and now includes many national weather agencies.

Computer models

Your weathercaster might have more charisma, but **computer models** are the building blocks of almost every weather forecast you'll ever get on TV or anywhere else. Modern meteorology would be impossible without them.

A computer model that predicts the weather is nothing more or less than a miniature replica of the atmosphere. Models are each made up of thousands of lines of computer code, but they behave more like dynamic enti-

Of hedgehogs and holidays

There's nothing like a gloomy winter to generate the longing for a good weather omen. In certain cultures, February 2 has traditionally filled that role for centuries. It's known in Christendom as **Candlemas Day** (formerly the Feast of the Purification of the Virgin). Sunshine on Candlemas Day has always signalled an ominous turn of events: "If Candlemas Day be fair and bright, Winter will have another flight/But if Candlemas Day bring clouds and rain, Winter is gone and won't come again." In the secular world, February 2 became **Groundhog Day**, named after the woodchuck (or badger, bear or hedgehog) whose shadow, if visible on that date, foretells six more weeks of winter. There's an astronomical basis to Groundhog Day that points to even more ancient roots. February 2 falls near the mid-point between the winter solstice and the spring equinox. In pagan cultures, such mid-point days were sacred, marked by rites and festivals. May Day (May 1) and Halloween (October 31) are also associated with ancient rites and culture. Halloween's Christian counterpoint, All Saints' Day (November 1), yields its own predictive power: "If All Saints' Day will bring out the winter, St. Martin's Day [November 11] will bring out Indian summer."

ties. Similar to an elaborate computer game, a model takes input and makes things happen. Cold fronts and low-pressure centres march across the face of a modelled Earth every day. These weather features aren't merely pulled out of storage to match the day's conditions; once a model is given today's weather, it actually produces tomorrow's from scratch, following the laws of atmospheric physics coded in the software. Some of the biggest weather features of recent years – such as the gigantic winter storm on the US east coast in March 1993, or the Christmas 1999 windstorm that pummelled France – came to life inside computer models a week or more before they materialized in the real world.

Waiting for computers

The idea of "calculating the weather", as **Frederik Nebeker** calls it in his book of the same name, didn't take hold until weather was measured by **numbers** rather than language. Before weather instruments came along in the seventeenth century, practically all observations of the weather were qualitative: "cloudy" skies, "heavy" rain, "strong" wind. The organized practice of recording and sharing weather data within countries wasn't in place until after 1850, and the global exchange of this data is a product of the twentieth century.

The seeds of a forecast revolution were sown in Bergen, Norway, around the time of the World War I by the same group of meteorologists that came up with the idea of cold and warm fronts (see "Where does the wind go?", p.23). **Vilhelm Bjerknes** and his colleagues devised a list of seven variables that controlled our weather. According to the Norwegians, if we had full knowledge of these seven factors, we could completely describe the weather across the entire planet. These elements were:

- **Wind speed** in all three directions (**east–west, north–south** and **vertically)**
- **Air pressure**
- **Temperature**
- **Moisture** by volume
- **Mass density,** or the amount of air in a given space

Computers weren't yet on the scene, so the Bergen group considered it impractical to extend these elements much more than a couple of days into the future. In the Bergen style of prediction – which led the field for decades, and is still influential – the forecaster tracks fronts, lows and other weather features and compares them to idealized life cycles that were rendered (quite elegantly) in three-dimensional drawings by the Norwegians. This approach relies heavily on a discerning eye and an aptitude for picturing where a front or low might go and how it might evolve.

© Bob Henson

Only a few years beyond writing on blackboards, TV weathercasters of the late 1980s were plotting maps on computer.

In the 1920s, the eclectic British scientist **L F Richardson** took the Norwegian ideas a step further. He came up with a set of seven equations that connected the seven elements above with the **rules of physics** laid down by Isaac Newton, Robert Boyle and Jacques Charles. If you could solve these seven equations, then Richardson believed you could not only describe the current weather, but extend it into the future. He envisioned a "forecast factory", where hundreds of clerks would carry out the adding, subtracting, multiplying and dividing needed to create a forecast by numbers. Richardson and his wife spent six weeks doing their own number-crunching in order to test his ideas on a single day's weather (see box, p.136).

Although Richardson's test was unsuccessful, his equations materialized again almost thirty years later in the world's first general-purpose **electronic computer**. The **ENIAC** (Electronic Numerical Integrator And Computer) was created in the US at the close of World War II. In order to show the machine's prowess and promise, its developers hunted for a science problem that could benefit from raw computing power. Weather prediction filled the bill. Richardson's equations were updated and translated into machine language, and on March 5, 1950, the ENIAC began cranking out the first computerized weather forecast. The computer took almost a week to complete its first 24-hour outlook. Later that spring, one-day forecasts were being finished in just about a day. The really good news was that, unlike Richardson's earlier test, nothing went horribly awry. In fact, ENIAC did a better than expected job of portraying large-scale weather a day in advance. Computer modelling was on its way, and in the years since, it's gone from strength to strength. Computer power, model sophistication and forecast accuracy have increased hand in hand.

How a model works

Imagine a 3-D lattice of **intersecting boxes**, with a tiny weather station in the centre of each box. Now extend that lattice over a whole continent, or the whole globe, and you have some idea of the framework used by a computer model.

To do its job, the model first breaks the atmosphere down into boxes that represent manageable chunks of air, far wider than they are tall. In a typical weather-forecast model, each chunk might be about 30km/18 miles across and anywhere from 8–800 metres/25–2500ft deep, with the sharpest resolutions closest to ground level. The model uses the physical equations above to track the weather elements at the centre of each box across a very short time interval. Two or three minutes of weather might be modelled in only a few seconds of computer time. At the end of each step forward, the data may be adjusted so that each box is in sync with adjacent boxes. Then the weather in the model moves ahead another couple of minutes. At regular intervals – perhaps every three to six hours within the model's virtual time – a set of maps emerges, used by forecasters to analyse the weather to come.

Today, just as in 1950, modelling the real atmosphere demands a stunning amount of computer horsepower. The weather changes every instant, and no human being or weather station can be everywhere around the world to monitor those changes 24 hours a day. The typical weather-forecast model is run on a **supercomputer** that can perform many billions of calculations each second, and it still takes an hour or so to finish its work. Global climate

Let's discuss this storm

For a glimpse into the minds of working meteorologists, the forecast discussions posted by the US **National Weather Service** each day are unbeatable. NWS forecasters use these to explain the reasoning behind their outlooks, acknowledging what is fairly certain and what might go wrong. Laden with jargon and sometimes sprinkled with weather-nerd humour, these discussions have never been intended for public consumption. For example, "ELY PD 2 CUD BE THE BEST CHC FOR CONV OMTNS THRU THE FCST PD" means that the early part of the second forecast period (the 12–24 hour window) could provide the best chance for convection (thunderstorms) over the mountains. Despite their semi-impenetrability, discussions were posted by a few universities on the Web starting in the mid-1990s, and the NWS soon added them to its own website. They remain little known outside the world of weather enthusiasts, although one did make the front page of the **Washington Post** (see "The surprise East Coast snow of 2000", p.138). A forecaster had expressed his dismay at the models' handling of the storm; the **Post** reporter found the discussion online and pulled a few words from it for his story. With more non-meteorologists now following their discussions, forecasters are easing up a bit on the jargon. You can read discussions for yourself on several of the major websites listed in Resources, p.385, or by going to the websites of local forecast offices listed on the main NWS portal, http://weather.gov.

models demand still more computer time, as we'll see in our primer on climate change (p.150).

Because high-speed computers are so expensive (tens of millions of US dollars for the best ones) and because their time is so valuable, there's tremendous pressure on modellers to be efficient. Most models that predict longer-term weather, up to ten or fifteen days out, span at least one hemisphere (northern or southern). The maps extracted from these models usually focus on a region of interest, plus enough surrounding territory to depict any weather features moving in from afar.

Where it all happens

Nearly all of the world's nations have their own weather services, but only a few of them operate full-blown global weather models. In these countries, modelling tends to be focused at a single location where computers can be concentrated. Some of these centres of action include:

- US **National Meteorological Center**, near Washington, DC
- The **Canadian Meteorological Centre**, just outside Montreal
- UK's **Met. Office** headquarters, based in Exeter
- Australia's **Bureau of Meteorology**, Melbourne
- Japan **Meteorological Agency**, Tokyo
- The co-operative **European Centre for Medium-Range Weather Forecasts**, Reading, England

Each of these centres has one or more of the world's top supercomputers, all produced by a handful of firms. Upgrades are frequent: in recent years the speed of computer processors has doubled about every eighteen months. This gives scientists the freedom to create more detailed and realistic models. A number of other countries, universities, multinational consortia and private companies operate various weather models that cover a region or nation rather than the entire globe.

It takes co-operation between countries to make continent-scale models happen. **Global agreements** that emerged in the twentieth century still guide the collection and sharing of weather data and the exchange of model output. Each model run is kicked off with a good first guess: the forecast for the present time pulled from the previous run. Then, data from a variety of sources – including ground-based weather stations, balloon-borne instruments, satellites and aircraft – are "assimilated" into the model. (Since eastward-flying aircraft often take advantage of the jet stream, this river of air is often one of the best-sampled regions of the upper atmosphere.) Weather balloons are traditionally launched at two key times each day: 00:00 and 12:00

Universal Time, which corresponds to midnight and noon at Greenwich, England. Ground-station reports are collected hourly at many locations, and more often than that by automated stations. As these reports come in, they're run through a sophisticated set of quality-control procedures.

Once all this raw material is fed into a model, it has to be sliced and diced into manageable portions. Each point in the model is assigned starting-point data that best match the surrounding observations, which may be 16km/10 miles apart in the Netherlands or 1600km/1000 miles apart over the ocean.

When models get into trouble

Even the best models have to make inevitable compromises if they're going to process the atmosphere's action quickly enough to be useful. A small mountain range may fall into a single chunk or two of model space, which means that all the detail and nuance of the peaks are translated into the equivalent of a giant concrete block. Instead of tracking realistic clouds, large-scale models rely on vastly simplified portrayals of the cloud types that reside within a chunk of air.

All this can obviously lead to inaccuracies. For instance, tropical forecasters have long had to watch for "**boguscanes**" – fictional hurricanes spun up by models too eager to build a tropical cyclone out of an inconsequential weather feature. Until it was improved in 2001, one major US model created dozens of boguscanes each year across the tropical Atlantic, far more than the actual number of hurricanes that occur. Models may also have trouble with shallow **cold air masses**, the kind that can slide a thousand kilometres southward from the Arctic in winter, but may extend upward less than a kilometre. These air masses once crept below a model's line of sight so cleanly that the model insisted the temperature of the lowest layer was above freezing when it was actually –10°C/14°F or colder. A new generation of regional models is packed with higher vertical resolution that largely alleviates this problem.

Most of the major recurring errors in a model are caught before, or shortly after, its release. But just as you might take the wrong medication for a cold and experience nasty side-effects, forecasters can make a model worse off when they tweak it to fix a recurring problem. Adding resolution tends to sharpen a model's picture of reality, but there's a catch: if each part of a model has a built-in amount of error, then making a model more complex may only multiply the uncertainty it holds. One solution is not to tamper with the model and simply account for boguscanes, missing cold waves or other problems as they occur. Another is to address some of these problems with a statistical massage.

Weather chaos

The early 1960s saw a boom in scientific studies, including meteorology, but in the background one particular sceptic was raising eyebrows. **Edward Lorenz**, a professor of meteorology at the Massachusetts Institute of Technology, discovered what still seems to be an outer limit to useful weather forecasts. The problem Lorenz found later became famous as "chaos". When a computer model begins to run, it has an incomplete view of the current weather, because it's starting with observations that may be separated by 160km/100 miles. Through a series of computer experiments, Lorenz found that small weather systems tucked out of view between the model points – as well as errors in the observations themselves – can grow with time as the model unfolds. Within a few days, the errors become large enough to jeopardize the entire forecast. To quote Lorenz's apt metaphor, which he used as the title of a talk, "Does a flap of a butterfly's wings in Brazil set off a tornado in Texas?" (His answer: perhaps. But so could all the wings of all the other butterflies – and, moreover, the flapping might actually prevent a tornado just as easily as it enables one to form.) One way forecasters get around the problem of chaos is to run a number of different models; the US runs several big ones at least four times a day. If all goes well, the errors will vary, while the bona fide weather features will show up in most or all of the models. Scientists may also introduce error on purpose. By running the same model twenty times or more, with tiny differences in the starting points each time, they can see which weather features are robust enough to appear in all of the variations. This approach, called ensemble modelling, is quickly becoming more feasible as computer power increases, and it's also being used in climate modelling (see p.128). Despite all of these tricks, the current thinking in the forecast world is that Lorenz was indeed right. Butterflies, and all the other things we can't specify, ensure that we will never exactly know what weather to expect at 4pm a month from today.

From model to forecast

You could take the raw data from a model and interpolate between boxes to produce a halfway acceptable forecast for your home. "Halfway" is the key word, however. Because of the limits of computer storage space and power, such a model can't specify all the local influences that might affect your weather, such as an inlet that keeps a coastal town warmer than its neighbours in winter and cooler in summer, or downtown buildings that generate enough heat to keep the nighttime air several degrees warmer than in the suburbs.

Many large-scale models are localized through a system called **Model Output Statistics**. MOS equations are designed to tweak the numbers provided by their parent large-scale model. Each MOS equation is built by comparing the parent model's forecasts of temperature, wind and the like, to the actual conditions observed in a given city, then identifying and accounting for the persistent biases. After every model run, MOS spits out a set of numbers that serves as a starting point in crafting a public weather forecast. MOS equations have been developed for more than one hundred towns and cities in the UK and over three hundred in the US.

It takes years before enough experience accumulates with a given model to build an MOS database for it. That's why MOS equations are typically based on older models, run alongside the state-of-the-art ones. The latter are used to scope out the overall weather situation and adjust the MOS data when needed. For instance, if the best models show that a rare snowfall is brewing at a sub-tropical location, the MOS equation might handle it poorly because it's never come across such a feature before (just as the snow itself might baffle the warm-blooded residents).

Putting it together

The maps and numbers generated by computer models and MOS are where objective guidance stops and the subjective skill of a human forecaster begins. It's up to this person to decide which words and numbers to pass on to the public or to a specialized audience.

As in a newsroom or a factory, the deadline pressure in a weather office can be tremendous – but not always. In places like the tropics or in seasons like mid-summer, days and even weeks can go by without a truly challenging forecast. MOS and other model guidance then become the order of the day, and a forecaster's main task is to make sure no errors or surprises are hidden in the model output. US forecaster **Leonard Snellman** came up with the phrase "meteorological cancer" in 1977, a few years after MOS was introduced. Snellman worried that complacency would allow errors to creep into a forecast and escape unchecked. His viewpoint resonated with many, although a general, gradual improvement in forecasting skill hints that the cancer Snellman feared isn't metastasizing on a widespread basis.

Whether public or private, every forecasting outfit has its own **lingo** and style. This helps establish a public identity; more importantly, it ensures that people aren't surprised too often by a term they've never heard before. A new phrase should – and typically does – mean something important, as it did in Oklahoma City, Oklahoma, on May 3, 1999. A gigantic tornado (soon to be the costliest in world history) was bearing down on the region that evening. To help convey the gravity of the situation to a twister-jaded populace, local NWS forecasters deviated from the usual "tornado warning" to declare a "tornado emergency".

Most wording decisions aren't made under such strain. Forecast entities might take years to hash out a seemingly trivial change in their terminology. It's because they're all too aware that people vary in how they interpret as simple a word as "sunny". In US forecasts, "partly sunny" and "partly cloudy" are interchangeable – they both indicate 30 to 60 percent of the sky will be cloud-covered. Yet when surveyed, a group of US university students actually thought "partly cloudy" pointed to a brighter outlook. To some of them, "mostly sunny" (10–30 percent cloud cover, according to the NWS)

Forecasting by the numbers

While employed as an ambulance-driver for the French army during World War I, **Lewis Fry Richardson** worked on a scheme that's still in use today for predicting weather by computer. Richardson devised **seven equations** that could be used to extend the current weather into the future. He put his ideas to the test with a grand experiment in 1922. Using data collected across western Europe a decade before – at 4am on May 10, 1910 – he calculated what the weather should have been at 10am on that day. It took Richardson six weeks to crunch the numbers for this six-hour forecast, and the results were abysmal. The pressure change he predicted for one point in Germany was a hundred times greater than the real thing. As author Frederik Nebeker points out, "It is ironic that this forecast, which is probably the forecast that consumed the greatest amount of labour by a single person, is no doubt one of the least accurate forecasts ever." Nonetheless, Richardson's test demonstrated to the world that one could – in principle, at least – carry out calculations and forecast the weather using more than intuition and rules of thumb that were standard in the 1920s. Richardson was a Quaker and conscientious objector, and he brought his mathematical bent to studies of human conflict ranging from arguments to wars. There was a lighter side to him as well. In describing how turbulence occurs, he wrote a classic one-sentence poem: "Big whirls have little whirls that feed on their velocity, and little whirls have lesser whirls and so on to viscosity."

meant that clouds would fill as much as 60 percent of the sky. In the UK, there aren't any such quantitative definitions, but the forecast lingo definitely tilts toward the cloudy side of the spectrum. The word "clear" isn't even in the Meteorological Office guidelines. "Bright" refers to a lightly overcast day with diffuse sunlight and perhaps some direct sun. (Of course, TV weathercasters and other forecast purveyors can always put their own spin on these officially sanctioned definitions.)

Reading between the lines

What kinds of forecasts are available out there? How solid are their claims of accuracy? What do you do when forecasts from different sources disagree? As consumers of weather information, it works to our advantage to ponder the forecasts we get and learn how to use them wisely. To paraphrase the Roman maxim *caveat emptor*, let the weather customer beware.

Forecasts for every time and place

The one- to two-day prediction that's so familiar to us as "the forecast" is only a sliver in a spectrum of outlooks that extend from a couple of hours to a year or more ahead. Here are a few of the most common types:

- **Nowcast** The briefest kind of official forecast, it's a quick take on what might happen in the next two hours or so. For instance, a 2pm nowcast might mention that storms with high wind are approaching your area and will arrive by 3pm. Nowcasts are a growing part of weather prediction in many countries, but unless they pertain to severe weather, you'll have a hard time finding them on radio or TV, since their short half-life doesn't mesh well with regularly slotted programming. Your best bet is to check the Internet site of your local forecast provider (see Resources, p.385). Or you can subscribe to one of the weather-alert services now available in many locations, especially in US high-tech corridors. The services pass on bulletins at almost any desired interval via cellphone, wireless modem, pager, fax, or proprietary device. Some countries provide forecasts, nowcasts and other information via a network of local radio stations accessible by dedicated, off-the-shelf receivers; in the US, for instance, there's NOAA Weather Radio (see Resources).

- **Severe-weatheralerts** Warnings for severe thunderstorms, flash floods or other short-fuse events operate on the same time scale as nowcasts, but they're much more likely to be aired through local media. In the US, life-and-death warnings are the one forecast area still considered the sacrosanct domain of government meteorologists. Competing interpretations of whether, say, a tornado is approaching could make the public more hesitant to seek safety (and might leave a private firm vulnerable to lawsuits). Every nation and every media outlet has its own opinion on which official warnings are most critical to put on air. Flash-flood warnings, for instance, aren't even issued by many European governments, and they're often ignored by commercial radio and TV elsewhere, even though flooding is the single deadliest storm hazard in many countries.

- **Short-rangeforecast** This is the 12- to 72-hour outlook that's been at the heart of weather prediction for many decades. The standard elements include high and low temperatures, cloud cover, the risk of rain or snow (either in words or in probabilities) and – especially for the first day or so of the forecast period – wind direction and speed. The last of these may be provided in kilometres or miles per hour, knots or as a numerical rating of force from the Beaufort wind scale, used in the venerable British shipping forecasts (see Resources, p.406). Some new features have been added to the standard forecast over the past decade or two. **Air-quality monitoring** has made it possible to give current and projected levels of pollutants, as well as pollen and other allergens. In the 1990s, **ultraviolet (UV) indices** were added to forecasts in more than thirty nations. Using a simple numbered scale these popular outlooks take into account the time of year and the level of cloudiness to give people a sense of the peak intensity of sunburn-inducing ultraviolet light.

- **Medium-rangeforecast** Covering from three to as long as ten days (even longer in some areas), this is the extended outlook that's an increasingly standard feature of weather reporting in newspapers and on TV. As of early 2006, official extended outlooks in **Canada** and **Britain** cover five days, including high and low temperatures and a word or two of broad-brush description (eg "mainly sunny"). **Australia** issues seven-day outlooks, although the final three days provide only verbal description of temperatures (eg "warm to hot") rather than numbers. And the **US National Weather Service** (NWS) provides full seven-day outlooks, including high and low forecasts. Not to be outdone, many local TV stations in the US now offer ten-day temperature

The surprise East Coast snow of 2000

The timing was almost perverse. On January 18, 2000, the US National Weather Service announced that its latest supercomputer for weather and climate models was officially online. Reaching five times the speed of its predecessor, increases in forecast accuracy were all but assured. Less than a week later, a snowstorm took shape over the eastern US in a far from straightforward fashion. The NWS's flagship **Eta** model kept the heaviest snow and rain just offshore, to the east of Washington DC, and other big coastal cities. Since this model often brought things too far west, forecasters downplayed the storm for the public – until the model changed its tune on the afternoon of Monday, January 23. The switch was too late to make the early-evening newscasts, but the late-night news spread the word that snow was on the way. No matter: the 30cm/12in that fell by morning threw the nation's capital into a frenzy. The drama was even more pitched in **Raleigh, North Carolina**. A day after mere flurries had been forecast, the city got some 51cm/20in of snow, more than had ever fallen in Raleigh over an entire month. A glimmer of light surfaced long after the snow had melted, however. Scientists testing a research model with similar resolution to the new NWS flagship model found that it might have been able to accurately predict a storm like this one, given enough good-quality starting data.

and precipitation outlooks. All this is a byproduct of the increasing foresight of computer models that now track large-scale weather features up to fifteen days. Simply because a model covers that extensive time-frame doesn't automatically mean it's trustworthy, of course. Even the best local forecasts of more than a week or so in advance are only a little better than a climatological estimate, if that. Nevertheless, the persistence of a locked-in pattern, or the emergence of a change in that pattern, may come to light in the models more than seven days in advance. If a big transition appears consistently over several days' worth of extended outlooks, it's a good sign that the change is likely to happen, even if the timing and intensity aren't yet certain.

▶ **Long-rangeforecast** Some governments and private firms issue **general outlooks** that extend more than a year in advance, relying on the influence of slowly evolving climate elements such as regions of warmer- or cooler-than-normal ocean temperatures. **El Niño** and **La Niña** (see p.114) are the premier examples of such forecasting tools. Once it's known an El Niño is unfolding, millions of people across the world can be given notice that floods or drought are more likely than usual in the months ahead. Perhaps because their output can be difficult to explain, long-range forecasts are seldom spotlighted in the media; they're easier to find on government Internet sites. Maps typically show the departures from average, by month or by season, expected in the forecast area for both temperatures and precipitation. The NWS uses a complex triad of probabilities (see box, p.140) to create such outlooks for overlapping three-month periods ranging up to fifteen months. While they're not much help for planning a wedding next year, such outlooks can make a huge difference to industry. For a utility company that generates and trades power each day, for instance, a degree or two of change in average temperature over a season can save (or burn up) millions of dollars. Seasonal forecasting has helped create a market for weather derivatives – a form of investment added to the Chicago Mercantile Exchange in 1999. Utilities, agricultural businesses and other firms purchase **weather derivatives** as insurance against a sharp temperature trend that might threaten their profits. If a season's climate deviates far from the expected range, the derivative pays off and the buyer may be rescued from financial ruin.

- **Climateprojections** A few research centres around the world are attempting to project global climate a hundred years or more into the future (see p.164). These aren't considered forecasts so much as **scenarios** – portraits of how climate might evolve if society grows and industrializes at a particular pace. No single projection can be taken as gospel, but these remain the best tools for gauging our possible impact on tomorrow's climate.

Comparison shopping

Before television weather came of age in the 1960s and 1970s, the public had to live with *the* forecast – the only one in existence, issued by the government. Now you can find numerous predictions through TV, radio, newspapers and the Internet (as well as pagers and other portable media). In many areas you can find differing outlooks among the four sources above, with each forecast prepared by a different meteorologist in the public or private sector. Factor in the multiple outlets in bigger cities for each form of media, and you could be facing literally dozens of forecasts valid for the same period.

Nearly all forecasts are at least partially derived from a common base of **government models**, so the differences between competing outlooks are often fairly minor, especially when the weather is tranquil. But when major changes are on the way, big discrepancies may be exposed, not only in the substance of competing forecasts but also in their timing. Professionals often go with a consensus of several different models in crafting their forecasts. This isn't such a bad course to follow, assuming the mixed forecasts consulted are all reasonable interpretations of a complex scenario. It's important to pay attention to when the forecasts are issued, however. All else being equal, the most recently issued forecasts should be weighed most heavily.

Consensus isn't necessarily the ideal option. Some forecasts are simply better than others. There are variations in procedures and skill levels even among seemingly uniform entities, such as the larger meteorological services. Only by paying attention to forecasts over time can you assess, if only in a gut-level way, whose predictions you trust the most. If you're really motivated, try scoring your local forecast sources for a few days on some easy-to-calculate variables like high and low temperature. (See below for more on assessing forecast accuracy).

Don't be dazzled by mere technology. In some places you can now get hourly forecasts of temperature or precipitation through a subscription service via cellphone, email, pager or the Web. While such services may be convenient, and they may in fact give you the most accurate forecasts for your area, their quality depends solely on the models and other data on which

A preview of next year's climate

If you can't get next week's weather right, how can you claim to predict anything a year in advance? It's the difference between **weather** and **climate** that makes long-lead forecasts possible. Actual storms can't be predicted a year out, but the longer-term features that affect climate across a season can sometimes be foretold with surprising accuracy, thanks in part to slow-changing **ocean temperatures**. In 1996 the US and Canada began issuing long-lead outlooks that specify, month by month, the likelihood of above- and below-average temperature and precipitation for three-month intervals. For instance, the set of thirteen outlooks issued in December 2005 would cover January, February, March 2006; February, March, April 2006; and so on to January, February, March 2007. Each map has contours showing where the climatic odds for above, near, or below-average conditions (each assumed to be 33 percent initially) have been shifted by forecasters. In an **El Niño winter**, for instance, the precipitation outlook might place a 50 percent contour across southern California. This means the odds of above-average rainfall have risen from 33 percent to 50 percent. As esoteric as all this might seem, it's the most concrete guidance available to the public on climatic conditions so far ahead. A similar system in Canada projected virtually all of the country to be warmer than average for the height of the 1997–98 El Niño. That outlook proved accurate for 88 percent of the nation. The US averages are drawn from a 30-year temperature base that was recently changed from 1961–90 to 1971–2000. Since the latter was a warmer period across North America, the very definition of "average" has moved up a notch.

they're based. No matter how finely a forecast is sliced and diced, increased precision alone doesn't improve accuracy.

What can possibly go wrong?

The official forecast in **New England** on the morning of September 21, 1938, was for cloudy, windy weather. Cloudy and windy it was – but nobody knew a severe hurricane would be tearing its way through the region by late afternoon. The storm was projected to move eastward and out to sea from its location off the North Carolina coast the previous night. But the forecasters of that time lacked data on the jet stream winds that steer hurricanes from above. In retrospect, we can now deduce that the jet was positioned in an unusual south-to-north orientation. It pulled the hurricane into Long Island and New England at the stunning rate of 113kph/70mph, one of the fastest motions on record for a tropical cyclone. Winds gusted to 299kph/186mph at Blue Hill Observatory near **Boston**, and catastrophic flooding resulted in **Providence, Rhode Island**. In all, some seven hundred people died.

Forecasts don't have to be this wildly off the mark to cause problems. Participants in Australia's 54th annual **Sydney-to-Hobart Yacht Race** were warned about gales as they embarked on December 26, 1998. The tailwinds got the race off to a fast start, but an unexpectedly strong storm – too compact to be well forecast by the models at first – cranked up just north of Tasmania. An hour into the race, the Bureau of Meteorology upgraded its outlook to a storm warning after a high-resolution model had finally revealed what was to come. Yet conditions were far worse than the yacht teams expected. As it pulled in westerly winds at near hurricane force, the fast-growing storm sent huge waves crashing almost directly into the existing swell. The result was not just a typical high sea – the average waves ran 9m/30ft – but a confused one, with occasional rogue waves that exceeded 15m/50ft. The nation's largest-ever marine rescue saved 55 people, but only a third of the 110 participating craft reached Hobart. Five boats sank, and six people drowned.

A failed prediction is every forecaster's worst nightmare. There's no way to tell in advance which forecasts are problematic, but there are clues that can help a layperson sniff out uncertainty hidden in the seeming certainty of a particular outlook. Here are a few of the most common forecast predicaments and how meteorologists deal with them:

▶ **Flip-flops** There is far more weather data available for some parts of the world than for others. When a system is developing in a data-sparse region, such as the **Atlantic** or (especially) the **Pacific**, computer models have a more difficult task projecting its future. Every six or twelve hours, the models take another stab at it. The projected scenario may shift dramatically from one model run to the next until the true character of the weather system becomes clear.

This is one source of what forecasters call **flip-flops**. The model guidance for next weekend may be screaming "cold" on Tuesday, then shift to "mild" by Tuesday night, and then "bone-chilling cold" on Wednesday. What should a forecaster do in such circumstances? The most common approach is to gradually shift the forecast in a plausible direction by looking at how several different models treat the situation. At all costs, forecasters want to avoid putting the public through the back and forth gyrations that the models inflict on the forecasters themselves. Thus, the weekend forecast might run cold on Tuesday and then a slight bit colder on Wednesday, disregarding the models' mild flip-flop on Tuesday night as well as the apocalyptic deep freeze projected in Wednesday's model run. If the public forecasts continue to get colder as time goes by, you can bet that Saturday might be still more bitter than they're acknowledging. Even when the models insist on an unusual event, it may take a forecaster some time to adjust the forecast so it reflects that event. This is partly to avoid the dreaded flip-flops and partly a recognition that model error goes down as the prediction period gets closer.

▶ **Either/or** Warm and cold fronts are better managed by the models than ever before, but not every parry and thrust of a front is predictable. **Cold fronts** (the more dramatic of the two) can be especially troublesome. The sheer density of the air behind a shallow, but intense, cold front can help it to roll across the countryside more quickly than a model may predict. Back-door cold fronts – those that ooze westward, against the upper-level flow – can also be tricky.

Once a mass of cold air is lodged in place, predicting its erosion is a challenge of another sort. Wintertime cold air may be dammed against mountains like the US Rockies or the German Alps for days on end, eventually to be dislodged by a burst of warmer air from above. The models might see the warm air coming, but be unable to tell whether it'll have the force to punch through the cold.

In a perfect world, forecasters would telegraph uncertainties like these in a forecast phrased something like this: "Tomorrow – a 60 percent chance of continued warm, muggy weather and a 40 percent chance of sharply colder conditions". But it's hard enough to deal with one forecast, let alone two or more for the same period. Until the public is ready to accept multiple-choice forecasts, the meteorological world is in no hurry to provide them. Typically, a forecaster will either go for broke – focusing on just one of the two or more options at hand – or else hedge his/her bets and issue a forecast somewhere between the alternatives. Instead of the muggy 30°C /86°F, or the chilly 10°C/50°F, the predicted high might be a lukewarm 20°C /68°F. From the standpoint of numerical accuracy, this is the safer forecast, though in a sense it's the less honest one.

There's no easy way for the consumer to spot an either/or forecast. However, any sharp contrast in air masses will eventually make its way into the forecast. It's wise to bear in mind that this changeover could happen a little sooner or later than predicted, and the transition could be sharper than indicated, especially when a cold snap is on the way.

▶ **Rain, snow and all that** If you consider how difficult it is to analyse how precipitation forms within a cloud, it's amazing that rain and snow forecasts are as good as they are. It's now rare that one will encounter precipitation in real life without some hint of its onset in the forecast. Probabilistic forecasting of rain and snow (see "Only a slight chance of confusion", opposite) allows people to decide for themselves whether a given chance of moisture is enough for a picnic to be cancelled or a hike to be postponed.

Forecasters approach model output on **precipitation** with a healthy dose of scepticism. While the physical equations within most models are quite skilful at moving air around, they're less able to make the fine distinctions between, say, air that's not quite saturated (90 percent relative humidity) and air that's fully saturated (100 percent RH). This helps lead to a depiction of rain or snow in the model that can be way off in location or intensity. As a rule, gentle, steady rains or snows are more reliably simulated than are the big events, such as intense winter storms, hurricanes and thunderstorms. The last of these depend on **convective effects** – where the air is unstable and buoyant, rather than merely pushed upward – and many models are notoriously poor at handling convection. Indeed, models traditionally "parameterize" thunderstorms, portraying their physical consequences in broad-brush strokes without even trying to simulate the storms directly.

Over time, forecasters have learned to look at the large-scale features in a model and infer from those the most likely patterns of rainfall or snowfall. In the past decade, **mesoscale models** have provided a narrower focus on features from 10 to 100km/6–60 miles across. In some cases, these models can depict thunder and snowstorms in a far more realistic fashion than ever before. A few experimental models can telescope their resolution down to as fine as 1km/0.6 miles in targeted areas. However, the heavy demand on computer time needed to produce such detail means that it may take years for a mesoscale model to move from the world of research into day-to-day forecasting.

Only a slight chance of confusion

It is as a direct result of Model Output Statistics (MOS) that the nifty forecast feature known as **probabilistic precipitation outlooks** exists. Statements like "a 70 percent chance of rain" now appear in forecasts around the globe. These were launched in the mid-1960s in the United States, where MOS was pioneered. As familiar as they seem now, probabilistic forecasts were quite the novelty when they debuted. MOS generates such numbers by calculating how often an expected weather scenario has produced a wet day in the past. Each facet of a given weather situation is fed into an equation that computes how likely it is that the overall weather picture will be a damp one. If the forecast is calling for a 30 percent chance of rain, this doesn't imply that 30 percent of the forecast area will see rain, or that it'll rain for 30 percent of the day – only that, for a given point in the forecast domain, there's a 3 in 10 chance it will rain sometime during the day. Nor do probabilities indicate how hard it might rain: a given forecast might show a low chance of a storm that could pour buckets should it materialize, or a high likelihood of drizzle that won't amount to much water. Human forecasters often bump these raw MOS numbers upward or downward based on a host of factors. You might still hear "a 100 percent chance of rain" here and there, although meteorologists now tend to avoid this invitation to embarrass themselves by simply saying "rain" if MOS decrees it's a certainty. In the US, when probabilities are below 20 percent but more than negligible, the number is typically omitted and the word "isolated" (showers, snow flurries or whatever) is substituted.

- **Snow** remains a particular challenge to forecast. This is especially true at the margins, when temperatures are just cold enough to support snow, or just warm enough not to. When you see a forecast of "rain mixed with snow", it's fair to assume that the situation could just as easily be all wet or all white. **Location** is another factor to consider. Snowfall – and rainfall, for that matter – can vary by more than 50 percent in as little as 1.6km/1 mile, particularly near mountains or large bodies of water where local effects come into play. Even if a forecast calls for a uniform amount of snow across your area, don't be surprised to find large variations across short distances.
- **When will it stop?** Predicting the end of a rainy or snowy period is one of the more clear-cut tasks in forecasting – except when a certain kind of storm has taken hold. In Los Angeles, weathercaster **George Fischbeck** popularized the phrase "cut-off low, weatherman's woe", and it hits the mark nicely. Especially at mid-latitudes, upper-level centres of low pressure can get pinched off from the jet stream that once carried them along. Think of a river's wide meander which, over many years, becomes a small lake as the river cuts a straighter course nearby. In the atmosphere, a **cut-off low** can form in a day or two, then spin virtually in place for a week or more. The result is a prolonged bout of unsettled weather in the low's vicinity and just to its east. Because there may be no clear "kicker" – an incoming system that pushes the low away – models and forecasters can't say for sure when the dreariness will end. If a cut-off low is giving you woe, it may well stick around for a day or two longer than forecasts indicate.

Making your own forecast

More than ever before, it's feasible for a layperson to craft their own weather forecast. The raw material used by the professionals – output from computer models – is available for many parts of the world on the Internet. Anyone can put together a home weather station at reasonable cost and see how their outlooks compare with reality. And, as always, the sights and sounds of weather on the move are here for all to interpret and enjoy.

What's typical, what's not

The best starting point in understanding your local weather is to pay close attention to it. Get to know the **average conditions** and how often and radically the weather departs from those averages. Mini-climatologies can be found in Weather around the World (p.170) for more than 200 cities. You can also find more data for these and many other locations by checking with your local weather service, or by following Web links for climate data listed in Resources, p.385.

You can build your own climatology by making regular observations at home. For more detail on the weather instruments below, see "Taking the weather's pulse" (p.30). Some of the basic components of a home weather station include:

- A **good thermometer**, ideally a "max–min" model that records daily highs and lows.
- A **barometer**, for measuring atmospheric pressure. You can keep this inside, since pressure isn't significantly different inside and outside a home (unless you live in a high-rise building).
- A **humidity-measuring** device, such as a sling psychrometer.
- An **anemometer**, for measuring wind speed.
- A **wind vane**, for sensing its direction.
- A **rain gauge**, which funnels rainfall into a narrow column so that fine-scale readings become visible. It should be placed in an open area as far from buildings as possible so that windblown rain won't be blocked.
- Where climate dictates, a **snowboard** – a small, flat board painted white to reflect sunlight and stay cool. Snowboards are placed in an open area and cleared off after every observation during a snowstorm.

These instruments can be purchased à la carte and placed in a well-ventilated shelter, which you can either build or buy ready-made. Several kinds of instruments are often bundled together as part of a **digital weather station**; these are more expensive, but far more convenient, than manual instruments. Some digital stations can upload their data directly into a personal

Davis Instruments

Davis Instruments is one of several US companies that caters to enthusiasts with all-in-one weather stations.

computer, making it easy to compile statistics on the climate at your home. In general, the quality of weather instruments goes up with price, but you can often save money by shopping around. A basic set of reliable, durable instruments could be put together by spending in the region of $100/£55. Basic all-in-one digital weather stations now sell for as little as $30/£16, while higher-quality packages go for $150/£80 and up.

Consider pooling your observations with other weather fans. Many local TV stations and newspapers across the US have informal networks of weather watchers who send in daily reports by phone or the Internet. Your nation may also have **co-operative weather observers** whose reports become part of the permanent climate record (see "Taking the weather's pulse", p.30). If you're a student, you and your school can take part in the GLOBE Program, which collects and posts environmental observations from all over the world (see www.globe.gov).

The verification game

When a forecaster claims to be "80 percent accurate", what does it actually mean? A whole sub-field of meteorology, forecast verification, is devoted to this question. The guru of verification was the late US researcher **Allan Murphy**, who claimed a good forecast needs to have consistency, quality and value. Imagine a set of forecasts that are always exactly 10°C/18°F too warm in the winter and 10°C/18°F too cool in the summer. On the surface, the quality of those forecasts is poor, but if one can spot the bias, their consistency and value is high. Many public and private forecasters score their predictions using statistical techniques that gauge the discrepancies between the actual and predicted numbers for high and low temperatures, chances of rain or snow, and the like. Some of these techniques emphasize total error; others place more weight on the bigger miscues. For instance, when a forecaster misses four high temperatures by 1° and the fifth by 16°, the average error is 4° per day. The same total error would accumulate for another forecaster who is 4° off on each of the five days. Which outlooks would you prefer? That depends on how you're using the forecast. Research suggests that people and industries tend to be most concerned about reducing the error on those rare days when the forecast is way off base. Some verification studies (including a few by motivated citizens) can be found on the Web. A few US TV stations even verify their work on the air by rewarding a random viewer when the previous day's forecast high is more than, say, 1.7°C/3°F off. Benchmarks like this make it possible to say that a forecast provider is 80 percent accurate – but the claim only makes sense if you know the criteria on which it's based.

Reading the sky

What the atmosphere intends for us is very often revealed in the **colour and texture of the sky**. Weather lore worldwide is packed with visual references to clouds and their appearance (see p.124). The images in the colour section will give you a feeling for the kinds of clouds that might pass over your area. Below are some tips to bear in mind as you watch the show unfolding above you:

- Strong mid-latitude storms typically send out **high clouds** well ahead of any rain or snow. A gradual lowering of the clouds, from cirrus or cirrostratus to altostratus, often heralds a wet period to come in the next day or so.
- **Cumulus** building on a spring or summer morning – especially if they appear as tall as they are wide – could spell heavy showers or thunderstorms later on. Watch for sharp-edged, cauliflower-shaped turrets: the crisper their appearance, the more well-defined the updraught that's feeding the storm.
- **Fast-moving** clouds reveal a strong upper-level wind, which hints at the potential for strong storm development in the near future. This is especially true if the low-level clouds have a northward component to their motion (southward in the Southern Hemisphere) and if high-level clouds are moving in a different direction from low clouds.
- **Poor visibility** can be a sign of moisture, pollution or both. If the air feels and smells unusually moist, an incoming storm may be able to produce heavier rain.

- All else being equal, **clear days** allow for a bigger temperature rise than cloudy ones, and clear, calm conditions lead to the coldest nights.

When out in the **wilderness**, it's especially crucial to notice your surroundings. Countless adventurers have lost their lives through unexpectedly bad weather. Keep in mind that some of the worst weather can be surrounded by some of the best. **Hurricanes**, for instance, produce a zone of subsiding air around their periphery: the result can be skies so azure and conditions so balmy that long-timers call it hurricane weather. Mountain climbers may be lured onto the slopes by unseasonably warm air pulled up in advance of an autumn or spring storm that brings bitter cold and snow hours later. If your weather seems unusually nice for the time of year, consider it added incentive to check the forecast before heading into the backcountry.

How to decipher the models

They aren't for everyone, but the **complex maps** cranked out each day by computer models provide a wealth of information. Many of the longer-range models cover the entire globe. Alas, the pressure for government weather services to make money from their modelling has prompted some to avoid putting their model output on the Web for all to see. Still, an amazing amount of guidance is on offer if you know where to look. Resources (p.385) lists some of the major websites that include model output.

It can take years of practice to learn how to read a model's maps quickly and skilfully. Below are some tips for beginners:

- Each model moves the weather ahead in tiny increments, and the results are depicted at **set intervals** – typically every 6, 12 or 24 hours. The starting time of most model runs is either 00.00 or 12.00 Universal Time (midnight or noon in Greenwich, England, and 7.00pm or 7.00am Eastern Standard Time in the US). It normally takes anywhere from one to six hours for a model run to be completed and posted on the Web. Some governments withhold the output from the Internet for a few extra hours in order to give a first look to paying customers only.

- Weather models analyse the atmosphere at several standard levels measured in **millibars or kilopascals** (units of pressure; 10mb=1kPA). The contours on each upper-level map show how high that pressure level sits above different points on the ground. "High heights", as they're called, correspond to a high-pressure centre aloft.

- The mid-point of the atmosphere is 500mb or 50kPa (roughly equivalent to 5.5km/3.4 miles). This is the level at which major weather systems are steered, although the strongest jet-stream winds are a bit higher up, around 250–300mb. A look at the **500mb map** reveals where big storms might go. Generally speaking, the wind at 500mb flows along the height contours, with lower heights to the left of the wind. Some sites include arrows or barbs to show the wind speed and direction. On other sites, you'll see blobs of **vorticity**,

© BBC Weather Centre

Many weathercasters such as Alex Deakin (BBC) employ surface weather maps that include fronts, isobars, and centers of high and low pressure.

a combined measure of wind speed and curvature that signifies the "spin" within a storm and, indirectly, its intensity.

- The **mean sea level pressure** (MSL) map is essentially the same as the standard surface map you see on TV, with highs, lows and fronts marching across the land. See "Where does the wind go?", p.23, for more on these weather makers.
- As you follow the models, notice how they **agree or disagree**. If a particular model suddenly changes its tune on an upcoming system, watch its next run closely to see if the change is consistent. Comparing different models for the same region and period is a great way to get a fuller picture of what might happen.

Chapter four
A primer on climate change

A primer on climate change

Only half a century ago, the most optimistic of scientists believed we might some day take control of the weather. Science, after all, had harnessed the atom and reined in some of the world's most destructive diseases. Why not, then, put the kibosh on hurricanes and tornadoes and order up the weather we like.

Little did those mid-century optimists suspect that the 1990s would turn out to be one of the warmest – perhaps *the* warmest – decade of the millennium. An ever-increasing volume of scientific evidence supports the idea that our industrialized culture is the prime reason for this recent global warming and that further warming is on the cards. Just as we once dreamed, we are now influencing the climate and, by extension, the weather – but, like Dr Frankenstein, we can't quite control the process that we've set in motion.

For a variety of reasons, all the drama and arguments surrounding global change have taken a long time to produce constructive action. Hopes were high in December 1997, when the world's governments met in Kyoto, Japan, to draft the world's first treaty for limiting emissions of greenhouse gases. In the summer of 2001, at a meeting in Bonn, Germany, 178 nations signed the protocol – but not the United States, which produces about a quarter of the world's greenhouse emissions. That critical no-show left it up to Russia to tip the scales by ratifying the protocol early in 2005, finally making it international law. However, the treaty will only be in force till 2012, and it will cut global emissions only by a few percent, if even that. How all of this might affect the weather we experience each day isn't clear, and that uncertainty is lukewarm comfort in a world that continues to heat up, decade by decade.

Below is a quick summary of the science and politics of global climate change. For a more comprehensive treatment, see **The Rough Guide to Climate Change.**

The basics

One of the most deceiving aspects of climate change is the aura of controversy surrounding it. In the US, in particular, many journalists have tended to portray global warming as an either/or proposition: either it's warming or it's not, either we're to blame or we aren't, either we know what's going to happen or we don't. By setting up these two-sided battles, the media granted disproportionate visibility to a small group of global-change "naysayers" (who are by no means a unified body) and to industries that have an understandable stake in the status quo

On the other side of this artificial divide are the vast majority of those scientists who gather climate data, sift through it for trends, and use computer models to simulate the atmosphere's future. Like the naysayers, these researchers are far from a unified group. However, they are in clear agreement on many aspects of the problem. There's no better proof of this than the reports of the **Intergovernmental Panel on Climate Change** (IPCC). Since 1988, the United Nations has sponsored this loose-knit body, which includes many hundreds of climate change scientists drawn from universities and laboratories worldwide. Every few years, the IPCC issues a comprehensive set of reports designed to help governments and other planners approach the questions posed by climate change. The first summary of its fourth major assessment was released in February 2007.

Some of the key points put forth in recent climate analyses:

- The **world's surface air temperature** has warmed by close to 0.8°C/1.44°F since the late 1800s. The figure is based on measurements from weather stations on land and from the 70 percent of Earth covered by ocean. For climate purposes, air temperature over the sea is usually estimated from water temperature; this tends to smooth out the short-term wiggles but retain the longer trends. Some places, like the Arctic, have warmed more than the average. Others have warmed less or even cooled slightly, such as the southeastern US. Satellites have inferred temperatures several miles up since the late 1970s. For years they showed less warming aloft than at the surface, but that difference has now been explained by instrument quirks and other factors.

- Beyond the temperature record, there is concrete evidence of warming in the world around us. Summer sea ice in the **Arctic** is barely half as thick as it was in 1950. **Mountain glaciers** around the globe have receded notably. **Sea level** has risen about 10–20cm/4–8in since 1900: most of this is due to sheer expansion of the warmed-up oceans rather than melted ice from glaciers. About a quarter of the world's **coral reefs** have been seriously damaged or destroyed in the last several decades, with rising ocean temperatures considered to be one of the main factors at work. The oceans themselves are also changing: as they absorb increasing amounts of carbon dioxide (CO_2) from the atmosphere, this makes the seawater less alkaline, posing a serious threat to many marine ecosystems.

- Industrial emissions of **greenhouses gases** – most importantly, **CO_2** – are almost certainly to blame for a good part of the past century's temperature increase. In 2007, the IPCC concluded that "most of the observed increase in globally averaged temperatures since the mid-20th century is very likely due to the observed increase in anthropogenic greenhouse gas concentrations." A few naysayers continue to argue that the twentieth-century warming may have occurred as a result of solar influences or other factors. We'll take a closer look at these arguments below.

▶ Temperatures will continue to go up. In 1990, and again in 1995, the IPCC projected that the global average might increase by about 1.5–4.5°C/2.7–8.0°F by the year 2100. In 2001 they expanded that range to 1.4–5.8°C/2.5–10.4°F. A follow-up study by IPCC scientists Tom Wigley and Sarah Raper projected a 50/50 chance that readings will fall near the centre of this range, or around 2.5–4°C/4–7°F. The 2007 IPCC projections changed little, with the best estimates of 21st-century warming at 1.8–4.0°C/3.2–7.2°F. The range reflects uncertainty in how much emissions will be controlled. These projections are drawn from computer models that have received their share of criticism. Imperfect and simplified as they are, these models are the only real tools at hand for gauging when and how global changes might become manifest. Indeed, every modern weather forecast is based on a computer model.

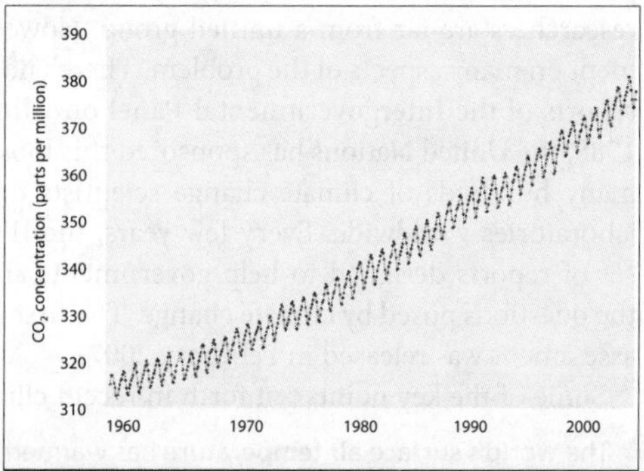

US Department of Energy

The trace of atmospheric carbon dioxide, measured in parts per million above Hawaii's Mauna Loa since 1957, shows a steady climb, plus yearly rises and falls caused by Northern Hemisphere plant cycles.

What's going on here?

The greenhouse effect itself is no cause for alarm. It's a natural part of our climate system, one that's been acknowledged for more than a century as something we need in order to survive. By holding in heat that would otherwise escape directly to outer space, greenhouse gases help to create a safe, cosy atmosphere. Without these gases, our average global temperature would be well below freezing. On the other hand, too much of a good thing can be counter-productive. Venus has an atmosphere that's 97 percent carbon dioxide. Coupled with its position closer to the sun, this keeps the Venusian surface atmosphere close to a broiling 460°C/860°F (see "Weather on other planets", p.372).

Ancient air samples taken from layers of ice in the **Arctic** and **Antarctic** reveal that the percentage of carbon dioxide in Earth's atmosphere has varied more or less in accordance with global temperature for over 100,000 years (see opposite). It's not yet proven whether changes in CO_2 caused these temperature changes, or vice versa; it appears that changes in either one can prompt a response in the other and trigger a self-reinforcing cascade of

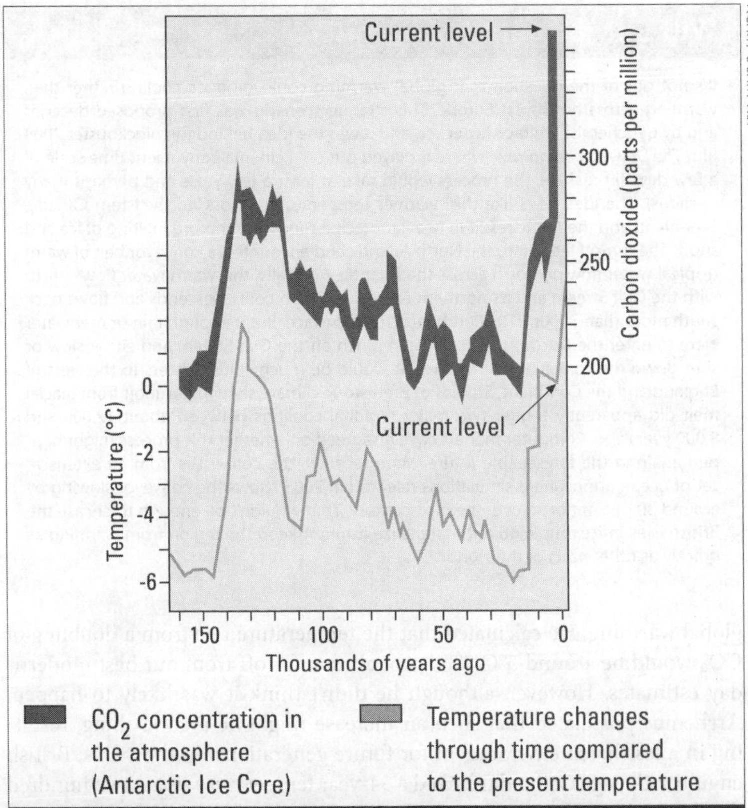

Global temperature has been closely coupled with the amount of carbon dioxide in the air for at least 150,000 years.

effects. Great sheets of ice covered much of present-day Canada and parts of Europe and Asia for more than 80,000 of these last 100,000 years, and CO_2 at that time was about 30 percent less prevalent in the atmosphere than today. Ice cores and other records indicate that carbon dioxide levels also dropped by several percent during the Little Ice Age, which caused noticeably more severe winters across Europe from the mid-1500s to the mid-1800s. Before industrial emissions began changing things, CO_2 made up an average of about 0.027–28 percent of the air, or 270–280 parts per million (ppm). This represents only about a hundredth of the world's carbon; much more is locked up in the oceans, in sediments and in fossil fuels that release CO_2 to the air as they're burned.

In 1895, Sweden's **Svante Arrhenius** was the first scientist to proclaim that the rise in carbon dioxide from fossil-fuel emissions would lead to

Will Europe get the big chill?

It's not out of the question that **global warming** could produce cooling rather than warming across northwest Europe. This strange scenario was first proposed decades ago by geochemist **Wallace Broecker**, and it was the idea behind the blockbuster 2004 film *The Day After Tomorrow*, where it played out on a cinema-convenient time scale of a few days (in real life, the process would take at least a few years and perhaps many decades). The idea goes like this: warmer temperatures across northeastern Canada, Greenland and the Arctic result in heavier precipitation and/or more melting of ice and snow. The run-off feeds into the North Atlantic and encounters a conveyor belt of warm tropical water flowing north across the Atlantic. Normally, this warm water flows north with the Gulf Stream and its northern extensions, then cools, descends and flows back south more than 3000m/10,000ft beneath the surface. But if enough rain or meltwater were to enter the North Atlantic, it could pinch off the Gulf Stream and either slow or shut down the conveyor belt. The result would be much colder waters to the west of England and the Continent. Studies of prehistoric climate show that runoff from glacial melt did apparently trigger two major regional coolings between about 12,000 and 8,000 years ago. Computer models have disagreed on whether this process might happen again in the foreseeable future. More recently, the consensus from an extensive set of ocean-atmosphere simulations released in 2005 shows the conveyor slowing by around 30 percent or so over the next century. That wouldn't be enough to encase the British Isles in ice, but should it materialize it might keep the region from warming as quickly as other parts of the world.

global warming. He calculated that the temperature rise from a doubling of CO_2 would be around 5°C/9°F – not all that far off from our best modern-day estimates. However, although he didn't think it was likely to happen, Arrhenius postulated that such an increase might be a good thing, resulting in a less harsh environment for future generations. In the 1930s, British engineer **George Callendar** linked a 54-year temperature rise at two hundred weather stations around the world to fossil-fuel emissions. Like Arrhenius, he saw this as a beneficial side-effect of industrial growth rather than a worrisome trend.

As an outgrowth of the International Geophysical Year in 1957–58, chemist **Charles Keeling** began carrying out regular CO_2 measurements on top of Mauna Loa in Hawaii using a new high-resolution sensor. Checking the data against CO_2 readings from the South Pole and elsewhere, Keeling and colleagues were able to produce one of the most famous, ominous graphs in science (see p.152). It shows an inexorable year-by-year rise in atmospheric carbon dioxide. Because the lowest few miles of the atmosphere are so thoroughly mixed, the Hawaii data reflects the state of carbon dioxide everywhere around the globe. The CO_2 rise is about 0.4 percent a year, from 316 ppm in 1958 to 378 ppm in 2004. As best we can tell, we now have the highest level of global atmospheric CO_2 in at least the past 600,000 years, and the level continues to rise.

There are other important greenhouse gases besides CO2. **Methane** is an industrial byproduct that's also emitted in surprisingly large amounts by rice paddies and by millions of belching cows. **Nitrous oxide** is another industrial gas with a powerful greenhouse effect. Even **water vapour** is a greenhouse gas – in fact, it's the most prevalent one (see box, p.156). Although several greenhouse gases are more potent on a molecule-by-molecule basis, there's so much carbon dioxide in the air that it has more overall impact than the other human-produced gases combined. Thus, many studies use CO_2 as a stand-in for the entire group.

The downs and ups of global warming

If scientists knew about the greenhouse effect a hundred years ago, and if it has been obvious for decades that atmospheric CO_2 is on the increase, why didn't we hear alarm bells ringing sooner? To begin with, the **temperature pattern** of the twentieth century threw scientists off the trail. The global average rose dramatically from the 1890s through the 1930s, as noted by Callendar, then began to level off. The decades of warming brought a spate of public attention – "Is the World Getting Warmer?" asked a *Saturday Evening Post* headline of 1950 – but the global average dropped from the late 1940s into the mid-1970s. This made many scientists wonder whether the human-enhanced greenhouse effect – at that stage a phenomenon generally accepted by researchers but still little known by the public – might be counteracted by the **cooling effects** of soot and grit in the air or by some other unknown element. With the help of a few eloquent spokespeople, books and magazine articles played up the risk of a "snow blitz" that could bring about a new Ice Age, just as researchers were beginning to trace the impact of previous ice ages through ocean sediments. A few scientists who expressed concern in the 1970s over global cooling later came to believe that **warming** was the bigger concern – only to find themselves dodging their earlier statements, flung back at them by global-change sceptics. Yet even during the height of the global-cooling media craze, many scientists continued to quietly study the long-term risks posed by greenhouse gases.

As we now know, the cooling of the mid-twentieth century was only one step backward in a forward-stepping process. Temperatures began to climb again in the 1980s, and the greenhouse effect became a household phrase in 1988. NASA scientist **James Hansen** testified before the US Congress that he was "99-percent confident" that the summer's abnormal heat was part of a trend related to global warming. A single hot summer doesn't equal climate change, of course, but Hansen's comments – based on years of climate modeling and other evidence – resonated in the steamy air of Washington, DC.

Muggy greenhouse

The most prevalent of our greenhouse gases is **water vapour**. It's responsible for about two-thirds of the natural greenhouse effect. **Carbon dioxide** is a much more powerful absorber, however, and so a small increase in CO_2 can have a wider impact. The important thing about water vapour is that it's an agent of positive feedback. As global temperatures rise, the oceans warm up and release extra water vapour. This water vapour can then absorb energy and, in turn, radiate some of it toward the ground, thus helping the **global temperature** to rise even further – which causes still more water vapour to be released. Eventually, it's believed, the extra moisture and heat would generate enough clouds to keep this process from becoming a runaway greenhouse effect, but how this safety valve would operate on a global level is unclear.

Things accelerated in the 1990s, which scored the five warmest years in the century-plus record up to that point. The trend was only briefly interrupted when the eruption of the Philippines' **Mt Pinatubo** produced the largest volcanic cloud of the century in June 1991. Millions of tons of volcanic debris were thrown into the stratosphere, blocking sunlight and cooling the globe. A few months after Mt Pinatubo blew its top, Hansen predicted that a multi-year global cooldown would be followed by more records in the late 1990s. Nobody else made quite as flat a prediction, but many quietly shared Hansen's belief, and it proved to be correct: the years 1997 and 1998 each set new global highs, and the worldwide trend played out in many local areas, especially at higher latitudes across the Northern Hemisphere. In the 1990s, **central England** saw about eight days per year whose average temperature was above 20°C/68°F. That's about double the rate observed in the preceding two hundred years. The first few years after 2000 have shown no sign of a planetary cooldown. The year 2005 was just as warm as the record-setting 1998 – and that was without any help from the gigantic 1997–98 El Niño, which gave global temperatures a temporary boost.

News you can use

Statistics are an abstract concept to most of us when we're freezing at the bus stop or over-heating in a room without air conditioning. How does global change translate into weather we can feel and see? The science of "downscaling" global warming to your backyard is still a fairly crude enterprise. Even what seems to be the most obvious conclusion – you'll be warmer more often – can't be taken for granted. Just as some places didn't share in the warming of the twentieth century, it's possible that a few spots (see p.154) may find themselves cooler on average when the year 2100 rolls around.

Even if forecasters can't tell you where every individual raindrop will fall, they can let you know with confidence that there's a better chance of rain tomorrow than there was today. This is the sort of information – probabilities, possibilities, risks – that scientists generate in order to help policy makers figure out how we should respond to the global change threat. Here are a few of the overall points to keep in mind. Some of them are widely accepted, while others are more controversial.

- Global warming won't do away with **cold snaps.** In the US, Arctic outbreaks often produce a rash of editorial cartoons and people on the street asking, "Whatever happened to global warming?" Although models tend to agree that **cold extremes** will become less intense in a greenhouse-warmed climate, this still allows for plenty of up-and-down variations from day to day – in other words, weather.
- Both **floods** and **droughts** may intensify. The reason for this apparent paradox is that, as temperatures go up, more water tends to evaporate from the ocean. That means, on average, more moisture is available in the air for making rain. **Global precipitation** is difficult to measure, since it's impossible to blanket the oceans with a network of fixed monitoring stations, and satellite measurements of rainfall are still being refined. However, the best estimates show an increase of as much as 5 to 10 percent in rain and snow amounts over Northern Hemisphere land areas in the past century. In the US, climatologist **Tom Karl** and colleagues have demonstrated that more of a given year's rainfall tends to fall in heavier bursts than it did a century ago. Similar results have been compiled for many other parts of the world, with a few noteworthy exceptions (such as southwest Australia, which has been drying out over recent decades). More intense rainfall suggests that further flooding or more severe flooding is possible, although many other factors are involved, such as how cities are built and how watersheds are managed. As for droughts, what little water remains in the soil is more likely to evaporate when it's warmer – thus, the dry ground gets even drier. Some computer models hint that the most likely place for increased summer drought in the decades to come is the interior of continents, such as central Europe or the US Great Plains.
- Overall, **warming** is likely to be more dramatic in the winter and at night. This was the pattern of the last century: winters warmed more than summers did, and over land the nighttime lows warmed almost twice as much as did the daytime highs. In part, this is probably a side effect of extra moisture and clouds in the air. These help keep temperatures from dropping as much as they otherwise would, especially on cold winter nights. One result already in evidence has been a lengthening of the **growing season** in many areas. A European survey published in the journal Nature found that by the mid-1990s plants were blossoming an average of six days earlier than they were in the 1960s, while autumn colours tended to arrive about five days later. At some high latitudes, the growing season is now more than two weeks longer than in the 1950s.
- **Global dimming** isn't likely to save the day. Since the mid-20th century, the amount of sunlight reaching land areas around the globe decreased by a substantial amount – 10 percent or more in some areas. The most likely culprits are airborne particles, or aerosols, emitted directly from cars, factories and homes or created as pollutants mingle with sunlight. Because our experience

What puts the greenhouse in gas?

All molecules of greenhouse gas – **ozone**, **carbon dioxide**, **water vapour** and the rest – have at least three atoms. These molecules capture and absorb radiation more easily than a two-atom molecule like nitrogen or oxygen, making them the prime culprits in global warming. If they're so effective, why don't these molecules catch energy on its way down rather than on its way up? Some incoming sunlight is indeed absorbed, but a large portion the heat we feel in the air actually comes up from the ground (try standing in a car park on a hot day and you'll get the picture). The energy that radiates upward is in the **infrared** – a longer wavelength range than most of the sunlight that warmed the ground in the first place. Since molecules of a three-atom gas can vibrate and rotate in sync with infrared radiation, they can absorb its energy much more easily than can two-atom gases. The greenhouse molecules can then emit energy in all directions, some of which goes towards the surface. Thus, you have the greenhouse effect: radiation comes in, but it can't all get out (very easily, anyhow).

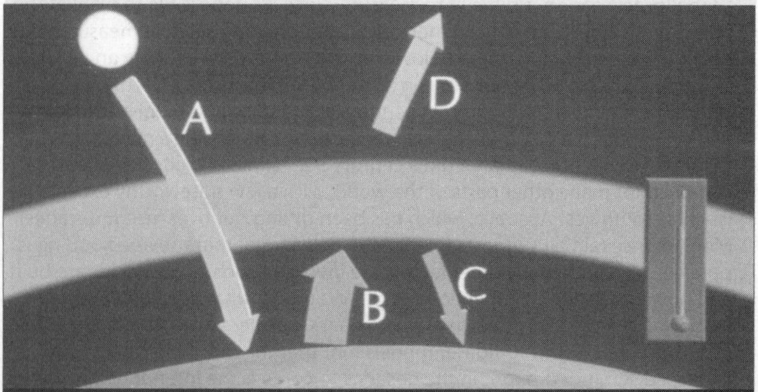

NCAR / UCAR / NSF

As sunshine reaches earth (A), some gets absorbed by the planet's surface, which then re-emits energy at a different wavelength (B). Some of this escapes (D); some gets absorbed and re-emitted by greenhouse gases (C), warming the planet.

of weather is dominated by daily and seasonal shifts among sunshine, cloudiness, and nighttime, it's doubtful that the world looks noticeably darker to the average person on the street. The dimming hasn't been enough to knock out global warming, either, given that temperatures have continued to rise since the 1970s. Some studies show the dimming effect has weakened since 1990, though, as cleaner-burning technology helps reduce the output of sulphates and other short-lived pollutants responsible for dimming. It's much easier to keep aerosols out of the air than it is to cut back on carbon dioxide, so we can expect dimming to take a back seat to warming on the global scale.

- **Hurricanes** appear to be gaining strength as global warming proceeds, but the specifics are still being debated. Today's global climate models don't have the resolution to track individual low-pressure centres in much detail, so they can't easily show whether such storms will change in number or strength as greenhouse gases increase. Despite 2005's bumper crop of Atlantic hurricanes (see p.85), there is little evidence that hurricane counts have increased globally or that they will rise in a warmer climate. However, a larger fraction of hurricanes worldwide has been reaching intense levels since the 1970s, according to several studies in 2005–06 that triggered a lively research and policy debate. Computer models hint that warmer oceans may cause a continued increase in strength among the most powerful hurricanes. The jury is still out on the significance of such a trend for society, given the uncertainties of hurricane landfall locations – but even if hurricanes don't get worse, the growth in coastal settlement across hurricane-prone areas will only increase the risk of devastation and injury.

- There's no particular reason to believe that **tornadoes** will increase or intensify. The upswing in tornado reports in the US and Europe across recent decades is believed to be mainly because more people are looking for **twisters**, and they're spotting more of the weak, short-lived ones that would have been missed in earlier times. Fortunately, the more violent tornadoes don't appear to be changing in frequency.

Wild cards

It's human nature to look for an easy solution to global warming. One such idea involves sprinkling tons of iron filings over the ocean to stimulate the growth of **plankton** that would consume **carbon dioxide**. Another involves spraying sun-blocking particles from aircraft into the stratosphere, mimicking the temporary cooling effects of strong volcanoes. Yet such mechanical escape hatches could be more trouble than they're worth. For example, it's unclear how the iron-filing approach would affect marine life and how long any benefits would last; and the cooling from a stratospheric shield wouldn't slow down the increase in atmospheric carbon dioxide or the acidification of oceans. Our planet's network of flora, fauna, oceans, ice and atmosphere is so integrated that surprises could easily result if we carry out global-scale tinkering, assuming that such a fix was even affordable. And as climatologist **Stephen Schneider** has pointed out, any unforeseen change in natural cycles could accentuate global warming just as easily as it might alleviate it.

It seems unlikely we'll escape a significant warming in the next century without making vast changes in the way we live. Even if we were to freeze the global output of greenhouse gases right away – a political and practical impossibility – we'd be in for some amount of warming, perhaps 0.5°C/0.9°F or more, simply due to the gases we've already emitted, their century-long lifespan in the atmosphere, and the extra heat already stored in our planet's oceans (see below). In that sense, the choice before global society is how much additional warming we want to try to prevent. A number of environmental and policy groups have settled on a target of no more than 2°C/3.8°F beyond the preindustrial global average. It's a goal they see as achievable through a gradual ramp-up in conservation, energy efficiency and new technologies, yet one low enough that it might prevent some of the more dire possibilities, such as wholesale extinctions or a collapse of the Atlantic conveyor-belt circulation (see p.154).

As we face the hand we've dealt ourselves, there are several wild cards to take into account:

- **The flip side of pollution** At the same time that greenhouse gases act to warm us up, other industrial emissions help cool us down. Mainly, these are **aerosols** – tiny particles of dust, smoke, soot, and other stuff spewed out from cars, homes and factories, as well as volcanoes and large-scale fires. Aerosols act more like a shield than a greenhouse. They tend to reflect more energy than they absorb, which keeps some of the sun's rays from warming up the atmosphere (although new research is finding that soot may absorb more energy than previously thought).

  The best guess is that aerosols cause about half as much cooling as greenhouse gases cause warming. But aerosols don't last long. Except for those thrown by volcanoes into the stratosphere (where they can cool global climate for one or two years), most aerosols stay in the air only a few days before falling out of the atmosphere. By contrast, the **greenhouse gases** produced by industry tend to stay airborne for decades. Because aerosols can't travel far during their short lifespan, their cooling tends to be concentrated near industrial centres, while most greenhouse gases are spread fairly uniformly around the entire world.

  Aerosols induce respiratory disorders and damage our health in general, with the result that most industrialized countries are seeking to curtail them. Ironically, this could allow global warming to proceed at a faster pace. On the other hand, nobody knows how much developing countries like China will rely on coal burning and other aerosol-spawning technologies in the next century. This could have a huge effect on how and where global warming might be partially offset by aerosol cooling.

- **What about the Sun?** Until recently we only had crude knowledge of how much the energy put out by the sun varied from year to year. New instruments are now achieving far more precise readings of **solar irradiance** (the amount of solar energy travelling to Earth) and how it varies. By correlating this data with visible **sunspots**, scientists have been able to use four hundred years of sunspot records to infer how the sun's output might have varied in the past. It now appears as if the cooldown in global temperature from about 1940 to 1970 coincided with a slight drop in solar energy. The temperature rises of 1900–1940 and 1970 into the twenty-first century seem to correspond to increases in solar energy.

  To the best of our knowledge, however, these changes in solar energy aren't nearly enough to fully explain the temperature rise of the past century. Overall, it's estimated that we've seen at least three times more warming than would have resulted from solar variations alone. In other words, even if the Sun helped to sculpt the uneven shape of our century's temperature trend, there's been a temperature climb well above and beyond this factor, especially since the 1970s.

- **Water and ice** The world's oceans are like an energy bank. Because it's so dense, water takes longer than air to warm up or to cool down. And the oceans have warmed up dramatically in the past hundred years. A study led by US oceanographer **Sydney Levitus** found far more warming than expected down to depths of 3000m/10,000ft, with half of the added energy stored above 300m/1000ft. Some of that heat may get transferred to the atmosphere in the coming years, whether through **El Niño** events (see p.114), **hurricanes** (see p.83) or some less dramatic means. Nobody knows how quickly the oceans might release the heat they've stored.

Ice, on the other hand, helps to keep us cool. Sunlight is reflected easily from a bright ice patch or snow field. All else being equal, the more ice cover there is, the less solar energy gets absorbed by Earth and its atmosphere. Mid-latitude **mountain glaciers** have shrunk dramatically in the past century, but they represent only a small fraction of Earth's ice cover. Arctic sea ice is a different matter. The summertime thickness of Arctic ice has gone down by some 40 percent in the last fifty years, and the normal summer retrenchment of the ice pack's surface area has grown more extensive: some models now predict virtually ice-free Arctic waters in summer by the 2040s. The record-setting zone of summer melt in 2005 exceeded that of the 1978–2000 average by an area roughly twice the size of Texas. Glaciers across southern Alaska and Greenland are also melting at an increasing clip, a factor not fully accounted for in models of sea-level rise. Most of Antarctica remains ice-covered, but some ice shelves lining the West Antarctic Ice Sheet have been decaying, and a few spectacular collapses have been observed – most notably the Larsen B ice shelf. This floating ice mass attached to the Antarctic Peninsula lost an area of ice the size of Delaware to the sea in March 2002. Most of Antarctica's ice is expected to remain intact through the 21st century and well beyond, but a substantial amount might be lost from the continent's western edges as warming proceeds.

▶ **Greenery** In 1992 a group of scientists teamed with the Western Fuels Association to produce *The Greening of Planet Earth*. This film – and a sequel, released in 1998 – made the case that greenhouse gas emissions might be a good thing: with increased carbon dioxide and warmer temperatures, plants would have a field day and society would benefit from it. As unyieldingly cheerful as the film's tone was, research has borne out some of its points. It does appear that extra carbon dioxide could be stimulating faster growth of some plants, although it's unclear what effect this might have on the plants' overall health and nutritive value and how other nutrients, such as nitrogen, might modify the plants' response. Also, the growing seasons in some places are becoming measurably longer. If agricultural interests were nimble enough to adapt to these changes, and if altered rainfall and snowfall patterns didn't get in the way, it's conceivable that the overall productivity of our planet's agriculture might increase – although, as climate boundaries shift, some countries could be winners while others would lose out. Right now it appears that these shifts would favour northern countries like Canada and Russia while adding climatic stress to the poverty-plagued tropics.

Natural **ecosystems** will face their own challenges. A 2004 review of studies spanning a fifth of Earth's land areas found that up to 37 percent of plant and animal species could face extinction by the year 2050 if emissions increase at a modest pace. More than half of all species could be at risk if emissions climb more dramatically. Some animals and insects may be able to adapt to warming by simply migrating with the climate, but other species – some forest inhabitants, for instance – may not be able to shift poleward or otherwise adapt quickly enough. Forests themselves play a large part in determining climate. Trees are usually darker than open fields, and therefore tend to absorb more sunlight – but trees also ingest large amounts of carbon dioxide. During Kyoto negotiations, lobbyists from some heavily forested countries successfully argued that nations should get credit not only for planting forests, but also for maintaining existing ones. This could mean less pressure to reduce emissions.

Now what?

The **Kyoto Protocol** won't come close to bringing the atmosphere back to its pre-industrial state – that would take centuries – and yet its goals are quite ambitious. By the year 2012, the original protocol called for an 8 percent cut in 1990 emission levels for the European Union, 7 percent for the US, and 6 percent for Japan and Canada, with lesser amounts for other nations. Such a plan might seem pie-in-the-sky, but there's a precedent for it. Following the discovery of the **ozone hole** above Antarctica (see opposite), it took only two years for the world's countries (via the UN) to ratify the **Montreal Protocol**, which successfully called for a gradual replacement of the chemicals that were responsible for ozone depletion. For Kyoto, the implications and costs are far more widespread and the opposition was far stronger. Nonetheless, the protocol in its final, ratified form calls for emissions from industrialized nations by the year 2012 (except for the US, which opted out) to be reduced by an average of 5.2 percent over 1990 values – a requirement somewhat watered down from the original Kyoto draft, but still unprecedented.

Some of the biggest uncertainties that obscure the climate-change picture are **human** ones. For example, how will changes in technology and lifestyle affect emissions? Even though the US and Europe produced the lion's share of emissions in the 1900s, the **developing world** will likely take on that role sometime in the next several decades, if only due to population growth. Much will depend on whether growing economies fall back on older, dirtier technologies or incorporate greener ones. With the US acting as a role model for the high-consumption lifestyle, many developing countries feel they have the right to raise their standard of living just as their industrialized predecessors did. By the same token, many policy-makers (especially in the US) are loath to alter economies based on fossil fuels only to watch other countries burn the fuel they saved. The result is a global political picture that makes emissions cuts difficult to implement.

As the world begins planning for the next diplomatic steps to follow Kyoto, watch for a complex set of political manoeuvrings, both within and between nations. In the meantime, with the threat of serious climate change looming ever larger, many individuals and groups are taking action now. Alternative energy sources are on the increase: wind and solar power, which are virtually emissions-free, are each increasing by roughly 30 percent per year. Research continues on more efficient forms of biofuels that are less reliant on petroleum-based farming techniques. Even nuclear power – which is free of carbon emissions, if environmentally problematic in other ways – is making a gradual comeback. The big challenge will be addressing the use of coal, oil and natural gas. As developing countries grow, the global use of fossil fuels continues to climb. Even with the possible arrival of "peak oil" in the next few years, there's more than enough fossil fuel to produce truly worrisome climate change. That's especially true for coal, which is why cleaner ways of burning coal are a topic of intensive research.

Even if we could preserve today's climate under glass, Earth's increasing population adds to the risk of death and destruction from the weather calamities we already know too well. Whether they set up housekeeping in paper-and-tin shacks or elegant mansions, people are flocking to homes near the **sea** – about two-thirds of the human race live within 160km/100 miles of a coastline – or in other weather-vulnerable locations. As political scientist, **Roger Pielke Jr**, has pointed out, once society has addressed these Achilles heels, we'll be better prepared for the future, no matter how severe climate change turns out to be.

The hole in the ozone and what we're doing about it

The **ozone hole** is more than a physical concept. It's a powerful symbol of how humans can wreak damage to the planet in ways we never dreamed. Be that as it may, the risks posed by the ozone hole appear to be on the decrease, and in many ways they pale next to the dangers connected to global warming.

In 1985 a team of British researchers was shocked to discover that levels of ozone in the stratosphere above Halley Bay, on the **Antarctic coast**, had plummeted in recent years. Up until then, hints from ground and satellite data had been either overlooked or discounted, largely because nobody expected to find such a dramatic change.

Ozone is a pollutant at ground level, harmful when we breathe it. However, the ozone layer that sits within the **lower stratosphere** (focused at about 25–40km/15–25 miles high) is a godsend. Even though it's a tiny fraction of the stratospheric air, it intercepts much of the **ultraviolet light** that can produce sunburns and skin cancer, damage our eyes, and cause all kinds of ill-health. **Stratospheric ozone** is measured from ground by a light beam that bounces up, hits the ozone and returns. By monitoring the strength of this signal, scientists can calculate the amount of ozone in an imaginary column that stretches above the measuring site. In 1985, the British team found that ozone had diminished from a near-normal level of 300 dobson units to about 200. Since then, values below 100 have been found. (The "hole" is defined as the area with dobson units below 220.)

A flurry of measurements from both ground instruments and satellites over the next few years clarified what was happening. Three factors conspire to form this patch of depletion – the ozone hole – that lasts for a few

Forecasts for the twenty-first century

The computer models that project global climate don't yet have the skill to issue regional outlooks with much confidence. This means that policy-makers don't work with forecasts per se. Instead, they evaluate "scenarios", examples of what could happen if things played out in a certain fashion. Researchers spin whole sets of scenarios under various assumptions: emissions will hold steady, they'll rise, they'll drop, etc. It's then up to policy-makers to decide which scenarios to plan for. Think of this as insurance: people often plan for potential calamities, like a car wreck or a home fire, that they don't necessarily expect will happen.

Research bodies around the world have taken a cue from the Intergovernmental Panel on Climate Change, or IPCC (see p.151), and produced reports that bring climate change into perspective for a single country or region. Below are summaries of two such studies.

▶ **United Kingdom.** The Hadley Centre, part of the UK Meteorological Office, produced the first comprehensive look at British climate-change scenarios in 1998. The 2002 update was dubbed UKCIP02, and a third update (UKCIPnext) is planned for later in the decade. The UKCIP02 scenarios are based on four IPCC projections of how global emissions might increase in the twenty-first century. Overall, the UK is projected to warm by the 2080s by an amount that ranges from 2.0°C/3.6°F in the lowest-emissions scenario to 3.5°C/6.3°F in the highest. The warming is somewhat greater in the summer and autumn than in winter and spring; in all scenarios the southeast warms more than the northwest. In general, winters moisten and summers dry out, with a slight overall drop in precipitation. Snowfall could decrease by anywhere from 30 to 90 percent by the 2080s, depending on the scenario.

As with their IPCC sources, none of the four UKCIP02 scenarios are judged as to their relative likelihood, because each depends on long-term societal choices that are virtually impossible to predict. For UKCIPnext, the probability of particular outcomes within each scenario will be assessed through ensemble modelling. This involves producing a number of model runs for each scenario, with tiny changes to the model starting points that capture a wide range of future climates and gauge how likely each might be.

▶ **United States.** The first US National Assessment, mandated by Congress and released in late 2000, was the result of nearly a decade of work. The assessment is still available on the Web; despite its age, it remains one of the most detailed looks at how US climate might change with global warming. Instead of looking at a range of scenarios from one model, the US assessment used one scenario each from two global climate models, one from the Hadley Centre and the other from Canada. The emissions levels in both are the same as the medium–high UK projection (increases of about 1 percent per year). The average US temperature is projected to rise a slight bit more rapidly than weeks during each Southern Hemisphere spring, from about September to November. The first ingredient is a special type of cloud, a **polar stratospheric cloud**, that only forms as winter temperatures fall below about −80°C/−112°F at high altitudes and latitudes. These clouds represent the stage for ozone depletion. The main players on that stage are the industrial chemicals known as **chlorofluorocarbons (CFCs)**. Starting in the 1920s, when they

the rest of the world's in the twenty-first century. An average of 0.28°C/0.5°F per decade is predicted by the Hadley model and 0.5°C/0.9°F by the Canadian model. As in the British projections, the US warming is most prominent during winter but present year round. In the already-steamy South, the average July heat index (a measure of heat and humidity combined; see Resources, p.402) is projected in the Canadian model to climb more than 11°C/20°F by 2100. Both models project big rises in precipitation for the southwestern deserts, with increases topping 80 percent in some spots. The models disagree on the eastern US: Hadley projects widespread increases in the 20–30 percent range, while the Canadian model dries out much of the southern tier by 10 to 30 percent. More recent models still disagree on how US precipitation might change, but in any event it's believed that the warmer US temperatures are liable to stimulate more intense and/or prolonged droughts. The Atlantic and Gulf coasts will be vulnerable to rising sea levels and storm surges that could jeopardize wetlands, while the levels of the Great Lakes are projected by some models to decline.

Succeeding the US National Assessment is the US Climate Change Science Program, which is centered on a series of 21 "synthesis and assessment" reports that will characterize the state of global-change research in various areas over the next few years. These reports are intended to clarify key questions about climate-change science and lay the groundwork for future research; they're not designed to provide specific regional projections in the style of the US National Assessment.

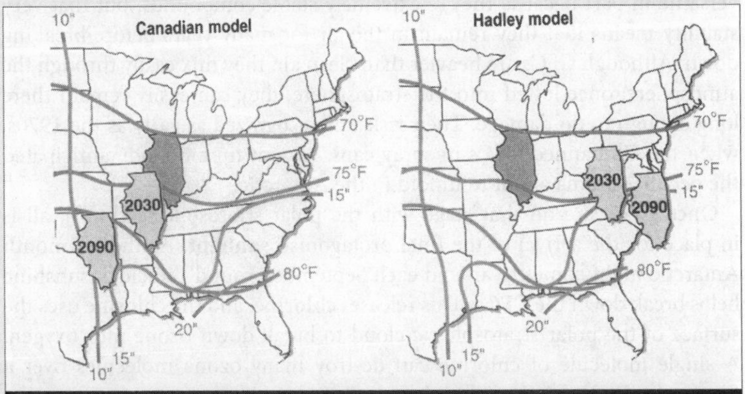

The Canadian climate model suggests that Illinois' climate might resemble that of eastern Oklahoma by the year 2090. The Hadley model indicates a more Carolinian regime might take hold.

What you can do about climate change

It's the biggest environment challenge our world has ever faced, but that doesn't mean global warming is insoluble. Here are a few ways you can make a difference.

▶ **Calculate your carbon.** Through a variety of online tools, you can determine how much carbon is produced by your daily activities – the perfect starting point for reducing your carbon footprint (see Resources, p.390).

▶ **Offset your emissions.** The steps below can help reduce the amount of carbon your activities consume. You can offset the remainder through a number of firms that invest in alternative energy, tree planting and other ways to reduce carbon output (see Resources). In this way, you're helping others to remove enough carbon to offset the amount you can't avoid emitting.

▶ **Make your home more energy efficient.** Compact fluorescent light bulbs produce the same amount of light with just 30 percent of the energy of traditional incandescents, and they last far longer. You can also save both money and energy by weatherstripping around doors and windows and beefing up insulation. Updating "energy hog" appliances, such as refrigerators and freezers, is another good idea. Also, try using surge protectors that allow you to shut down remote-controlled gadgets that silently drain power while apparently turned off. If you have the choice, opt for renewable sources of electricity: a solar-powered hot water heater is one good way to start.

▶ **Shop smartly.** If you buy food that's organic and/or locally grown, you can limit the amount of fossil fuel needed to grow and move them around. Also, many containers (cans, bottles, and the like) and printed materials (newspapers, magazines, catalogues) can be

were developed for refrigeration, CFCs have been used in spray cans, air conditioners, computer-manufacturing plants and many other places. CFCs are versatile in part because they're extremely stable compounds, but that very stability means that they remain in the air for many years before breaking down. Although CFCs are heavier than clean air, they mix easily through the atmosphere; once lofted into the stratosphere, they can easily remain there long enough to do damage. Their risk was recognized as early as the 1970s, when the US banned CFCs in spray cans. At that time nobody anticipated the peculiar drama about to unfold in the Antarctic.

Once CFCs are on that stage with the polar stratospheric clouds, all is in place for the arrival of the final protagonist: **sunlight.** As the six-month Antarctic night comes to an end each September, round-the-clock sunshine helps break down the CFCs. This releases **chlorine**, and the chlorine uses the surface of the polar stratospheric cloud to break down ozone into **oxygen**. A single molecule of chlorine can destroy many ozone molecules over a few weeks. By November of each year, the stratosphere has warmed up, the clouds disappear, and the ozone hole closes once more.

If the ozone hole covered the whole planet, we'd have to wear sunscreen and hats for weeks at a time. Fortunately, the hole has never extended much recycled. And if everyone brought a reusable bag when shopping, we could cut back dramatically on the estimated 500 billion plastic bags that are used and tossed each year.

▶ **Find the best ways to get around.** Transportation is the fastest-growing major source of greenhouse gases. If you're a frequent driver, make your next car as fuel-efficient as you can afford, and strive for higher efficiency right away. Keeping your tyres at their proper inflation, starting and stopping gently, and driving at moderate speeds all help. Consider taking fewer trips but making them longer. Trains or boats often use a great deal less carbon per mile than airplanes or automobiles. And air travel puts the greenhouse gases at higher altitudes, where they are several times more powerful. (A typical return flight from Europe to California emits the same amount of carbon per person as the typical UK car does in an entire year.) Rough Guides regards travel, overall, as a global benefit, while recognizing that we all have the responsibility to reduce our carbon footprints. Together with Lonely Planet and other concerned partners in the travel industry, Rough Guides encourages you to "fly less and stay longer", and we support a carbon offset scheme run by Climate Care (www.climatecare.org). Please take the time to view our website (www.roughguides.com/climatechange) and see how you can make your next trip climate neutral.

▶ **Spread the word.** If you feel strongly about climate change, put your concern into action by working for lower-carbon practices at your school or workplace and in your community. Another way to make an impact is by putting any investments you may have in companies that take global warming seriously. And, of course, you can contact your legislature or join any of the numerous activist groups trying to turn the tide of climate change.

beyond Antarctica, although it has encroached on southern Chile. Southern Australia and New Zealand, while outside the hole per se, have seen ozone reductions of more than 10 percent at times.

Why isn't there an Arctic ozone hole?

In fact, there is, although it's not as powerful and reliable as its southern counterpart. For a variety of reasons, including the effect that extensive **mountains** across North America and Eurasia have on air circulation, the wintertime vortex that forms over the **Arctic** is less stable than its Antarctic counterpart. This limits the growth of polar stratospheric clouds and helps keep a bona fide ozone hole from forming. Still, the springtime depletion over the Arctic grew to levels of 60 percent by 2000 (comparable to the Antarctic's percentage drop), thanks largely to increasingly colder Arctic temperatures (see overleaf).

A related concern – distinct from the ozone hole – is the weaker but broader and more persistent **ozone depletion** of some 5 to 10 percent across much of the globe compared to the temporary drops of 50 percent or greater in the ozone hole itself. Some of this worldwide depletion is likely due to the yearly dispersal of the ozone hole and the mixing of that ozone-depleted air around the globe.

Before long, the ozone hole should be on the mend. The 1987 **Montreal Protocol**, orchestrated by the United Nations and ratified with amazing speed, called for a gradual rampdown in CFC manufacture, with a virtually complete ban by 1995. In place of CFCs are substitutes such as **halochlorofluorocarbons**, which are far less likely to break down and release chlorine. There is a black market for CFCs, which will help keep some emissions leaking into the air, but global CFC emissions have dropped significantly since the late 1990s, and chlorine concentrations in the stratosphere have also begun to show signs of a decrease.

Then there's global cooling...

One puzzle remains. Scientists thought we'd see recovery in ozone levels by the year 2000, yet that year's hole was the strongest yet, and through 2005 the ozone hole's extent varied little from its 1990s average. The problem, oddly enough, seems to be **global cooling** – in the stratosphere, that is. Even as the global average of surface temperature has hit record highs here at ground level, it's been dropping to **record lows** in the stratosphere. This is because ozone itself is a greenhouse gas. By removing it from the stratosphere, we've been helping that layer to cool down dramatically plus allowing more energy to get down to the lower atmosphere, where we live. Another cause of the chill is – ironically – the increase in carbon dioxide, which acts as a cooling agent high in the stratosphere. These cold temperatures have induced polar stratospheric clouds to form more readily and last longer. This, in turn, provides a sturdier stage for CFCs to continue causing damage – in the Arctic as well as the Antarctic – even as the culprits go down in numbers.

Thus, although we're doing the right things in the long run, nature has added a delay of what could be several decades to the ozone layer's recovery. It's likely to be 2050 or beyond before ozone levels are back near normal, and with global warming set to produce major changes in the global atmosphere, it's possible that the ozone layer won't completely return to its pre-1980s state. For now, a small but significant amount of extra ultraviolet light (a few percent's worth) will continue to reach the parts of our planet where most people live. That could trigger a measurable rise in skin cancer rates, as well as some harder-to-predict impacts on other species.

Chapter five
Weather around the world

Weather around the world

Whether you're looking for unusual weather or (more likely) hoping to avoid it, a broad understanding of the world's climate is invaluable. You'll find plenty of background on what actually makes weather happen in the earlier sections of this book. In this particular chapter, the practical conditions you're most likely to encounter in major countries and cities dotted throughout the globe are discussed. Vital statistics – including annual temperature range and rainfall levels – can be found at a glance on the city graphs supplied here for more than 200 destinations worldwide. For a more general discussion of world climate zones, turn to p.43.

Even without human intervention, Earth's climate evolves over decades, centuries and millennia. Bear in mind, as you digest the information to follow, that some regions are more variable than others, and it's a rare day that conforms strictly to the norm. The graphs for each spotlighted city also reveal some of the extremes in temperature observed over the past few years – scenarios that you might want to keep in the back of your mind, if not at the front of your itinerary, as you travel. (See also the "Global extremes" map on pp.172–173). If you need more specific detail for a given location, you should consult the climate-related websites listed in the Resources section, p.385.

How to use the city climate charts

The information in the charts to follow is intended to serve as a guideline, rather than an absolute guarantee, of what you might expect to encounter at a given place.

- **Precipitation** The graphic top left shows the average amount of precipitation (including rainfall and melted snowfall) recorded per month. In general, there's about a 10:1 ratio between the depth of snow and the amount of water in it. This ratio can run lower for a very wet snow or higher than 30:1 for the dry snow observed in extreme cold or at higher altitude. In desert climates, the average precipitation may be a blend of one unusually wet month and many bone-dry ones, so the "typical" month will see less rain than the numerical average might indicate.

- **Number of wet days** The figures on top of each bar of the month-by-month precipitation graph show the average numbers of wet days per month. Note the threshold value near the top of the chart (for example, "# days > 1mm"). This threshold (usually 1mm/0.04in) can be considered the amount of rain or melted snow that would constitute a wet day – just a bit more than a sprinkle or flurry. Because of data limitations, the threshold for a few stations is "0mm". This means that even trace amounts of rain or snow are included – so for these locations, you may want to reduce the number in each month by a few days (two or three will work well in most places) to get an idea of how many days might be more substantially wet. A few other cities have thresholds of 2.5mm/0.1in; for these, you may want to add a wet day or two per month.
- **Thunder and snow** There's no threshold for these values. If the weather observer heard a clap of thunder or saw a snowflake, that day goes down as thundery or snowy. As with the 0mm threshold for precipitation above, expect only a sub-set of these days to give you any real trouble. You can gauge the potential for heavy snow by looking at the average temperatures and precipitation totals for a given month. If the month tends to run near or below freezing, with large amounts of precipitation and a large number of snow days, be prepared. On the other hand, if the precipitation values are small despite a large number of snow days, you're more likely to encounter a dusting than a heavy snowfall. Climate-change alert: warming may push down the number of snow days dramatically in places where winter temperatures average close to freezing, such as northern and central Europe (witness their almost-snowless winter of 2007).
- **Sunrise/sunset** The numbers indicated were calculated for the 15th day of each month, using a US Naval Observatory online calculator at http://aa.usno.navy.mil/data/docs/RS_OneYear.html. Adjustments were made for Daylight Savings Time as practised in each country as of early 2002. Several websites show the DST start and stop dates for each country, as well as any recent changes. One such site is the Daylight Savings Time page of WebExhibits (webexhibits.org/daylightsaving/i.html).
- **Temperature** Both averages and extremes are shown for each month. Periods of record for the extremes appear at the bottom of each temperature chart (eg "21-yr high"). These indicate how many years you might expect to go before encountering the highest or lowest readings on the chart. Keep in mind that a record could be broken any time that conditions permit.
- **Humidity** The average relative humidity is shown for a specified afternoon hour in each month (typically 1500, or 3pm). By comparing this number to the average afternoon highs, you can get a sense of the comfort level you might expect (or endure). Note that the effect of a given relative humidity jumps greatly as the temperature goes up. You can use a heat-index chart (p.402) to get a quantitative sense of this effect.

Global extremes

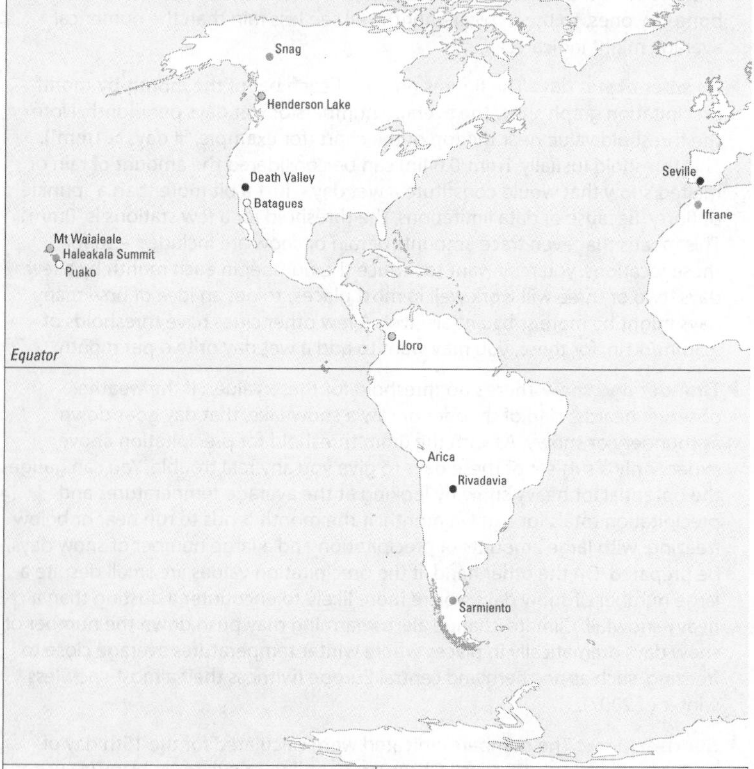

PACIFIC ISLANDS 42°C/108°F, Tuguegarao, Philippines (April 29, 1912) -10°C/14°F, Haleakala Summit, Maui, Hawai'i, US (Jan 2, 1961) 11,684mm/460", Mt Waialeale, Kauai, Hawai'i, US 227mm/8.93", Puako, Hawai'i, Hawai'i, US	**NORTH AMERICA** 57°C/134°F, Death Valley, California, US (July 10, 1913) -63°C/-81°F, Snag, Yukon, Canada (Feb 3, 1947) 6502mm/256", Henderson Lake, BC, Canada 305mm/1.2", Batagues, Mexico
SOUTH AMERICA 49°C/120°F, Rivadavia, Argentina (Dec 11, 1905) -33°C/-27°F, Sarmiento, Argentina (June 1, 1907) 13,299mm/523.6" (est.), Lloro, Colombia 0.8mm/0.03", Arica, Chile	**AFRICA** 58°C/136°F, El Azizia, Libya (Sept 13, 1922) -18°C/-11°F, Ifrane, Morocco (Feb 11, 1935) 10,287mm/405", Debundscha, Cameroon <0.25mm/<0.1", Wadi Halfa, Sudan

The figures above, which reflect readings through the end of 2006, are provided courtesy of the US National Climate Data Center. Precipitation is shown as annual averages over periods ranging from 9 to 59 years. Since these depend in part on

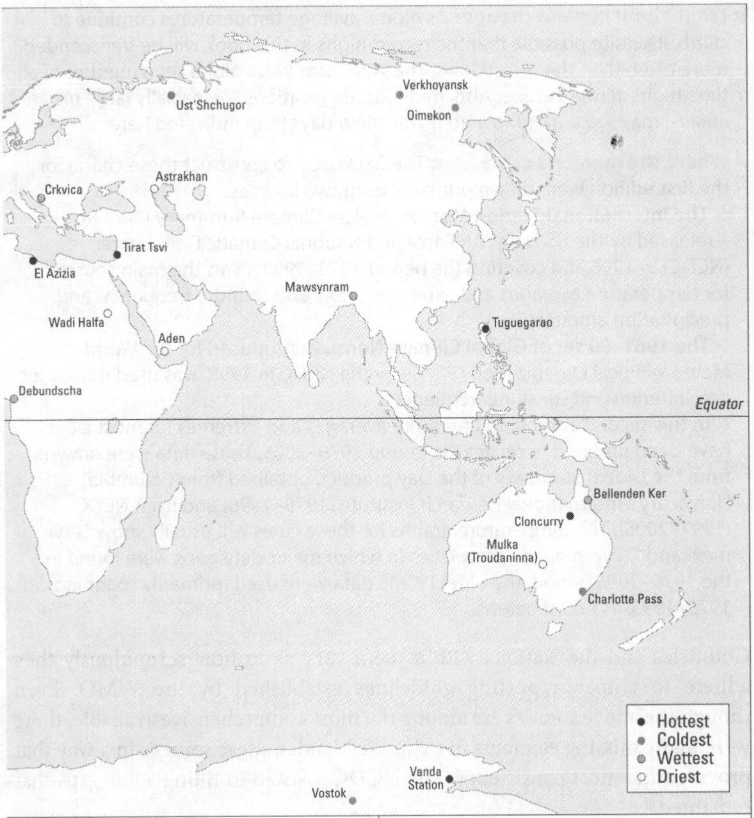

EUROPE	ASIA/ MIDDLE EAST
50°C/122°F, Seville, Spain (Aug 13, 1881) -67°C/-89°F, Ust'Shchugor, Russia (date/year unavailable) 4648mm/183", Crkvica, Bosnia-Herzegovina 163mm/6.4", Astrakhan, Russia	54°C/129°F, Tirat Tsvi, Israel (June 21, 1942) -68°C/-90°F, Verkhoyansk, Russia (Feb 7, 1892) and Oimekon, Russia (Feb 6, 1933) 11,872mm/467.4", Mawsynram, India 46mm/1.8", Aden, Yemen
ANTARCTICA	**AUSTRALIA**
15°C/59°F, Vanda Station, Scott Coast (Jan 5, 1974) -89°C/-129°F, Vostok (Jul 21, 1983) Precipitation averages are problematic in Antarctica due to constant blowing and resettling of snowfall	53°C/128°F, Cloncurry, Queensland (Jan 16, 1889) -23°C/-9°F, Charlotte Pass, NSW (June 29, 1994) 8636mm/340", Bellenden Ker, Queensland 103mm/4.05", Mulka (Troudaninna), SA

measurement techniques, some uncertainty is inherent. The figures are listed as follows: hottest, coldest, wettest, driest.

▶ **What about climate change?** As global average temperatures continue to climb, it's quite possible that the record highs in this book will be transcended more often than the record lows. The 2003 heat wave alone set a number of all-time highs across Europe. Also, mid-latitude locations – especially large urban areas – may see a trend toward fewer snow days than indicated here.

▶ **Where the numbers came from** The data used to construct these charts for the first edition were drawn primarily from two sources:

The International Station Meteorological Climate Summary (ISMCS) – released by the US Navy, Air Force and National Climatic Data Center (NCDC) in 1996 and covering the period 1973–1993 – was the main source for temperature averages and extremes, snow and thunder frequency, and precipitation amounts.

The 1961–90 set of Global Climate Normals, produced for the World Meteorological Organization (WMO) by the NCDC in 1998, was used mainly for precipitation and sunshine frequency.

In this second edition, temperature averages and extremes for most cities have been updated to reflect the period 1979–2005. These data were drawn from the Global Summary of the Day product, obtained from Columbia University's International Research Institute (1979–1996) and from NCDC (1997–2005). The temperature graphs for these cities will usually show "27yr max" and "27yr min." For a few cities in which major data gaps were found in the 1979–2005 period, the older ISCMS data were used (primarily spanning 1973–1993, as noted below).

Countries and the stations within them vary as to how scrupulously they adhere to climate-reporting guidelines established by the WMO. Even though the above sources are among the most comprehensive available, there were many missing elements in each. We blended these sources in a way that provided the most consistency. The NCDC assisted in filling a few gaps that remained.

Most of the ISMCS data spans the period from 1973 to 1993, although some of the records for average precipitation go back much further. While the 1973–93 period is somewhat shorter than the 1961–90 period used in the WMO data set, it's a more up-to-date interval. A similar philosophy applies to the use of 1979–2005 temperature data in this second edition. Globally, temperatures were considerably cooler in the 1960s than in the 1990s, and the warmer readings have continued into the first decade of the twenty-first century. Average temperatures in this edition have risen by 0.6–1.2°C/1–2°F for many months in many cities. And many of the extreme highs and lows you'll find in this second edition are dramatically warmer than those in the first edition – sometimes by 10°C/18°F or more. This is especially true for extreme midwinter lows.

North America

North America's mercurial climate has something for everyone. Severe hurricanes, killer tornadoes, crippling snowstorms, torrential downpours, withering drought and a fair share of nice weather, too – all pay frequent visits to this prosperous, yet vulnerable, land that stretches from the **Canadian Arctic** to tropical **Mexico**. Hollywood has bombarded the world with films of extraordinary American weather, painting a picture of the continent as one big meteorological melodrama. In reality, most short-term visitors will experience nothing more ominous than a few brisk wind shifts, some bursts of rain or snow, and perhaps a thunderstorm. Moreover, because North America – especially the **United States** – is so weather-obsessed and media-saturated, you're likely to hear if any truly threatening weather is in the offing.

The main reason for the frequent and dramatic changes in North America's weather is the region's unique geography. Most continents have mountain ranges arranged in a variety of orientations, but only the Americas have substantial north–south ranges that run the entire length of both continents. This barrier tends to block marine air flowing in from the Pacific on the westerly winds that encircle the globe at mid-latitudes. Within North America, only the west coasts of the US and Canada have climates akin to the more muted regimes of Europe, where Atlantic air dominates.

With the Pacific influence so restricted, you might expect the rest of North America to be a parched and barren domain, like the Australian Outback. Actually, the eastern half of the US and Canada is by and large a fairly moist place. Forests cover a good deal of the landscape outside cities, and almost every spot in the eastern US gets more rain than London, Beijing or Cape Town. The sub-tropical **Gulf of Mexico** is in a perfect position to moisten southerly winds that feed into storms sweeping across the US and southern Canada. And the western Atlantic is warmed by the north-flowing **Gulf Stream**. This allows coastal storms – even in winter – to tap into a source of near-tropical warmth well into the mid-latitudes.

It's a good 4800km/3000 miles across Canada from **Vancouver to Halifax**. Much of that trek is across glacier-levelled prairie and woodland. The great flatlands that stretch from central Canada south to the Gulf coast are a super-highway for bitter Arctic northerlies that pour down cold air each winter. In the summer, the traffic flow reverses, and southerly winds scream north, carrying warmth and moisture (the US state of Kansas is named after the indigenous Kanza tribe, the "people of the south wind"). When these competing air masses fight it out, especially in the spring, intense thunderstorms and tornadoes often result.

The western mountains of North America spill across Mexico's western half, with a vast plateau extending to the east. The high terrain mixes with the low latitude to produce a pleasing combination of ample sunshine and warm temperatures that rarely reach scorching levels, except in the extreme northwest. Rainfall tends to be focused during the six-month wet season, from about May to October, when most of Mexico falls under the influence of tropical trade winds. Hurricanes can boost the rainfall total and wreak havoc along the south and east Mexican coastlines – as well as along the US east coast from Texas all the way to New England.

Canada

The world's second-largest country, **Canada** has the world's most extensive coastline, fronting the Atlantic, Pacific and Arctic Oceans. This provides plenty of raw material for a varied climate, much like that of the United States, but colder and drier on average. Storm systems often spin up along the nation's southern tier (especially in **Alberta**) before diving south of the border toward the **Great Lakes**. Another big factor is **Hudson's Bay**, which cuts well into the eastern half of Canada. Frozen for roughly half of the year, the bay encourages polar cold to extend much further south than it otherwise would. Permafrost edges closer to the equator here than anywhere else in the world. Only the maritime provinces of the far southeast benefit from the northerly extensions of the Atlantic's Gulf Stream current; further up Canada's east coast is the cold Labrador Current. Despite this icy picture, there are, in fact, many places and times where Canadians take their heavy coats off and visitors can comfortably marvel at the scenery. The western prairie states often thaw out in mid-winter; the marine cities of **Vancouver** and **Halifax** aren't much colder than Seattle or Boston, respectively; and most of the nation enjoys long, sunny, warm days in July. In fact, **southern Ontario** can be decidedly muggy at times.

Canada is one of the world's most sensitive barometers of climate change. The average temperature in recent years has sometimes run more than 1.5°C/2.7°F above the long-term national average, with the warming especially pronounced in winter. Erratic thaws have complicated life for those who depend on the yearly melting-and-freezing cycle, from indigenous people to polar bears.

Eastern Canada

Many visitors to Canada end up inside the broad southeastern belt that holds a majority of the nation's 30 million people, from southern Ontario across **Quebec** to the **Maritime Provinces**. This zone – tucked between the Great

Lakes, the **Gulf of St Lawrence** and the **Atlantic Ocean** – has a four-season climate much like New England's. Cities well inland, such as **Ottawa**, struggle to get above freezing during the depths of winter, while the average January highs for the lakeside city of **Toronto** and the coastal city of **Halifax** are closer to 0°C/32°F. Along with coastal gales and bright, bitterly cold days, winter can bring a few rounds of freezing rain as well as snow. The cold hangs on well into spring, and fog often cloaks the maritime reaches, but summer and early autumn are quite pleasing. Lows typically stay above 10°C/50°F and highs approach 26°C/79°F, with conditions a touch warmer as you move in from the Atlantic. A few intense tornadoes have been known to strike southern Ontario, the so-called "banana belt" of Canada (although no actual bananas grow there).

Further east, the island of **Newfoundland** has a climate similar to **Nova Scotia**'s, albeit cloudier and damper. **Labrador** is even colder in winter and tends to be chilly and cloudy in summer (the snow often doesn't melt until late May). On rare occasions, hurricanes that quickly curve up from the western Atlantic or the US East Coast can bring high wind and heavy rain across the Maritimes. Weather changes are frequent throughout the year across eastern Canada, and precipitation is remarkably well distributed. Across the populated areas, a typical month brings about 50–80mm/2–3in of moisture falling on about 10 to 15 days.

The Prairie Provinces

Canada's northernmost cities of consequence – **Winnipeg**, **Saskatoon**, **Regina**, **Calgary** and **Edmonton** – are spread out across the open lands that separate the Rocky Mountains from the forests and bogs around Hudson's Bay. Much like the steppes of Ukraine and southern Russia, this prairie has rich soil and very long, warm summer days well suited to growing grain; in fact, the eastern prairies can get somewhat muggy in mid-summer. There are just enough summer showers and thunderstorms to keep the land fertile (although they also bring an occasional tornado and, especially in Alberta, frequent rounds of hail). Prairie winters alternate between howling northerlies and mild, but ferocious, westerlies blowing off the Rockies. Blizzards are a very serious business – visibility can drop to near zero – but with moisture limited, the snowfalls are seldom as heavy as in the worst East Coast storms. Averaged across the year, this is the sunniest part of Canada.

British Columbia

Much like Oregon and Washington, **British Columbia** features a stark contrast between a wet, mild coastline and a drier basin-and-range climate further inland. **Vancouver** is only a couple of degrees cooler than Seattle, but lacking the shield of the Olympic Peninsula, it gets almost twice as much rain

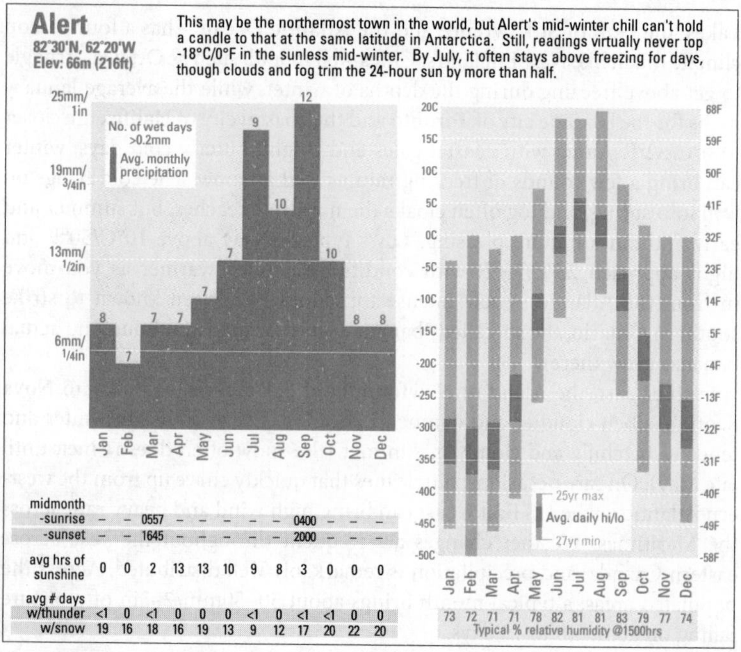

(not so much more often as more intensely). The west slopes of the coastal mountains and the higher peaks inland get huge amounts of wintertime snow and rain, but the east slopes are far drier. British Columbia feels the effects of both **El Niño** (producing a tendency toward dryness) and **La Niña** (higher odds of abnormally heavy rain and snow). Unusually mild winters and dry summers have boosted the risk of forest fire in recent years across the interior.

Northern Canada

Northern Canadians deal with some of the most inhospitable weather of any populated place on the globe. In January the average temperature from **Yukon** to the coast of **Hudson's Bay** runs from –25°C/–13°F to –35°C/–31°F. However, conditions actually become comfortable in the near-continuous sunlight of summer. Interior sections climb above 16°C/60°F on most summer days, and highs above 35°C/95°F have occurred. Near the Arctic Ocean, the barely-above-freezing water keeps the typical July day below 16°C/60°F, and clouds and fog are frequent. If that's too warm for you, head north across the hundreds of islands spread between Canada and Greenland. At latitude 82.5°N, you'll find **Alert**, the world's northernmost permanently inhabited settlement, where about 100 weather and climate researchers coexist with Arctic foxes and polar bears.

Calgary

51°06'N, 114°01'W
Elev: 1084m (3556ft)

Sitting next to the Rockies, sunny Calgary has a climate of wide temperature swings much like Denver's. Winter days often warm well above freezing on dry, sometimes fierce west winds, and the bitter cold snaps don't feel as penetrating as they do farther east. The mild, stormy summers can bring hail and rare tornadoes.

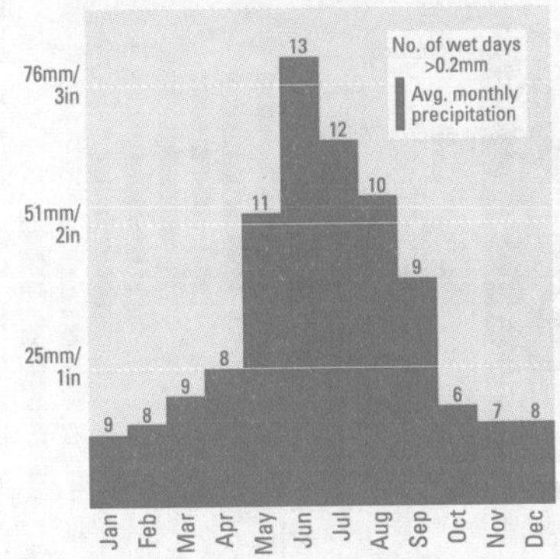

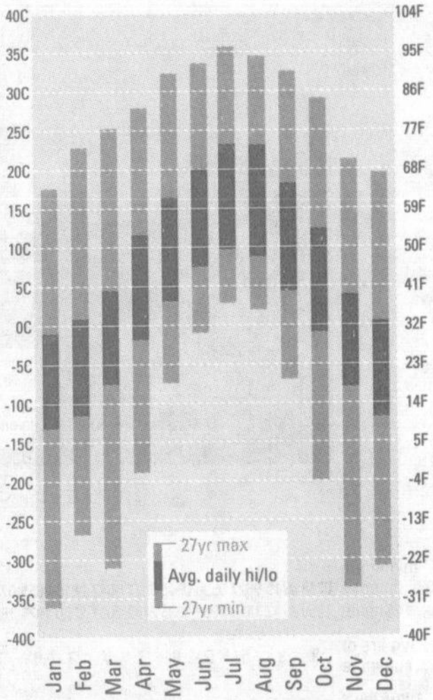

	Jan	Feb	Mar	Apr	May	Jun	Jul	Aug	Sep	Oct	Nov	Dec
midmonth -sunrise	0833	0749	0650	0641	0546	0521	0539	0623	0711	0800	0753	0833
-sunset	1658	1752	1841	2032	2120	2153	2145	2057	1950	1843	1648	1629
avg hrs of sunshine	4	5	6	7	8	10	10	9	7	6	4	3
avg # days w/thunder	<1	0	<1	1	3	7	7	5	2	<1	<1	0
w/snow	14	13	15	9	4	<1	0	<1	3	6	12	14

	Jan	Feb	Mar	Apr	May	Jun	Jul	Aug	Sep	Oct	Nov	Dec
Typical % relative humidity @1700hrs	59	58	55	42	44	44	44	46	47	47	60	61

Churchill

58°45'N, 94°04'W
Elev: 29m (94ft)

A visit to Churchill calls for careful packing. If you're whale-watching in mid-summer, expect chilly nights and cool days, but bring a T-shirt (it can get warm) and a coat (there might be frost). Expect a shower every third day or so. During the polar bear migrations of mid-autumn, it's already as cold as mid-winter in Germany.

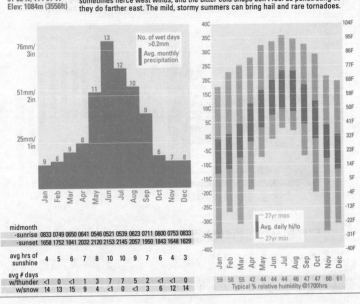

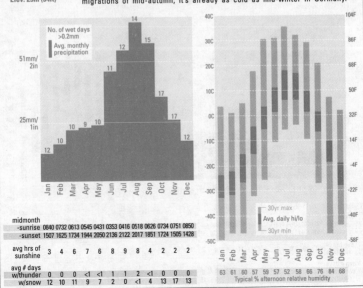

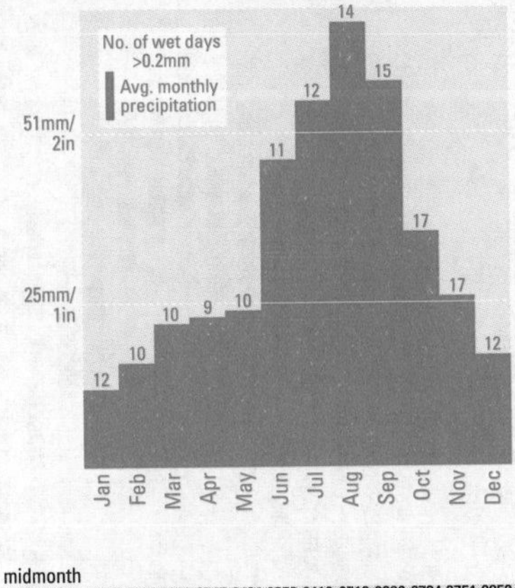

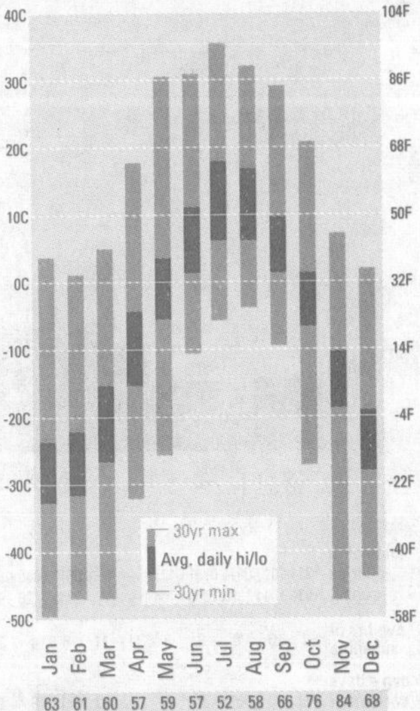

	Jan	Feb	Mar	Apr	May	Jun	Jul	Aug	Sep	Oct	Nov	Dec
midmonth -sunrise	0840	0732	0613	0545	0431	0353	0416	0518	0626	0734	0751	0850
-sunset	1507	1625	1734	1944	2050	2136	2122	2017	1851	1724	1505	1428
avg hrs of sunshine	3	4	6	7	6	8	9	8	4	2	2	2
avg # days w/thunder	0	0	0	<1	<1	1	1	2	<1	0	0	0
w/snow	12	10	11	9	7	2	0	<1	4	13	17	13

	Jan	Feb	Mar	Apr	May	Jun	Jul	Aug	Sep	Oct	Nov	Dec
Typical % afternoon relative humidity	63	61	60	57	59	57	52	58	66	76	84	68

Halifax

44°53'N, 63°30'W
Elev: 145m (476ft)

Part of the Maritime Provinces with good reason, Halifax gets over twice as much rain as London in a typical month. In winter, both heavy rain and snow are possible. Summers are appealing, with a shower roughly every third day and heat or cold rare. Halifax often stays mild into early October, but winter chill can drag into May.

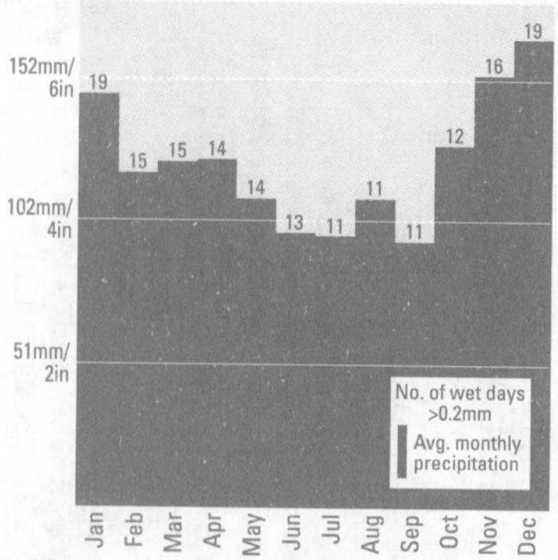

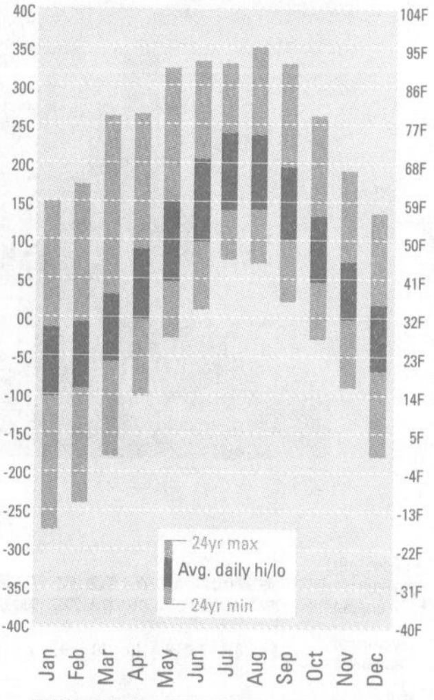

midmonth -sunrise	0748	0715	0627	0630	0546	0527	0542	0616	0653	0729	0711	0745
-sunset	1659	1742	1820	1959	2036	2102	2057	2020	1925	1829	1646	1633
avg hrs of sunshine	3	3	5	5	6	7	7	7	6	5	3	3
avg # days w/thunder	<1	<1	<1	<1	1	3	3	2	1	<1	<1	<1
w/snow	22	18	16	8	2	<1	0	0	0	1	9	19

76 72 67 67 66 65 65 67 69 73 78 79
Typical % relative humidity @1700hrs

Inuvik

68°18'N, 133°29'W
Elev: 68m (223ft)

Summers are short and intense in this high-latitude town. From June through to August, most days top 10°C/50°F. In July, afternoons can go well above 20°C/68°F, although chilly fogs can suddenly roll in off the Arctic and a shower may cool things off. The pitch-dark winters are frigid and lightly snowy.

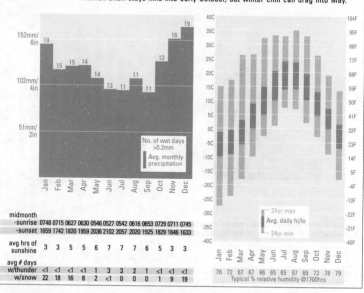

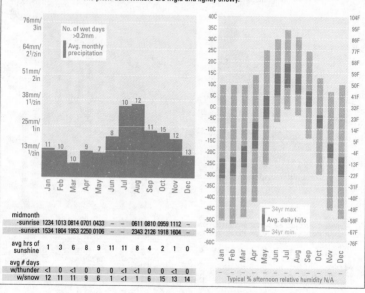

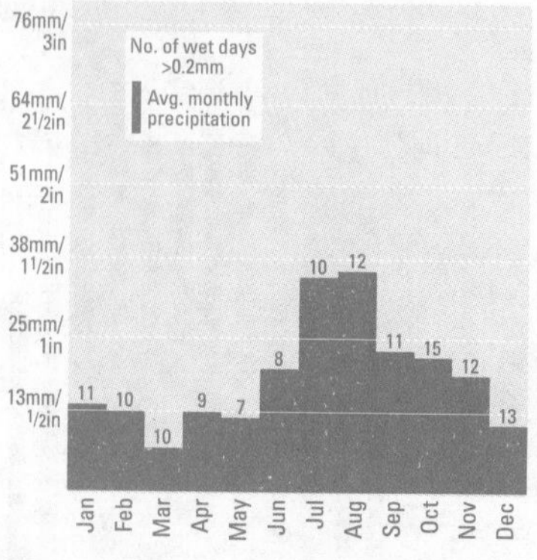

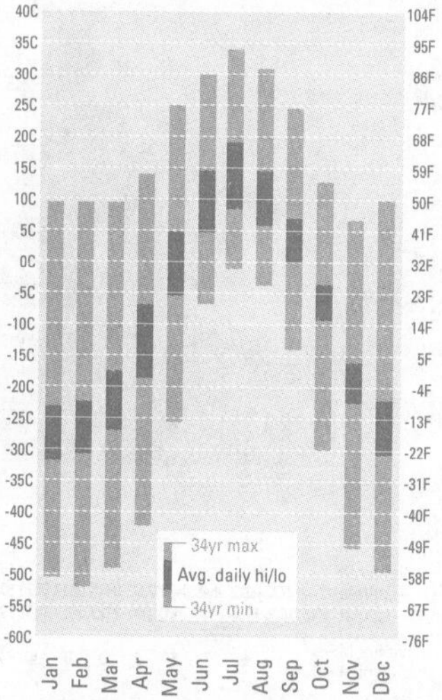

midmonth -sunrise	1234	1013	0814	0701	0433	–	–	0611	0810	0959	1112	–
-sunset	1534	1804	1953	2250	0106	–	–	2343	2126	1918	1604	–
avg hrs of sunshine	1	3	6	8	9	11	11	8	4	2	1	0
avg # days w/thunder	<1	0	<1	0	0	0	<1	<1	0	0	<1	0
w/snow	12	11	11	9	6	1	<1	1	6	15	13	14

– – – – – – – – – – – –
Typical % afternoon relative humidity N/A

Montreal

45°28'N, 73°45'W
Elev: 36m (118ft)

Although it's a good distance from the Great Lakes, Montreal is actually damper than Toronto. Summers are mild, but winter digs in deeply. January is as cold here as in Moscow, with heavier snow and the risk of ice storms. Late spring and early autumn can still get nippy, but there are lots of bright, crisp days.

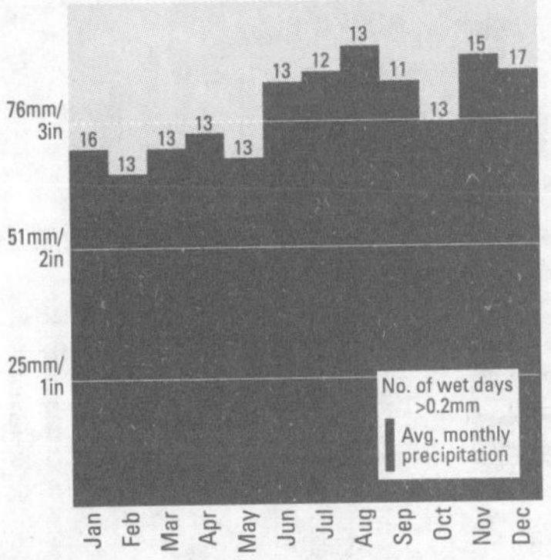

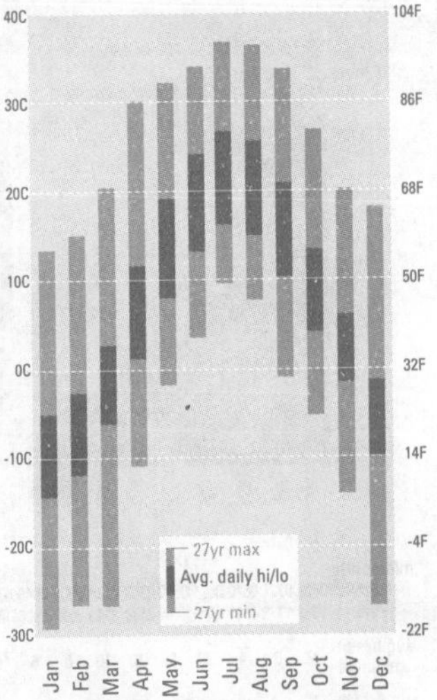

	Jan	Feb	Mar	Apr	May	Jun	Jul	Aug	Sep	Oct	Nov	Dec
Typical % relative humidity @1600hrs	69	64	61	55	53	56	56	59	63	64	71	72

midmonth												
-sunrise	0731	0657	0608	0610	0525	0506	0521	0556	0633	0711	0654	1728
-sunset	1638	1722	1801	1941	2019	2045	2041	2003	1906	1810	1625	1612
avg hrs of sunshine	3	4	5	6	7	8	9	8	6	5	3	3
avg # days w/thunder	<1	<1	<1	1	2	5	6	6	2	1	<1	<1
w/snow	22	17	13	6	<1	0	0	0	<1	2	11	21

Toronto

43°40'N, 79°38'W
Elev: 173m (567ft)

Summers are much like those across northern Europe. Expect mild to warm days, with periodic cool spells and heat waves and a shower or storm about twice a week. Early autumn can be gorgeous, apart from early cold snaps. Spring is often sunny but slow to warm up. Winters are cold with light snows (heavy on occasion)

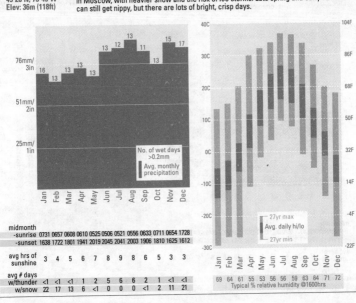

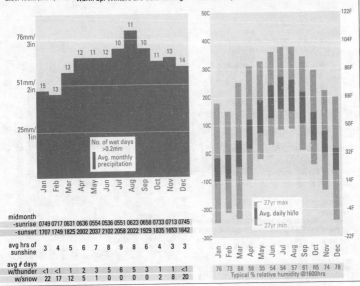

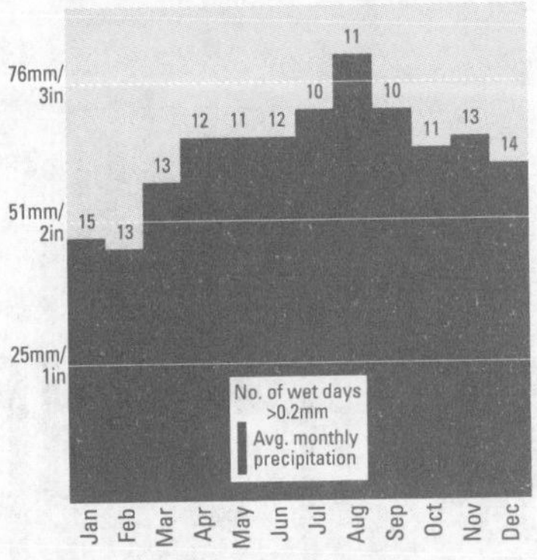

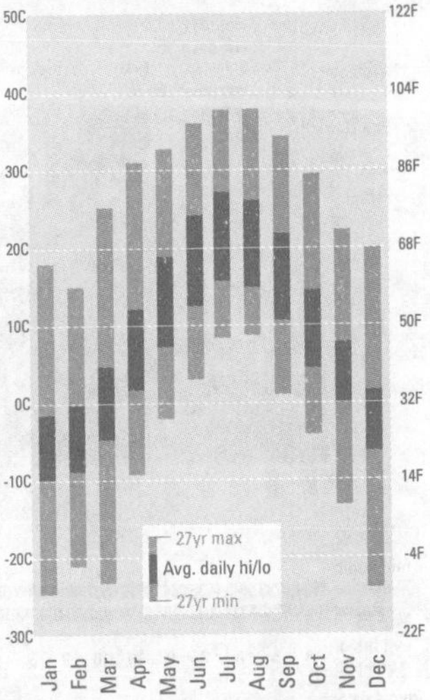

	Jan	Feb	Mar	Apr	May	Jun	Jul	Aug	Sep	Oct	Nov	Dec
Typical % relative humidity @1600hrs	76	73	68	58	55	54	54	57	61	65	74	78

midmonth												
-sunrise	0749	0717	0631	0636	0554	0536	0551	0623	0658	0733	0713	0745
-sunset	1707	1749	1825	2002	2037	2102	2058	2022	1929	1835	1653	1642
avg hrs of sunshine	3	4	5	6	7	8	9	8	6	4	3	3
avg # days w/thunder	<1	<1	1	2	3	5	6	5	3	1	1	<1
w/snow	22	17	12	5	1	0	0	0	0	2	8	20

Vancouver

49°11'N, 123°10'W
Elev: 2m (7ft)

It rains in Vancouver almost two out of three winter days, and weeks can pass without sunny spells. As in Seattle, Arctic cold and snow are rare, but fierce when they do strike. Summers compensate with strings of gloriously mild, sunny days, although nights are often cool. Serious heat and thunder occur once or twice a year.

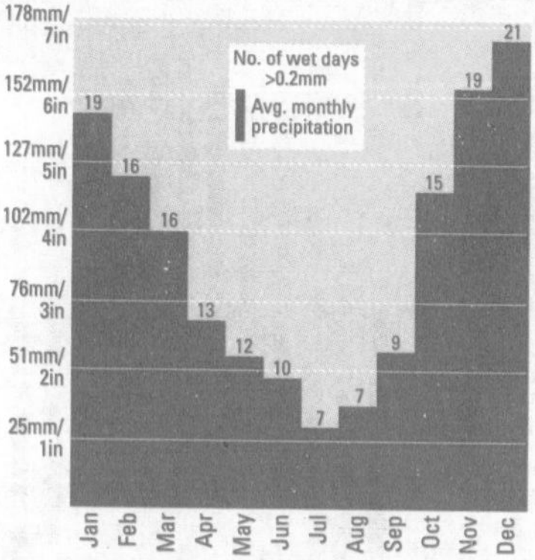

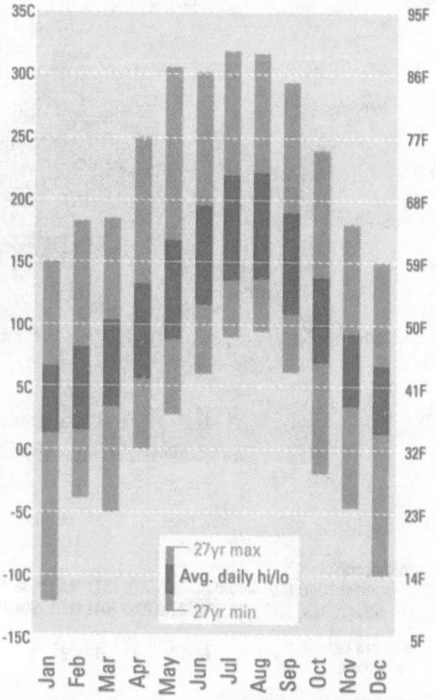

	Jan	Feb	Mar	Apr	May	Jun	Jul	Aug	Sep	Oct	Nov	Dec
midmonth -sunrise	0802	0721	0626	0621	0530	0507	0524	0605	0649	0734	0723	0801
-sunset	1643	1733	1818	2005	2049	2120	2113	2029	1926	1822	1631	1615
avg hrs of sunshine	2	3	4	6	8	8	10	8	6	4	2	2
avg # days w/thunder	<1	<1	<1	<1	<1	1	1	1	1	<1	<1	<1
w/snow	5	3	2	<1	0	0	0	0	0	<1	2	4
Typical % relative humidity @1600hrs	82	77	71	66	63	62	61	63	69	78	80	82

Winnipeg

49°54'N, 97°14'W
Elev: 239m (784ft)

For sheer contrast between winter and summer, this prairie city beats even Minneapolis. Mid-summer is close to ideal, with loads of sunshine; the humidity is seldom high, although gusty storms often roll through. May and September are only a bit cooler and just as bright. Mid-winter brings bursts of snow and windy, finger-freezing cold.

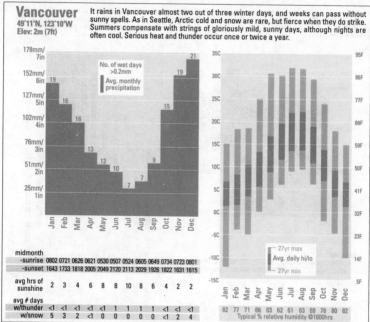

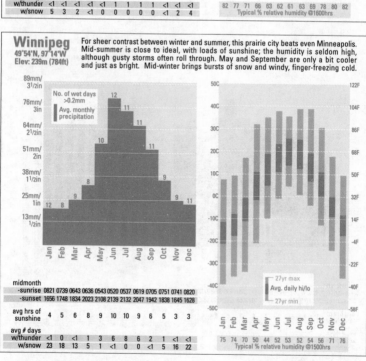

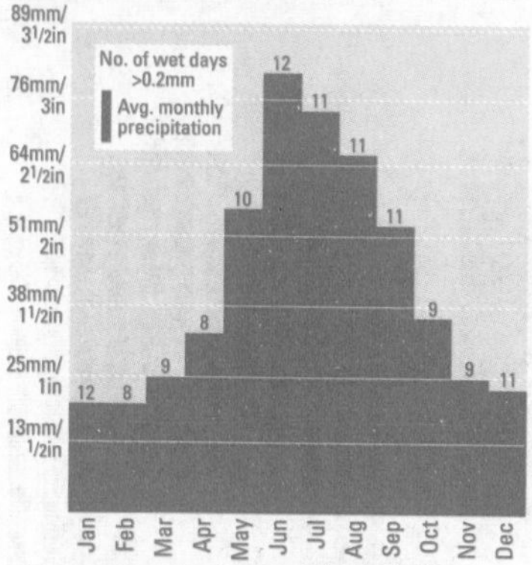

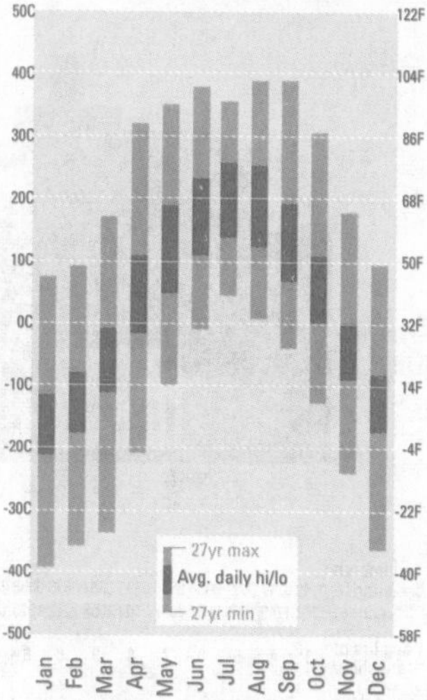

	Jan	Feb	Mar	Apr	May	Jun	Jul	Aug	Sep	Oct	Nov	Dec
midmonth -sunrise	0821	0739	0643	0636	0543	0520	0537	0619	0705	0751	0741	0820
-sunset	1656	1748	1834	2023	2108	2139	2132	2047	1942	1838	1645	1628
avg hrs of sunshine	4	5	6	8	9	10	10	9	6	5	3	3
avg # days w/thunder	<1	0	<1	1	3	6	8	6	2	1	<1	<1
w/snow	23	18	13	5	1	<1	0	0	<1	5	16	22
Typical % relative humidity @1500hrs	75	74	70	50	44	52	53	52	54	56	71	76

© Gregory Thompson/Inclouds.com

In the Swiss Alps, as with other mountain ranges, thick clouds may blanket the lower elevations while the high peaks bask in sunlight

© Weatherwise/Mashahiro Kajikawa

Iridescent clouds paint the sky above Kiruna, Sweden

© Bob Henson

Severe storms rumble across the US prairie

© Bob Henson

Afternoon sunlight can produce a rainbow uncommonly low to the horizon (as shown here), while sunset rainbows arc high in the sky

For more details on cloud types, see the box on p.415

The wispy fragments of cirrus are a familiar sight worldwide

The ragged tops of a stratocumulus sheet bubble above Maui

A setting sun behind cumulus can produce vivid crepuscular rays

Air circulating around California's Mt Shasta produced this eerie cloud

© Kenneth D Langford

Mountains can produce lenticular (lens-shaped) clouds in circular or linear fashion

© Rob Henson

Mammatus loom over the Texas prairie from beneath a storm

© Caspar Ammann

A thin rain shower and its virga produce a sublime Swiss rainbow

© Greg Stumpf

A vast sheet of altocumulus fills the sky at sunset

© Carlye Calvin

A setting sun paints these mountain wave clouds perched above a Colorado lake

© Gregory Thompson/InClouds.com

This slender, slow-moving tornado lurked near the town of Limon, Colorado, for almost half an hour

Mammatus hangs over the southern flank of this severe thunderstorm over the US High Plains

City lights that struck uniformly aligned crystals produced this set of ice pillars in the frigid North Dakota night

Shadows beneath cumulus speckle the South Pacific east of Australia

© Kenneth D Langford

The line between fog and sunlight can be razor-sharp, especially in areas near a coastline

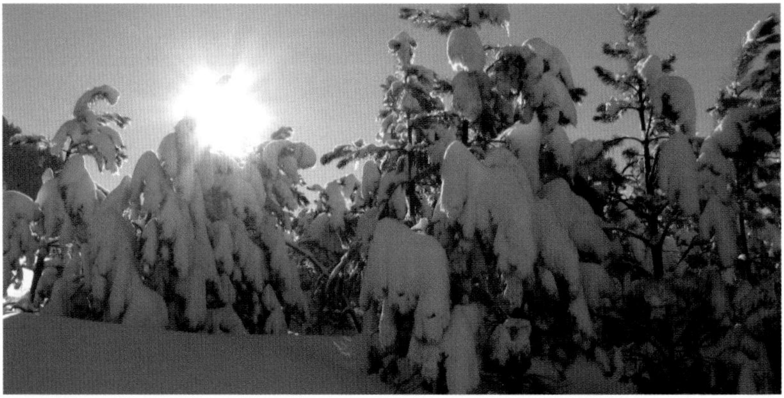

© Bob Henson

Fresh snow glints in the low-hanging sun of a winter afternoon in the Rockies

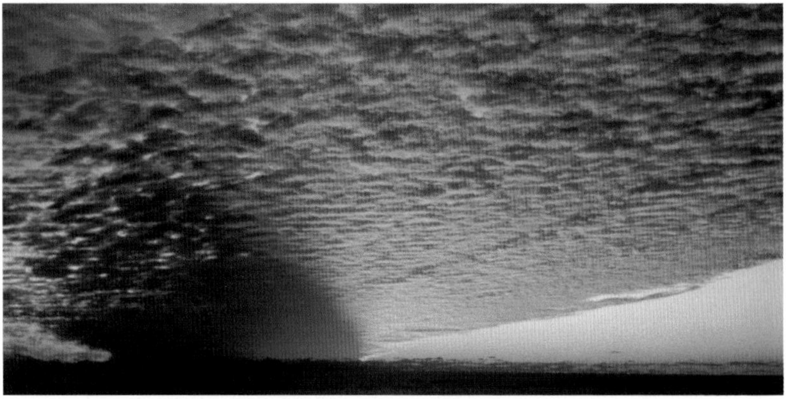

© Weatherwise/James Doughty

This majestic sheet of altocumulus is split by the light of the setting sun off Baja California

© Elva Adams

Not all ominous storm clouds are deluge-producers. This behemoth near Onslow, Australia, produced damaging wind but more dust than rain.

© Carlye Calvin

A double rainbow arcs over the Front Range of Colorado. In between the two bows is the dim belt known as Alexander's dark band.

© Howard B Bluestein

Lightning from a dramatic thunderstorm splits the night sky. Some storms can produce hundreds of cloud-to-ground strikes per hour.

© Gregory Thompson/InClouds.com

Clouds need ice crystals to form a halo, but iridescent clouds like these can be produced through either crystals or water droplets

© Carlye Calvin

A fiery sunrise illuminates altocumulus and cirrostratus clouds. Dust and other airborne particles can shift the sky's hue.

© Kenneth D Langford

When seen from a mountaintop or an aeroplane, a deck of low clouds may produce an "undercast"

Acapulco

16°45'N, 99°45'W
Elev: 5m (16ft)

Few beach towns are as reliably torrid as Acapulco. Even with a sea breeze, most days are tropical by anybody's standard. Temperatures rarely drop below 20°C/68°F even in winter, and 22°C/72°F is a cool night during the extremely sultry summer. It's mostly dry from November–May. Hurricanes usually – but not always – remain offshore.

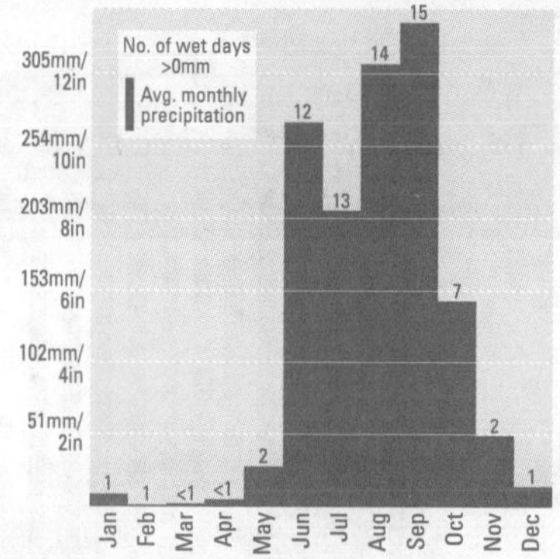

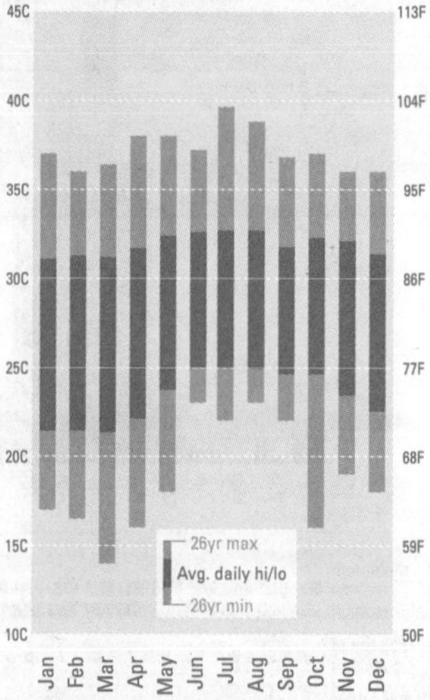

	Jan	Feb	Mar	Apr	May	Jun	Jul	Aug	Sep	Oct	Nov	Dec
Typical % relative humidity @1500hrs	62	62	62	65	65	65	66	66	69	66	64	64

	Jan	Feb	Mar	Apr	May	Jun	Jul	Aug	Sep	Oct	Nov	Dec
midmonth -sunrise	0711	0705	0647	0624	0608	0606	0614	0623	0627	0632	0643	0700
-sunset	1826	1841	1849	1855	1903	1913	1916	1904	1841	1818	1804	1808
avg hrs of sunshine	9	9	9	9	8	7	7	7	6	8	9	8
avg # days w/thunder	<1	<1	<1	<1	2	5	7	7	6	4	1	<1
w/snow	0	0	0	0	0	0	0	0	0	0	0	0

Cancun

21°02'N, 86°53'W
Elev: 5m (16ft)

Balmy air rules for most of the year at Cancun. Cold fronts from North America sometimes make it here – especially in December and January, visitors should be prepared for a coolish, cloudy day or two. The hurricane threat peaks from August through October. The Yúcatan gets drier and less humid toward the northwest.

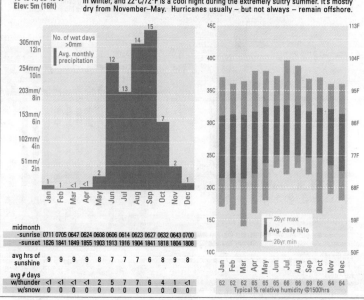

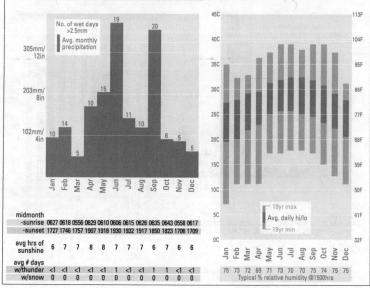

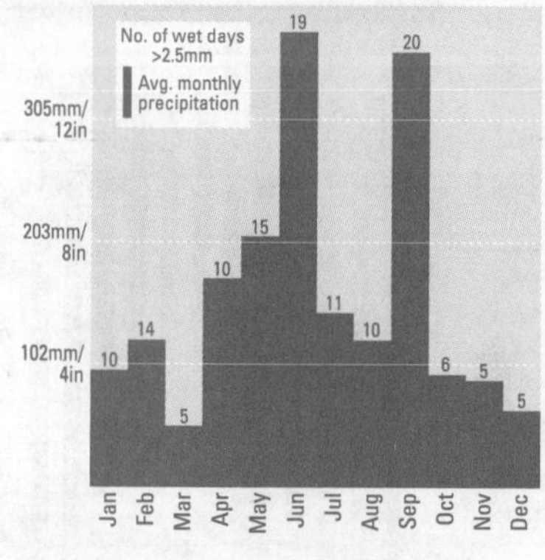

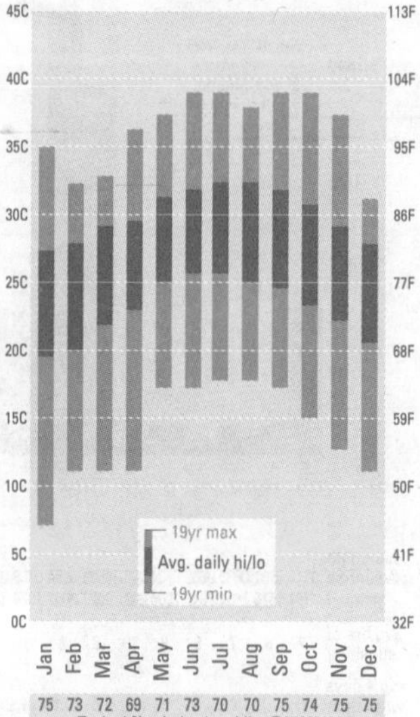

	Jan	Feb	Mar	Apr	May	Jun	Jul	Aug	Sep	Oct	Nov	Dec
Typical % relative humidity @1500hrs	75	73	72	69	71	73	70	70	75	74	75	75

	Jan	Feb	Mar	Apr	May	Jun	Jul	Aug	Sep	Oct	Nov	Dec
midmonth -sunrise	0627	0618	0556	0629	0610	0606	0615	0626	0635	0643	0558	0617
-sunset	1727	1746	1757	1907	1918	1930	1932	1917	1850	1823	1706	1709
avg hrs of sunshine	6	7	7	8	8	7	7	7	6	7	6	6
avg # days w/thunder	<1	<1	<1	<1	<1	1	<1	<1	1	1	<1	<1
w/snow	0	0	0	0	0	0	0	0	0	0	0	0

Guadalajara

20°31'N, 103°19'W
Elev: 1526m (5005ft)

On the fringe of the Central Plateau, Guadalajara is open to more external weather influences than Mexico City. Spring is dry and hot; afternoon and evening rains strike with a vengeance from June–September. The dry winter sees big 24-hour temperature swings; any given day could be toasty or chilly. Last century's one big snow hit in 1997.

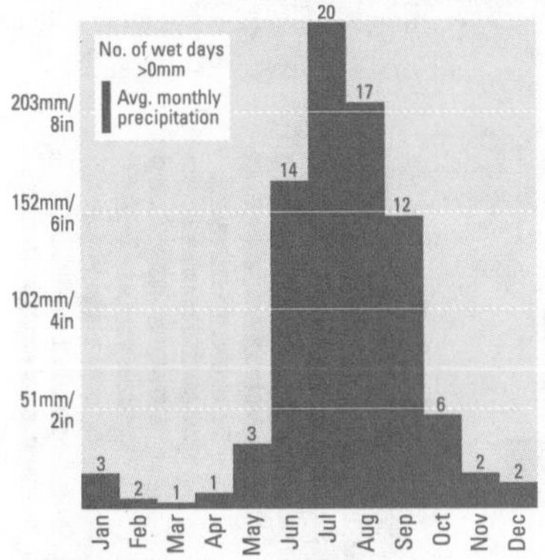

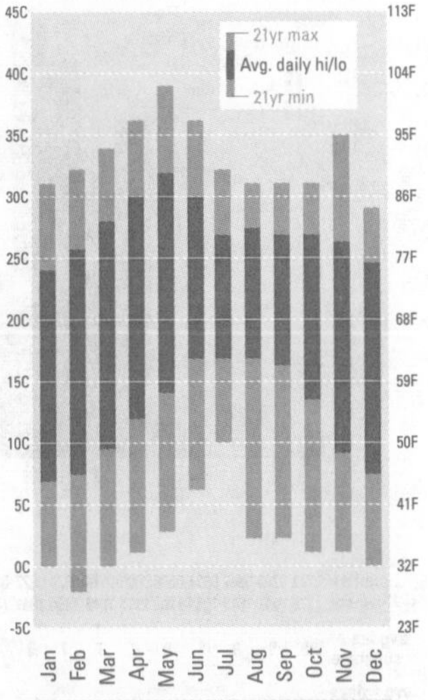

	Jan	Feb	Mar	Apr	May	Jun	Jul	Aug	Sep	Oct	Nov	Dec
midmonth -sunrise	0732	0723	0702	0735	0716	0713	0721	0733	0741	0748	0703	0722
-sunset	1834	1852	1903	2012	2023	2035	2037	2023	1956	1929	1813	1815
avg hrs of sunshine	7	9	9	9	9	7	6	7	6	7	8	6
avg # days w/thunder	<1	<1	<1	1	2	9	14	12	8	3	<1	<1
w/snow	0	0	0	<1	0	0	0	0	0	0	0	0
Typical % relative humidity @1700hrs	36	29	22	21	24	41	58	59	58	48	39	39

Guaymas

27°58'N, 110°56'W
Elev: 27m (89ft)

Guaymas manages to out-swelter Acapulco in summer, when the Gulf of California is at its toastiest. There's a shower every third or fourth day from July–September. The gulf's still warm in autumn, but rains slacken and temperatures drop. Dry mid-winter days are usually warm rather than hot, and nights are cool but never cold.

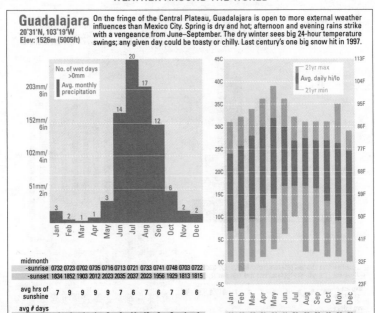

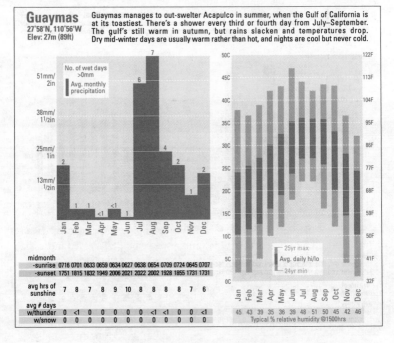

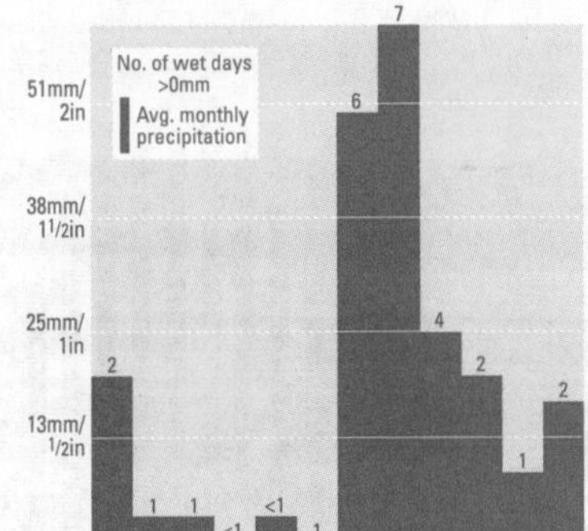

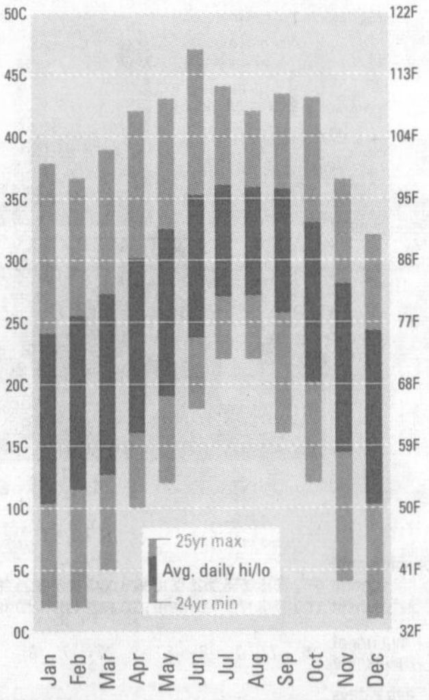

	Jan	Feb	Mar	Apr	May	Jun	Jul	Aug	Sep	Oct	Nov	Dec
midmonth -sunrise	0716	0701	0633	0659	0634	0627	0638	0654	0709	0724	0645	0707
-sunset	1751	1815	1832	1949	2006	2021	2022	2002	1928	1855	1731	1731
avg hrs of sunshine	7	8	7	8	9	10	8	8	8	8	7	6
avg # days w/thunder	0	<1	0	0	0	0	0	<1	<1	0	0	<1
w/snow	0	0	0	0	0	0	0	0	0	0	0	0
Typical % relative humidity @1500hrs	45	43	39	35	36	39	48	51	50	45	42	46

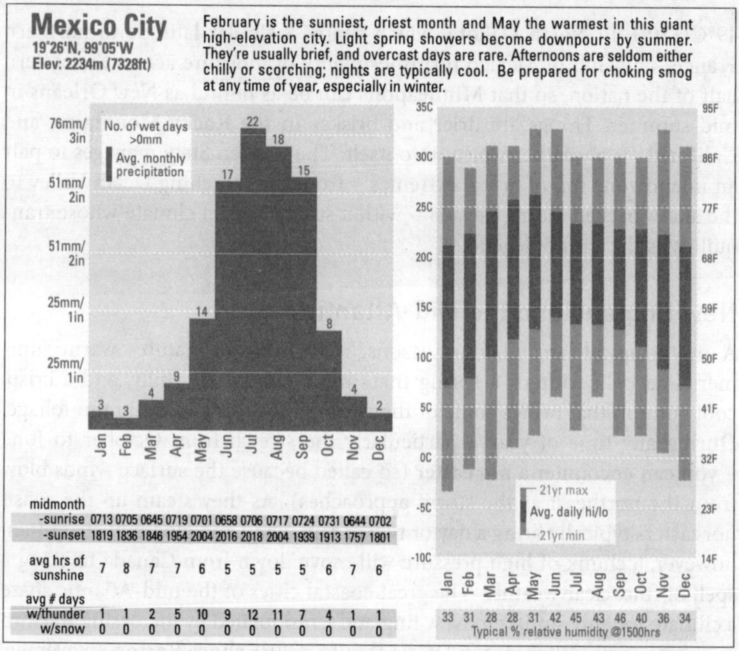

to **Hermosillo** – whereas the western Baja gets most of its scant rain in the winter. Once every few years a hurricane limps into the gulf, seldom causing major damage, except to fishing fleets. Cool waters make the west Baja coast almost hurricane-proof.

The United States

Entire shelves of books have been written about **United States** weather. In part, that's because the country is both sprawling and populous. The nation's agricultural roots and high mobility also play into its obsession with the atmosphere. But the weather itself deserves some credit. Europeans who settled America from east to west were progressively amazed by the spectrum of conditions they encountered. Not many nations can boast a temperature spread of –62°C/–80°F to 57°C/134°F, as well as a place where it rained more than 1000mm/39in in less than a day and a spot where it snowed more than 2500cm/1000in in a single winter.

The contiguous 48 US states are planted firmly in the temperate zone, so you could encounter pleasing temperatures – or uncomfortable cold or heat – almost any place. The people of **Miami Beach** watched snowflakes fly across the surf in January 1977, and temperatures have soared above

49°C/120°F in **North Dakota**, which shares a similar latitude to northern France. The Gulf of Mexico furnishes plenty of moisture across the eastern half of the nation, so that **Minneapolis** can be as humid as **New Orleans** in mid-summer. Things are drier and brisker in the **Rocky Mountains**, and **California** is a weather world unto itself. The Golden State manages to pair an inland zone full of brutal extremes – from the scorching Death Valley to the snow-packed Sierra Nevada – with a sunny coastal climate whose tranquillity is the stuff of legend.

New England and the Mid-Atlantic

A proto-typical land of four seasons, **New England** features warm summers and cold winters, a spring that's often sunny but chilly, and a crisp, cool autumn that brings some of the world's most dramatic autumn foliage. During any time of year – particularly the stretch from October to June – you can encounter a **nor'easter** (so called because the surface winds blow from the northeast as the storm approaches). As they steam up the coast, nor'easters typically bring a day or two of rain, wind and gloom. Just as often, however, a chunk of high pressure will move down from Canada to bring a spell of fine, clear weather. The great coastal cities of the mid-Atlantic share a climate roughly similar to New England's, modulated by increasing warmth as you go south. **Washington DC**, is about 5°C/9°F above **Boston** on average, enough to make its summers almost sub-tropical but its winters noticeably less bitter, with only about half as much snow as in Boston.

East Coasters often refer to typical summer weather with three H's – hazy, hot and humid. The worst heat waves, with temperatures that can top 38°C/100°F, occur as drier west winds descend from the Appalachian mountains that parallel the coast from **Maine** to **Alabama**. Points at elevation are a few degrees cooler than those nearer the shore. Winter's cold can be brutal from upstate New York into northern New England, where lows can drop to –34°C/–30°F or worse. The major coastal cities rarely drop below –18°C/0°F, but in any winter they could experience a blizzard with gale-force winds and over 30cm/12in of snow that shuts down airports and highways for a day or more. Hurricanes and intense tornadoes are very infrequent this far northeast, but either can be devastating. Even near-misses from tropical storms can cause dangerous rip-tides and swells along the coast. All this travail aside, the big northeastern cities of the US get plenty of fine weather, especially in spring and autumn, when temperatures are often at their most moderate and the air at its clearest.

The South

Only mid-winter completely escapes the heat and humidity that define Southern climate. Average highs from **Louisiana** to **Georgia** exceed 26°C/79°F from May into October and sometimes top 35°C/95°F in midsummer. Some relief can be found along the coast, where sea breezes kick in on most hot afternoons, or in **the Appalachians**, which extend as far south as northeast Alabama and north Georgia. The long Southern summer is punctuated by frequent showers and thunderstorms and the occasional tropical storm or hurricane, which can dump enormously heavy rains. Winter brings more variety, with a good helping of bright, clear weather – often pleasingly mild, sometimes bitingly cold – and everything from thunderstorms to snow. (Winter rains here tend to be heavier during El Niño years.) The belt from **Arkansas** to **the Carolinas** is prone to occasional ice storms. Tornadoes occur year-round across the South, but they are most likely in late winter and early spring, and middle-of-the-night twisters are a particular threat. **North Florida** is a climatic extension of the Deep South, while south of **Orlando**, the regime is tropical. Here, freezes are almost nonexistent, and the sun is intense even in winter. Florida's rainfall is focused on near-daily summer downpours with vivid lightning. From June through November, some of Florida's rain falls in tropical cyclones, which strike the state once per year on average (though the state endured a record-smashing seven hurricanes in 2004 and 2005).

The Great Lakes and Midwest

The heartland of America has a continental climate that imports elements from every direction. September through October offer the best chance of comfortable weather: summer's worst heat is gone, but serious chill has yet to arrive. Winters here can get as cold or colder than in the Northeast (often with severe wind chill) but the snowfall tends to fall in quick shots rather than in huge dumps. The exception is along the south and east shores of the **Great Lakes**, where cold air passing over the relatively warm lakes in autumn and early winter can trigger enormous lake-effect snow squalls. These can generate more than 100cm/39in in a day or two along the coast and almost nothing just an hour's drive inland. Warm spells in winter tend to be a bit drier and sunnier here than in the Northeast. Afternoons in late spring and summer can bring intense thunderstorms, especially along the tier just south of the Great Lakes from **Iowa** and **southern Minnesota** to **Ohio**. On the shores of the lakes themselves, summer temperatures are moderated, leading to strings of sunny, warm days broken only on occasion when a heat wave builds in. Mid-summer humidity can be surprisingly high across the agricultural **Midwest**, as the croplands trap and evaporate moisture. Combined with the heat radiating off buildings and pavements, the humidity can keep temperatures in cities like **St Louis** and **Chicago** above 26°C/79°F for several nights on end.

The Great Plains

It was once dubbed the Great American Desert by European settlers, but the land from **North Dakota** to **Texas** is actually a steppe climate – a transition zone between the moist regime further east and the mountains and true deserts to the west. Summers can be scorchingly hot across all of the plains, but the humidity is generally lower as you head northwest. Autumn is the most pleasant season here. Sharp wintertime cold fronts – called "blue northers" in Texas – can push sub-freezing weather from the Canadian border to the Gulf Coast in less than two days. Equally fast warm-ups often sweep in, especially just east of the Rockies, where modified Pacific air can replace Arctic air in a flash: temperatures have been known to jump more than 22°C/40°F in minutes. The **northern Plains** are markedly colder than their counterparts to the south, but the difference in rainfall from east to west is just as important in Great Plains climate. In southeast Texas, **Houston** averages 1200mm/47in of rain a year, while the Rio Grande Valley town of **Del Rio** gets only about half that amount. With no major mountain ranges to steer the air flow, the plains are a playground for low-pressure centres that spin up just east of the Rocky Mountains and draw in air from all quarters. Winds often gust to more than 64kph/40mph across the heart of the plains, and it's a rare day when there is no noticeable breeze at all. In a dry spring, dust storms can tinge the sky copper and even drop a thin layer of grit, an echo of the disastrous Dust Bowl storms of the 1930s. When there's moisture and a front approaching, spectacular thunderstorms and deadly tornadoes may dance across the landscape.

The Rocky Mountains

Temperature extremes – season-to-season and hour-to-hour – are the defining feature of the vast **Rocky Mountains.** Thanks to the altitude and the relative lack of moisture, a day's temperature spread can easily span 20°C/36°F. Much of the interior West falls within the **Great Basin** – an irregular plateau with winter snowfall that's usually fairly light and brief summer storms that can trigger flash floods. Arctic air rarely flows into the basin, but once trapped there, it can produce weeks of frigid smog in cities like **Boise** and **Salt Lake City**. Summers in the Great Basin are gloriously bright and warm by day and cool by night. In the vast deserts and sprawling cities of **Arizona, southern Nevada** and **southeast California**, summers are blisteringly hot and dry, except for the few weeks (usually in July and August) when the summer monsoon – a weaker version of India's – brings down temperatures slightly but adds humidity. Cities astride the east slopes of the Rockies, such as **Denver**, have a more prairie-like climate, with thunderstorms in the summer, frequent cold blasts in the winter and occasional rounds of strong wind.

Without much moisture to temper the transition seasons, autumn and spring can run hot or cold in the Rockies.

The West Coast

In 1981, author Joel Garreau proposed forming a separate state called Ecotopia – the sliver of coastline from near Los Angeles to British Columbia. From a meteorological point of view, it makes perfect sense to split off this green, moist strip from the more arid eastern parts of California, Oregon and Washington. The exception is the skyscraping Sierra Nevada mountain range, which captures enough Pacific moisture for legendary snowfalls. On the Pacific coast is a far different land: a Mediterranean climate zone that seldom gets muggy or frigid. Summers are bone-dry here – it virtually never rains along the **California coast** from late June into September – but the air is softened by marine moisture, and fog is a frequent companion on spring and summer mornings. The world-famous fogs of **San Francisco** result largely from Pacific air flowing through the Golden Gate (the narrow strait for which the famous bridge is named). California's **Central Valley** is a former inland sea with more seasonality than the coast. Across California, winter is the wet season, when hills turn green and days of downpour can be interspersed with spells of sunshine (even in a typical mid-winter, only about one out of three days in San Francisco sees rain). Further north, the winter rains are typically more persistent: after two weeks of solid overcast, weathercasters may announce when "sun breaks" are expected. Summers here are as reliably bright as in LA, although the latitude keeps this area a few degrees cooler. Most winter storms along the West Coast are rather gentle by eastern standards, but there are exceptions. Extended wet periods can cause flooding, high winds may knock out power, and when Arctic air sneaks over the Continental Divide, western Washington can get pummelled by ice or heavy snow all the way to the coast. During **El Niño** winters, the rains usually increase in California while the Pacific Northwest is prone to drought. **La Niña** reverses this picture: **Seattle** and **Portland** often get even more winter rain than usual, while southern California may miss out.

Alaska

Alaska's climate is far more similar to that of its next-door neighbour, Canada, than to that of the lower 48 US states. Summers bring the fabled midnight sun to Alaska, along with daytime temperatures that often exceed 21°C/70°F in the interior. Even in summer, however, most of Alaska dips below 10°C/50°F almost every night, and showers are frequent, with a few thunderstorms thrown in. In winter, cold air masses often build up over a stretch of days in **eastern Alaska** and northwest Canada, as if resting before their plunge south. **Fairbanks**, near the centre of the state, can stay below

The great American thermometer race

Roadside kitsch is a US tradition, and the country's oversized weather extremes are a natural source of material. At least two tornado "museums" sprang up during the 1990s, for instance. On your way to Death Valley, California – the hottest spot in the Americas – it's difficult to miss the world's largest **thermometer display**. Erected next to a restaurant in **Baker, California**, it stands 41m/134ft in honour of Death Valley's all-time high of 57°C/134°F. Lights illuminate the current temperature in steps of 5.6°C/10°F, so the hotter it is, the more you have to crane your neck. Baker's thermometer bested the previous world champion, a display 6.7m/22ft-high located in **International Falls, Minnesota** – the "Ice Box of the Nation". The city takes its self-proclaimed title seriously, with a week-long festival each January that includes a 10km/6.2-mile road race. International Falls is hardly an ice box in summer, when temperatures can soar well above 32°C/90°F. However, the average January high is barely –12°C/10°F, and lows can dip below –40°C/–40°F (although such extreme cold has become more scarce in recent years; in December 2005, only five nights chilled below the –18°C/0°F mark).

–40°C/–40°F for days, and ice fog (exacerbated by smoke and car exhaust fumes) makes matters worse. The **south coast** – an enormous arc that runs from the **Aleutians** to the **Panhandle** – tends to be dry and cool in summer and much milder than the rest of the state in winter. Pacific storms that slam into the Panhandle can bring near-biblical amounts of rain or snow. The bogs and tundra of **northwest Alaska** are unrelentingly cold in the winter and often cloudy and foggy in summer, especially along the coastlines. The air is too cold for much moisture, so it seldom rains or snows heavily. **Barrow**, the state's northernmost settlement, actually qualifies as a desert, receiving only about 1000mm/4in of rain and melted snow a year.

Hawaii

The 50th US state delivers on its tropical promise: trade winds, sunshine and reliable temperatures. There's only a slight warm-up from winter to summer, and the typical range between day and night throughout the year is only about 6°C/10°F. What makes the difference in Hawaiian weather is location. Because of the persistent northeast trades, the **eastern sides** of the islands experience heavy rainfall – **Hilo**, on the Big Island, averages 3300mm/130in a year – while the west shores, such as the Big Island's **Kona Coast**, are perpetually parched. Even in Hilo, it doesn't rain constantly: showers are actually focused during the evening and overnight hours. (Emissions from the Big Island's two volcanoes sometimes mix with fog to produce a hazy compound that locals call "vog".) **Honolulu**, on the southeast coast of Oahu, has a nearly rain-free summer and a winter with just enough rain to keep things green. **El Niño** tends to make winter and spring in Hawaii drier than usual. Although the waters surrounding Hawaii are a little too cool for hurricanes to thrive, one manages to straggle through every few years.

Anchorage

61°10'N, 150°01'W
Elev: 35m (114ft)

Though tempered somewhat by the nearby Pacific, Anchorage still manages some impressive temperature swings. Mid-winter may bring a string of above-freezing days or a spell of severe cold; summers average cool but can turn distinctly warm or chilly. Winter snows and summer rains are seldom heavy, and the sun breaks through on many days.

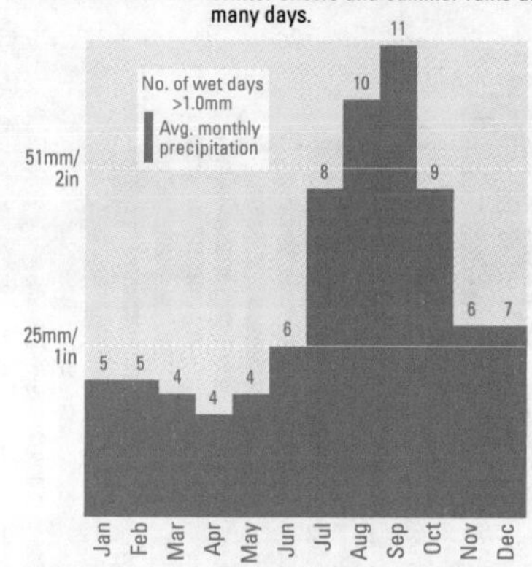

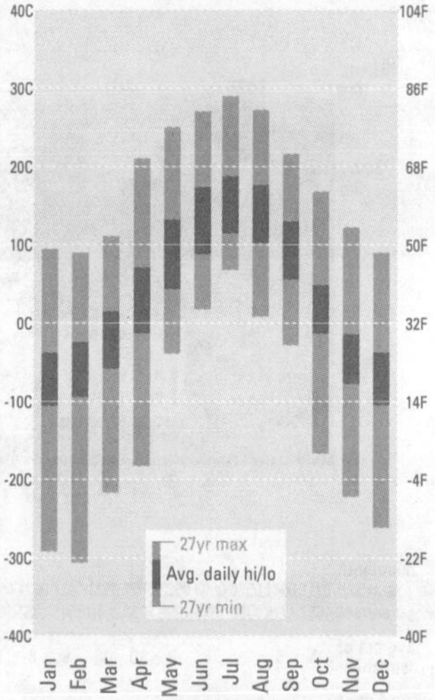

	Jan	Feb	Mar	Apr	May	Jun	Jul	Aug	Sep	Oct	Nov	Dec
midmonth -sunrise	0957	0842	0717	0639	0513	0422	0452	0609	0727	0843	0906	1009
-sunset	1623	1747	1902	2123	2242	2340	2318	2158	2022	1848	1623	1541
avg hrs of sunshine	3	4	6	8	9	9	8	7	5	4	3	2
avg # days w/thunder	0	0	0	0	<1	<1	<1	<1	<1	0	0	0
w/snow	17	14	13	9	1	0	0	0	1	9	15	19

	Jan	Feb	Mar	Apr	May	Jun	Jul	Aug	Sep	Oct	Nov	Dec
Typical % relative humidity @1500hrs	73	67	57	54	50	55	62	64	64	67	74	76

Atlanta

33°39'N, 84°26'W
Elev: 308m (1010ft)

The highest of the ten largest US cities, Atlanta sits just above the South's worst summer heat. It's still very warm and humid, with frequent thunder. Sun and rain alternate year round. Autumn is the driest time; winter can be damp and chilly. Tornadoes may occur any time of year, the worst from winter into spring.

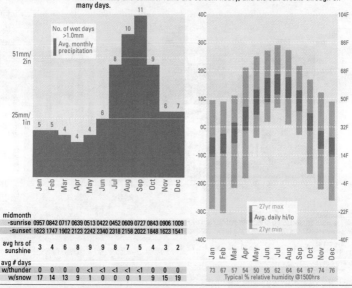

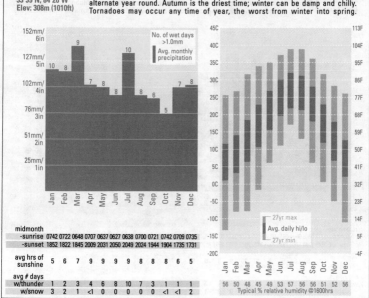

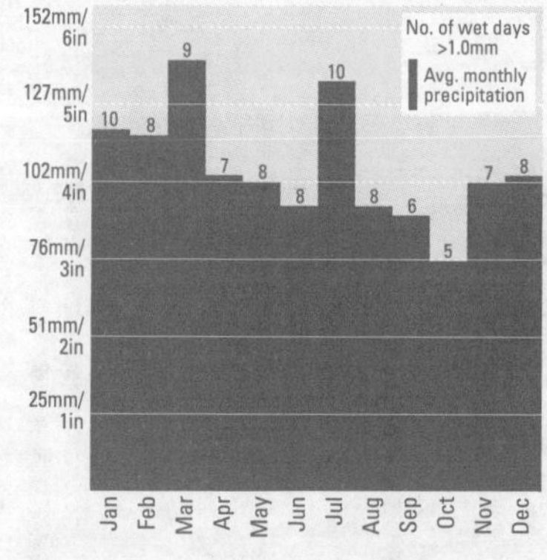

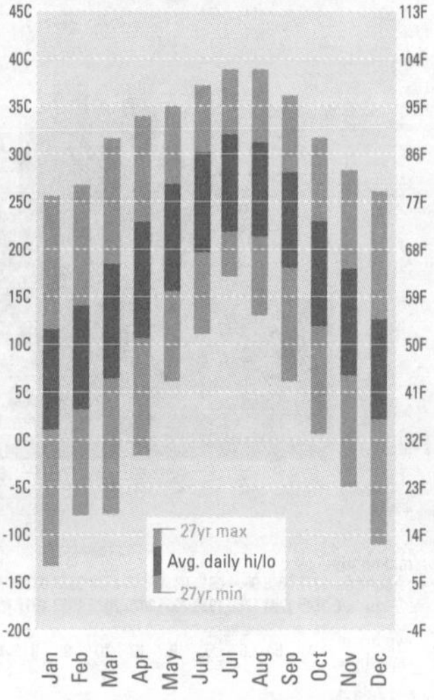

	Jan	Feb	Mar	Apr	May	Jun	Jul	Aug	Sep	Oct	Nov	Dec
midmonth -sunrise	0742	0722	0648	0707	0637	0627	0638	0700	0721	0742	0709	0735
-sunset	1852	1822	1845	2009	2031	2050	2049	2024	1944	1904	1735	1731
avg hrs of sunshine	5	6	7	9	9	9	9	8	8	8	6	5
avg # days w/thunder	1	2	3	4	6	8	10	7	3	1	1	1
w/snow	3	2	1	<1	0	0	0	0	0	<1	<1	2

	Jan	Feb	Mar	Apr	May	Jun	Jul	Aug	Sep	Oct	Nov	Dec
Typical % relative humidity @1600hrs	56	50	48	45	49	53	57	56	56	51	52	56

Boston

42°22'N, 71°02'W
Elev: 6m (20ft)

This city's four-season climate includes many glorious autumn days followed by windy winter spells of cold rain or snow. Every few years a blizzard dumps over 30cms/12 inches. Cool Atlantic waters can postpone spring for weeks and take the edge off summer heat, although west winds sometimes usher in a batch of sultry air.

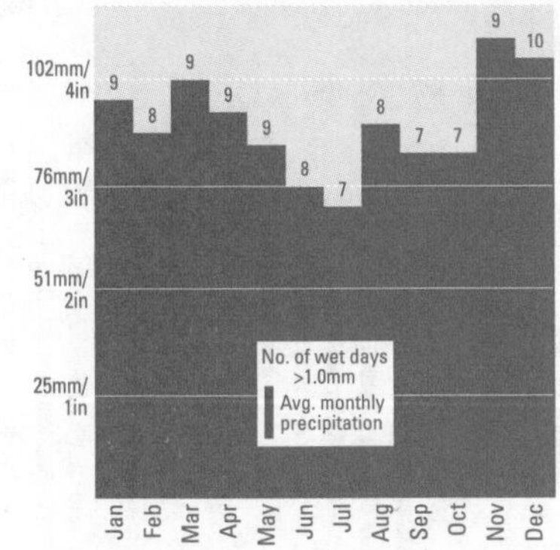

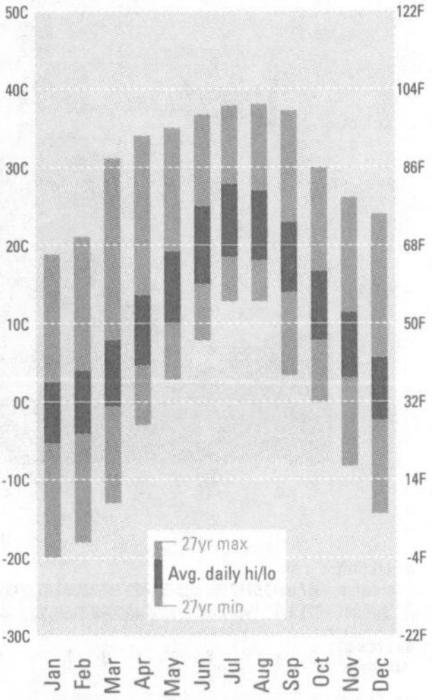

	Jan	Feb	Mar	Apr	May	Jun	Jul	Aug	Sep	Oct	Nov	Dec
midmonth -sunrise	0711	0641	0556	0603	0523	0507	0521	0551	0624	0657	0635	0706
-sunset	1637	1716	1750	1926	1959	2023	2019	1945	1854	1802	1622	1612
avg hrs of sunshine	5	6	7	8	9	10	10	9	8	7	5	5
avg # days w/thunder	<1	<1	1	1	2	3	4	3	1	1	<1	<1
w/snow	13	11	9	3	<1	0	0	0	0	<1	3	11
Typical % relative humidity @1600hrs	58	56	57	56	58	58	57	60	61	59	61	60

Chicago

41°59'N, 87°54'W
Elev: 205m (674ft)

In true Midwestern fashion, Chicago offers roller-coaster weather, as daily temperatures bounce above and below average. Lake breezes help cool off summer heat; a few nights may hang above 26°C/79°F. Once winter sets in, Chicago can be sharply colder – or warmer – than Moscow, as wind-whipped snow alternates with mild spells.

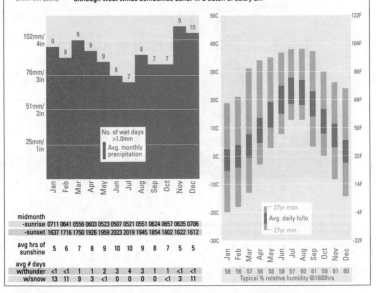

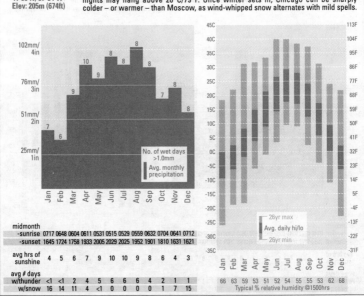

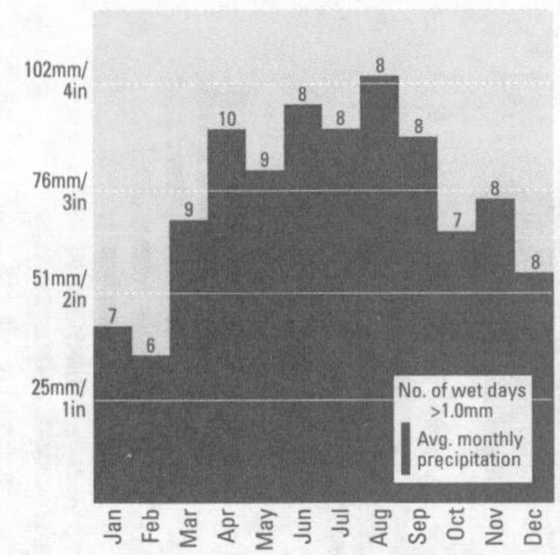

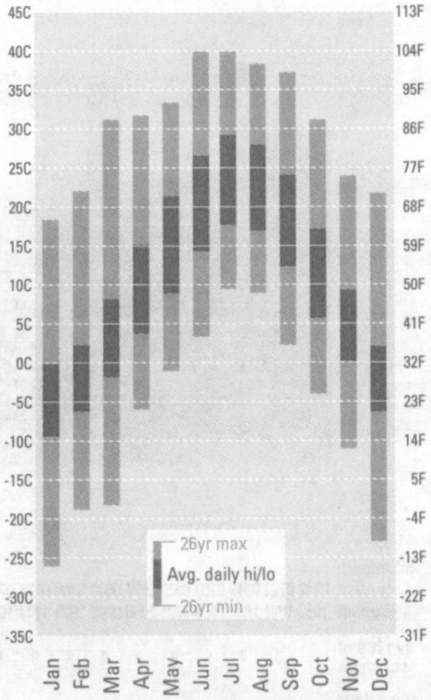

	Jan	Feb	Mar	Apr	May	Jun	Jul	Aug	Sep	Oct	Nov	Dec
midmonth -sunrise	0717	0648	0604	0611	0531	0515	0529	0559	0632	0704	0641	0712
-sunset	1645	1724	1758	1933	2005	2029	2025	1952	1901	1810	1631	1621
avg hrs of sunshine	4	5	6	7	9	10	10	9	8	6	4	3
avg # days w/thunder	<1	<1	2	4	5	6	6	6	4	2	1	1
w/snow	16	14	11	4	<1	0	0	0	0	1	7	15
Typical % relative humidity @1500hrs	66	63	59	53	51	52	54	55	55	53	62	68

Dallas

32°44'N, 96°58'W
Elev: 151m (495ft)

Dallas resembles an oven from July into September, as highs can top 38°C/100°F for days on end. Spring brings strong winds and violent thunderstorms; tornadoes are a serious threat here. Winter is variable but often bright and mild; snow or freezing rain may coat highways now and again. Autumn is the least windy, most temperate season.

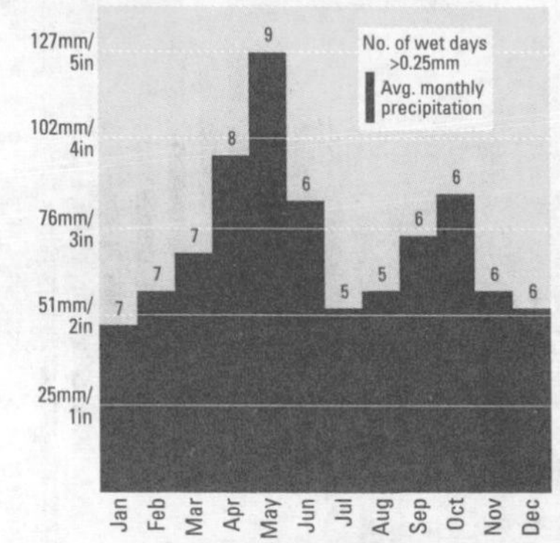

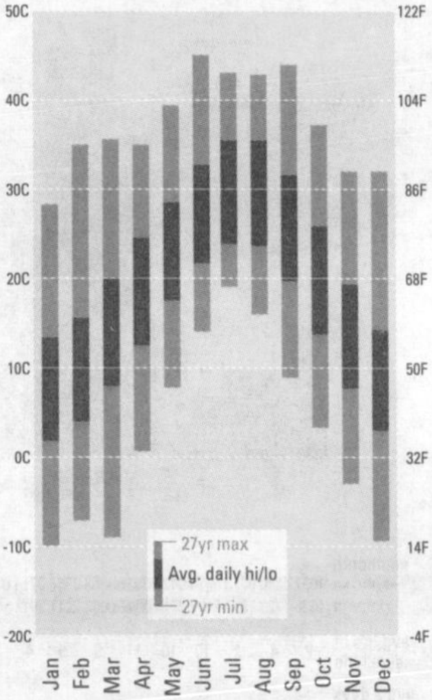

	Jan	Feb	Mar	Apr	May	Jun	Jul	Aug	Sep	Oct	Nov	Dec
Typical % relative humidity @1500hrs	53	51	47	49	51	48	43	41	46	46	48	51

	Jan	Feb	Mar	Apr	May	Jun	Jul	Aug	Sep	Oct	Nov	Dec
midmonth -sunrise	0730	0711	0638	0658	0629	0620	0631	0651	0712	0732	0658	0723
-sunset	1745	1813	1836	1958	2020	2037	2037	2013	1934	1855	1727	1723
avg hrs of sunshine	6	6	7	8	8	10	11	10	8	7	6	6
avg # days w/thunder	1	2	5	7	8	6	5	5	4	3	2	1
w/snow	4	2	1	<1	0	0	0	0	0	<1	<1	1

Denver

39°46'N, 104°52'W
Elev: 1612m (5286ft)

Temperatures at Denver's altitude can soar by day and plummet by night. Summers feature hot days and mild nights, with afternoon storms frequent (some bearing large hail). Winter days can top 18°C/64°F in between prolonged stretches below freezing; snows tend to melt quickly but can be heavy. Spring is often damp and autumn usually dry.

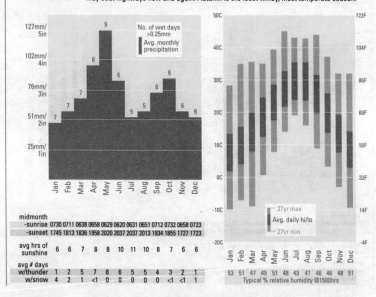

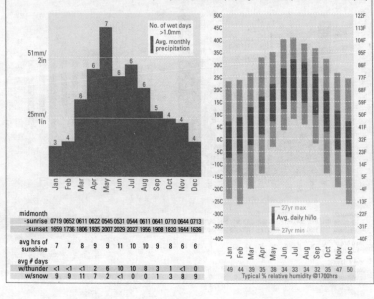

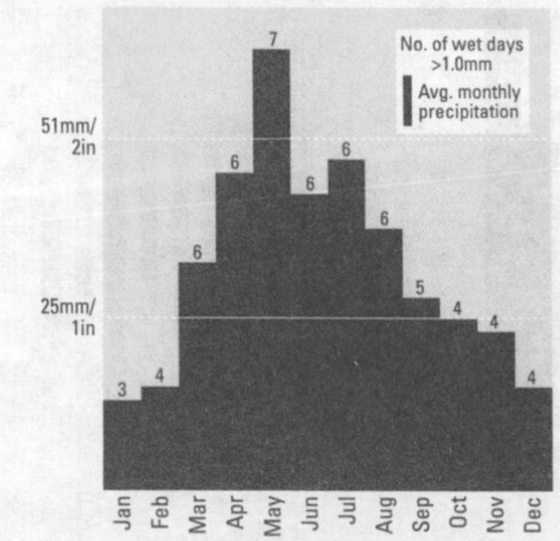

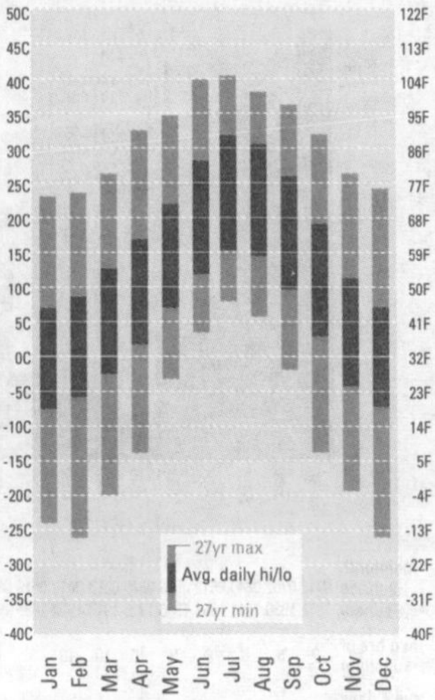

	Jan	Feb	Mar	Apr	May	Jun	Jul	Aug	Sep	Oct	Nov	Dec
Typical % relative humidity @1700hrs	49	44	39	35	38	34	33	34	32	35	47	50

	Jan	Feb	Mar	Apr	May	Jun	Jul	Aug	Sep	Oct	Nov	Dec
midmonth -sunrise	0719	0652	0611	0622	0545	0531	0544	0611	0641	0710	0644	0713
-sunset	1659	1736	1806	1935	2007	2029	2027	1956	1908	1820	1644	1636
avg hrs of sunshine	7	7	8	9	9	11	10	10	9	8	6	6
avg # days w/thunder	<1	<1	<1	2	6	10	10	8	3	1	<1	0
w/snow	9	9	11	7	2	<1	0	0	1	3	8	9

Fairbanks

64°49'N, 147°52'W
Elev: 133m (436ft)

This ultra-continental climate can be hard to dress for – especially in spring and autumn, when you may need a T-shirt and parka in the same week. Fairbanks' most brutal winter cold, complete with ice fog, can last for days, but summer has long stretches of sun and mildness.

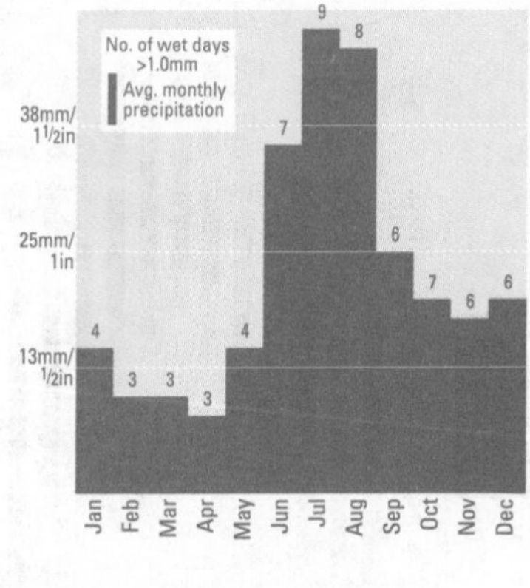

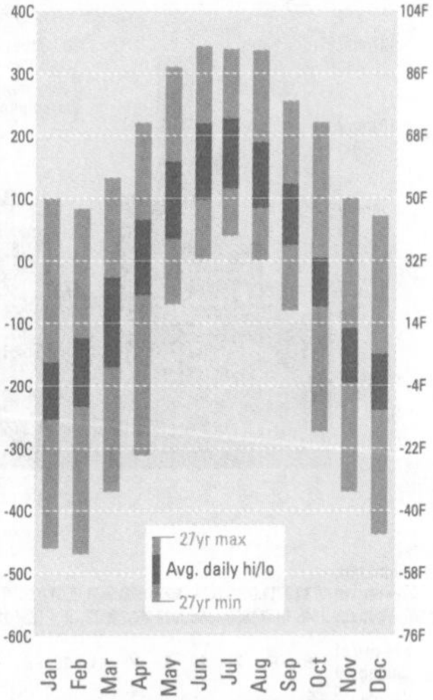

	Jan	Feb	Mar	Apr	May	Jun	Jul	Aug	Sep	Oct	Nov	Dec
midmonth -sunrise	1027	0850	0710	0616	0429	0302	0353	0538	0714	0844	0927	1052
-sunset	1536	1723	1852	2129	2310	0042	0002	2211	2018	1829	1544	1441
avg hrs of sunshine	2	4	7	10	10	11	9	5	4	3	2	1
avg # days w/thunder	1	<1	0	<1	<1	3	3	1	<1	<1	<1	<1
w/snow	16	14	12	9	2	0	0	<1	4	17	18	18

Jan	Feb	Mar	Apr	May	Jun	Jul	Aug	Sep	Oct	Nov	Dec
70	64	54	46	38	44	51	55	56	69	74	72

Typical % relative humidity @1500hrs

Honolulu

21°20'N, 157°55'W
Elev: 2m (7ft)

Weather rarely interferes with life in this tropical town. It's dependably warm by day and mild at night, with temperatures a touch lower in winter than in summer. Expect a few showers in the cooler months, although sunshine is plentiful. Island rainfall is much heavier toward the east and lighter to the west.

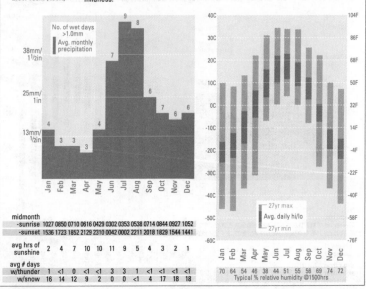

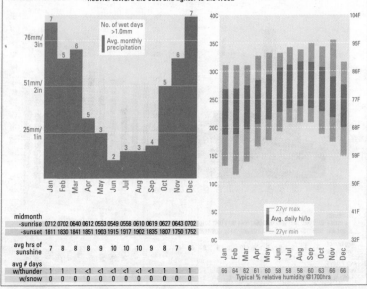

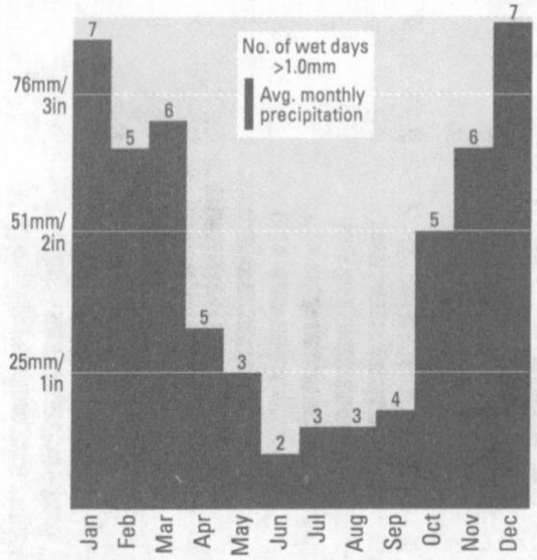

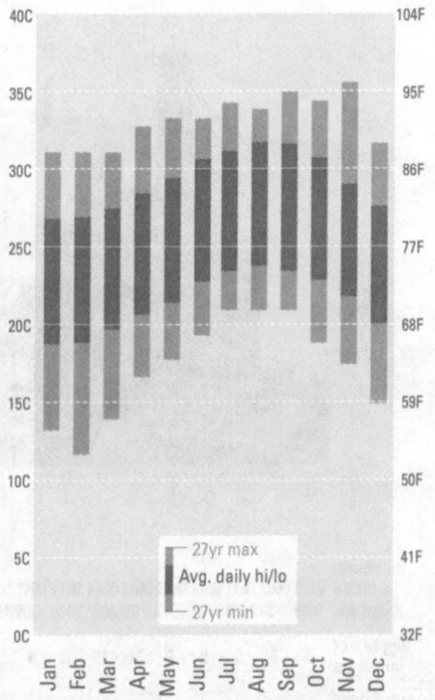

	Jan	Feb	Mar	Apr	May	Jun	Jul	Aug	Sep	Oct	Nov	Dec
midmonth -sunrise	0712	0702	0640	0612	0553	0549	0558	0610	0619	0627	0643	0702
-sunset	1811	1830	1841	1851	1903	1915	1917	1902	1835	1807	1750	1752
avg hrs of sunshine	7	8	8	8	9	10	10	10	9	8	7	6
avg # days w/thunder	1	1	1	<1	<1	<1	<1	<1	<1	1	1	1
w/snow	0	0	0	0	0	0	0	0	0	0	0	0

Jan	Feb	Mar	Apr	May	Jun	Jul	Aug	Sep	Oct	Nov	Dec
66	64	62	61	60	58	58	58	60	63	66	66

Typical % relative humidity @1700hrs

Las Vegas

36°05'N, 115°10'W
Elev: 659m (2162ft)

The summer heat in Vegas rivals that of Phoenix, although the monsoon humidity and storms are a bit less prevalent here. In mid-winter, Las Vegas tends to be noticeably cooler than Phoenix. Expect a few windy, cool days and nippy nights; even a dusting of snow is possible. Sunshine predominates year round.

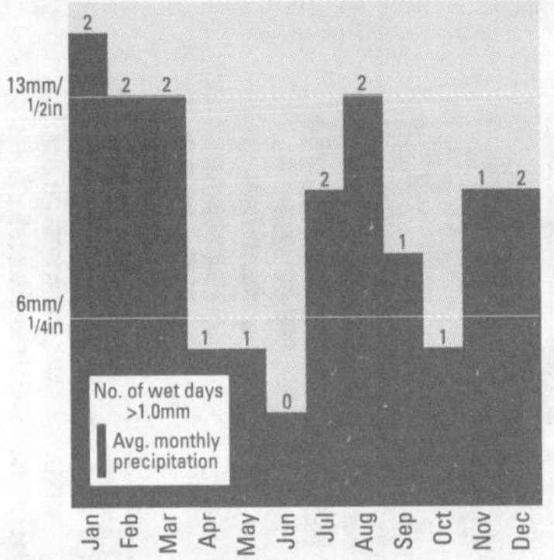

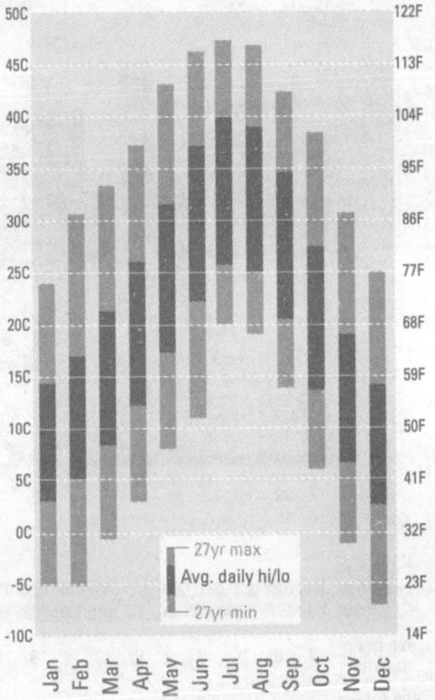

	Jan	Feb	Mar	Apr	May	Jun	Jul	Aug	Sep	Oct	Nov	Dec
midmonth -sunrise	0751	0728	0652	0708	0635	0623	0635	0659	0726	0748	0717	0744
-sunset	1750	1822	1848	2014	2040	2059	2058	2031	1948	1905	1733	1728
avg hrs of sunshine	8	9	10	12	13	13	13	12	11	10	8	8
avg # days w/thunder	<1	<1	<1	<1	1	1	4	4	2	<1	<1	<1
w/snow	1	<1	<1	<1	0	0	0	0	0	0	<1	1
Typical % relative humidity @1600hrs	32	26	21	15	13	10	14	16	15	18	26	31

Los Angeles

33°56'N, 118°23'W
Elev: 30m (100ft)

Temperatures vary widely across town in LA's fabled summer sun. Beaches may be 25°C/77°F even as valleys to the north and east top 38°C/100°F. Winter rains can trigger flash floods, especially during El Niño, with picture-perfect weather after the skies clear. Fierce Santa Ana winds bring hot, dry weather a few days each year. Smog is worst in the summer.

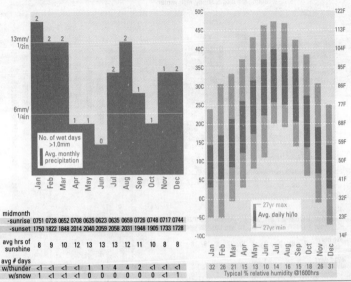

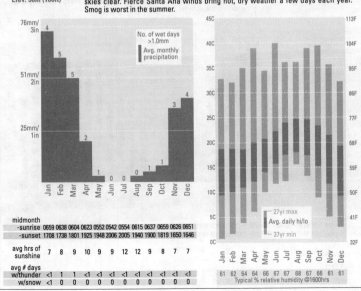

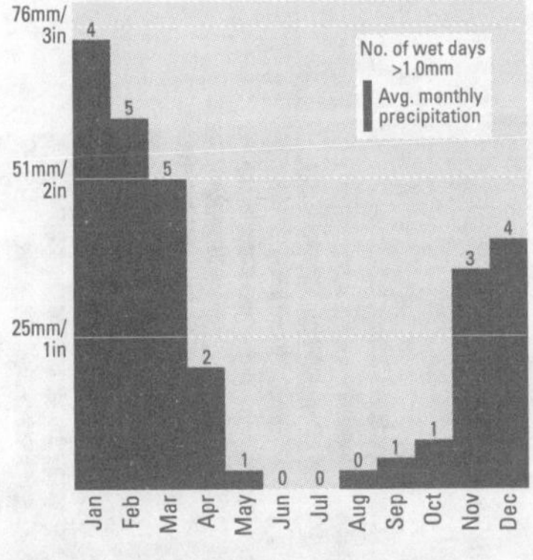

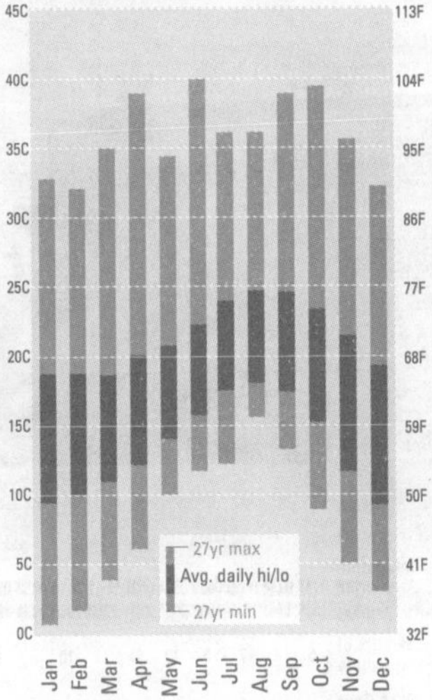

	Jan	Feb	Mar	Apr	May	Jun	Jul	Aug	Sep	Oct	Nov	Dec
midmonth -sunrise	0659	0638	0604	0623	0552	0542	0554	0615	0637	0659	0626	0651
-sunset	1708	1738	1801	1925	1948	2006	2005	1940	1900	1819	1650	1646
avg hrs of sunshine	7	8	9	10	9	9	12	12	9	8	7	7
avg # days w/thunder	<1	1	1	<1	<1	<1	<1	<1	<1	<1	<1	<1
w/snow	<1	0	0	0	0	0	0	0	0	0	0	0
Typical % relative humidity @1600hrs	61	62	64	64	66	67	67	68	67	66	61	61

Miami

25°48'N, 80°18'W
Elev: 4m (12ft)

Miami combines ample sunshine with twice the rainfall of western Europe. Most of it falls in afternoon torrents during the sultry stretch between May and October. Tropical systems, including hurricanes (a real threat here), add to the rain from August onward. Only a few chilly spells and wet periods occur during the mostly sunny, mild winter.

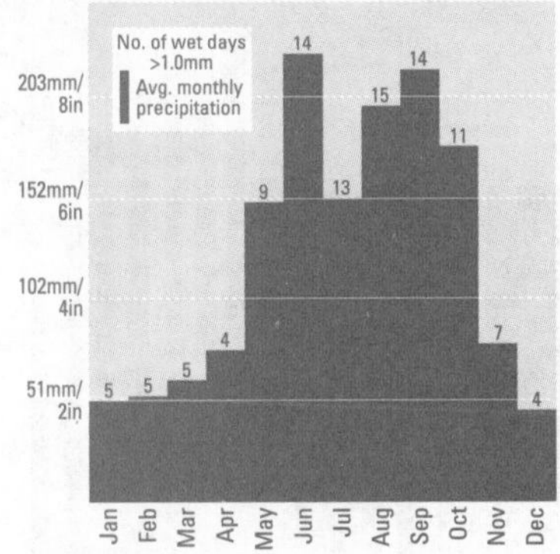

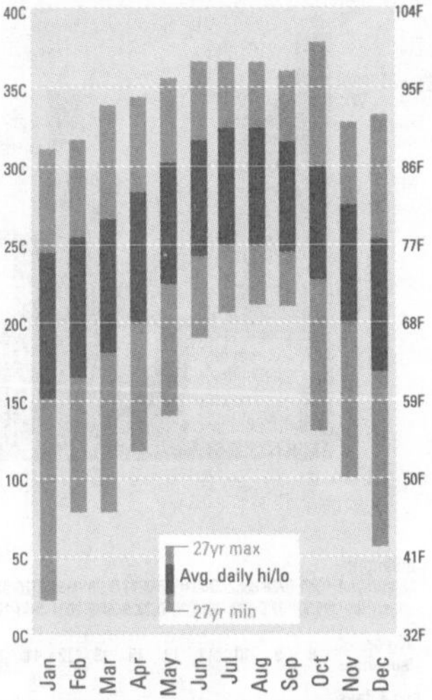

	Jan	Feb	Mar	Apr	May	Jun	Jul	Aug	Sep	Oct	Nov	Dec
midmonth -sunrise	0710	0656	0631	0658	0636	0629	0639	0654	0707	0720	0639	0700
-sunset	1752	1815	1830	1944	2000	2014	2015	1957	1925	1854	1732	1733
avg hrs of sunshine	7	8	9	10	10	10	10	9	9	8	7	7
avg # days w/thunder	1	1	2	3	6	12	15	16	12	5	1	1
w/snow	0	0	0	0	0	0	0	0	0	0	0	0

Jan	Feb	Mar	Apr	May	Jun	Jul	Aug	Sep	Oct	Nov	Dec
60	58	57	57	61	68	66	67	69	65	63	60

Typical % relative humidity @1600hrs

Minneapolis

44°53'N, 93°13'W
Elev: 254m (834ft)

Few cities outside of eastern Asia have more dramatic continental climates. Minneapolis is often the coldest big US city, sometimes failing to rise above –18°C/0°F for days; people navigate downtown through enclosed walkways. Spring and autumn are quite varied. Aside from spells of intense humidity and strong thunderstorms, summer can be a treat.

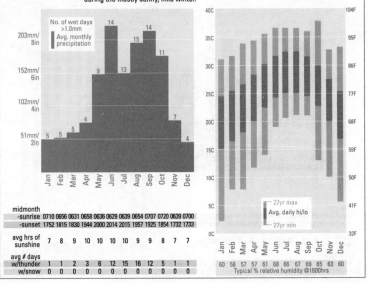

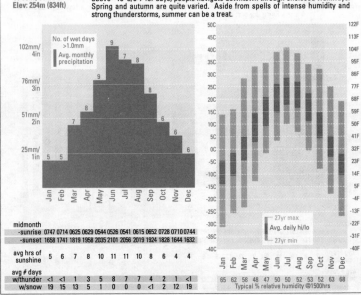

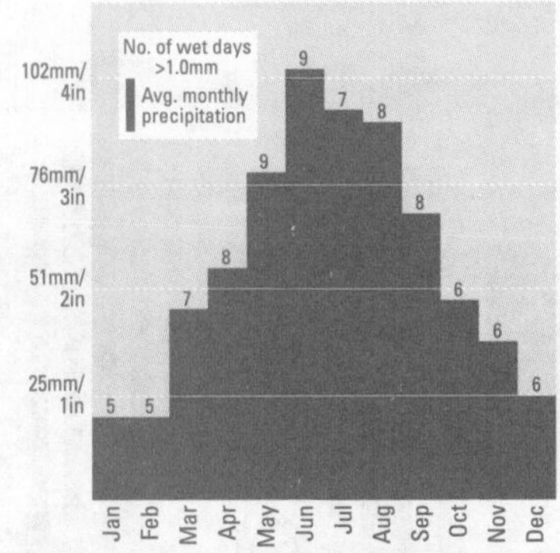

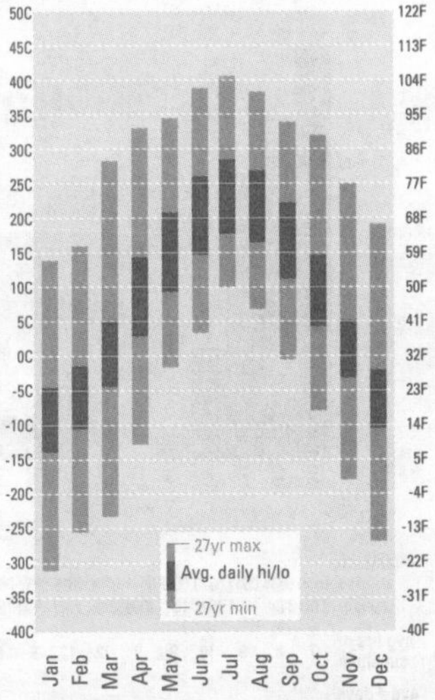

	Jan	Feb	Mar	Apr	May	Jun	Jul	Aug	Sep	Oct	Nov	Dec
midmonth -sunrise	0747	0714	0625	0629	0544	0526	0541	0615	0652	0728	0710	0744
-sunset	1658	1741	1819	1958	2035	2101	2056	2019	1924	1828	1644	1632
avg hrs of sunshine	5	6	7	8	10	11	11	10	8	6	4	4
avg # days w/thunder	<1	<1	1	3	5	8	7	7	4	2	1	<1
w/snow	19	15	13	5	1	0	0	0	<1	2	12	19

Jan	Feb	Mar	Apr	May	Jun	Jul	Aug	Sep	Oct	Nov	Dec
65	62	58	48	47	50	50	52	53	52	63	68

Typical % relative humidity @1500hrs

New Orleans

29°59'N, 90°15'W
Elev: 1m (4ft)

Humidity rules this sub-tropical town. Summer nights seldom dip below 21°C/70°F. Early spring and late autumn are mild, bright, and often less humid. A damp winter day can feel quite raw; Mardi Gras weather can range from clammy to gorgeous. The hurricane threat is greatest from August–October. (As Katrina made clear, much of New Orleans sits below sea level.)

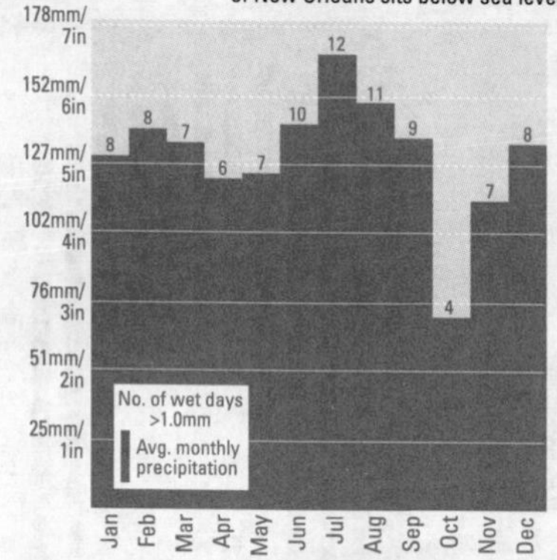

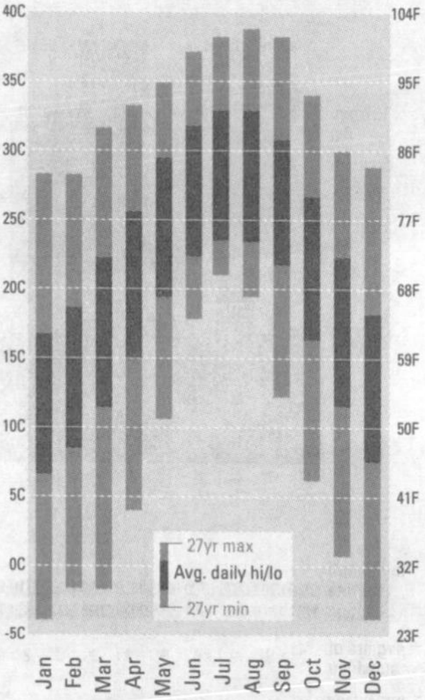

	Jan	Feb	Mar	Apr	May	Jun	Jul	Aug	Sep	Oct	Nov	Dec
midmonth -sunrise	0658	0641	0611	0634	0608	0600	0610	0628	0646	0703	0626	0649
-sunset	1724	1750	1809	1928	1947	2003	2004	1942	1906	1830	1705	1703
avg hrs of sunshine	5	6	7	8	9	9	8	8	8	8	6	5
avg # days w/thunder	2	3	4	5	6	10	16	13	7	2	2	2
w/snow	<1	<1	<1	0	0	0	0	0	0	0	0	<1
Typical % relative humidity @1500hrs	63	59	57	57	58	61	66	65	63	56	59	62

New York

40°47'N, 73°58'W
Elev: 40m (131ft)

New York's climate is notably diverse. Summer days are typically hazy, hot and humid, with occasional thunderstorms providing a break. Autumn and spring are variable but often beautiful. Cold winter winds whistle through Manhattan, and snow can pile up to more than 30cm/12 inches. Many snowfalls turn into rainstorms.

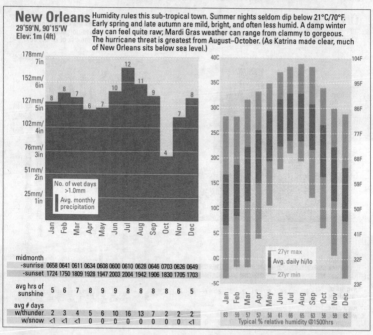

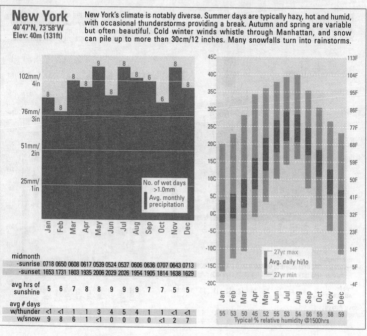

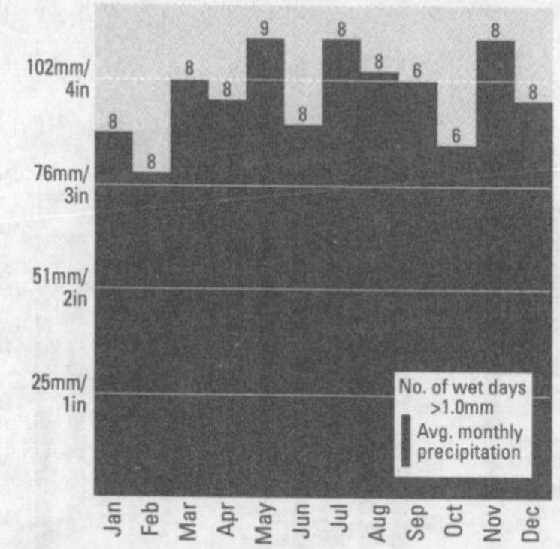

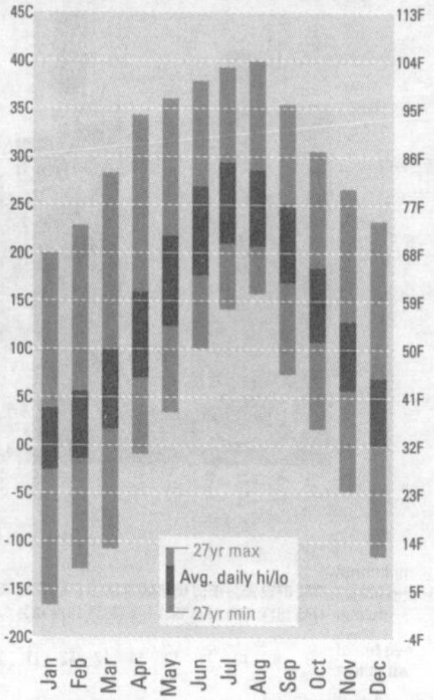

	Jan	Feb	Mar	Apr	May	Jun	Jul	Aug	Sep	Oct	Nov	Dec
midmonth -sunrise	0718	0650	0608	0617	0539	0524	0537	0606	0636	0707	0643	0713
-sunset	1653	1731	1803	1935	2006	2029	2026	1954	1905	1814	1638	1629
avg hrs of sunshine	5	6	7	8	8	9	9	9	7	7	5	5
avg # days w/thunder	<1	<1	1	1	3	4	5	4	1	1	<1	<1
w/snow	9	8	6	1	<1	0	0	0	0	<1	2	7
Typical % relative humidity @1500hrs	55	53	50	45	52	55	53	54	56	55	58	59

Orlando

28°26'N, 81°20'W
Elev: 28m (91ft)

This city's theme parks are drenched by more than sun: showers and storms pass through on nearly half of summer afternoons (central Florida leads the US in lightning). Mid-winter brings a few hard freezes and damp days but otherwise gorgeous weather. Early spring and mid-autumn escape the worst summer heat and the risk of all-out cold.

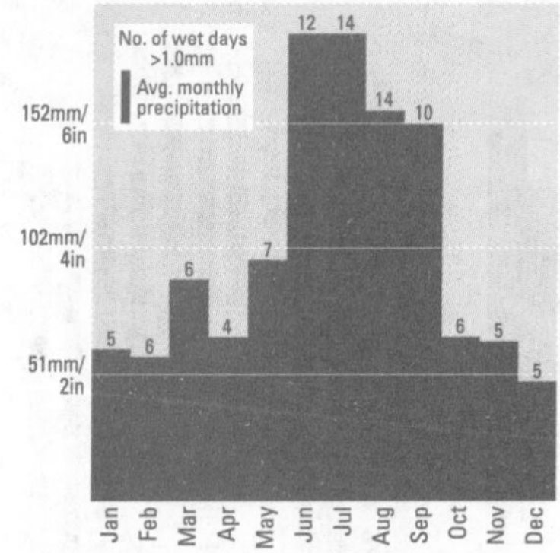

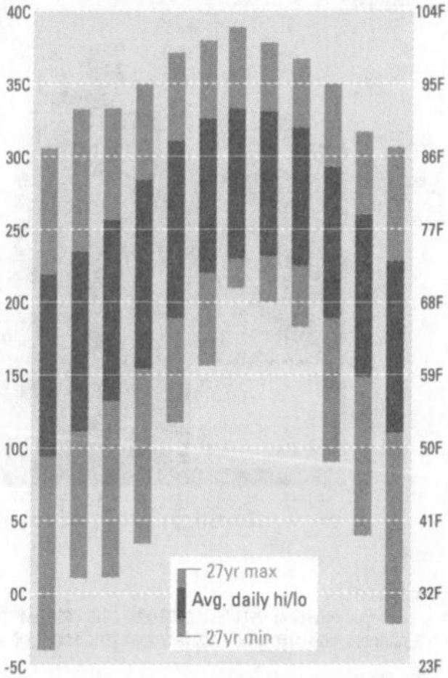

	Jan	Feb	Mar	Apr	May	Jun	Jul	Aug	Sep	Oct	Nov	Dec
midmonth -sunrise	0719	0703	0635	0700	0635	0628	0638	0655	0710	0726	0648	0710
-sunset	1751	1816	1834	1951	2009	2024	2025	2004	1930	1856	1732	1731
avg hrs of sunshine	7	8	9	9	9	9	9	9	8	7	7	7
avg # days w/thunder	1	2	3	3	8	16	19	19	11	3	1	1
w/snow	<1	0	0	0	0	0	0	0	0	0	0	0
Typical % relative humidity @1600hrs	54	50	48	47	541	61	64	66	65	59	58	58

Phoenix

33°26'N, 112°01'W
Elev: 338m (1110ft)

Autumn and early spring are good times to avoid Phoenix's heat, the most intense of any big city in the Americas. Summer begins scorching and dry; by July, lightning-laced storms and uncomfortable humidity arrive. The desert nights turn quite cool in winter (frosts are possible), but only a few breezy, cloudy fronts interrupt the sunshine.

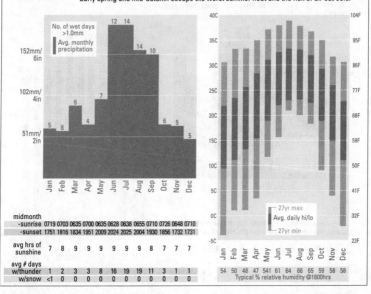

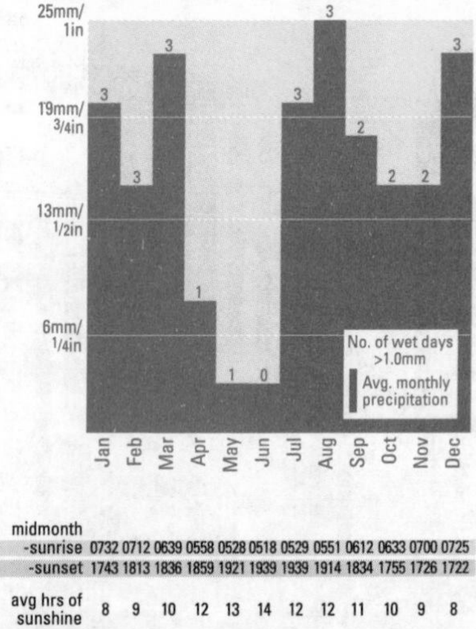

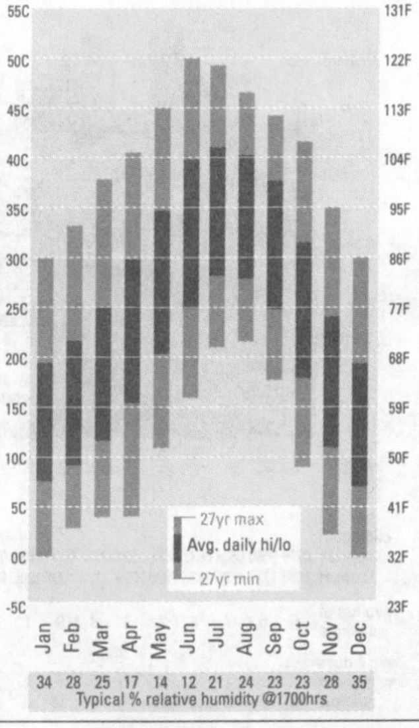

	Jan	Feb	Mar	Apr	May	Jun	Jul	Aug	Sep	Oct	Nov	Dec
midmonth -sunrise	0732	0712	0639	0558	0528	0518	0529	0551	0612	0633	0700	0725
-sunset	1743	1813	1836	1859	1921	1939	1939	1914	1834	1755	1726	1722
avg hrs of sunshine	8	9	10	12	13	14	12	12	11	10	9	8
avg # days w/thunder	<1	1	1	1	1	1	6	7	4	1	1	<1
w/snow	<1	<1	0	0	0	0	0	0	0	0	0	<1
Typical % relative humidity @1700hrs	34	28	25	17	14	12	21	24	23	23	28	35

San Francisco

37°46'N, 122°26'W
Elev: 39m (128ft)

San Francisco's weather can vary from block to block. Fingers of chilly, foggy Pacific wind often wrap around the city's western hills while downtown remains sunny. The marine air is scarcely warmer – and sometimes cooler – in summer than in winter. Gusty spells of rain arrive from November–April, with brief heat waves from April–October.

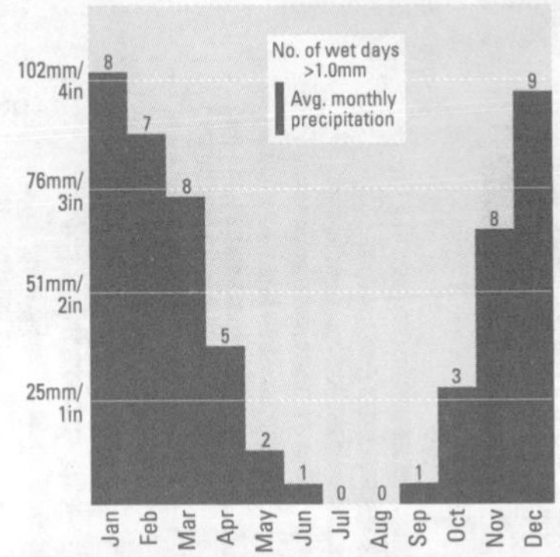

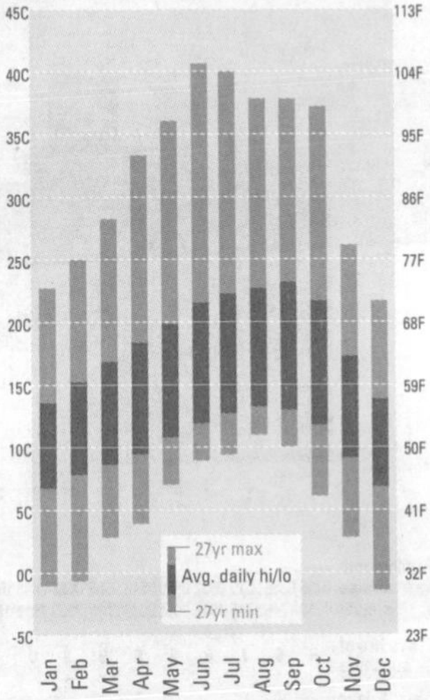

	Jan	Feb	Mar	Apr	May	Jun	Jul	Aug	Sep	Oct	Nov	Dec
midmonth -sunrise	0724	0659	0621	0635	0600	0547	0600	0625	0652	0718	0650	0718
-sunset	1715	1749	1817	1945	2013	2033	2031	2002	1918	1832	1658	1652
avg hrs of sunshine	6	7	9	10	10	10	10	9	9	8	6	5
avg # days w/thunder	<1	1	<1	1	0	<1	1	1	<1	1	<1	0
w/snow	<1	<1	<1	0	0	0	0	0	0	0	0	<1
Typical % relative humidity @1600hrs	66	64	62	59	59	59	59	61	58	59	63	67

Seattle

47°27'N, 122°18'W
Elev: 137m (450ft)

Scandinavians founded Seattle, but Londoners may feel at home here. Temperatures in the two cities are similar year round; Seattle summers provide a good deal more sunshine and little rain. Winter's moisture is frequently gentle, although howling wind and downpours sometimes strike. Heavy snows occur only during rare Arctic blasts.

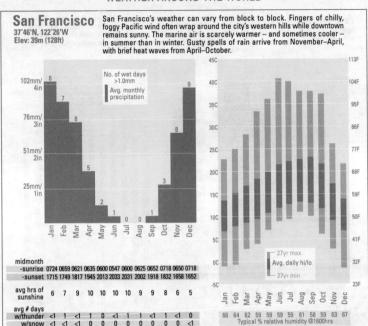

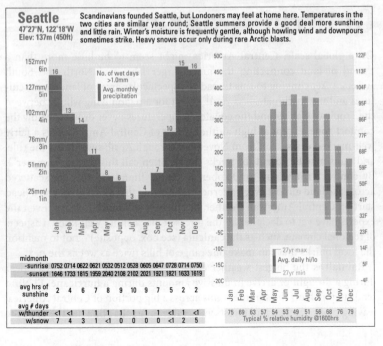

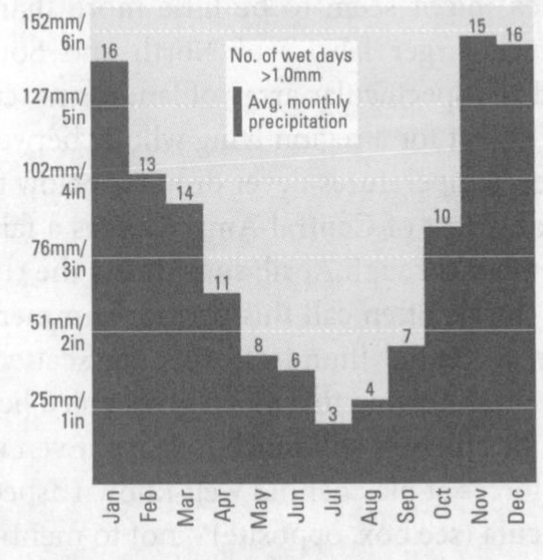

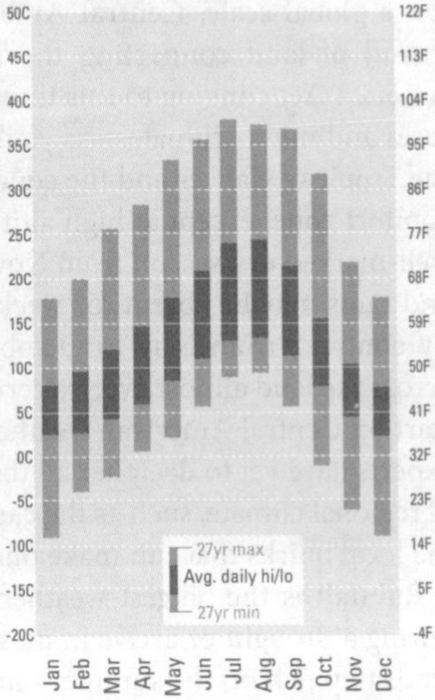

	Jan	Feb	Mar	Apr	May	Jun	Jul	Aug	Sep	Oct	Nov	Dec
midmonth -sunrise	0752	0714	0622	0621	0532	0512	0528	0605	0647	0728	0714	0750
-sunset	1646	1733	1815	1959	2040	2108	2102	2021	1921	1821	1633	1619
avg hrs of sunshine	2	4	6	7	8	9	10	9	7	5	2	2
avg # days w/thunder	<1	<1	1	1	1	1	1	1	1	<1	1	<1
w/snow	7	4	3	1	<1	0	0	0	0	<1	2	5
Typical % relative humidity @1600hrs	75	69	63	57	54	53	49	51	56	68	76	79

Washington DC

38°51'N, 77°02'W
Elev: 3m (10ft)

Built on a swamp, Washington reveals its origins during the torrid summer. Highs can top 32°C/90°F for days on end, and many locals head for the hills or the seashore. Autumn is more pleasing. Winter days can be grey but are seldom bitterly cold; most years see a few snowfalls, usually light. Thunderstorms spice up the bright, breezy spring.

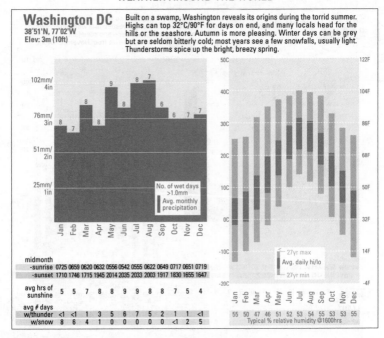

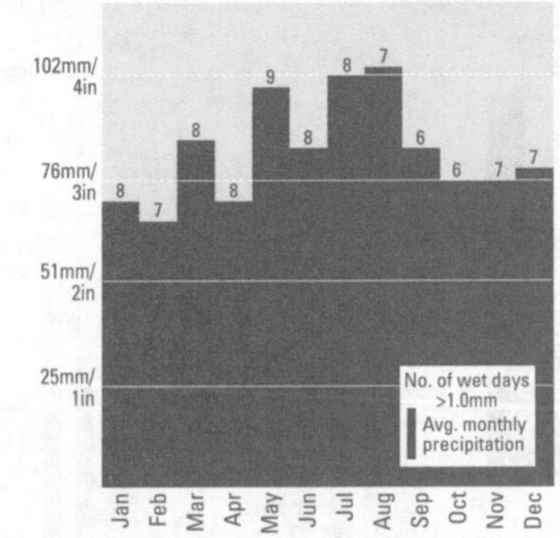

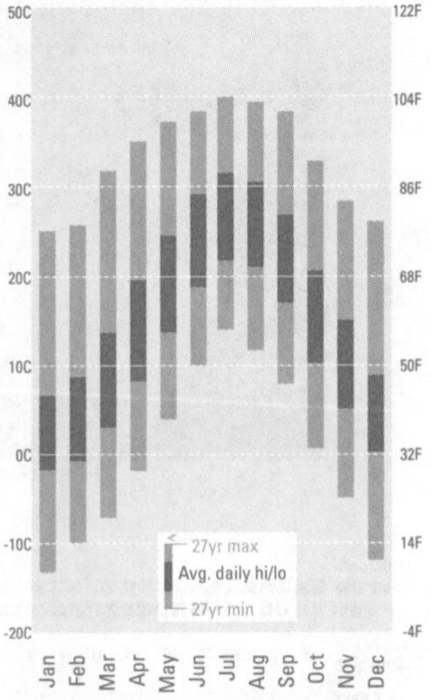

	Jan	Feb	Mar	Apr	May	Jun	Jul	Aug	Sep	Oct	Nov	Dec
midmonth -sunrise	0725	0659	0620	0632	0556	0542	0555	0622	0649	0717	0651	0719
-sunset	1710	1746	1715	1945	2014	2035	2033	2003	1917	1830	1655	1647
avg hrs of sunshine	5	5	7	8	8	9	9	8	8	7	5	4
avg # days w/thunder	<1	<1	1	3	5	6	7	5	2	1	1	<1
w/snow	8	6	4	1	0	0	0	0	0	<1	2	5
Typical % relative humidity @1600hrs	55	50	47	46	51	52	53	54	55	53	53	55

Central America

On a global scale, **Central America** might seem to be little more than a thread of land connecting the much larger fabrics of North and South America. Yet clinging to this thread is a spectacular array of landforms, cultures and micro-climates. As you'd expect for a region lying wholly between the Tropic of Cancer and the equator, temperatures never drop far below the comfort zone, except at high altitude. Most of Central America sees a fairly well-marked dry season from November through April: sunshine is the rule, and rains may be absent for weeks (locals often call this period "summer"). By contrast, from May to October, torrential thunderstorms are scattered across the land almost every afternoon. Knowing this much gives you a head start on Central American weather, but there is still much to learn. Even the experts have yet to decipher all the forces at play behind well-known aspects of regional climate, such as the **canicula** (see box, opposite) – not to mention the local quirks that can make one range of hills twice as wet as the next.

Rainfall is the biggest weather concern here. It may virtually disappear during a drought or arrive in massive amounts during a **hurricane. El Niño** tends to pinch off the summer rains across a big portion of Central America, leaving crops parched and forests vulnerable to fire. **La Niña** has the opposite

The wet season's "little summer"

If you visit Central America or south and east Mexico during the heart of the wet season, you may find the heavens failing to open up as expected. The afternoon rains typically slacken across much of the region for two or three weeks during July or August, producing the **canicula** or **veranillo** (literally, the little summer, since the wet season itself is referred to as winter). The air is still humid and the sky may be filled with puffy clouds, but a week or more might pass between daytime storms. However, in some regions the canicula has little effect on the nighttime rains that are also part of the wet season. At many sites, the total rain for July or August is only two-thirds the amount registered in May or June, thanks to the canicula. Ironically, the **Caribbean coast** from **eastern Nicaragua** to **Panama** can get some of its heaviest downpours even as the canicula is playing out just to the west. The break in the rains along the Caribbean doesn't usually hit until August or September, and the following upsurge may extend into the New Year. Some of the heaviest rains of the year may follow the canicula: spells of dampness called **temporals** that can last a week or longer.

effect, usually producing a higher frequency of garden-variety rains as well as Caribbean hurricanes. Since hurricanes typically move with a westward component while in the tropics, they virtually never strike the Pacific coast head on. However, the **Pacific slopes** are equally – or more – at risk of flooding than towns facing the Caribbean, because a hurricane approaching from the east can pull Pacific moisture up and into the mountains. In October 1998, **Hurricane Mitch** slid from the Caribbean into the north coast of **Honduras** and wrung out colossal amounts of rain – as much as 900mm/35in – across the country's highlands. Mitch's winds had weakened dramatically by the time it made landfall, but three days of incessant rains produced catastrophic flooding and mudslides that killed more than 11,000 people across the region (most of them on the Pacific side) and devastated the Honduran economy.

The islands arcing around the north and east sides of the Caribbean sweat out a few hurricane scares of their own each year, but otherwise their climates are enviably placid. **Cuba** and the **Bahamas** are close enough to North America to get a few distinct cold fronts each winter. Elsewhere, island residents, like those throughout much of Central America, are most affected by the ebb and flow of wet and dry seasons, with little in the way of temperature leaps or tumbles to upset the equanimity.

Belize

Cays and coral reefs are one of the main tourist draws here, so the heat and humidity that rule **Belize** almost year-round aren't necessarily a drawback. The dry season is shorter than usual for Central America. Summer and autumn deliver a downpour about every alternate day, and in winter a few

Belize

17°32'N, 88°18'W
Elev: 5m (16ft)

You can get a damp day any time of year in this steamy coastal city. The odds are lowest in March and April, after the occasional cool, damp northers of mid-winter are gone but before summer and autumn downpours set in about every other day. Try to avoid soggy, hurricane-prone September and October.

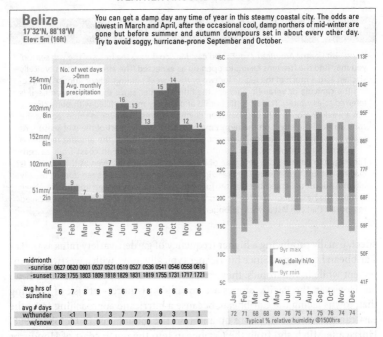

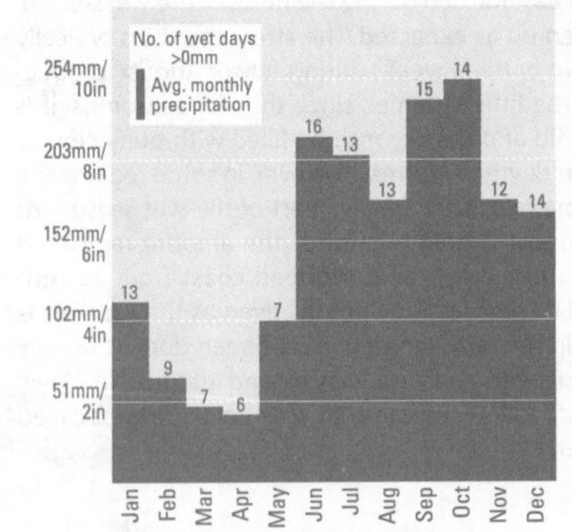

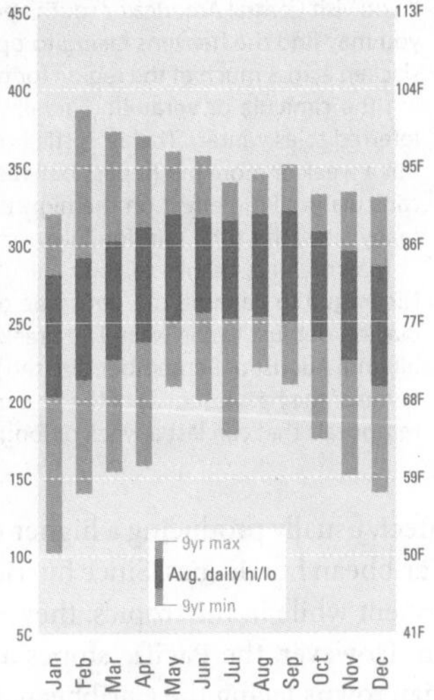

	Jan	Feb	Mar	Apr	May	Jun	Jul	Aug	Sep	Oct	Nov	Dec
midmonth -sunrise	0627	0620	0601	0537	0521	0519	0527	0536	0541	0546	0558	0616
-sunset	1739	1755	1803	1809	1818	1829	1831	1819	1755	1731	1717	1721
avg hrs of sunshine	6	7	8	9	9	6	7	8	6	6	6	6
avg # days w/thunder	1	<1	1	1	3	7	7	7	9	3	1	1
w/snow	0	0	0	0	0	0	0	0	0	0	0	0
Typical % relative humidity @1500hrs	72	71	68	68	69	76	75	74	75	76	74	74

fronts push south from the Gulf of Mexico, bringing damp and coolish weather that can tarry for a day or two. The warm-season rains usually come and go quickly, but expect a fair amount of cloudiness. A *canicula* often interrupts the rains for a few days in late July or August. The sultriness that pervades Belize is most intense as you move inland from the sea. Rains are heavier as you go south, ranging from over 1000mm/39in in the far north to more than 4000mm/157in along the seaward slopes of the **Maya Mountains.** Your best bet for sustained sunshine and dryness is from February to early May, with heat and humidity building through that span. The **hurricane threat** is greatest from August through October.

Costa Rica

Nicaragua

For all the renowned ecological diversity of this small nation, its meteorological diversity is worth noting as well. This is a country where micro-climates reign supreme. The combination of an **irregular coastline**, the presence of **two contrasting bodies of water** nearby, and the extremely **varied topography** leads to a plethora of local wind and rainfall patterns that are nearly impossible to summarize. For what it's worth, you can expect pleasantly mild conditions at moderate altitudes (such as **San José**), with nights a touch cool in the dry season, but afternoons seldom very hot, even in the wet mid-summer. The **lowlands** are, unsurprisingly, warmer and more humid. The heat is a bit more intense on the Pacific side than on the Atlantic, sometimes topping 35°C/95°F in the spring across **Guanacaste** and **Puntarenas**. Although some pockets of **Costa Rica**, such as the **Valley of the General**, are drier than others, "dry" is a relative term in this well-watered country. You can expect frequent afternoon downpours across most of the country between May and early November. The *canicula* dries things out in many areas for a couple of weeks at some point in July or August. On the Pacific slopes and plains and near the Caribbean coast, the rains usually diminish in September and October, then pick up again in November and continue into January or beyond. Indeed, almost every month sees ample rainfall near the Caribbean; much of it falls at night, though. It's dry nearly everywhere else from February through April, with more sunshine than in the summer. The heat gets intense across the lowlands as spring approaches.

To the north, **Nicaragua** has its own share of climatic regimes. Heat and humidity rule the **eastern lowlands**: ample rain falls year-round, and warm-season deluges are especially common on the southeast coast. (**Hurricanes** are also a definite risk along the Nicaraguan coast, especially from September into November.) The **northwest highlands** provide weather much like that

Puntarenas

9°58'N, 84°50'W
Elev: 3m (10ft)

If you're descending from the hills, steel yourself for thick, tropical air. Afternoons are especially hot toward the end of the dry season, from February into April. Late-day rains from May through October are as heavy as in San José but a little less frequent. The veranillo is most likely in July.

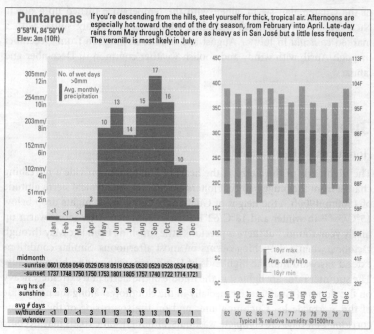

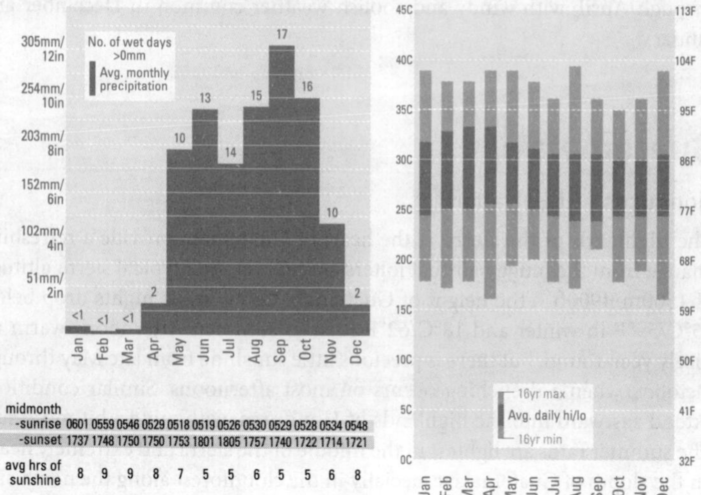

	Jan	Feb	Mar	Apr	May	Jun	Jul	Aug	Sep	Oct	Nov	Dec
midmonth -sunrise	0601	0559	0546	0529	0518	0519	0526	0530	0529	0528	0534	0548
-sunset	1737	1748	1750	1750	1753	1801	1805	1757	1740	1722	1714	1721
avg hrs of sunshine	8	9	9	8	7	5	5	6	5	5	6	8
avg # days w/thunder	<1	0	<1	3	11	13	12	13	13	10	5	1
w/snow	0	0	0	0	0	0	0	0	0	0	0	0
Typical % relative humidity @1500hrs	62	60	62	66	74	77	77	78	79	79	76	70

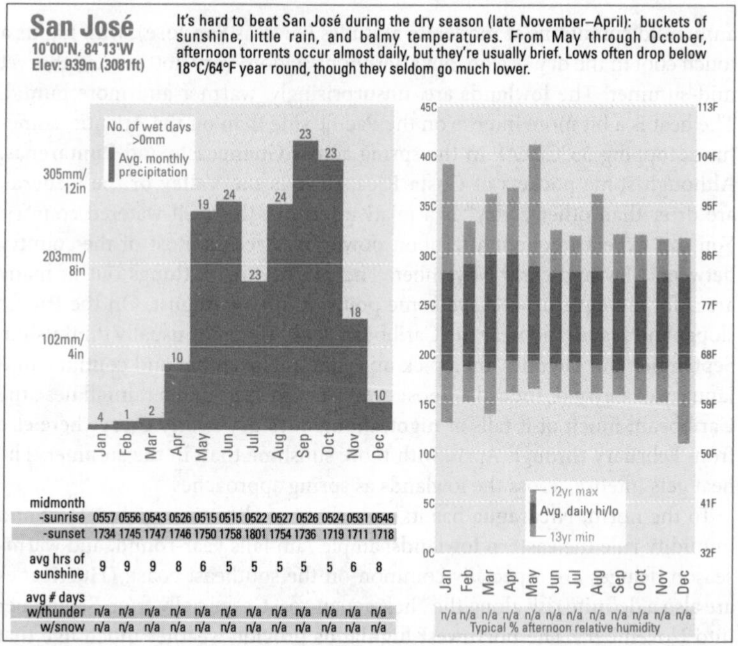

of the Guatemalan sierra, albeit with the temperature turned up a notch. Nicaragua's touristed west, including **Managua**, is reliably sultry, with a marked *canicula* in July or August and a bone-dry season from December through April, with windy and coolish weather common in December and January.

Guatemala

Honduras | El Salvador

The **highlands of the sierra** at the heart of **Guatemala** provide a refreshing change from the muggy air that loiters at sea level. At a typical sierra altitude of 1500m/4900ft – the height of **Guatemala City** – most nights drop below 15°C/59°F in winter and 18°C/62°F even in summer. Afternoons warm up nicely year-round, but there is precious little sunshine from late May through October, when a drenching occurs on most afternoons. Similar conditions extend eastward into the highlands of **Honduras**, with quite a bit more sun. The summer rains are lightest in the middle of the sierra and extremely heavy on the slopes of Guatemala, especially in the cloudforest along the north side of the **Sierra de los Cuchumatanes** and on the south side of the volcanic corridor. Few days are complete washouts, although some roads may become

Guatemala City

14°35'N, 90°31'W
Elev: 1499m (4917ft)

Altitude and clouds help keep the sierra air refreshingly mild and not too humid year round. To avoid the touristed core of the dry season (December–March), try either end; afternoons will be warmer and sunnier in April than in November. Expect lots of thunder and relatively little sun during the wet season.

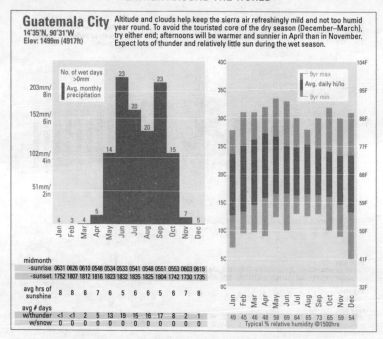

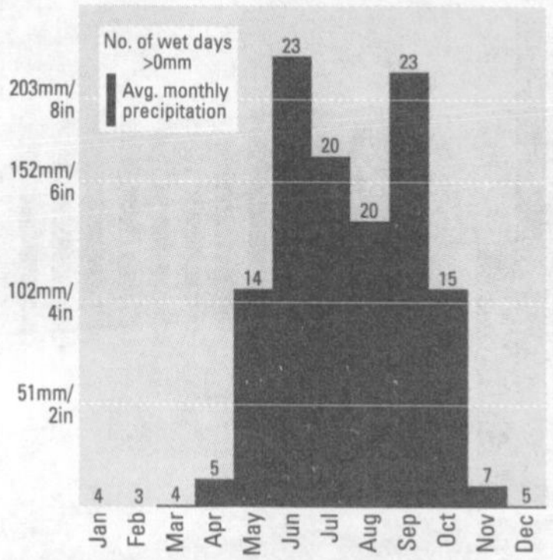

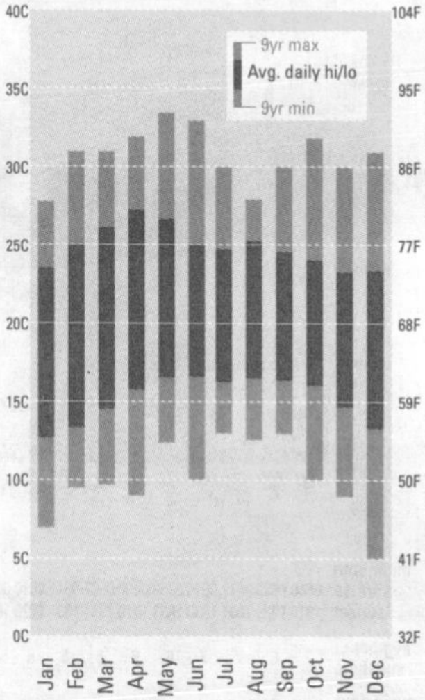

	Jan	Feb	Mar	Apr	May	Jun	Jul	Aug	Sep	Oct	Nov	Dec
midmonth -sunrise	0631	0626	0610	0548	0534	0533	0541	0548	0551	0553	0603	0619
-sunset	1752	1807	1812	1816	1823	1832	1835	1825	1804	1742	1730	1735
avg hrs of sunshine	8	8	8	7	6	5	6	6	5	6	7	8
avg # days w/thunder	<1	<1	2	5	13	19	15	16	17	8	2	1
w/snow	0	0	0	0	0	0	0	0	0	0	0	0
Typical % relative humidity @1500hrs	49	45	46	48	58	69	64	65	73	65	59	54

difficult to navigate as the wet season unfolds. The low-lying **rainforests of Petén** are sultry year round, except for sea breezes and a few shots of cool air in winter from northers that can bring rain as far south as the highlands near **Coban**. The Pacific coast gets progressively hotter and more humid as you move south toward **El Salvador**. Even at San Salvador's altitude of over 680m/2230ft, high temperatures usually hit 30°C/86°F during the wet season and often exceed 34°C/93°F when it's rainless, with humidity to boot. From November through April, most areas – except for those on and just in from the Caribbean coast – see periods of rain less than once in 2 or 3 weeks on average, with the low-latitude sun broken by scattered clouds.

Panama

Like a microcosm of Central America, **Panama**'s length is far greater than its breadth. The country snakes its way from west to east, putting the Atlantic to its north and the Pacific to its south (if you're sailing from Europe to Asia, you actually end up further east after going through the **Panama Canal**). The intersection of the prevailing winds with this contorted topography makes for a number of local climatic effects, with the big picture somewhat more uniform. Panama's low latitude means that it gets relatively few seasonal

Colón

9°13'N, 79°33'W
Elev: 12m (39ft)

Anticipating the temperature here is easy: it's sultry year round, and a tiny bit more humid than it is on coastlines further south. October and November are the wettest months; the rains often slog into December. From late January to early April, there's plenty of sun and rainfall is scanty (even lighter as you go south).

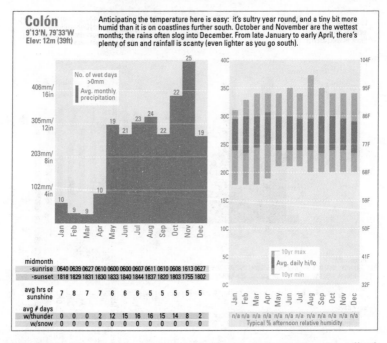

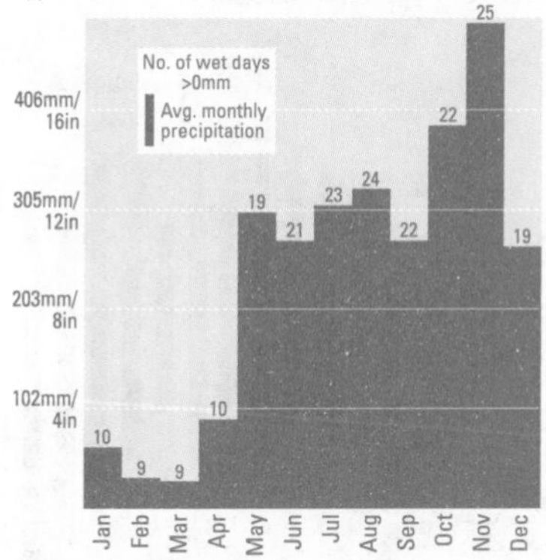

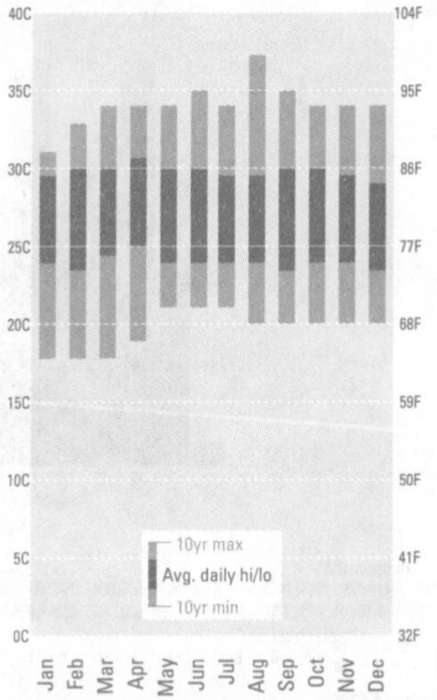

	Jan	Feb	Mar	Apr	May	Jun	Jul	Aug	Sep	Oct	Nov	Dec
midmonth -sunrise	0640	0639	0627	0610	0600	0600	0607	0611	0610	0608	1613	0627
-sunset	1818	1829	1831	1830	1833	1840	1844	1837	1820	1803	1755	1802
avg hrs of sunshine	7	8	7	7	6	6	6	5	5	5	5	5
avg # days w/thunder	0	0	0	2	12	15	16	16	15	14	8	2
w/snow	0	0	0	0	0	0	0	0	0	0	0	0

Jan	Feb	Mar	Apr	May	Jun	Jul	Aug	Sep	Oct	Nov	Dec
n/a	n/a	n/a	n/a	n/a	n/a	n/a	n/a	n/a	n/a	n/a	n/a

Typical % afternoon relative humidity

variations in temperature, even compared to, say, Costa Rica. Overall, it's consistently warm and humid at sea level. The worst heat builds toward the end of the dry season, which extends roughly from January to April (except on the soggy northwest coast, where northers can trigger heavy rain even in mid-winter). In general, the west and east ends of the country are the wettest regions, and the heaviest rains fall in October and November, although not from hurricanes – they aren't a serious risk this close to the equator. At higher points along the **Cordillera Central**, expect cool nights and mild days, as in Costa Rica's highlands. The main variation at altitude is between the soggy, cloudy wet season and the sunny mid-winter.

South America

Bone-dry deserts overlooking tropical oceans, perpetual windswept chill, hushed tropical splendour – **South America** isn't lacking in climatic variety. As opposed to the chameleon-like seasonal shifts found in Asia and North America, South America specializes in more-uniform climates that maintain their bold, distinctive hues throughout the year.

Despite the "South" in its name, a good fraction of South America lies in the Northern Hemisphere. The continent's gracefully tapered form is at its widest just south of the equator, accommodating the stunningly broad **Amazon Valley** along more than 2400km/1500 miles from headwaters to Atlantic. This is one of the world's lushest areas: the **rainforest** and its ecological diversity thrive in an environment where rainfall is generous and temperatures seldom get much below 20°C/68°F or much above 35°C/95°F. The northernmost reaches of South America are also quite moist, with the perplexing exception of the arid, scrubby Caribbean coast (see Venezuela, p.223).

If the Amazon is the heart of South America, the continent's backbone is the improbably long chain of mountains that runs nearly the entire length of the Pacific coast. The volcano-studded **Andes** are actually a set of several parallel chains, with many peaks extending above 4000m/15,700ft from **western Colombia** to **central Chile**. Together, the Andes serve as a barrier that keeps cool Pacific air bottled up against the west slopes, while heat and humidity rule the tropical east. The long, gentle curves of the Andean spine combine with distinctive ocean currents to produce jaw-dropping weather contrasts from one stretch of shoreline to another. On the Colombian coast, it rains as much as anywhere else in the Western Hemisphere. Further south, where **Peru** meets the Pacific, it hardly rains at all. The only thing that can alter this regime dramatically is a famous pair of offspring from the Pacific. During **El Niño**, the waters off Peru can warm by more than 6°C/10°F, triggering rain across normally dry areas but encouraging drought across northern South America. The cooler-than-normal waters of **La Niña** have the opposite effect – they tend to reinforce the patterns that make one spot run dry and another wet, producing weather that's "normal, only more so".

South America narrows as it extends into the southern mid-latitudes, so there isn't quite enough continent to produce the kind of bitter cold one finds across Canada or Russia. Instead, the oceanic westerlies maintain a cool, breezy grip. Storm systems in this persistent flow dump enormous amounts of rain and snow on the west slopes of the southern Andes. On the other side are the bright, thundery pampas of **Argentina** and, further south, the barren, chilly uplands of **Patagonia**.

Argentina

If any part of South America approaches the four-season regime familiar to Americans and Europeans, it's the heart of **Argentina**. Most of the country gets recognizably warm in summer and cool in winter, with clear-cut transitions in between. What's more, the bulk of Argentina manages a great deal of sunshine without becoming a desert. Some spots average 100 clear days a year and 8 hours of sun a day. When it does rain, it's often in the form of showers or thunderstorms that sweep through quickly.

Much of Argentina's population is clustered along the **Paraná River** and on the **Rio de la Plata**, which snakes its way close to the country's eastern border, reaching the Atlantic just past **Buenos Aires**. These rivers are a useful meteorological boundary. Eastward, toward **Uruguay** and eastern **Paraguay**, rainfall and humidity increase. Buenos Aires itself is fairly sultry in mid-summer, though much less oppressive than Rio de Janeiro. To the north is the **Gran Chaco**, a scrubby lowland that extends into northwest Paraguay. It's the hottest spot in South America: parts of the Gran Chaco reach 45°C/113°F almost every summer (winters can be quite cool, though).

West and south of the Paraná lies the heartland of Argentina and its endless fields of treeless **pampas**. Sunshine rules during the summer, with hot days (sometimes scorchingly so) and mild nights. On many afternoons, thunderstorms form just off the Andes and move eastward. These can be quite intense – large hail is a perennial threat, especially in and around the **Sierras de Córdoba** – but rainfall is seldom extreme. Especially in spring, temperatures may soar across the pampas courtesy of the *zonda,* a dusty *föhn*-type wind. Winters average on the cool side, but temperatures can gyrate from well below freezing to readings above 26°C/79°F. The highlands of the **Puna de Atacama** experience equally dramatic temperature swings, but shifted toward the cold side; above 4000m/13,000ft, it's unlikely to exceed 20°C/68°F any time of year. Lightning-laced storms dot the Puna de Atacama in summer; they're less common as you move south along the eastern Andes.

Summer afternoons are quite warm across the lowlands as far south as the **Rio Negro**, but once you enter **Patagonia** the averages begin to drop and the clouds and wind become more frequent. Winters tend to be fairly cloudy across all of Patagonia. Summers bring periods of sunshine across the north, but it's actually the cloudiest, gustiest season across the extreme south. The Argentinian part of Patagonia is far drier than its Chilean counterpart: many spots qualify as desert, with less than 200mm/4in of rain or melted snow each year. Although summer is the dampest time overall, any month can bring some moisture. There's a bit more temperature range here than on the Chilean side. As far south as **Sarmiento**, many summer afternoons top 26°C/79°F, while most winter nights fall below freezing. It's colder as you move south or gain elevation: some protected valleys above 4000m/13,000ft have been known to dip below –20°C/–4°F.

Buenos Aires

34°49'S, 58°32'W
Elev: 20m (66ft)

The climate here is similar to that of Sydney, with a few more seasonal extremes tossed in. Summer days are fairly warm and humid, with thunder about once a week. Offshore winds can deliver extreme heat. Winters are cool and only slightly drier. Nights may dip near freezing, but snow's a novelty.

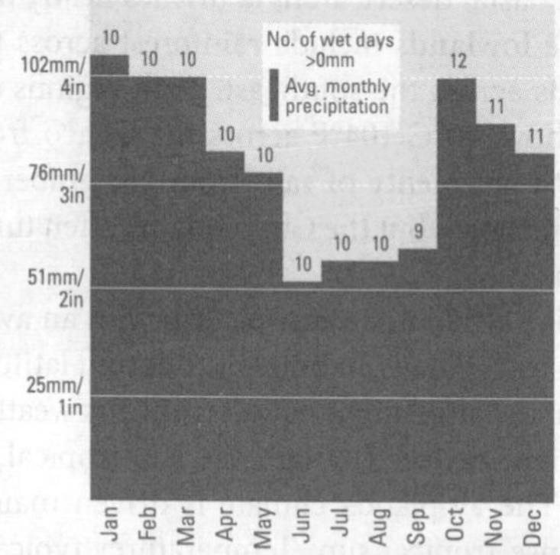

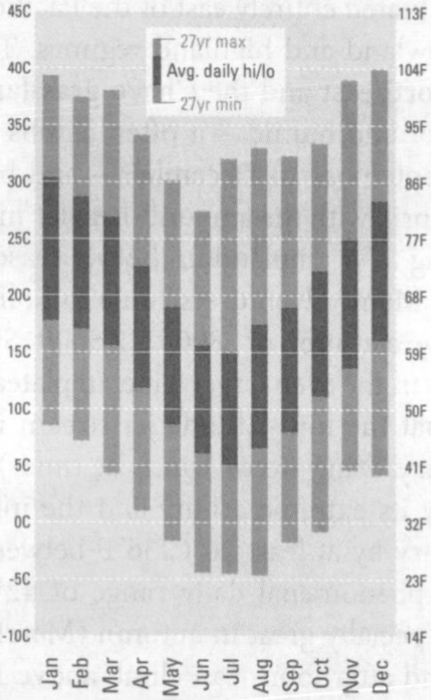

	Jan	Feb	Mar	Apr	May	Jun	Jul	Aug	Sep	Oct	Nov	Dec
midmonth -sunrise	0557	0628	0653	0718	0741	0800	0759	0734	0654	0612	0541	0535
-sunset	2010	1948	1912	1830	1759	1749	1801	1823	1845	1908	1937	2004
avg hrs of sunshine	8	8	7	6	5	4	4	6	6	7	8	8
avg # days w/thunder	6	5	4	3	3	2	2	3	2	5	5	7
w/snow	0	0	0	0	<1	<1	<1	<1	<1	0	0	0
Typical % relative humidity @1500hrs	47	52	54	59	60	62	62	57	53	56	53	48

Córdoba

31°19'S, 64°13'W
Elev: 474m (1555ft)

Most of Córdoba's moisture arrives in the summer, when strong afternoon storms every three or four days may drop hail as well as rain. The air is warm but not terribly humid in summer; rare bouts of intense heat are usually dry. Winter brings little rain and a good deal of sun, but occasional sharp cold.

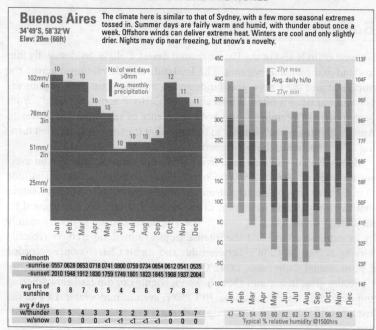

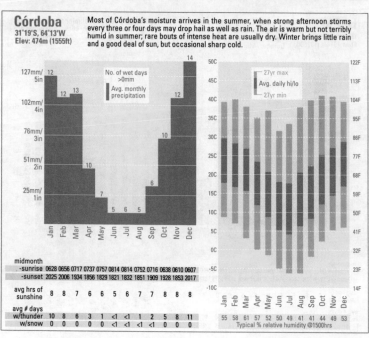

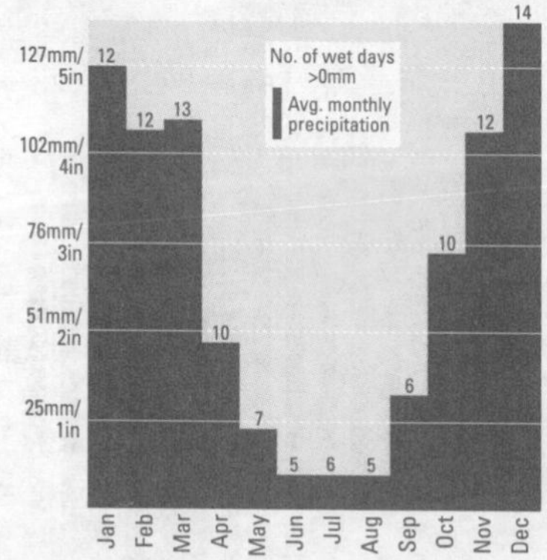

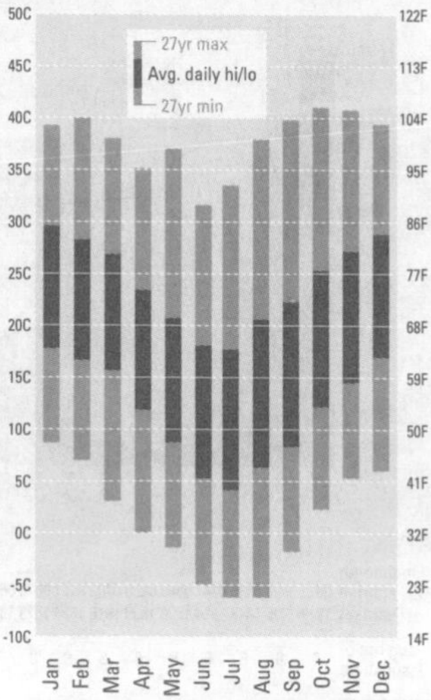

	Jan	Feb	Mar	Apr	May	Jun	Jul	Aug	Sep	Oct	Nov	Dec
midmonth -sunrise	0628	0656	0717	0737	0757	0814	0814	0752	0716	0638	0610	0607
-sunset	2025	2006	1934	1856	1829	1821	1832	1851	1909	1928	1853	2017
avg hrs of sunshine	8	8	7	6	6	5	6	7	7	8	8	8
avg # days w/thunder	10	8	6	3	1	<1	<1	1	2	5	8	11
w/snow	0	0	0	0	0	<1	<1	<1	<1	0	0	0
Typical % relative humidity @1500hrs	55	58	61	57	52	50	49	41	41	44	49	53

Bolivia

Situated entirely east of the Pacific coastal desert, **Bolivia** divides neatly into lowland and highland regimes. The **lowlands** include rainforest across the **northeast** and the **Chaco grasslands** across the southeast. Both regions are hot and humid – it often climbs above 40°C/104°F across the Chaco from September to December – and both get plenty of rain from November to April, with the amounts heavier in the north but the Chaco plains often turning to swampland in the wet season.

Many adventurers come to Bolivia's **altiplano**, a vast plateau with an average elevation of 3800ft/12,500ft. Since it is lower in both altitude and latitude than the even larger Tibetan plateau, the altiplano is cut off from any weather that the mid-latitude jet stream might deliver (though the sub-tropical jet may bring fierce winds at times). The altiplano's climate is driven mainly by its extreme aridity and the intense tropical sun. Temperatures typically vary by at least 20°C/36°F between day and night; **Charaña** once recorded a phenomenal daily range of 42°C/75°F. These day-to-night contrasts are especially great in autumn (March–May). Nights can dip far below freezing and afternoons can climb above 10°C/50°F, even in mid-winter. **La Paz** sits just below the altiplano's north edge, helping it to remain milder on the cold-

La Paz

16°31'S, 68°11'W
Elev: 4014m (13166ft)

The world's highest capital city descends from the adjacent altiplano, where winter nights often plunge to –10°C/14°F. Frost is uncommon, snow rarely sticks, and winds are lighter than on the altiplano itself. Expect a few noisy thunderstorms anytime from October–March. It's cool year round.

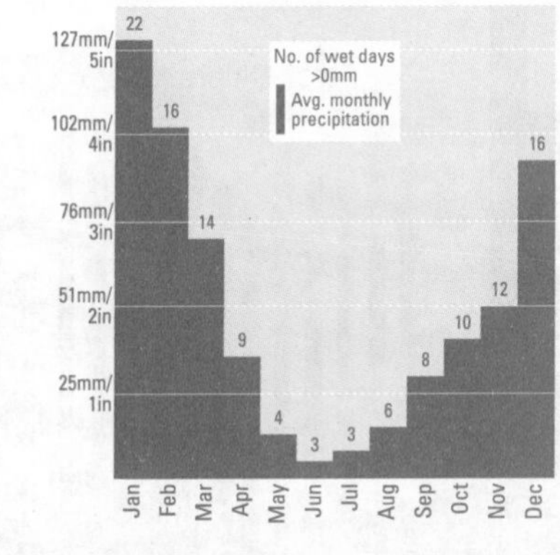

midmonth	Jan	Feb	Mar	Apr	May	Jun	Jul	Aug	Sep	Oct	Nov	Dec
-sunrise	0612	0628	0636	0641	0649	0659	0702	0651	0628	0605	0551	0555
-sunset	1912	1905	1847	1824	1809	1808	1816	1824	1828	1832	1844	1901
avg hrs of sunshine	5	6	5	6	8	8	8	9	8	7	6	5
avg # days w/thunder	16	13	12	9	2	1	<1	2	4	10	12	15
w/snow	1	<1	1	1	1	2	2	3	3	2	1	1

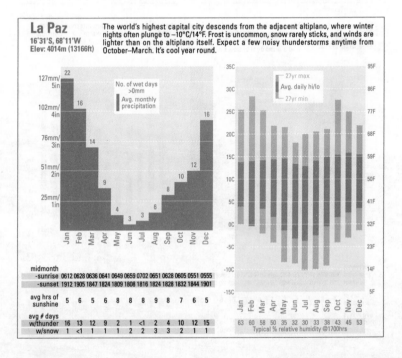

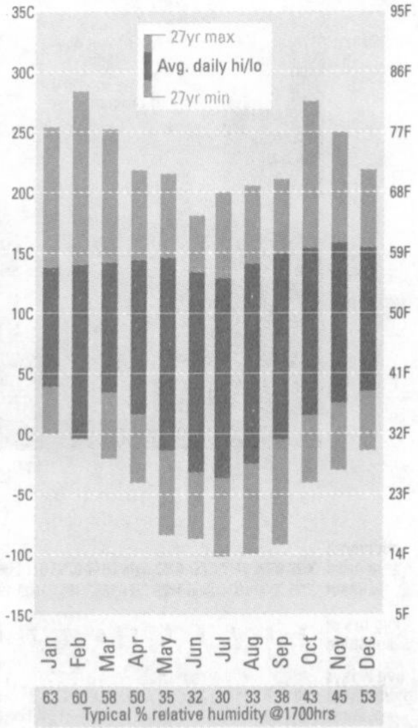

est nights. Winter days on the altiplano tend to be crystal-clear, with only a handful of light snows each year. The mild summer afternoons often bring a quick, intense round of thunder, sometimes dropping sleet or snow that quickly disappears.

Brazil

Roughly as large as the United States or China, **Brazil** is by far the world's largest nation straddling the equator. For all its size, however, Brazil has relatively few weather surprises, thanks to its low latitude and the absence of any huge mountain ranges.

Amazonia and Mato Grosso

The northwest third of Brazil is dominated by the vast **Amazon basin**. The entire basin gets at least 1500mm/60in of rain a year, usually in the form of heavy afternoon thunderstorms and a few rounds of nighttime rain. The Amazon's rains are heaviest on either end – at the headwaters and near the Atlantic coast – and somewhat lighter (this being relative) across **western Pará state**. The sea breeze often triggers storms by afternoon near the coast; these drift westward, dumping late-night downpours on western Pará and sometimes reaching **Manaus** in the wee hours. Across the **Guiana Highlands** of the far north, the rains tend to be concentrated from April to September; along and just south of the Amazon, they're heaviest from November to May, but a downpour can occur there in any month.

Lows across **Amazonia** are usually within a few degrees of 23°C/73°F and highs are in the neighbourhood of 31°C/88°F. The humidity is predictably high, although it dips a little during the dry season. The temperature may take a rare tumble between June and September with the arrival of a *friagem* – a cool front from Argentina. Once or twice in a typical year, a *friagem* makes it into upper Amazonia, sometimes pushing as far east as Manaus. It can bring two or three days of overcast and temperatures that might dip below 20°C/68°F (or even lower, the further south you are).

The seasons are a bit sharper in the **Mato Grosso**, the vast lowlands – and swamplands, during the wet season – that extend south from the Amazon and west from the Central Plateau. Here, the summer showers and thunderstorms aren't quite as heavy as on the Amazon, and during the May-to-September dry season, weeks may pass without a drop of rain. Even then, a *friagem* might bring a day of wintertime drizzle, and after it clears, temperatures can plummet. Even at latitude 15°S, it's been known to dip near freezing. As spring comes on, the Mato Grosso starts to roast. By October, most days top 30°C/86°F, and a few spots can hit 40°C/104°F during late spring and summer.

East Coast

Seasonal shifts are gentle along the stunning shoreline where southeast Brazil meets the Atlantic. From **Vitória** to **Porto Alegre**, the summer months (including the time of Rio's Carnaval) are made for beachgoing. It's hot and humid, with a thunderstorm about every third or fourth day at **Rio de Janeiro** and **São Paolo** and a bit less often further south, where it's somewhat less muggy. The southernmost coast, around Porto Alegre, usually dips below 20°C/68°F even in mid-summer. Winter brings a succession of cool fronts that wrap around the coast, tempering the tropical heat. Their impact is strongest toward the south. Mid-winter in Porto Alegre is similar to that in Los Angeles, with a few prolonged bouts of rain and coolish temperatures, while the coast from Rio northward typically gets little more than clouds with each front. Even in July, most afternoons in Rio top 25°C/77°F.

Contrary to the usual tropical pattern, the easternmost tip of Brazil – including the coast from **Salvador to Recife** and **Cabo de São Roque** – gets most of its rain in the autumn and early winter, from March into August. There's still a good deal of sunshine during those wet months. The showers are least frequent and far lighter from October to December. Sultriness reigns throughout the year.

Plateau

Brazil's **Central Plateau** offers a respite from the intense coastal heat and humidity. The altitudes are modest, but they make a difference. In **Brasilia** (just over 1000m/3300ft), afternoons typically stay close to 25°C/79°F in both winter and summer. Most evenings dip below 16°C/60°F from May to August. In winter, the plateau is sunnier, drier and cooler than the coast. The western plateau gets frequent afternoon thunderstorms during the summer, while the rains are lighter but a bit more sustained further east. The southernmost highlands, fronting the Atlantic, are the only ones that get substantial rain in winter; at the highest points, you might see a snowflake.

Nordeste

West of **Fortaleza**, the coastal downpours arrive in summer and early autumn, typically from about January to May, with light showers the rule otherwise. Both the wet and the dry seasons get wetter as you move toward **Belém**. Moving inland, you'll find one of Brazil's most distinctive climatic regions, the **Nordeste**. From **northern Bahia** to the **Rio Grande do Norte**, rainfall is puzzlingly sparse – less than 1000mm/39in per year in general, and

Belém

1°23'S, 48°29'W
Elev: 16m (53ft)

Moisture pervades this delta city near the mouth of the Amazon. The surrounding swamps keep humidity at steam-room levels. Almost every day from January–May brings a spell of rain, and the pace slackens only slightly in the so-called dry season. Expect hot afternoons and sultry nights year round.

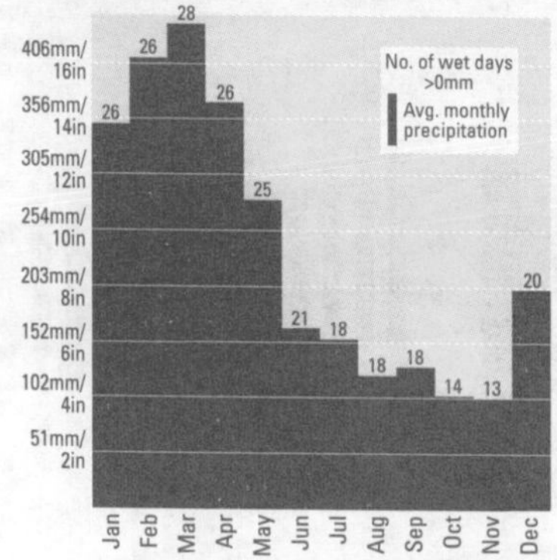

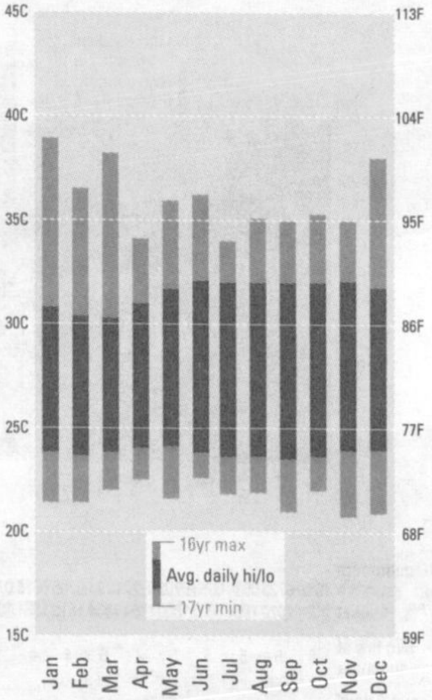

	Jan	Feb	Mar	Apr	May	Jun	Jul	Aug	Sep	Oct	Nov	Dec
midmonth -sunrise	0618	0623	0619	0612	0609	0613	0618	0616	0606	0556	0553	0603
-sunset	1829	1833	1826	1816	1812	1816	1821	1820	1812	1804	1804	1815
avg hrs of sunshine	4	4	3	4	6	8	8	8	8	7	7	6
avg # days w/thunder	10	10	9	11	11	10	9	10	10	7	6	10
w/snow	0	0	0	0	0	0	0	0	0	0	0	0

	Jan	Feb	Mar	Apr	May	Jun	Jul	Aug	Sep	Oct	Nov	Dec
Typical % relative humidity @1500hrs	80	81	82	81	73	65	64	66	70	70	70	74

Brasilia

15°52'S, 47°56'W
Elev: 1061m (3480ft)

This capital city is distinctly milder than most of Brazil's other population centres. There's a sharp wet season in summer, with sunshine interrupted by thundery downpours roughly every other day from October–March. Most afternoons are comfortably warm year round; nights can get nippy in the dry season.

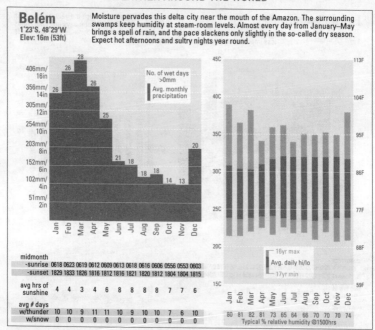

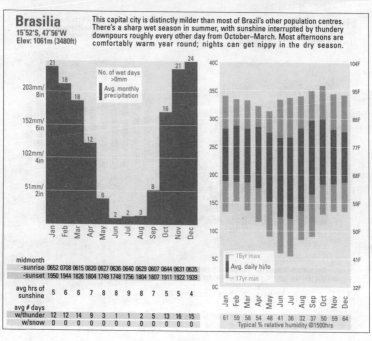

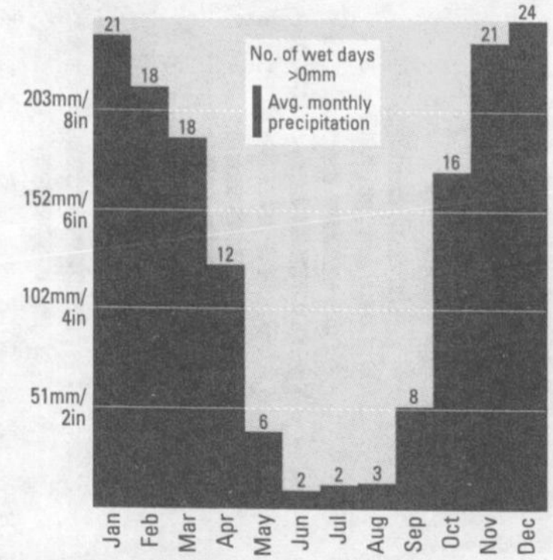

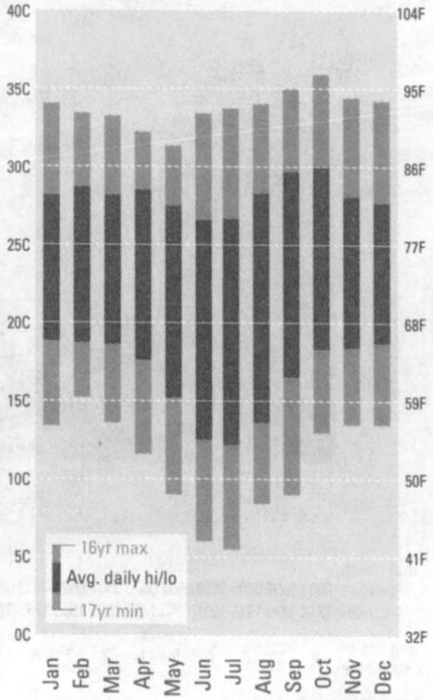

	Jan	Feb	Mar	Apr	May	Jun	Jul	Aug	Sep	Oct	Nov	Dec
midmonth -sunrise	0652	0708	0615	0820	0627	0636	0640	0629	0607	0644	0631	0635
-sunset	1950	1944	1826	1804	1749	1748	1756	1804	1807	1911	1922	1939
avg hrs of sunshine	5	6	6	7	8	8	9	8	7	5	5	4
avg # days w/thunder	12	12	14	9	3	1	1	2	5	13	16	15
w/snow	0	0	0	0	0	0	0	0	0	0	0	0

	Jan	Feb	Mar	Apr	May	Jun	Jul	Aug	Sep	Oct	Nov	Dec
Typical % relative humidity @1500hrs	61	59	58	54	48	41	36	32	37	50	59	64

Foz do Iguaçu

25°44'S, 54°28'W
Elev: 270m (886ft)

Wet and dry seasons aren't too sharply marked here – through the year, about one in four days is wet, with heaviest rainfall in summer. Expect warm days and cool nights in autumn and winter, with only a few freezes or hot days. Summer heat is a shade less intense than Rio's, with thunder less frequent.

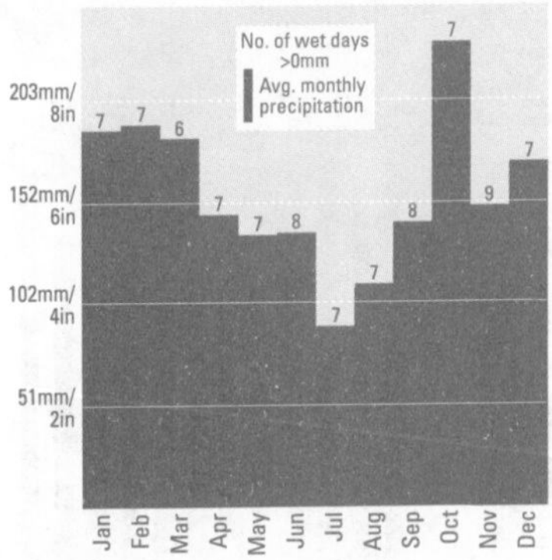

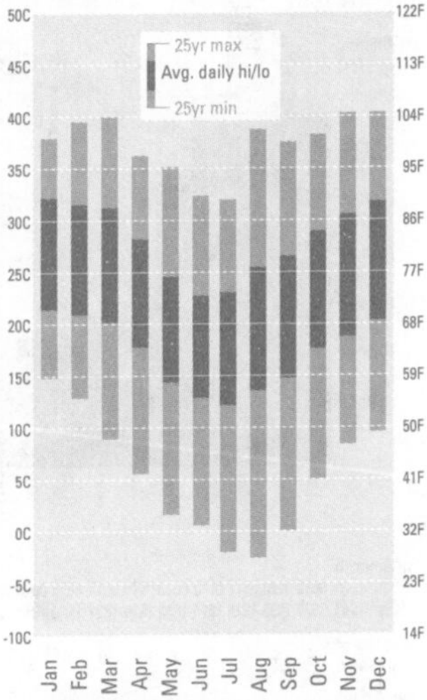

	Jan	Feb	Mar	Apr	May	Jun	Jul	Aug	Sep	Oct	Nov	Dec
midmonth -sunrise	0700	0723	0639	0653	0708	0722	0724	0706	0635	0703	0641	0641
-sunset	2034	2020	1854	1822	1800	1754	1804	1819	1831	1944	2004	2025
avg hrs of sunshine	6	5	5	5	5	4	5	4	4	5	7	7
avg # days w/thunder	5	6	4	3	3	3	2	3	4	5	6	5
w/snow	0	0	0	<1	0	0	0	0	0	0	0	0
Typical % relative humidity @1500hrs	58	59	59	61	63	66	61	60	57	57	57	55

Manaus

3°09'S, 59°59'W
Elev: 84m (276ft)

This is one of the less rainy points along the Amazon, though it still gets many downpours. There's a distinct dry season from July–September, when sunshine increases and it rains only about once a week on average.One or two cloudy cool fronts may bring temperatures down a notch in the dry season.

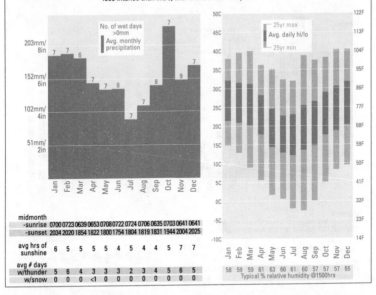

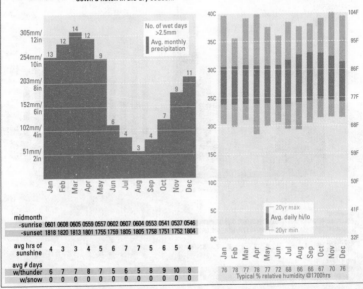

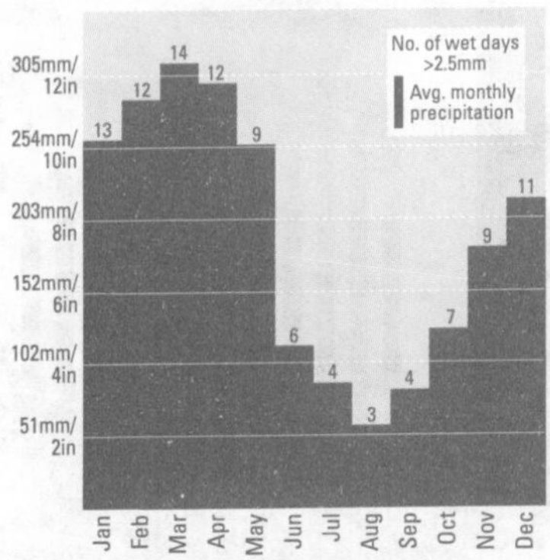

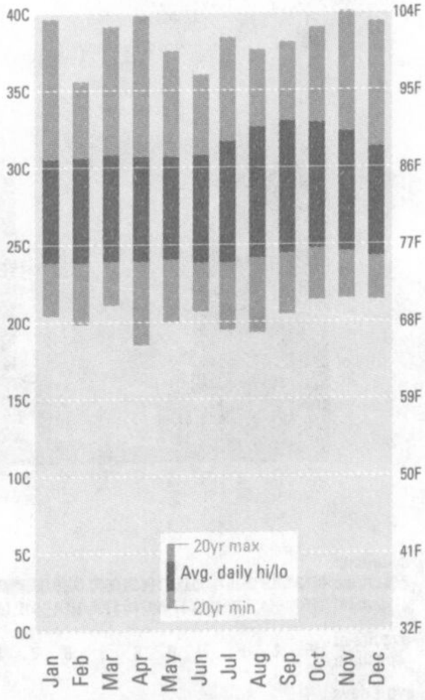

	Jan	Feb	Mar	Apr	May	Jun	Jul	Aug	Sep	Oct	Nov	Dec
midmonth -sunrise	0601	0608	0605	0559	0557	0602	0607	0604	0553	0541	0537	0546
-sunset	1818	1820	1813	1801	1755	1759	1805	1805	1758	1751	1752	1804
avg hrs of sunshine	4	3	3	4	5	6	7	7	5	6	5	4
avg # days w/thunder	6	7	7	8	7	5	6	5	8	9	10	9
w/snow	0	0	0	0	0	0	0	0	0	0	0	0
Typical % relative humidity @1700hrs	76	78	77	78	77	72	68	66	66	67	70	76

Recife

8°04'S, 34°51'W
Elev: 19m (62ft)

If you're planning to visit Brazil during the standard dry season, bring an umbrella to Recife anyway. This stretch of the Atlantic coast gets showers most days from April–August. At other times, the rains are lighter and more sporadic. It's muggy year round, though a bit less so as the wet season fades.

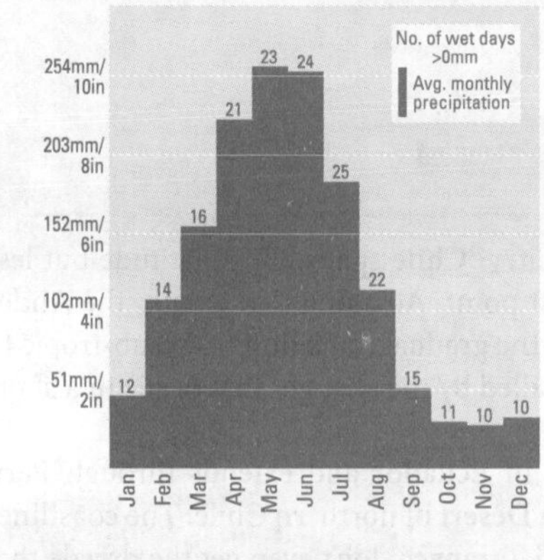

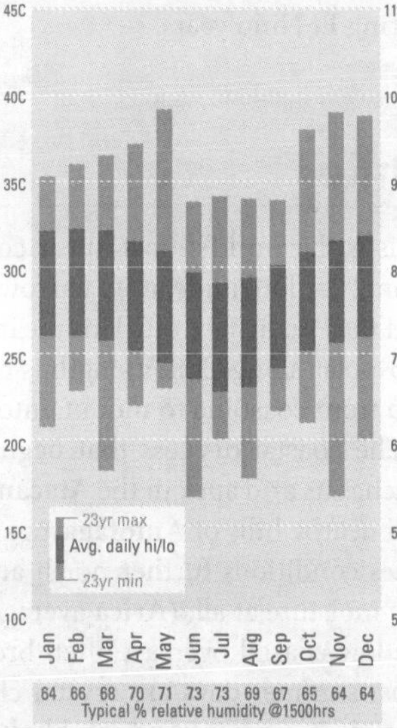

	Jan	Feb	Mar	Apr	May	Jun	Jul	Aug	Sep	Oct	Nov	Dec
midmonth -sunrise	0613	0623	0524	0522	0523	0530	0535	0529	0513	0557	0550	0557
-sunset	1845	1844	1733	1717	1708	1710	1716	1719	1716	1814	1819	1832
avg hrs of sunshine	8	7	7	6	6	6	5	3	7	8	9	8
avg # days w/thunder	1	1	2	3	1	<1	<1	<1	0	<1	<1	1
w/snow	0	0	0	0	0	0	0	0	0	0	0	0

Rio de Janeiro

22°49'S, 43°15'W
Elev: 6m (20ft)

The weeks around Carnaval are some of the sultriest in Rio, with a thunderstorm every few days and occasional spikes of dry heat from the interior. Winter is still on the warm side, with just as much sun but less humid nights. A lingering winter front can bring a day or two of coolish dampness.

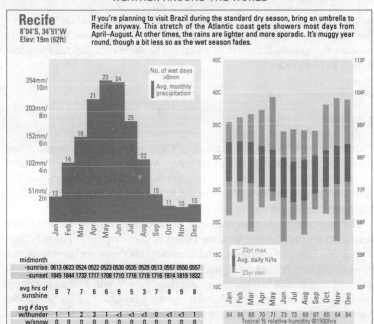

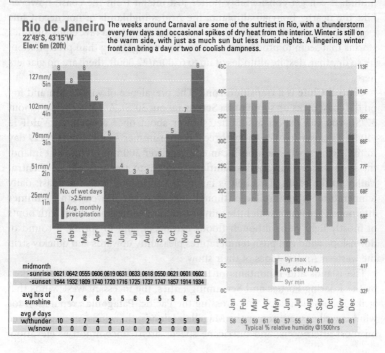

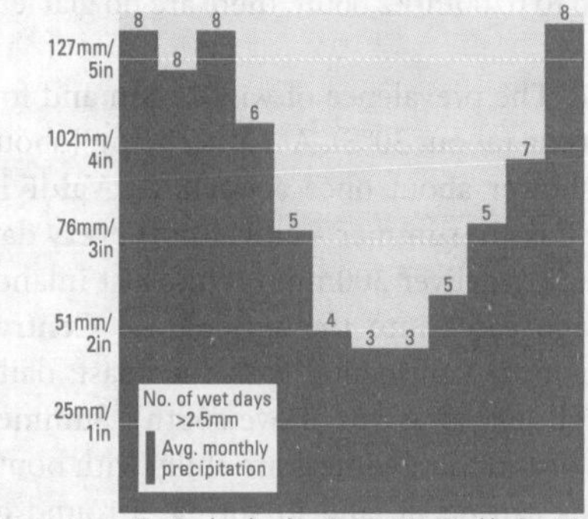

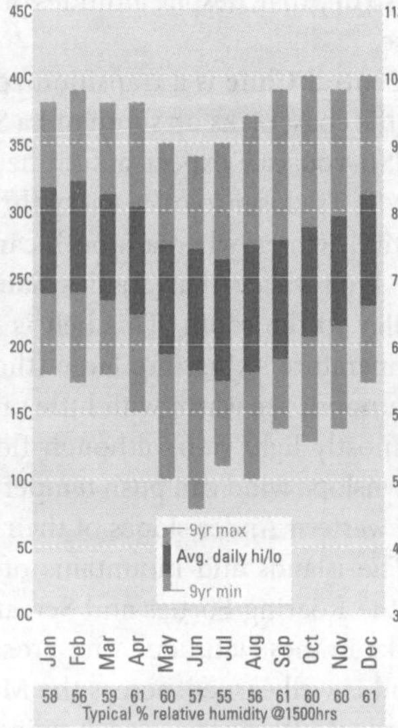

	Jan	Feb	Mar	Apr	May	Jun	Jul	Aug	Sep	Oct	Nov	Dec
midmonth -sunrise	0621	0642	0555	0606	0619	0631	0633	0618	0550	0621	0601	0602
-sunset	1944	1932	1809	1740	1720	1716	1725	1737	1747	1857	1914	1934
avg hrs of sunshine	6	7	6	6	6	5	6	6	5	5	6	5
avg # days w/thunder	10	9	7	4	2	1	1	2	2	3	5	9
w/snow	0	0	0	0	0	0	0	0	0	0	0	0

below 500mm/19in from **Petrolina** northeast toward the coast. Moreover, the Nordeste's rainfall is exceedingly erratic – it's common for a year to produce far more or far less than average. The worst droughts tend to occur during **El Niño** years.

Chile

This is the world's most linear country: **Chile** spans 38° of latitude but less than 2° of longitude at its narrowest point. All along that length, the Andes and the Pacific face off. This means the gradual transition from sub-tropic to sub-polar along Chile's length is rivalled by the changes that occur in a short trip from coastline to mountaintop.

The coastal dryness that begins in **Ecuador** and extends through **Peru** reaches its arid apex in the **Atacama Desert** of northern Chile. The coastlines and nearby hills of **Antofagasta** and **Tarapacá** don't even get the drizzle that eases conditions further north and south: this region can go years without any moisture at all. (**Arica** averages less than 1mm/0.04in a year.) In winter, the low coastal overcast often breaks, leaving a bright, but cool, afternoon. Most summer days are crystal clear and just warm enough to feel slightly humid. It can get considerably hotter just inland from the coast, however. Like the nearby Bolivian altiplano, the **Andes highlands** of northern Chile experience huge daily temperature ranges year round and raw, chilly summer storms that seem to produce more thunder and lightning than precipitation. It's so dry that, despite altitudes of up to 6700m/22,000ft, there are no glaciers here.

Central Chile is a transition zone. The prevalence of winter rain and fog on the coast picks up south of **La Serena** (about 30°S). At **Concepción** (about 37°S), you can expect a summer shower about once a week. At **Valdivia** (40°S) it's closer to once every three days in summer – and almost every day in mid-winter, when a month can easily deliver 300mm (10in). Just inland, the high valley that cradles **Santiago** eases into the agricultural **Central Valley** to the south. This belt is far more continental than the coast: daily temperature swings are large (though less so as you move south). Summer afternoons are warm with little rain or thunder; winters are chilly with bouts of mostly light rain, although floods can occur, and in spring, a round of downslope wind can push temperatures pleasantly upwards and quickly strip the western Andes slopes of their snow.

The islands and mountains of far southern Chile are perpetually raked by the Roaring Forties and Screaming Fifties, the incessant west winds that strike in these latitudes after crossing the Pacific. **Patagonia**'s reputation for wind is well earned: across the **Magellan Islands**, the winds reach gale force on most afternoons, and hurricane speeds aren't uncommon at elevation.

Easter Island

27°09'S, 109°25'W
Elev: 47m (154ft)

Far enough into the Pacific to avoid the desiccation of the Chilean desert, Easter Island gets a generous amount of rain throughout the year (very little thunder and no hurricanes, though). It's marginally wetter and warmer from December–May. The dry-season nights may feel quite cool.

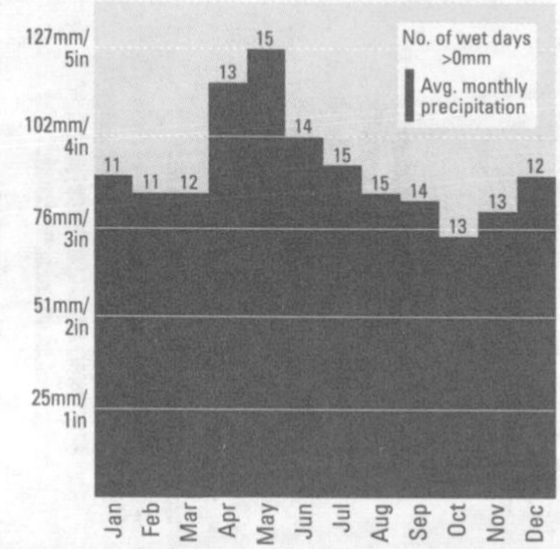

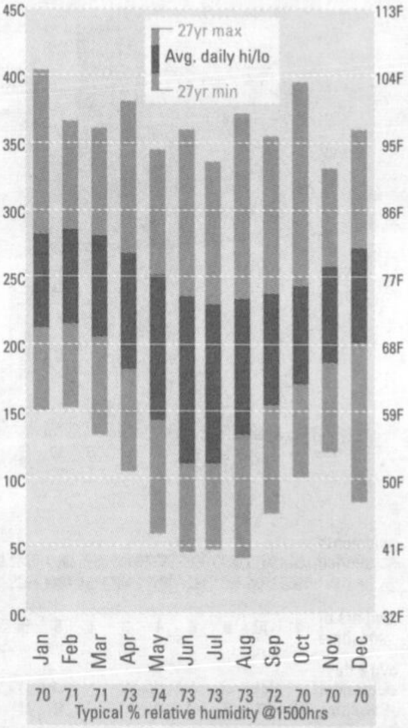

	Jan	Feb	Mar	Apr	May	Jun	Jul	Aug	Sep	Oct	Nov	Dec
midmonth -sunrise	0737	0802	0719	0734	0750	0805	0806	0748	0715	0742	0719	0718
-sunset	2117	2102	1934	1901	1837	1831	1841	1857	1911	2025	2046	2108
avg hrs of sunshine	n/a	n/a	n/a	n/a	n/a	n/a	n/a	n/a	n/a	n/a	n/a	n/a
avg # days w/thunder	<1	<1	<1	<1	<1	<1	<1	<1	0	<1	<1	0
w/snow	0	0	0	0	0	0	0	0	0	0	0	0

Punta Arenas

53°00'S, 70°51'W
Elev: 37m (121ft)

Temperatures seldom rise much above 16°C/60°F across far south Patagonia. The incessant wind makes it feel even chillier, and sunshine tends to be sporadic. It's cloudiest and windiest in summer. Light rain or snow can fall any time of year. Expect conditions to vary drastically across short distances and with altitude.

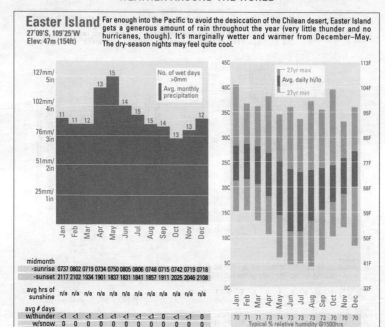

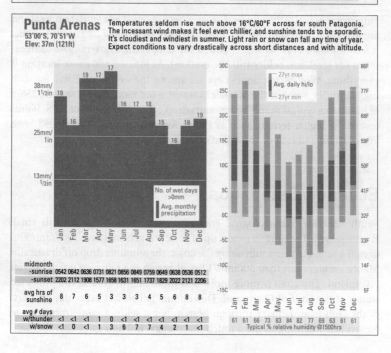

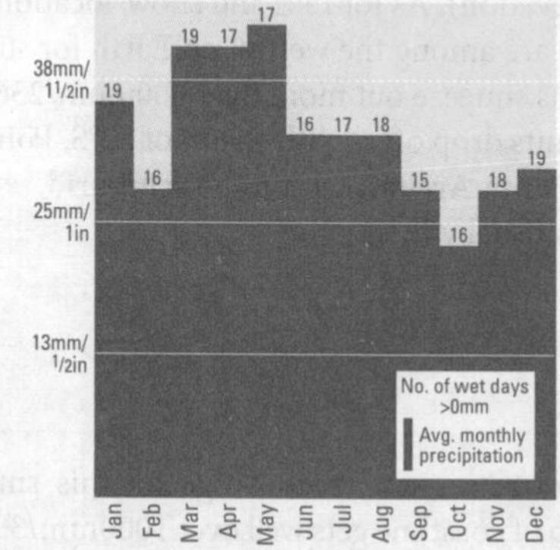

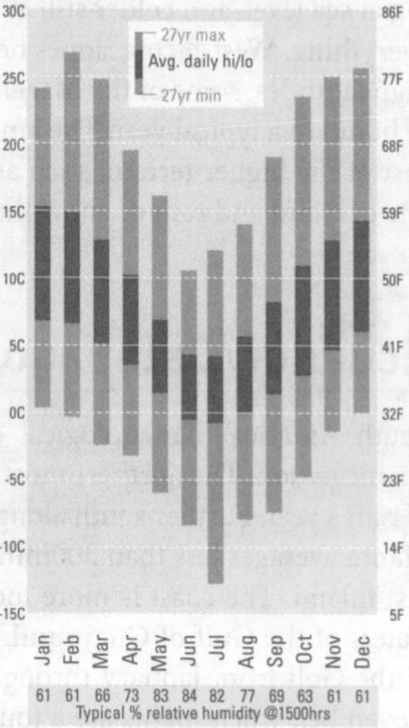

	Jan	Feb	Mar	Apr	May	Jun	Jul	Aug	Sep	Oct	Nov	Dec
midmonth -sunrise	0542	0642	0636	0731	0821	0856	0849	0759	0649	0638	0536	0512
-sunset	2202	2112	1908	1577	1658	1631	1651	1737	1829	2022	2121	2206
avg hrs of sunshine	8	7	6	5	3	3	3	4	5	6	8	8
avg # days w/thunder	<1	<1	<1	1	0	<1	<1	<1	<1	<1	<1	<1
w/snow	<1	0	<1	1	3	6	7	7	4	2	1	<1

Santiago
33°23'S, 70°47'W
Elev: 574m (1555ft)

Nestled in a basin, Santiago gets more ups and downs in temperature than the moist Pacific coast nearby. Nights are cool year round, although they seldom drop much below freezing. Gusty downslope winds can cause daytime readings to soar. The Santiago basin suffers serious air pollution, especially in the winter.

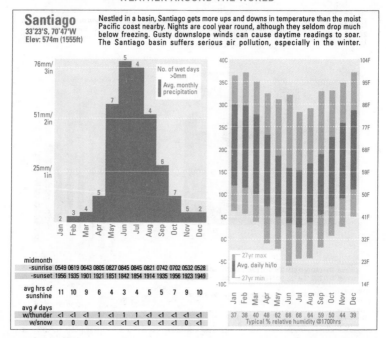

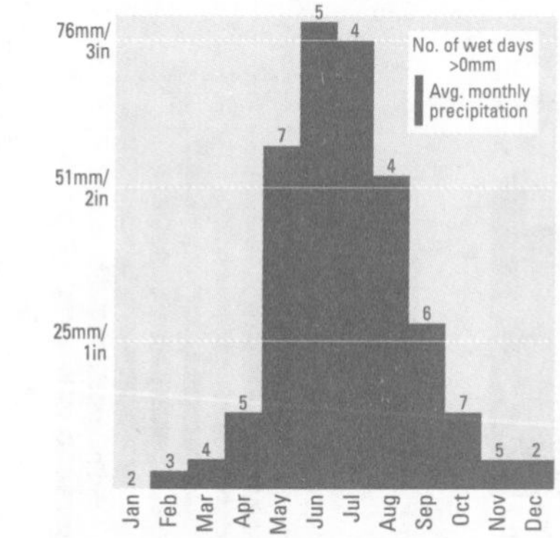

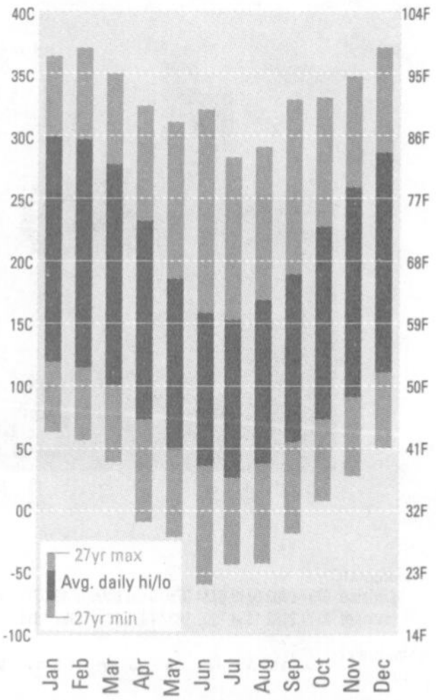

	Jan	Feb	Mar	Apr	May	Jun	Jul	Aug	Sep	Oct	Nov	Dec
midmonth -sunrise	0549	0619	0643	0805	0827	0845	0845	0821	0742	0702	0532	0528
-sunset	1956	1935	1901	1921	1851	1842	1854	1914	1935	1956	1923	1949
avg hrs of sunshine	11	10	9	6	4	3	4	5	5	7	9	10
avg # days w/thunder	<1	<1	<1	1	<1	1	1	<1	<1	<1	<1	<1
w/snow	0	0	0	<1	<1	<1	<1	0	<1	<1	0	<1
Typical % relative humidity @1700hrs	37	38	40	48	62	68	68	64	59	50	44	39

Summer is actually the windiest season here. Even at sea level, temperatures are low enough in summer (average highs are well below 20°C/68°F) so that it feels almost as raw as winter (when the temperature hovers close to freezing at sea level, and colder still at elevation). As for rain and snow, location is everything. West-facing slopes here are among the wettest on Earth for such high latitudes; some of the Magellans squeeze out more than 6000mm/236in of liquid in a typical year. The amounts drop off slightly south of 52°S. Points east of the higher terrain, such as **Punta Arenas**, tend to be semi-arid, with lots of cloud and rain or snow that's usually light.

Ecuador/Galapagos

South America's climatological contrasts seem to converge on this small Pacific nation. The northernmost tip of coastline gets well over 1000mm/39in of rain a year. Further south along the coast, the amounts drop off drastically. **Manta** averages less than 300mm/12in, although the rains are much heavier just inland. The coast is more moist and more humid around the protected waters of the **Gulf of Guayaquil**. Downpours fall almost every day on parts of the Gulf from January through March, but it seldom rains from June to November, when nights are a touch less muggy. **El Niño** can boost the rain-

Guayaquil

27°58'N, 110°56'W
Elev: 9m (30ft)

Just east of the coastal desert, Guayaquil gets a healthy dose of rain from January into April. It's usually bone-dry from June into December. It's fairly warm and humid in the dry season, more so when the rains arrive. Expect the micro-climate to vary wildly on a short trip in almost any direction.

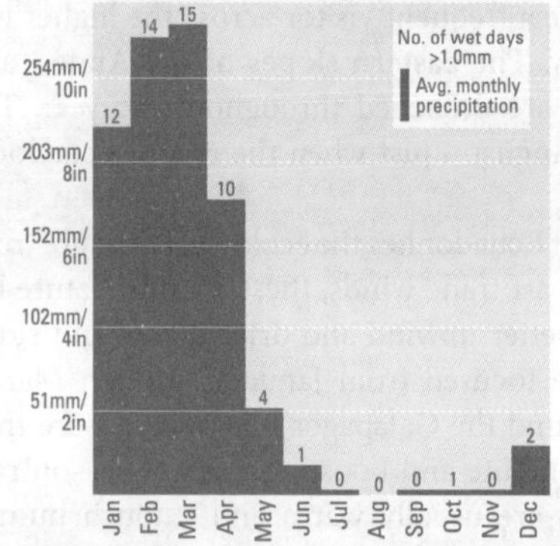

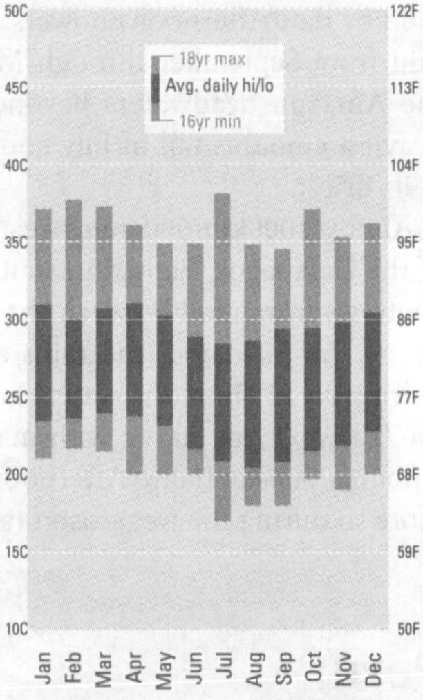

	Jan	Feb	Mar	Apr	May	Jun	Jul	Aug	Sep	Oct	Nov	Dec
midmonth -sunrise	0622	0628	0625	0618	0615	0620	0625	0623	0612	0601	0558	0607
-sunset	1836	1839	1832	1821	1816	1820	1826	1825	1818	1810	1811	1822
avg hrs of sunshine	3	4	5	5	5	4	4	4	5	4	4	4
avg # days w/thunder	1	3	4	4	2	1	<1	<1	<1	<1	1	<1
w/snow	0	0	0	0	0	0	0	0	0	0	0	0

	Jan	Feb	Mar	Apr	May	Jun	Jul	Aug	Sep	Oct	Nov	Dec
Typical % relative humidity @1600hrs	60	64	61	62	61	62	60	57	56	57	55	54

Quito

0°09'S, 78°29'W
Elev: 2812m (9223ft)

Though it's got a reputation for rain, Quito is merely a touch wetter than other highland points to the north and well to the south. About every third day brings a shower or thunderstorm from October–May, with a slight drop-off in November–December. Readings are cool year round; extreme heat and cold are rare.

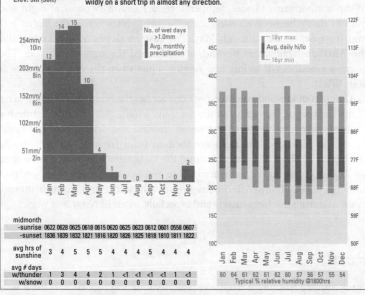

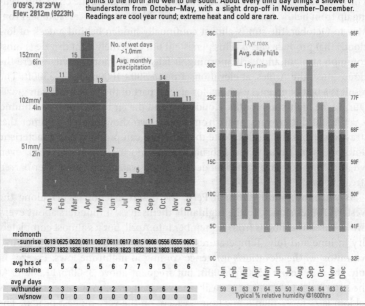

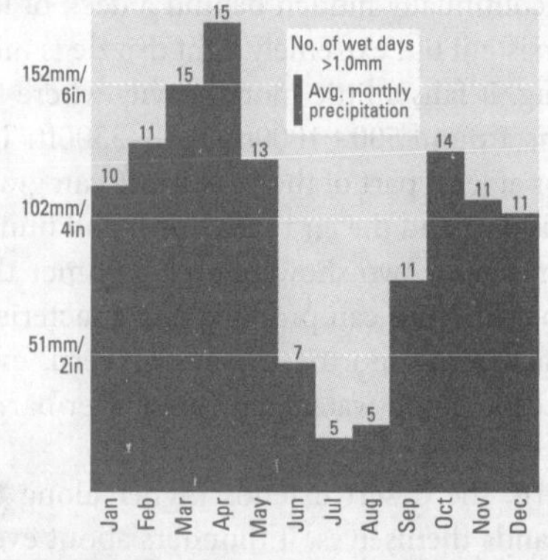

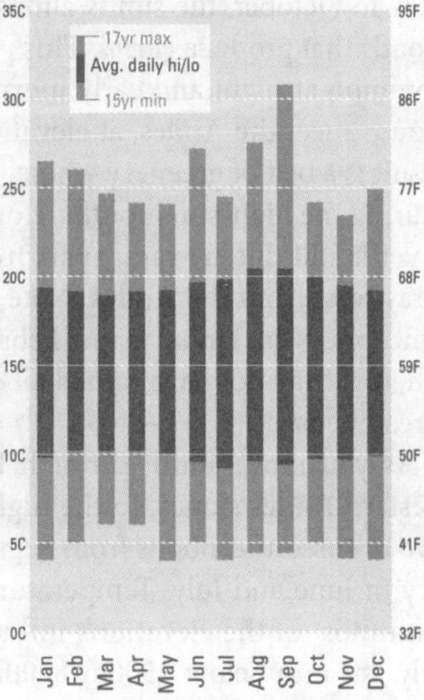

	Jan	Feb	Mar	Apr	May	Jun	Jul	Aug	Sep	Oct	Nov	Dec
midmonth -sunrise	0619	0625	0620	0611	0607	0611	0617	0615	0606	0556	0555	0605
-sunset	1827	1832	1826	1817	1814	1818	1823	1822	1812	1803	1802	1813
avg hrs of sunshine	5	5	4	5	5	6	7	7	9	5	6	6
avg # days w/thunder	2	3	5	7	4	2	1	1	5	6	4	2
w/snow	0	0	0	0	0	0	0	0	0	0	0	0

	Jan	Feb	Mar	Apr	May	Jun	Jul	Aug	Sep	Oct	Nov	Dec
Typical % relative humidity @1600hrs	59	61	63	67	64	55	50	49	56	64	63	62

fall totals considerably; during the severe 1982–83 event, Guayaquil more than doubled its wet-season average of 1000mm/39in. Across the **highlands** of the northernmost Andes, temperatures are normally chilly at night and mild by day. Afternoon showers are a frequent visitor across the higher terrain from September through May. The **eastern slopes of the Andes**, and **the Amazon headwaters** beyond, are drenched throughout the year. The heaviest amounts fall in July and August – just when the rest of Ecuador is at its driest.

Almost 1000km/600 miles west of Ecuador lies the ecological treasure trove of the **Galapagos**. Bathed in southeast trade winds, these islands feature big contrasts in rainfall between the wetter upwind and drier downwind sides. As on the mainland, the rains are focused from January through March. During an **El Niño**, the waters around the Galapagos may warm more than 6°C/10°F, playing havoc with marine life and boosting the wet-season rain amounts substantially. Afternoons are usually warm and a touch humid, more so during the wet season (and especially when El Niño is at work).

Peru

For a country fronting a tropical ocean, it's amazing that **coastal Peru** is as dry as it is. Some towns along the Pacific go years without measurable rain. The desert air, shaped by cold waters moving north along the coast and welling up from below, tends to be humid, yet mild, for its tropical latitude. From May to October, the sun is almost continually hidden behind a deck of low clouds that produce *garua*. This persistent but extremely light drizzle is most common at night and early morning. It falls a little more heavily where the clouds meet the Andes, at elevations around 500– 1000m/1650–3300ft. The result is a belt of greenery where, for at least part of the year, cattle can graze. During the high-sun months, clouds part and the air turns warm and humid – yet it still doesn't rain, apart from one or two showers each summer that stray west from the Andes. Note that **El Niño** can produce uncharacteristic downpours in January and February across northern parts of Peru, even where it may not have rained all decade; flood waters may pour over barren streambeds.

As you move **south** through Peru, the desert extends higher along the western Andes slopes. In the **highlands** themselves, it thunders about every two or three afternoons from September into April, but it's almost completely dry in June and July. Temperatures cool only gradually with height at first, with most of the elevation's influence coming in nighttime lows. Freezes in July are rare below 2500m/8200ft, but frequent above 3500m/11,500ft. At the latter height, it can just as easily snow as rain, even in summer (the snow seldom lasts more than a few hours, except on the very highest peaks). The

Lima
12°00'S, 77°07'W
Elev: 13m (43ft)

Although it sits astride the Pacific, Lima gets only about as much rainfall as Cairo, Egypt. Summers are sunny (a shower here is big news) and oddly humid for such a dry town. From May–November, there is virtually no sun, except toward the end of a day of cool overcast and ultra-light *guara* drizzle.

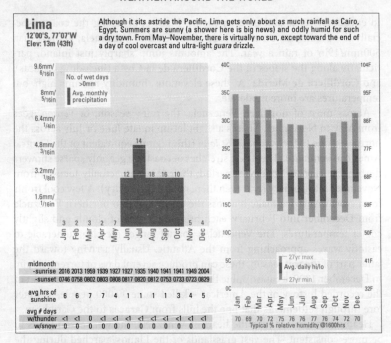

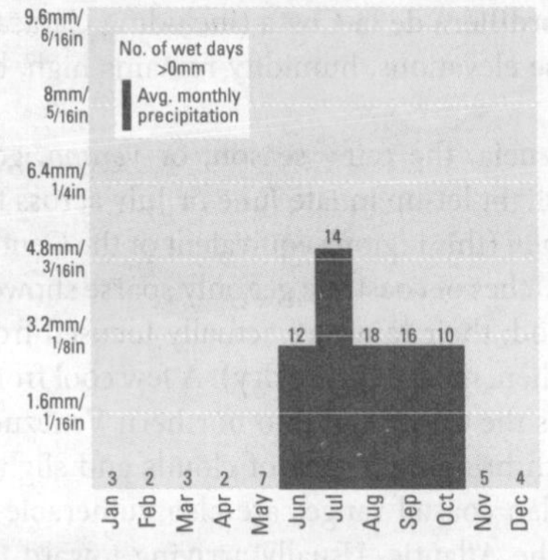

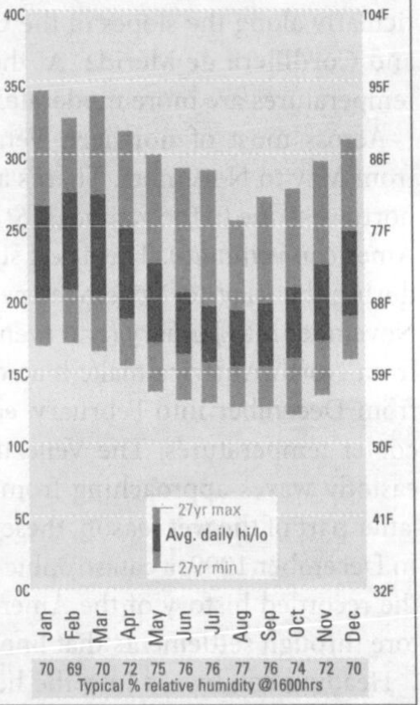

	Jan	Feb	Mar	Apr	May	Jun	Jul	Aug	Sep	Oct	Nov	Dec
midmonth -sunrise	2016	2013	1959	1939	1927	1927	1935	1940	1941	1941	1949	2004
-sunset	0746	0758	0802	0803	0808	0817	0820	0812	0753	0733	0723	0829
avg hrs of sunshine	6	6	7	7	4	1	1	1	1	3	4	5
avg # days w/thunder	<1	<1	0	<1	<1	<1	<1	<1	<1	<1	<1	<1
w/snow	0	0	0	0	0	0	0	0	0	0	0	0

Andes gradually descend into Peru's two eastern flanks: the **Madre de Dios** of the southeast, where rains are focused in the October-to-May wet season, and the **Amazon headwaters** of the northeast, where afternoon deluges occur almost every day throughout the year. There's only a slight easing of the Amazonian downpours around July, when the monthly average rainfall at **Iquitos** is at its lowest: close to 150mm/6in.

Venezuela

Colombia | French Guiana | Guyana | Suriname

The usual ingredients of tropical climate – sharp moisture contrasts, temperatures that hang steady – come together in intriguing ways through the belt of countries across the north end of South America.

The Caribbean coast of **Venezuela** and **Colombia** is famous for its oddly dry climate, with striking differences in rainfall across short distances. Although the coast stays warm and humid year round, the easterly trade winds tend to parallel the shoreline, rather than blowing onshore. That – along with some large-scale climate mechanisms still being unravelled – makes the coast from **Río Caribe** westward to northern **Colombia** far more arid than one might expect. Instead of rainforest, you're more likely to encounter scrubland or even cactus. Many spots along the coast, especially the peninsulas on either side of the **Gulf of Venezuela**, get less than 500mm/19in of rain a year. The amounts jump sharply just inland, particularly along the slopes of the **Cordillera de la Costa** (including Caracas), and **Cordillera de Mérida**. At these elevations, humidity remains high, but temperatures are more moderate.

Across most of **northern Venezuela**, the rainy season, or *verano*, goes from May to November. There's a slight let-up in late June or July across the northwest due to the *verano de St John* (this region's equivalent of the Central American *veranillo*). The driest stretches of coastline get only sparse showers during the May-to-November period; their rains are actually focused from November into January (and even then, most days are dry). A few cool fronts from North America make it across the Caribbean into northern Venezuela from December into February, each bringing a spell of clouds and slightly cooler temperatures. The Venezuelan coastal ranges are also vulnerable to easterly waves approaching from the Atlantic. Usually arriving toward the latter part of the wet season, these can produce several days of torrential rain. In December 1999, a catastrophic flash flood – the deadliest weather event in the recorded history of the Americas – killed some 30,000 people, as water tore through settlements that line the hills from **Caracas** to the Caribbean.

Heading southward into the heart of Venezuela, the wet season tends to be more consistent. The vast grasslands of the **Llanos**, parched during the months of low sun, are inundated by torrential thunderstorms on most days from April to November. Even heavier rains fall across the **Guiana Highlands**, where the altitude yields somewhat milder temperatures. The **Amazon lowlands** extend from Brazil into southeast Colombia and southernmost Venezuela; here, you should be prepared for sultry conditions and a possible heavy thunderstorm most any time of year. The heaviest rains fall from April to July. Annual totals top 3000mm/118in across much of this area. Rainfall is also extensive as you move from Venezuela into the forests of **Guyana**, **Suriname** and **French Guiana**. The intertropical convergence zone (ITCZ) sweeps through this region twice each year, following the sun. The ITCZ tends to produce near-daily torrential storms as it shifts north (roughly April to July) and less-intense, but more-prolonged, nighttime rains as it passes to the south (November–February). Except on the easternmost Guiana Highlands, these countries are sweltering year round.

If it's rain you seek, you'll be hard-pressed to beat the **Pacific coast** of Colombia, which ranks with northeast India as one of the world's soggiest spots. As moist southwest winds – contrary to the usual easterly trades – encounter the Andes, they dump buckets of rain throughout the year along the coastal plain and adjacent slopes. Some points around **Quibdó** average more than 10,000mm/390in a year, and in 1936 the city itself scored a record for the Americas with 19,839mm/782in. There's method to this moist mad-

Spinning fiction out of Colombian climate

The fractured mix of coastline and mountain across northernmost Colombia and Venezuela is laced with small-scale weather regimes that violate the normal rules of climate. As the childhood home of novelist Gabriel García Márquez, this land was fertile ground for the blend of magic and realism that permeates his writing. García Márquez grew up in Aracataca, a town between the east coast of the Gulf of Venezuela and the compact Sierra Nevada de Santa Marta. Because the cool oceanic inversion is so strong along the Colombian coast, trade winds can't ascend the Sierra and produce the usual heavy rains on the windward side. Instead, they wrap around the mountains and converge on the leeward side, making the region around **Aracataca** more than four times wetter than the **Guajira Peninsula** (less than 160km/100 miles to the north). As geographer Gary Elbow has noted, it only took a dash of literary licence for García Márquez to write in *One Hundred Years of Solitude* about a rain that lasted "four years, eleven months, and two days" – or to depict the Guajira Peninsula in his novella Eréndira as a place with "miserable and burning streets where the goats committed suicide from desolation when the wind of misfortune blew".

ness. Generally, the slopes get afternoon thunderstorms and lighter nighttime rains. The lower elevations typically get a period of light, steady rain from sundown to sun-up, with the bulk of the day frequently bright and dry – a convenient window for intrepid visitors who want to see the wettest spot in this half of the world.

Caracas

10°30'N, 66°53'W
Elev: 835m (2739ft)

This city above the Caribbean can sizzle, but it seldom does. Most summer afternoons are merely warm, with a shower about every other day from June–October. It can pour as late as December on rare occasions. With the exception of an occasional cloudy, drizzly cool spell, winters are bright, mild, and dry.

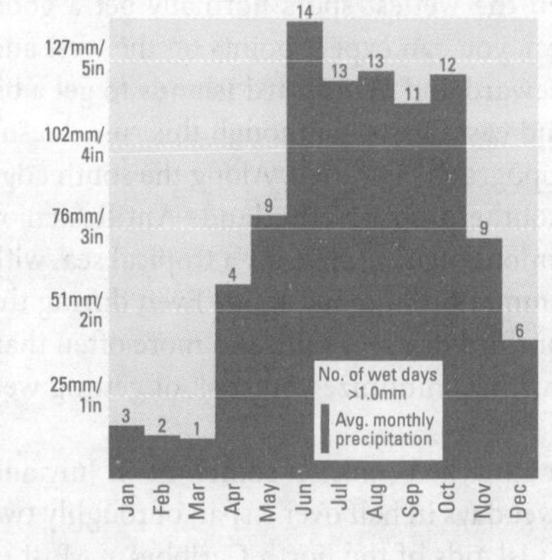

	Jan	Feb	Mar	Apr	May	Jun	Jul	Aug	Sep	Oct	Nov	Dec
midmonth -sunrise	0650	0648	0635	0617	0606	0606	0613	0618	0617	0616	0623	0637
-sunset	1824	1836	1838	1838	1842	1850	1854	1846	1828	1810	1801	1808
avg hrs of sunshine	7	8	8	6	6	6	7	7	7	7	7	7
avg # days w/thunder	<1	<1	<1	1	1	1	1	3	3	3	2	<1
w/snow	0	0	0	0	0	0	0	0	0	0	0	0

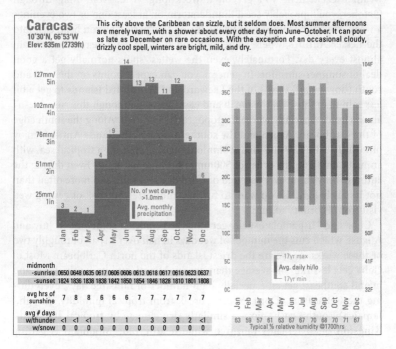

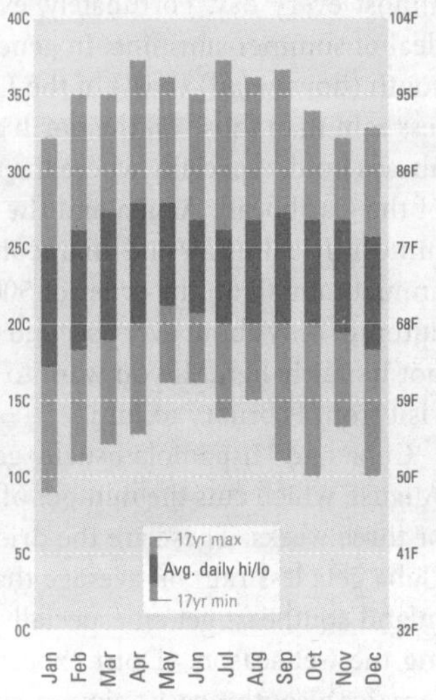

	Jan	Feb	Mar	Apr	May	Jun	Jul	Aug	Sep	Oct	Nov	Dec
Typical % relative humidity @1700hrs	63	59	57	61	63	67	67	70	68	70	71	67

Caribbean/Atlantic Islands

Aruba | Bahamas | Bermuda | Cuba | Dominican Republic | Haiti | Jamaica | Netherlands Antilles

The homogeneous cultures that extend across much of the **Caribbean** have been shaped in part by a beneficent climate. Across this great arc of islands, temperatures gradually rise and fall from pleasantly warm levels in January to readings by July that are definitely tropical but seldom scorching. The most distinct transitions occur across the westernmost islands of **Cuba** and **Hispaniola** (including **Haiti** and the **Dominican Republic**), as well as the islands of the **Bahamas**. **Havana** and **Nassau** often drop below 16°C/60°F in January, but seldom below 10°C/50°F. The rare damp days in mid-winter are often the coolest. A few winter days also see rainfall along the northeast coasts of the Dominican Republic and **Puerto Rico**, where trade winds – reinforced by cool high pressure over the Atlantic – push moisture up steep mountain slopes. Across the flat Bahamas, the cool spells tend to be less moist, and their likelihood decreases as you head from northwest to southeast.

Across most of the Caribbean, the core of the dry season extends from February to April. During this span, a stout trade wind may be the most weather excitement you get on a week-long visit. From May through October, you'll probably see a thunderstorm on at least one or two evenings a week. Storms become more likely during this period as you head east into the **Lesser Antilles**, where some highland stations get a heavy dose of rain almost every day. Fortunately, even the wettest spots normally get a good deal of summer sunshine. In general, you can expect points on the west and south (downwind) coasts of the **Leeward** and **Windward Islands** to get a bit less rain than those on the north and east coasts – although this rule doesn't always hold, especially where the topography is varied. Along the south edge of the Caribbean, **Aruba** and the southernmost **Netherlands Antilles** enjoy uniformly balmy air and an uncommonly arid climate for a tropical sea, with annual rainfall on the order of 500mm/19in in many spots. Even during the autumn, only about every second or third day sees rain, and more often than not it's fairly light. If you want to further minimize your risk of getting wet, visit from February to July.

Cuba and Hispaniola usually get a marked *canicula* sometime in July and August, which cuts the number of wet days in half over a span of roughly two or three weeks. These are the driest islands of the north Caribbean. Most of Cuba gets less rain on average than nearby Miami, and its south coasts and inland southeast get an especially high number of rain-free days, even during the wet season. (Don't expect a break from tropical humidity, though.) **Jamaica**'s south coast enjoys a similarly dry climate by tropical standards. **Kingston** often remains rain-free while nearby Blue Mountain is getting

Basseterre

17°17'N, 62°44'W
Elev: 9m (30ft)

This west-coast town gets considerably less rain than Martinique's eastern highlands. A downpour could strike every second or third day from June through November, when heat, humidity, and hurricane risk peak. Rains only gradually increase during the pleasingly balmy period from January to May.

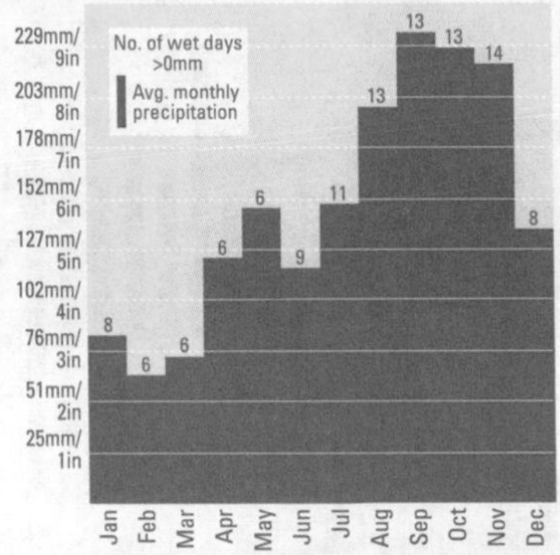

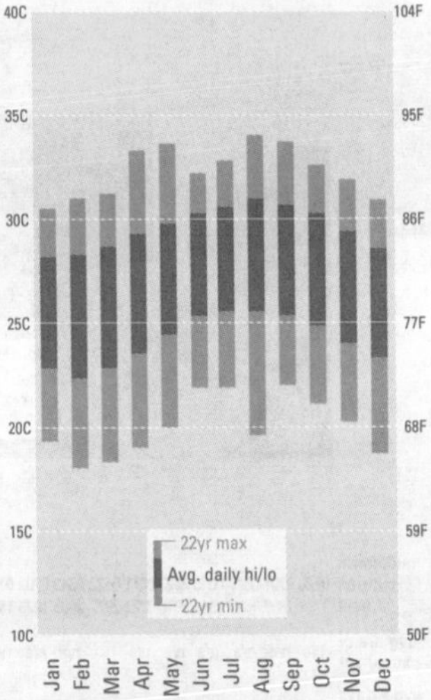

	Jan	Feb	Mar	Apr	May	Jun	Jul	Aug	Sep	Oct	Nov	Dec
midmonth -sunrise	0644	0638	0619	0555	0539	0537	0545	0554	0559	0604	0616	0633
-sunset	1757	1813	1821	1827	1836	1846	1849	1837	1813	1749	1735	1739
avg hrs of sunshine	8	8	8	8	8	8	8	8	7	7	7	7
avg # days w/thunder	0	<1	0	0	<1	<1	<1	1	1	<1	<1	0
w/snow	0	0	0	0	0	0	0	0	0	0	0	0

	Jan	Feb	Mar	Apr	May	Jun	Jul	Aug	Sep	Oct	Nov	Dec
Typical % relative humidity @1700hrs	71	70	70	71	74	73	73	74	74	76	77	74

Fort-de-France

14°36'N, 61°00'W
Elev: 4m (13ft)

Expect at least one or two rainstorms per week in Martinique no matter when you visit. Downpours interrupt sunshine almost daily from June–January. The island's northeast side is a bit drier in winter but prone to particularly heavy rains from west-moving systems (including hurricanes) during the wet season.

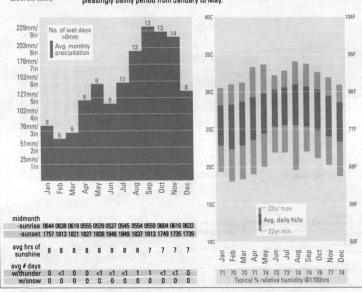

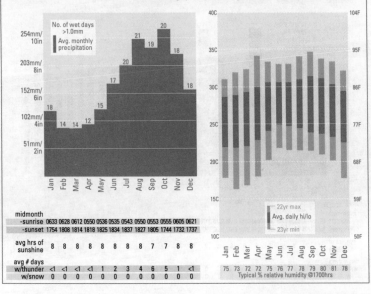

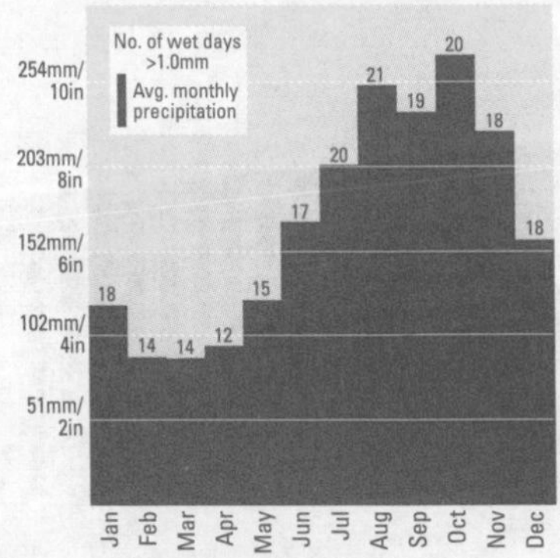

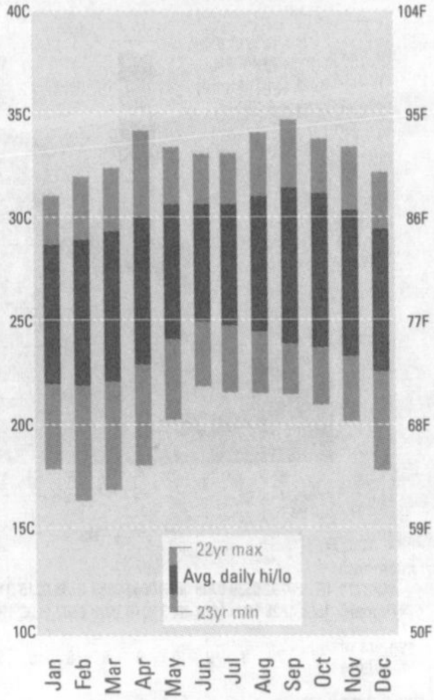

	Jan	Feb	Mar	Apr	May	Jun	Jul	Aug	Sep	Oct	Nov	Dec
midmonth -sunrise	0633	0628	0612	0550	0536	0535	0543	0550	0553	0555	0605	0621
-sunset	1754	1808	1814	1818	1825	1834	1837	1827	1805	1744	1732	1737
avg hrs of sunshine	8	8	8	8	8	8	8	8	7	7	8	8
avg # days w/thunder	<1	<1	<1	<1	1	2	3	4	6	5	1	<1
w/snow	0	0	0	0	0	0	0	0	0	0	0	0

	Jan	Feb	Mar	Apr	May	Jun	Jul	Aug	Sep	Oct	Nov	Dec
Typical % relative humidity @1700hrs	75	73	72	72	75	76	77	78	79	80	81	78

Hamilton

32°22'N, 64°40'W
Elev: 34m (110ft)

Like eastern North America, Hamilton is a well-watered place, with healthy rains through the year and thunder possible in any month. All the moisture helps keep temperatures mild (albeit a bit sticky in summer), and there's plenty of sun. Hurricanes and hybrid storms may threaten in summer and autumn.

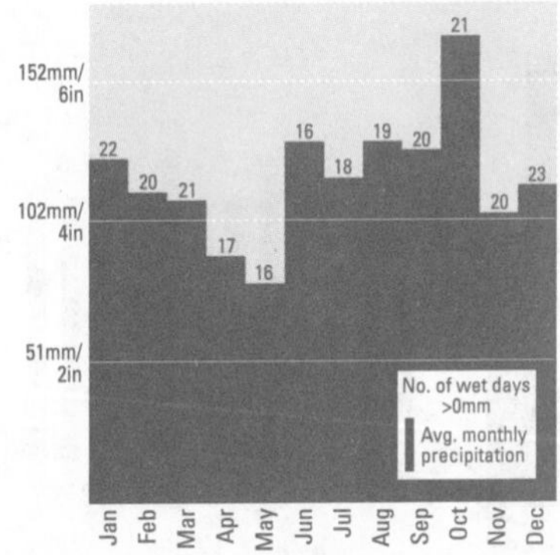

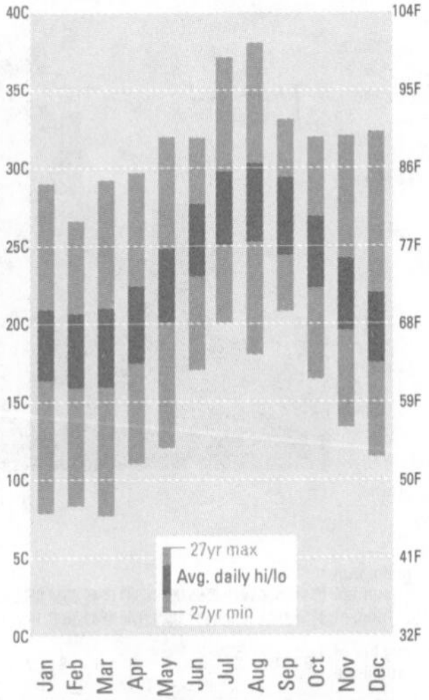

Jan	Feb	Mar	Apr	May	Jun	Jul	Aug	Sep	Oct	Nov	Dec
69	69	68	68	72	75	72	71	71	71	69	68

Typical % relative humidity @1700hrs

	Jan	Feb	Mar	Apr	May	Jun	Jul	Aug	Sep	Oct	Nov	Dec
midmonth -sunrise	0720	0701	0629	0650	0621	0611	0622	0643	0702	0722	0648	0713
-sunset	1736	1805	1827	1948	2010	2027	2027	2003	1925	1846	1718	1715
avg hrs of sunshine	n/a	n/a	n/a	n/a	n/a	n/a	n/a	n/a	n/a	n/a	n/a	n/a
avg # days w/thunder	3	3	4	3	3	4	8	9	7	5	3	2
w/snow	0	0	0	0	0	0	0	0	0	0	0	0

Havana

22°59'N, 82°24'W
Elev: 59m (194ft)

Havana isn't an arid town, but it's notably drier than Miami. The wet season is broken by a sharp veranillo in July or August, and even in September two out of three afternoons are dry. Summers are sultry and winter days are pleasant, although a few fronts bring clouds and cool breezes, especially in January.

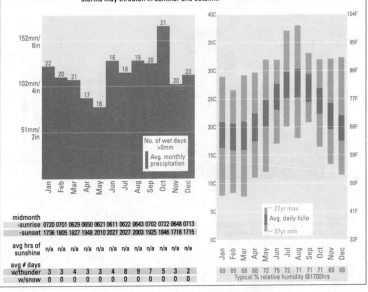

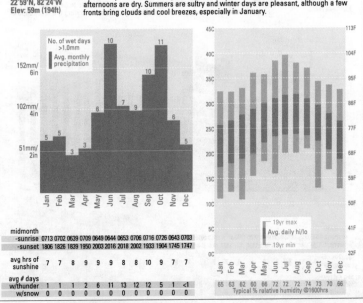

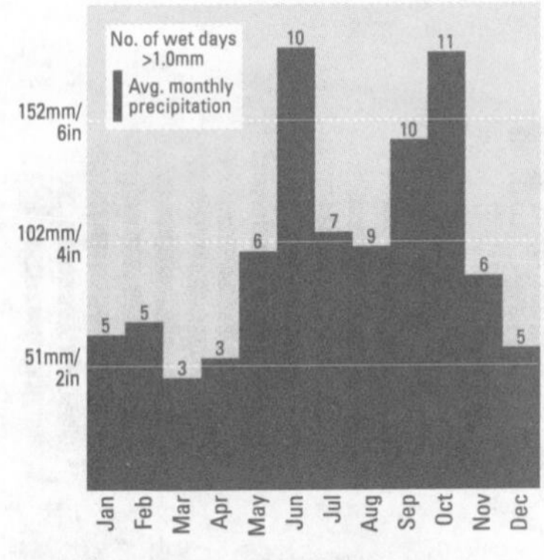

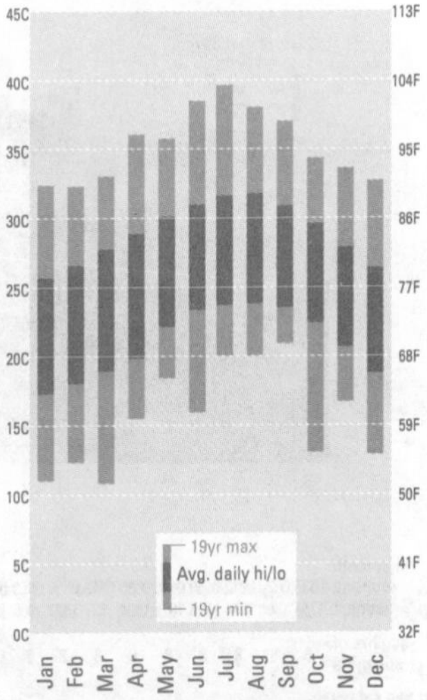

Jan	Feb	Mar	Apr	May	Jun	Jul	Aug	Sep	Oct	Nov	Dec
65	63	62	60	66	72	72	72	74	73	70	66

Typical % relative humidity @1600hrs

	Jan	Feb	Mar	Apr	May	Jun	Jul	Aug	Sep	Oct	Nov	Dec
midmonth -sunrise	0713	0702	0639	0709	0649	0644	0653	0706	0716	0726	0643	0703
-sunset	1806	1826	1839	1950	2003	2016	2018	2002	1933	1904	1745	1747
avg hrs of sunshine	7	7	8	9	9	9	8	8	10	9	7	7
avg # days w/thunder	1	1	1	2	6	11	13	12	12	5	1	<1
w/snow	0	0	0	0	0	0	0	0	0	0	0	0

Kingston

17°56'N, 76°47'W
Elev: 9m (30ft)

Mountains shield Kingston from the prevailing trade winds and help keep thunderstorms just over the horizon. It seldom rains more than once or twice a week, outside of a hurricane. This city endures intense heat and humidity year round. Head for the hills or the beach for relief.

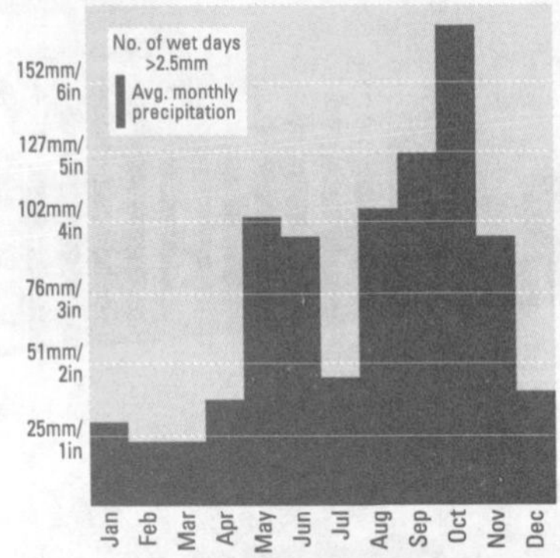

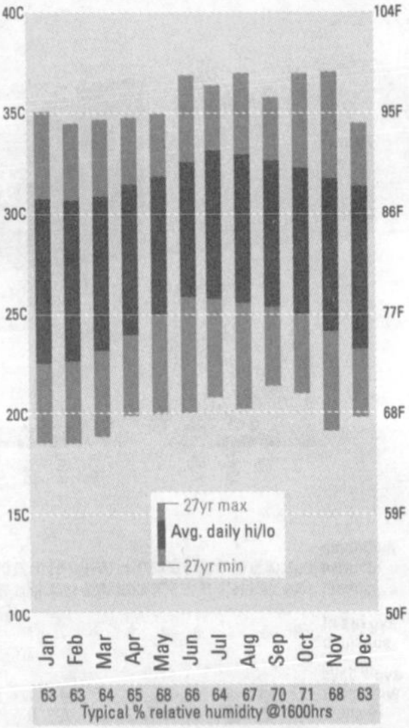

	Jan	Feb	Mar	Apr	May	Jun	Jul	Aug	Sep	Oct	Nov	Dec
midmonth -sunrise	0641	0634	0615	0551	0534	0532	0540	0549	0555	0601	0613	0630
-sunset	1752	1808	1817	1824	1833	1844	1846	1833	1809	1745	1730	1734
avg hrs of sunshine	8	9	9	9	8	8	9	9	8	7	8	8
avg # days w/thunder	<1	<1	<1	1	3	4	5	9	10	9	4	1
w/snow	0	0	0	0	0	0	0	0	0	0	0	0

Nassau

25°03'N, 77°28'W
Elev: 7m (23ft)

Winter in Nassau provides lots of warm afternoons, though an occasional brisk north wind arrives with clouds and readings below 16°C/60°F. The days are hottest and thunderstorms most prevalent (every two or three afternoons) from May–October. The long hurricane season peaks in September.

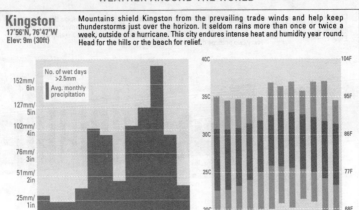

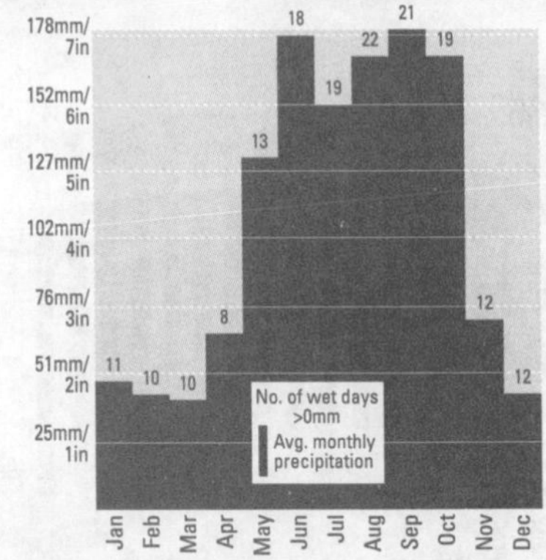

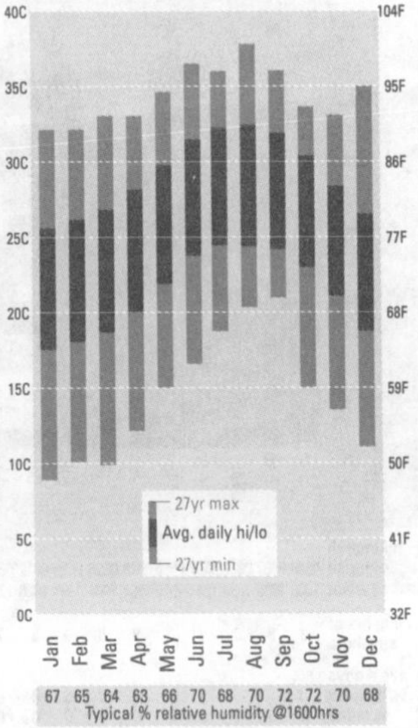

	Jan	Feb	Mar	Apr	May	Jun	Jul	Aug	Sep	Oct	Nov	Dec
midmonth -sunrise	0657	0644	0619	0648	0625	0620	0629	0644	0656	0708	0627	0647
-sunset	1742	1804	1819	1932	1947	2001	2002	1945	1914	1843	1722	1723
avg hrs of sunshine	7	8	8	9	9	8	9	9	7	7	7	7
avg # days w/thunder	1	1	1	2	4	9	12	14	10	4	1	1
w/snow	0	0	0	0	0	0	0	0	0	0	0	0

Port of Spain

10°35'N, 61°21'W
Elev: 15m (49ft)

At close to 10°N, Trinidad gets little dramatic temperature change through the year. The wet season extends from May–December, with showers on most days but plenty of sun as well. Winter nights are a touch less humid, with rain only every two or three days.

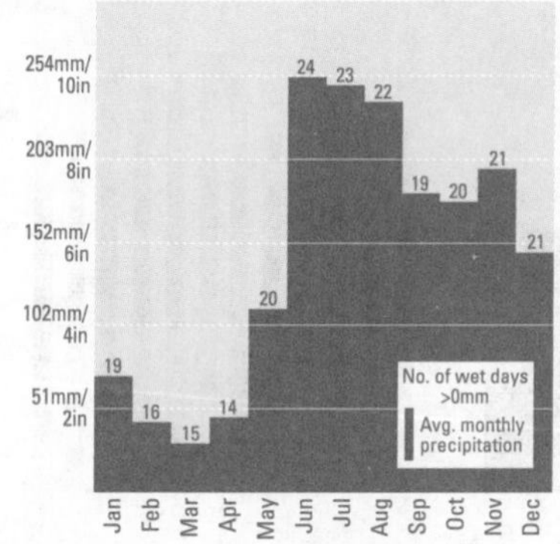

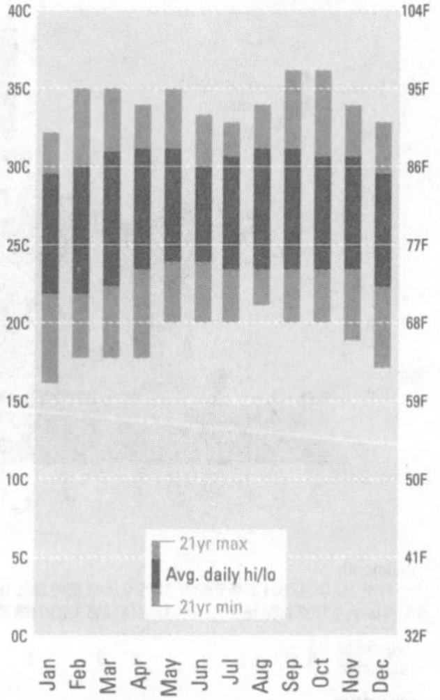

	Jan	Feb	Mar	Apr	May	Jun	Jul	Aug	Sep	Oct	Nov	Dec
midmonth -sunrise	0628	0626	0613	0555	0543	0544	0551	0556	0555	0554	0601	0615
-sunset	1802	1813	1816	1816	1820	1828	1832	1824	1806	1748	1739	1746
avg hrs of sunshine	8	8	8	8	8	6	7	7	7	7	7	7
avg # days w/thunder	<1	<1	<1	<1	1	3	5	6	7	6	3	1
w/snow	0	0	0	0	0	0	0	0	0	0	0	0
Typical % relative humidity @1700hrs	70	67	64	65	70	75	75	76	76	78	79	76

San Juan

18°26'N, 66°00'W
Elev: 3m (9ft)

Although San Juan faces the North Atlantic, winter chill is rare. Nights never dip much below 20°C/68°F. Even during the driest period (February–March), some rain falls every third day or so. Wet-season showers and storms and the occasional hurricane can trigger serious flooding. It's mostly sunny year round.

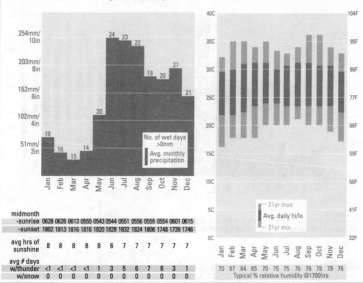

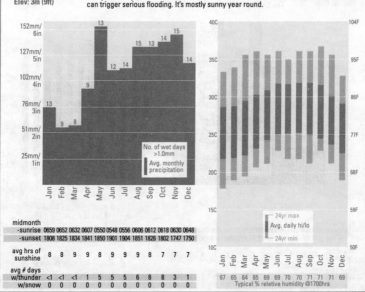

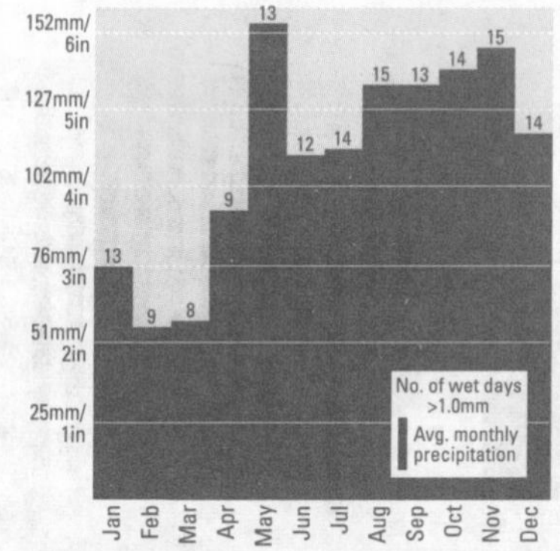

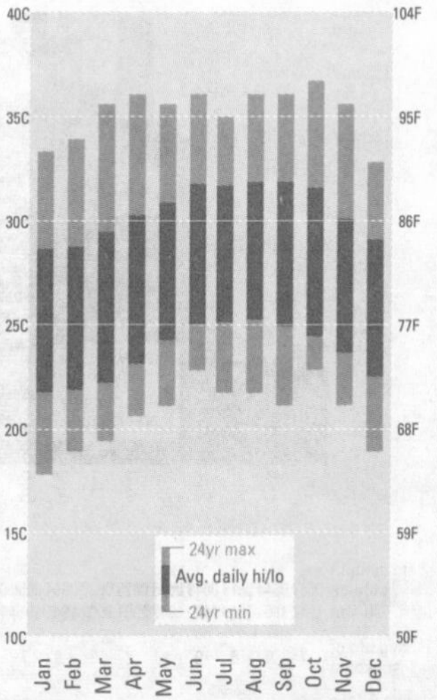

	Jan	Feb	Mar	Apr	May	Jun	Jul	Aug	Sep	Oct	Nov	Dec
midmonth -sunrise	0659	0652	0632	0607	0550	0548	0556	0606	0612	0618	0630	0648
-sunset	1808	1825	1834	1841	1850	1901	1904	1851	1826	1802	1747	1750
avg hrs of sunshine	8	8	9	9	8	9	9	9	8	7	7	7
avg # days w/thunder	<1	<1	<1	1	5	5	5	6	8	8	3	1
w/snow	0	0	0	0	0	0	0	0	0	0	0	0
Typical % relative humidity @1700hrs	67	65	64	65	69	69	70	70	71	71	71	69

Santo Domingo

18°26'N, 69°53'W
Elev: 14m (56ft)

The wet season follows the typical Caribbean pattern, with a downpour about every third day from May–November. Northers that bring coolness and moisture to the north coast of Hispaniola in mid-winter seldom do much here; it rains lightly perhaps once a week or so. Sunshine is ample year round.

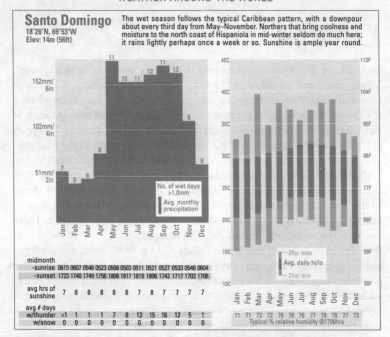

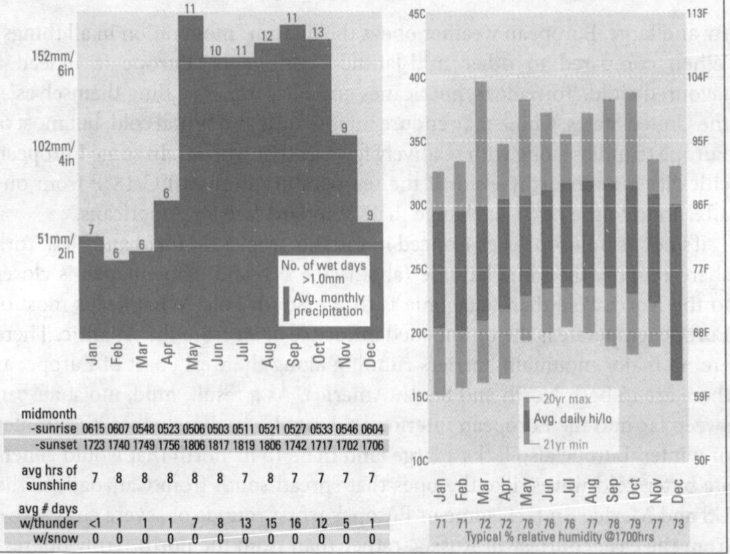

drenched. Haiti, lying downwind from moisture-intercepting mountains, is distinctly drier than the Dominican Republic. Some parts of interior Haiti get well under 1000mm/39in a year, and **Port-au-Prince** soaks in sunshine year round.

Although its weather resembles that of the Caribbean during the warm, humid summer, **Bermuda** – at 32°N – barely qualifies as a sub-tropical island. It's far enough north so that mid-latitude storms can brush by from autumn through spring. Thus, a decent rain may fall any time of year; spring is a bit drier than the other seasons. Wintertime temperatures are dependably cool to mild, with little more risk of a truly cold night than in the Bahamas.

The official **Atlantic hurricane season** runs from June through November. Climatology tends to favour some areas over others as the season unfolds. Some of the worst storms emerge off Africa and develop as Cape Verde hurricanes; these are most likely from August through September and can threaten any part of the region. The Caribbean itself tends to become active later in the season, producing hurricanes that may approach Hispaniola or the northern parts of Central America from September into late October. The Bahamas and, in particular, Bermuda are sometimes hit by hybrid systems late in the season. Part tropical, part mid-latitude, these hybrids seldom cause the widespread destruction that full-fledged hurricanes can produce, but they may still put a damper on a holiday with heavy rain and wind.

Europe

By and large, European weather obeys the dictum "moderation in all things". When compared to other mid-latitude continents, **Europe** is indeed a favoured child. Tornadoes, hurricanes and blizzards may fling themselves at the United States; China may endure intense heat and brutal cold; but most of Europe trundles along with relatively few weather worries. In some European cities it's a noteworthy event if the temperature drops 10°C/18°F from one afternoon to the next – a change that's standard fare for Americans.

It's not as if Europe is ensconced in the sub-tropics. **Madrid** and New York share roughly the same latitude (about 40.5°N), and **Copenhagen** is closer to the North Pole than Patagonia is to the South Pole. What keeps most of Europe temperate is the unimpeded influence of air from the **Atlantic**. There are no major mountain barriers running along the west coast of Europe, as there are in both North and South America. As a result, mild, moist air can sweep far into the European interior on a regular basis, whether it's summer or winter. Europe also lacks a large land mass to its north that would generate bitter cold waves, like the ones that spread south from Canada into the US and Mexico. In fact, some of Europe's worst rounds of severe chill arrive from the east, courtesy of **Russia**, rather than from the north. Truly oppressive heat is rare across northern and central Europe, although it's a different story south of the Alps. As the Mediterranean warms up in summer, countries from **Spain** to **Greece** typically swelter. Occasional bursts of Saharan air, borne on dusty, irritating south winds, can also ratchet up the temperature across large parts of the continent.

With the Atlantic exerting such a big influence, you might expect Europe to be awash in water. Northern and central Europe do get rain and/or snow throughout the year, and certainly the climate will seem moist enough to someone who visits **Amsterdam** in November. When you add up the amounts, however, the rainfall across great stretches of Europe is surprisingly paltry. Most of the continent north of **the Alps**, from **France** across **Germany** to Russia, averages less than 1000mm/39in of rain and melted snow per year. The figures for **London** and **Paris** are even lower, in the region of 600mm/24in – less than the averages for Sydney, New York or San Francisco. In another part of the world, this wouldn't be enough to sustain the kind of agriculture and flora that predominate across Europe. However, the continent's high latitude makes the summer sun too weak to parch plants easily, and the typically high humidities and extensive cloud cover help keep the rain from evaporating too quickly. Also, rather than pouring and running off, much of Europe's rain falls in drizzles, mists or sprinkles, allowing it to be well absorbed by the ground. (The city graphs in this book show the number of days on which rain exceeds a small threshold. Especially in Europe, visi-

A thousand winds, a thousand names

Do Mediterranean countries actually get more kinds of localized **wind** than other parts of the world? It's a near-impossible claim to verify, but southern Europeans certainly have an ample supply of wind names. The Mediterranean is one of the planet's most favoured regions for **low-pressure** formation. As these lows twirl eastward, they pull breezes around mountains and hills and help create a panorama of recognized local winds. The **föhn** (a warm current that howls down Swiss slopes) has become a worldwide synonym for wind that heats up as it blows downslope (see "Other windstorms", p.96). When a **cold front** careens over the mountains of Slovenia into the Adriatic Sea, the howling north wind that results is a **bora.** A hot, dry, sometimes-dusty wind from the Sahara that gains speed as it crosses the sea is called a **scirocco** in Italy. Light northerlies that hug the west coasts of Turkey and other southeast Europe countries, especially in summer, are called **etesians.** Each of these wind types may go by different names from place to place. For instance, the **mistral** that haunts the Cote d'Azur is a type of bora, and the scirocco is called a **leveche** when it blows across southeastern Spain.

tors can expect several additional days with sprinkles in any given month that's wet to begin with.)

Exceptions make the rule, and European weather has its share. The western coastlines of the **British Isles** and **Scandinavia** – where it rains both hard and long – are infamously soggy. The Alps produce a cornucopia of effects on the weather surrounding them. Spain – with mountains and plateaus scattered across it and two bodies of water nearby – has micro-climates galore. Even the more uniform regimes of, say, Belgium and The Netherlands include frequent day-to-day variations, as weak Atlantic systems brush past. Although the recipes for European weather may be dominated by vanilla rather than Tabasco, one learns to appreciate subtleties in the mix.

Austria

Liechtenstein

The broad-brush picture of weather in **Austria** is fairly straightforward; it's the delicate strokes that can throw a visitor. On the large scale, Austria features a Swiss-style mountainous regime to the west and a central European lowland climate to the east. The **Tyrol region** of western Austria – and **Liechtenstein** just beyond – is crisscrossed by narrow valleys where conditions may be quite different from those at altitude. North-facing slopes and valleys tend to be cool and cloudy in winter, but every few days a strong *föhn* and its screaming winds drives the temperature as high as 10°C/50°F and beyond in spots. The southern Tyrol and **Tauern slopes** can be surprisingly bright in winter, and ski resorts experience frequent strings of sunny days as well. Meanwhile, the **Danube valley** is often buried under a shallow dome

Innsbruck

47°15'N, 11°20'E
Elev: 581m (1960ft)

Sunshine and a few downpours mark summer in this Austrian valley. It's usually quite comfortable, though night chill is not unusual. Autumn is drier, but winter brings frequent spells of low cloud and cold on sharp bise winds. The gloom is chased out by mild föhn winds every few days.

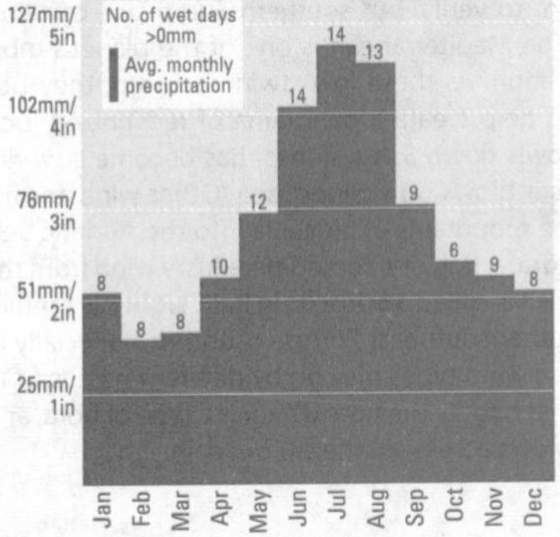

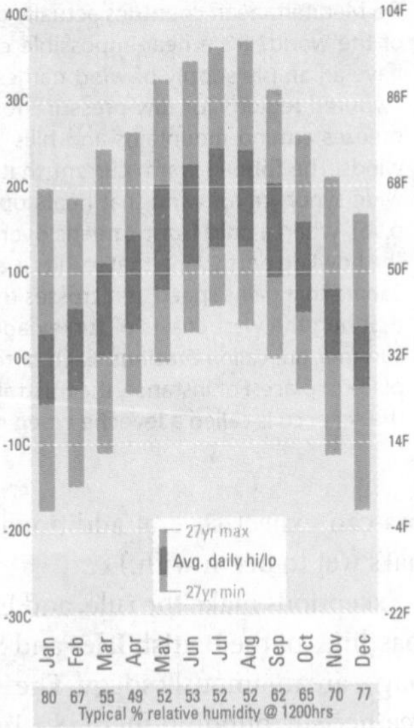

	Jan	Feb	Mar	Apr	May	Jun	Jul	Aug	Sep	Oct	Nov	Dec
midmonth -sunrise	0757	0720	0628	0627	0539	0518	0534	0611	0652	0733	0718	0755
-sunset	1651	1738	1820	2003	2044	2112	2107	2027	1927	1828	1640	1625
avg hrs of sunshine	2	4	5	6	6	6	7	6	6	5	3	2
avg # days w/thunder	<1	0	<1	<1	3	5	7	5	2	<1	<1	0
w/snow	n/a	n/a	n/a	n/a	n/a	n/a	n/a	n/a	n/a	n/a	n/a	n/a

Vienna

48°07'N, 16°34'E
Elev: 190m (623ft)

Shielded by mountains, Vienna gets only a few days of rain or snow in any month. Summer days are warm but seldom very humid, and sunny afternoons can be quite mild as early as April or as late as October. In winter, Vienna can experience clouds, chill, and snow flurries even when nearby slopes are sunny.

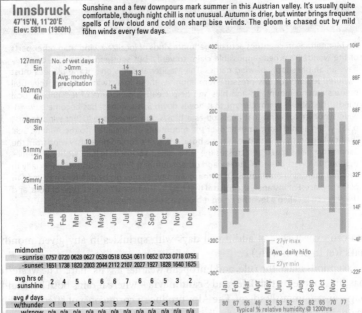

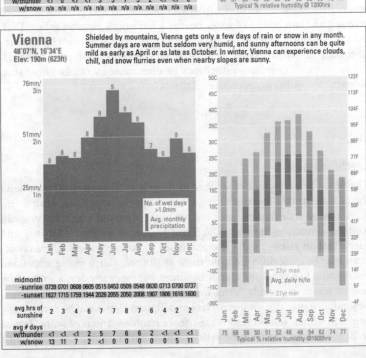

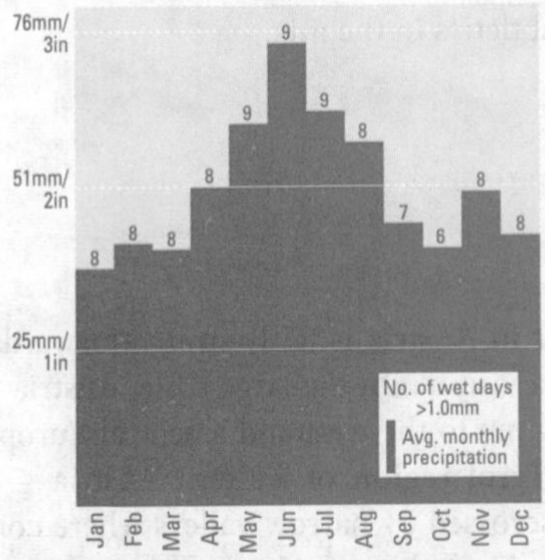

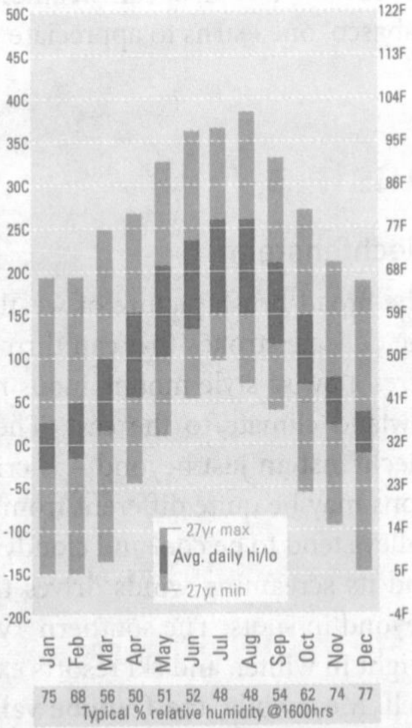

	Jan	Feb	Mar	Apr	May	Jun	Jul	Aug	Sep	Oct	Nov	Dec
midmonth -sunrise	0739	0701	0608	0605	0515	0453	0509	0548	0630	0713	0700	0737
-sunset	1627	1715	1759	1944	2026	2055	2050	2008	1907	1806	1616	1600
avg hrs of sunshine	2	3	4	6	7	7	8	7	6	4	2	2
avg # days w/thunder	<1	<1	<1	2	5	7	6	6	2	<1	<1	<1
w/snow	13	11	7	2	<1	0	0	0	0	0	5	11

of persistent clouds and chill that seldom breaks from December through February. It's colder and foggier still in the southeastern lowlands; the average mid-winter lows in **Graz** are typically below those of even Warsaw or Stockholm. Nights across Austria tend to stay chilly into May, but the afternoons by then can be extremely pleasant outside of a shower every couple of days. Fed by the ample summer sun, thunderstorms are fairly frequent across the country, with the rainfall a bit lighter in the northeast than elsewhere. Valleys are often the sunniest, calmest spots, as storms gather by midday around the higher terrain. For the driest, most temperate weather at altitude and below, September and early October are unrivalled.

Belgium

Luxembourg

Belgium is a land of gentle transitions meteorologically as well as culturally. The weather of **Flanders** is very similar to that of the adjacent Netherlands: cool winters, milder summers (especially as you move inland) and, in all seasons, frequent – through seldom intense – rain. As you move south into **Wallonia**, the climate becomes a little more continental. It's a touch cooler

Brussels

50°54'N, 4°32'E
Elev: 58m (190ft)

Belgian weather is easy to plan for. Expect a dose of rain every two to three days, more showery in summer and more drizzly in winter. Severe summer heat waves and winter cold waves are both uncommon. There's little major difference across the months from late spring to early autumn.

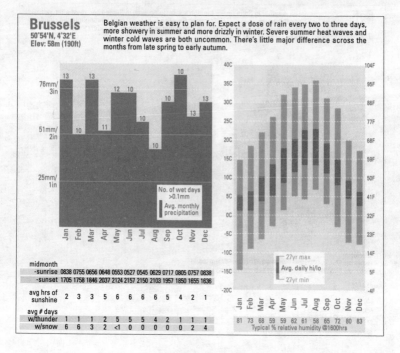

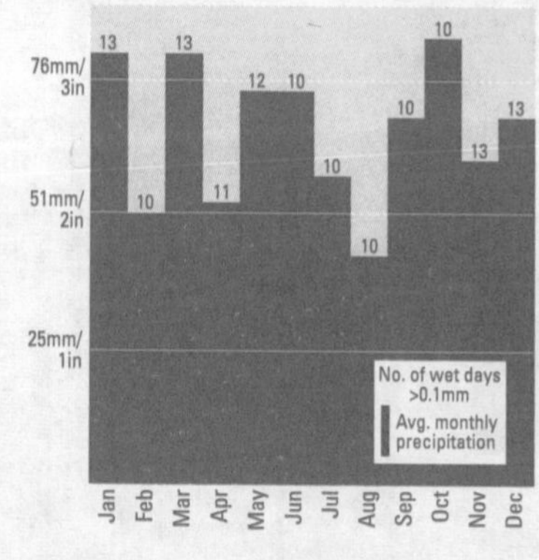

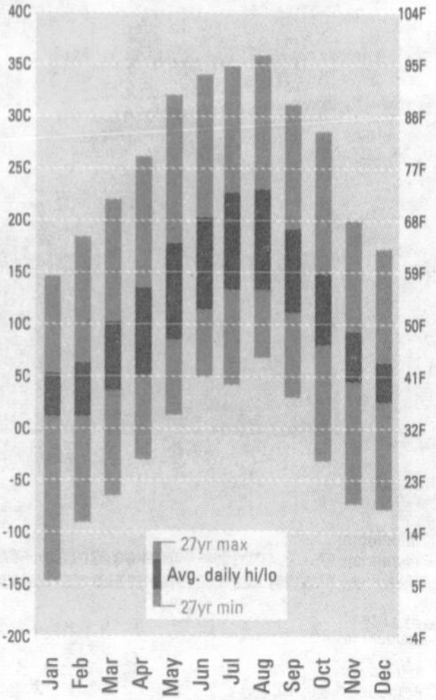

	Jan	Feb	Mar	Apr	May	Jun	Jul	Aug	Sep	Oct	Nov	Dec
midmonth -sunrise	0838	0755	0656	0648	0553	0527	0545	0629	0717	0805	0757	0838
-sunset	1705	1758	1846	2037	2124	2157	2150	2103	1957	1850	1655	1636
avg hrs of sunshine	2	3	3	5	6	6	6	6	5	4	2	1
avg # days w/thunder	1	1	1	2	5	5	5	4	2	1	1	1
w/snow	6	6	3	2	<1	0	0	0	0	0	2	4

	Jan	Feb	Mar	Apr	May	Jun	Jul	Aug	Sep	Oct	Nov	Dec
Typical % relative humidity @1600hrs	81	73	68	59	59	62	61	58	65	72	80	83

in winter, and there's a day or two less rain, but the difference isn't much to write home about. The higher hills of the **Ardennes** and **Luxembourg** do run about 2–3°C/3–5°F cooler on average than central Belgium. The bulk of Ardennes, especially the **Hautes Fagnes**, also stay cloudier and get roughly 300m/12in more annual moisture than points north and west.

Bulgaria

Two mountain ranges and two broad valleys make up the setting for **Bulgaria**'s four-season climate. The **Balkans** run east–west through the nation's heart, and they shield the southern lowlands somewhat from outbreaks of winter cold. Along the southern border, the **Rhodope Mountains** help prevent the hot, humid air of Greece from sweeping northward. It's still warm here in the summer, with a big temperature jump on sunny afternoons (followed every few days by a thunderstorm). **Sofia** is a few degrees cooler than the low valleys to its east. Autumn brings moderate temperatures and the driest weather of the year, while winter is usually chilly and cloudy, with light snows likely at Sofia and other higher points. Late spring is the wettest

Sofia

42°39'N, 23°23'E
Elev: 595m (1952ft)

The weather in Sofia ripens through the summer: even in September, most afternoons are still warm, and rain is infrequent. Expect a light snow cover throughout the cloudy, chilly winter, with frequent fog in December and January. Thunderstorms drop their heaviest rains in May and June.

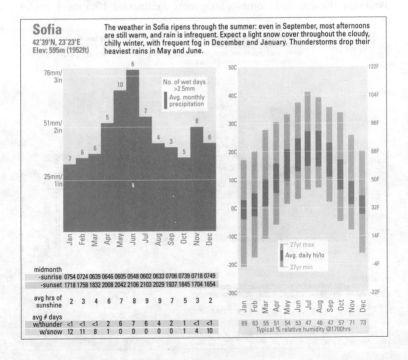

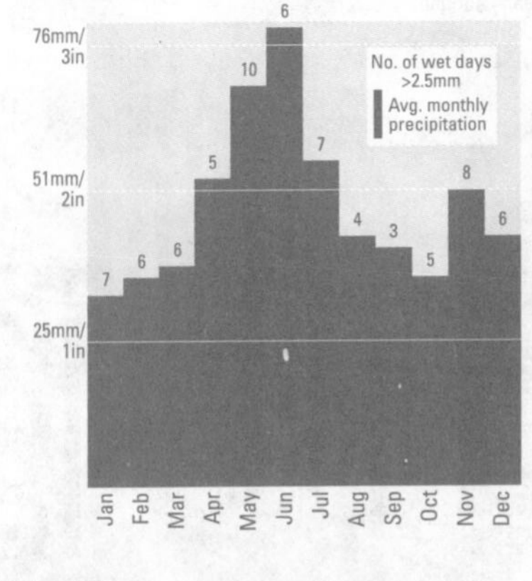

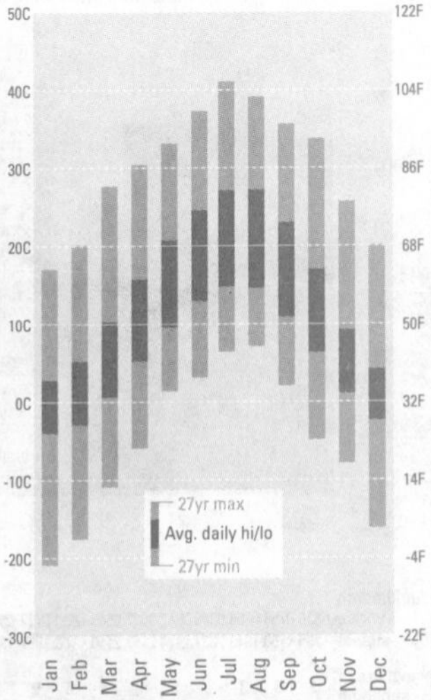

	Jan	Feb	Mar	Apr	May	Jun	Jul	Aug	Sep	Oct	Nov	Dec
midmonth -sunrise	0754	0724	0639	0646	0605	0548	0602	0633	0706	0739	0718	0749
-sunset	1718	1758	1832	2008	2042	2106	2103	2029	1937	1845	1704	1654
avg hrs of sunshine	2	3	4	6	7	8	9	9	7	5	3	2
avg # days w/thunder	<1	<1	<1	2	6	7	6	4	2	1	<1	<1
w/snow	12	11	8	1	0	0	0	0	0	1	4	10
Typical % relative humidity @1700hrs	69	63	55	51	54	53	47	46	47	57	71	73

season across much of the country. Along the breezy **Black Sea coast**, rain and snow are fairly light but well distributed throughout the year.

Czech Republic

The mountains that encircle much of the **Czech Republic** help to shield it from the coldest of the air that flows west from Russia. Nonetheless, the country tends to remain on the chilly side during the winter throughout the lowlands, and colder still in the mountains. A dusting of snow may fall every 2 or 3 days in winter, although the snowpack comes and goes a bit more frequently here than it does in the heart of Poland. The spring days warm quickly, and by May most afternoons are pleasantly mild, perhaps after a nippy morning. Summer variations depend on where you are relative to the mountains. Storms often develop close to the higher peaks and march east, giving downpours in some places and leaving others high and dry. On the whole, though, temperatures are comfortable and sunshine is plentiful. A frequent downslope flow ensures that **Brno** averages 4°C/7°F warmer than **Prague** during the summer. The occasional heat wave can push readings

Prague

50°06'N, 14°15'E
Elev: 361m (1184ft)

This is the land of the winter sprinkle or flurry: every two or three days see a dab of moisture, but it's rarely heavy. Even in summer, the average temperatures are no higher than Warsaw's, but there's less risk of a bona fide Arctic outbreak in winter. Spring gets thundery by May; early autumn is drier.

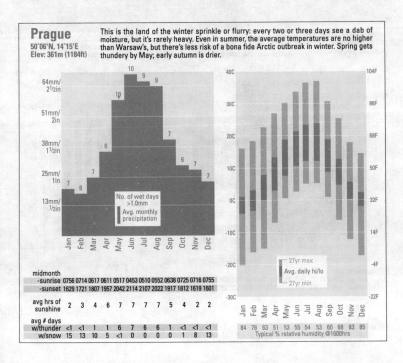

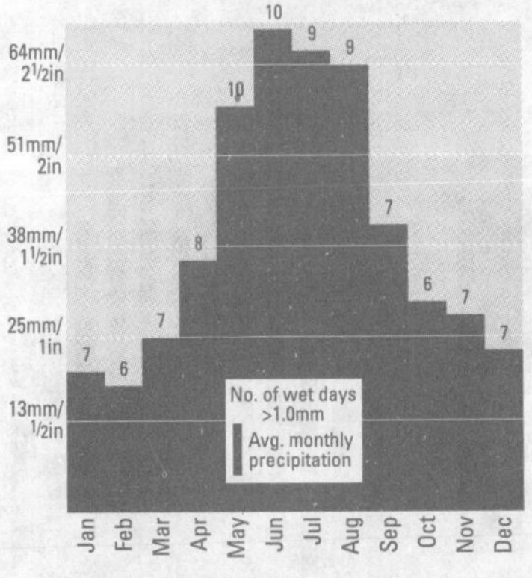

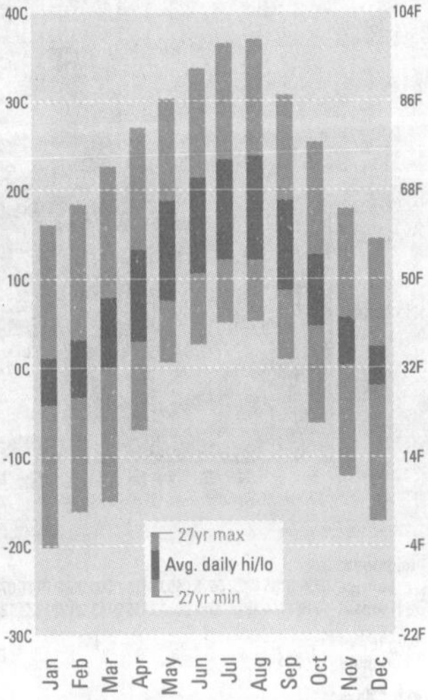

	Jan	Feb	Mar	Apr	May	Jun	Jul	Aug	Sep	Oct	Nov	Dec
midmonth -sunrise	0756	0714	0617	0611	0517	0453	0510	0552	0638	0725	0716	0755
-sunset	1629	1721	1807	1957	2042	2114	2107	2022	1917	1812	1619	1601
avg hrs of sunshine	2	3	4	6	7	7	7	7	5	4	2	2
avg # days w/thunder	<1	<1	1	1	6	7	6	6	1	<1	<1	<1
w/snow	15	13	10	5	<1	0	0	0	0	1	8	13

84	76	63	51	53	55	54	53	60	68	83	85

Typical % relative humidity @1600hrs

above 30°C/86°F nationwide; most spots have record highs above 35°C/95°F. As the air stabilizes in autumn, clear and settled weather can take hold for days on end, making September and October fine times to visit.

Denmark

Except for its border between Jutland and Germany, **Denmark** is essentially an island nation with a climate to match. As in The Netherlands and southern Sweden, you can expect some form of light precipitation almost every other day year-round, although the rain or snow is seldom very heavy. Atlantic gales pound the west coast of Jutland in the autumn and winter, balanced by occasional frigid intrusions from the Arctic or Siberia. Summer days across Denmark are mild, with coolish evenings in the northwest European mold. The absence of a large land mass helps prevent the occasional heat waves of summer from reaching the intensity of nearby countries. Here, you can enjoy the clear skies and sunshine of a hot spell without quite as much sizzle. Late summer and early autumn bring the best chance of heavier showers and thunderstorms. By late October, it's usually quite grey and gloomy. May and

Copenhagen
55°38'N, 12°40'E
Elev: 5m (16ft)

Summer is the clear winner in Copenhagen's climate. Even in July, average highs struggle to top 20°C/68°F, but the many hours of sunshine help it feel warmer. The shoulder seasons are nearly as mild. From late November into April, light snow and rain take turns; it's often damp and chilly but seldom frigid.

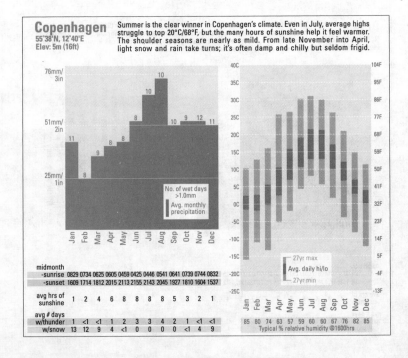

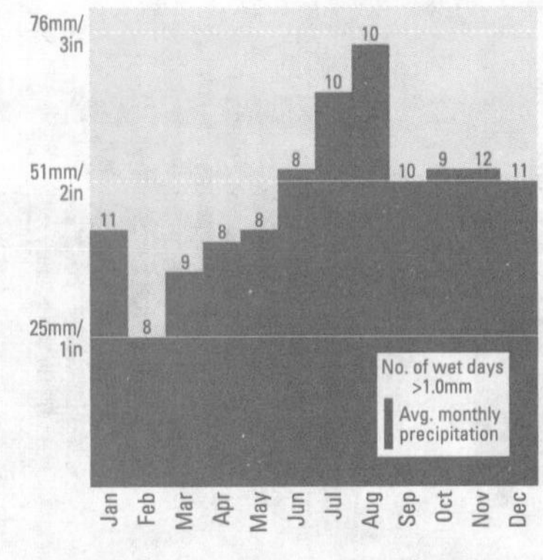

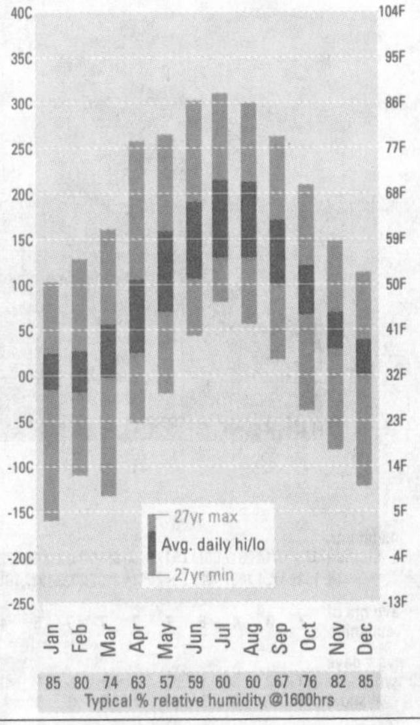

	Jan	Feb	Mar	Apr	May	Jun	Jul	Aug	Sep	Oct	Nov	Dec
midmonth -sunrise	0829	0734	0625	0605	0459	0425	0446	0541	0641	0739	0744	0832
-sunset	1609	1714	1812	2015	2113	2155	2143	2045	1927	1810	1604	1537
avg hrs of sunshine	1	2	4	6	8	8	8	8	5	3	2	1
avg # days w/thunder	1	<1	<1	1	2	3	3	4	2	1	<1	<1
w/snow	13	12	9	4	<1	0	0	0	0	<1	4	9

June tend to be mild and sunny across Denmark and distinctly drier than the months that follow.

Finland

Lake-strewn **Finland** has a maritime feel, although the huge air masses that build over continental Russia are never too distant. The ebb and flow between these influences is the main driver of Finnish climate. The **Atlantic** and the high latitude keep things relatively cool in summer, with few big temperature differences across the country. It seldom gets much above 26°C/79°F anywhere in Finland; the warmest ever recorded across **Lapland** in mid-summer is on a par with that in **Helsinki**. Showers and some thundery spells dampen the landscape every two or three days, with plenty of sunshine otherwise. It doesn't take long for autumn to feel wintry, especially across the north. By January, Siberian cold usually muscles out Atlantic air, freezing over the Gulfs of Bothnia and Finland. Even the **southwest Finnish coast** – the closest approach to open water in mid-winter – is nearly as frigid as Moscow on average. There's at least a dab of snow (usually light) on most

Helsinki
60°19'N, 24°58'E
Elev: 56m (184ft)

Despite its latitude, Helsinki in July averages only about 3°C/5°F cooler than Paris. The short, sunny summers do get variety: some heat, a few chilly nights, and occasional thunder. Autumn turns rainy before the snow and cold set in. The brighter spring warms quickly in May once the ice and snow melt.

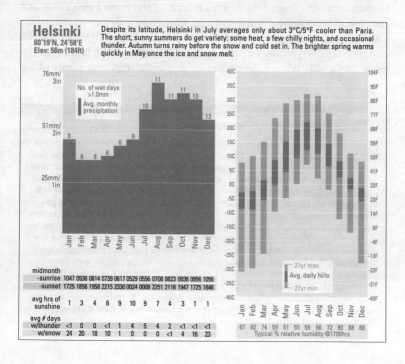

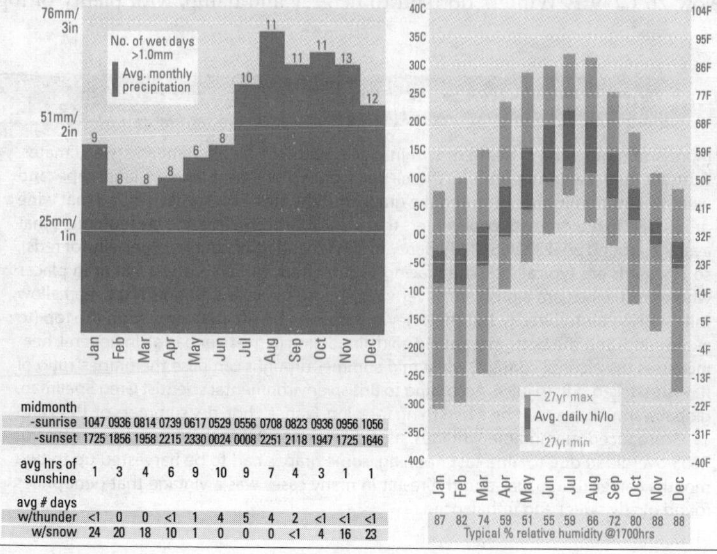

	Jan	Feb	Mar	Apr	May	Jun	Jul	Aug	Sep	Oct	Nov	Dec
midmonth -sunrise	1047	0936	0814	0739	0617	0529	0556	0708	0823	0936	0956	1056
-sunset	1725	1856	1958	2215	2330	0024	0008	2251	2118	1947	1725	1646
avg hrs of sunshine	1	3	4	6	9	10	9	7	4	3	1	1
avg # days w/thunder	<1	0	0	<1	1	4	5	4	2	<1	<1	<1
w/snow	24	20	18	10	1	0	0	0	<1	4	16	23

	Jan	Feb	Mar	Apr	May	Jun	Jul	Aug	Sep	Oct	Nov	Dec
Typical % relative humidity @1700hrs	87	82	74	59	51	55	59	66	72	80	88	88

days, and temperatures struggle to get above freezing. As you move north, the cold is more entrenched and the snowfall a bit lighter, but the snowpack itself is deeper due to less thawing and a longer snow season. Lapland gets the worst of it, although most of Finland has dipped below –30°C/–22°F at least once. The biting cold spells of winter become crisp and sunny during spring, Finland's driest period. The snow melts from south to north and inland during April and May, and temperatures become comfortable enough to enjoy the sunshine without a heavy jacket. Finland experiences a bit more fog than neighbouring countries: up to eight days a month at some coastal cities and towns, with the lowest likelihood in spring.

France

Although it's not an island like its neighbour across the English Channel, **France** is bordered on three sides by water. The **Atlantic** and **Mediterranean** help to moderate the temperature swings that otherwise might occur at France's latitude. The ragged northwest coast of **Brittany** and **Normandy** has a climate not too dissimilar to much of Great Britain, a little milder and wetter than London's. Toward the heart of France – **Paris** to the **Loire Valley** and east toward **Strasbourg** – you'll find a bit more variety. Despite its reputation for sultriness (and apart from rare events like the brutal 2003 heat wave), **central France** is typically very pleasant in mid-summer, with highs typically below 26°C/79°F. Winters do tend to be grey and damp, with plenty of fog

How does weather shape France's wines?

More and more **wine** is being grown in places with hot Mediterranean-style climates, Australia and California among them. But France's unique blend of landscape and weather – its *terroir* – has nurtured the grape for centuries. Experts have found that wine grows best in regions with cool winters that are a little above freezing and summers that average around 20–22°C/68–72°F. Plenty of **sun** is needed, of course (especially for reds), so vineyards are typically situated facing south, often with an eastward bent in places where west winds are a problem. River valleys close to oceans (Rhone, Napa, etc) allow **mild marine air** to filter in, but the vineyards need to be situated away from the top (to avoid wind) and the bottom (to avoid cold air pooling in the river basin). In general, heat increases the alcohol content, while mid-summer drought can raise the grapes' ratio of flavour-packed skin to juice. According to British environmental scientist Greg Spellman, global warming may not be a bad thing for wine. France's hot, dry summers of 1990 and 1995 produced memorable vintages, as did 2005 and 2006. However, the summer of 2003 overdid it: due to ultra-fast ripening, some grapes had to be harvested up to two months earlier than usual, and the result in many cases was a vintage that oenophiles found overly sweet and unbalanced.

from late autumn onward. Occasional Siberian onslaughts can cause readings to plummet well below 0°C/32°F for several days.

Since the land rises to the south as one approaches the **Massif Central** and **the Alps**, the winters here can be just as cold as further north, although the distance from the sea allows for brighter weather in summer. Summer storms can be intense throughout the easternmost provinces. Towns at higher elevation, such as **Le-Puy-en-Velay** in the Massif and **Grenoble** in the Alps, can be nippy on summer evenings and downright frigid in January. They compensate with strong daytime warm-ups on sunny days, especially toward **Provence**. Even in winter, warm winds descending **the Pyrenees** can bring a spell of days above 20°C/68°F to the **Gascogne region**, including **Toulouse**.

Summer on France's **Mediterranean coast** is almost rain-free, with stray showers along parts of the coast about once per week. **Marseilles**' July temperatures are almost identical to those in Los Angeles, without the blasts of 40°C/104°F heat that sometimes bake LA. Winter on the **Cote d'Azur** (including **Monaco**) is equally easy to take: temperatures normally reach 10°C/50°F even in January, and it only rains about every third or fourth day. The fly in the ointment is the infamous **mistral**, a bitter, grating north wind – sometimes exceeding 100kph/60mph – that descends from the Alps. It's

Bordeaux
44°50'N, 0°42'W
Elev: 61m (200ft)

Winters in Bordeaux are typically as mild as those on the Cote d'Azur, although there's more rain and less sun. Most mid-summer days are on the mild side – a bit warmer than Paris – although an east wind descending the Massif Central can bring toasty heat.

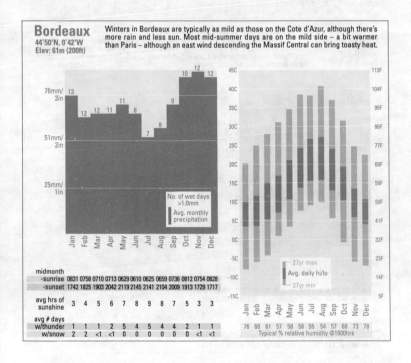

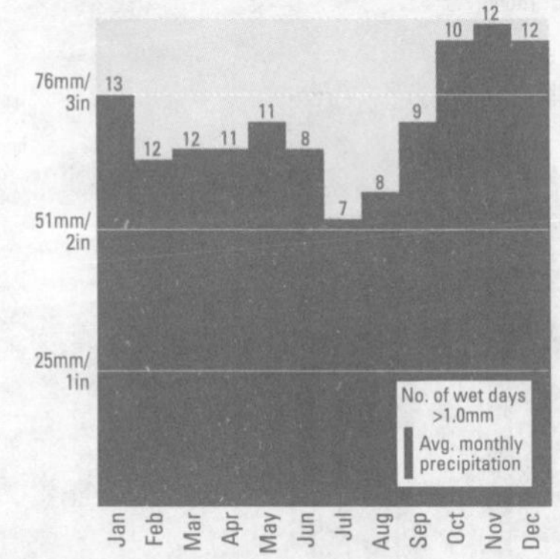

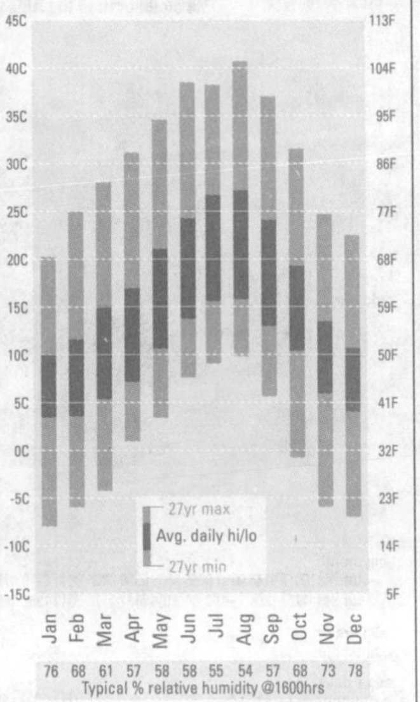

	Jan	Feb	Mar	Apr	May	Jun	Jul	Aug	Sep	Oct	Nov	Dec
midmonth -sunrise	0831	0758	0710	0713	0629	0610	0625	0659	0736	0812	0754	0828
-sunset	1742	1825	1903	2042	2119	2145	2141	2104	2009	1913	1729	1717
avg hrs of sunshine	3	4	5	6	7	8	9	8	7	5	3	3
avg # days w/thunder	1	1	1	2	5	4	5	4	4	2	1	1
w/snow	2	2	<1	<1	0	0	0	0	0	0	<1	<1

Cherbourg

49°59'N, 1°28'E
Elev: 138m (453ft)

If you want French culture with British weather, this is the place. Summers are a bit cooler than London's, and winters are similarly chilly, with occasional stout gales, especially on this peninsula. Winter rains set in early here (October), but they also slacken early. Though cool, March and April often have fine spells.

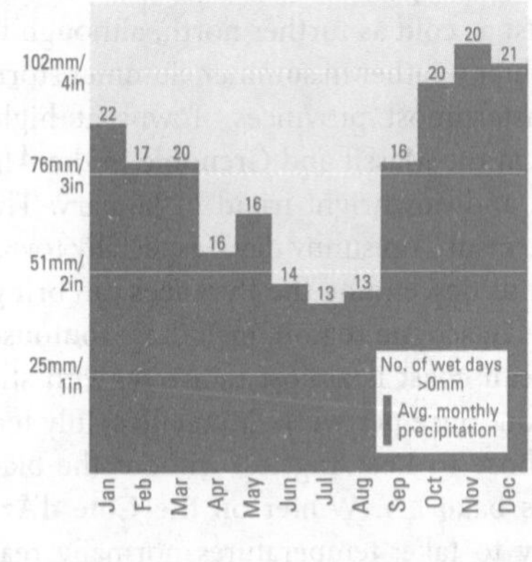

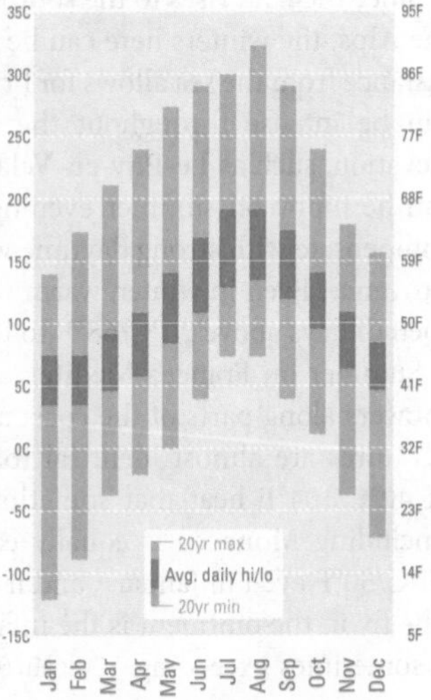

	Jan	Feb	Mar	Apr	May	Jun	Jul	Aug	Sep	Oct	Nov	Dec
midmonth -sunrise	0847	0805	0708	0702	0609	0545	0601	0643	0730	0816	0806	0846
-sunset	1721	1812	1859	2048	2133	2205	2158	2113	2008	1904	1710	1653
avg hrs of sunshine	2	3	4	6	7	7	7	7	6	4	3	2
avg # days w/thunder	1	<1	<1	1	1	1	2	1	1	1	1	1
w/snow	3	3	2	1	<1	0	0	0	0	0	<1	1
Typical % relative humidity @1600hrs	82	78	76	71	71	71	72	72	74	77	79	82

Lyon

45°43'N, 4°57'E
Elev: 201m (659ft)

More showery and thundery than Paris, Lyon gets a greater number of days with rain even as it racks up more sunshine than the City of Light. Cold winter air can wedge into the Rhone Valley and produce prolonged periods of fog. The nearby Massif Central is cloudier than the Alps year round.

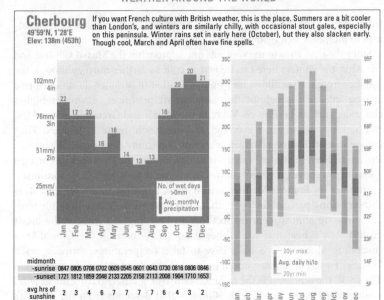

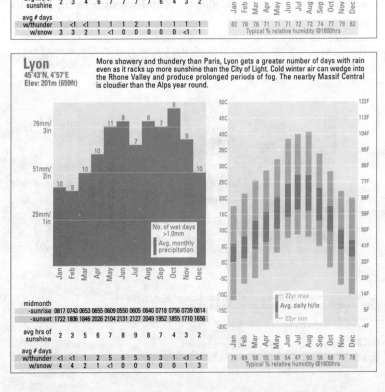

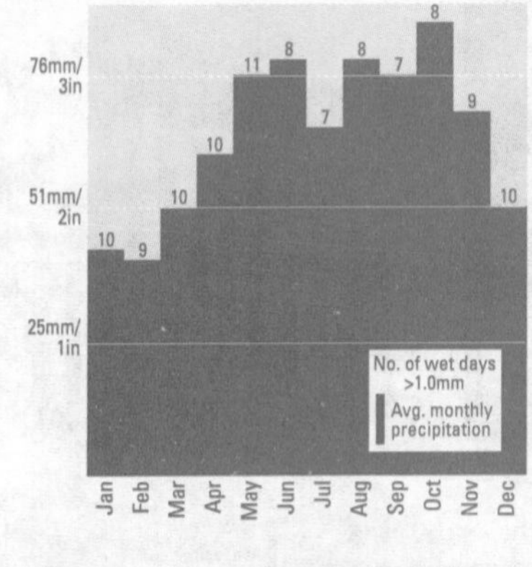

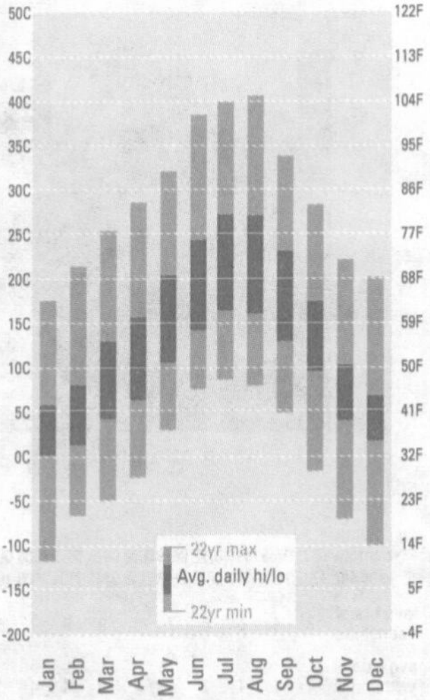

	Jan	Feb	Mar	Apr	May	Jun	Jul	Aug	Sep	Oct	Nov	Dec
midmonth -sunrise	0817	0743	0653	0655	0609	0550	0605	0640	0718	0756	0739	0814
-sunset	1722	1806	1846	2026	2104	2131	2127	2049	1952	1855	1710	1656
avg hrs of sunshine	2	3	5	6	7	8	9	8	7	4	3	2
avg # days w/thunder	<1	<1	1	2	5	6	5	5	3	1	<1	<1
w/snow	4	4	2	1	<1	0	0	0	0	0	1	3
Typical % relative humidity @1600hrs	76	69	58	55	56	54	47	50	56	68	75	78

Marseille

43°27'N, 5°14'E
Elev: 36m (118ft)

As annoying as a mistral can be, the northeast winds help keep Marseille brighter than northern France in winter. Fog and truly cold days are rare. Summer days are warm and nights moderately humid. Thunder is most likely on the fringes of summer; the heaviest storms strike in early autumn.

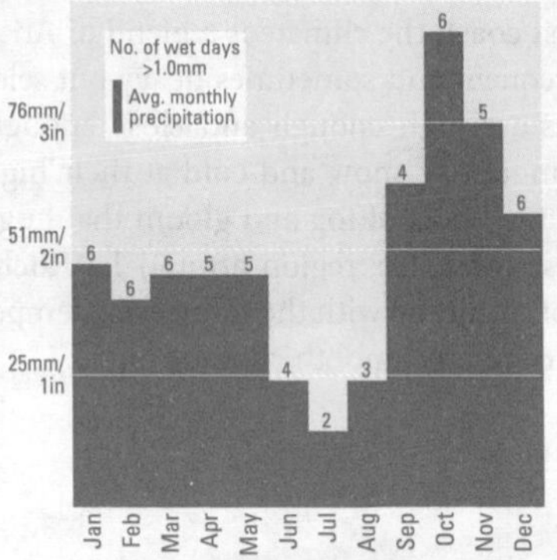

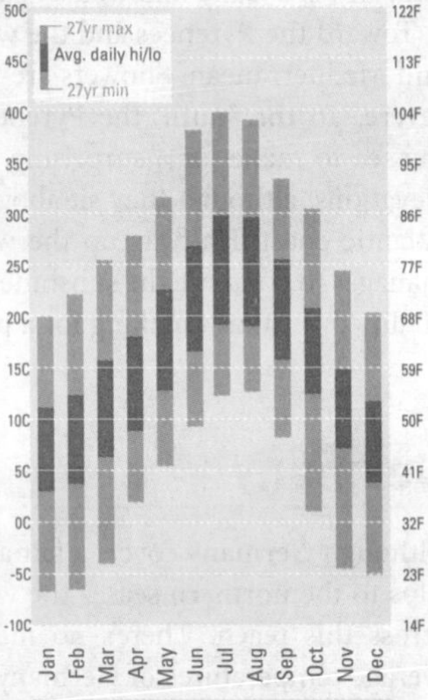

midmonth	Jan	Feb	Mar	Apr	May	Jun	Jul	Aug	Sep	Oct	Nov	Dec
-sunrise	0809	0738	0652	0657	0615	0558	0612	0644	0718	0753	0732	0805
-sunset	1728	1809	1845	2022	2056	2121	2118	2043	1950	1856	1715	1703
avg hrs of sunshine	5	6	7	8	9	11	12	11	8	7	5	5
avg # days w/thunder	1	1	1	1	2	3	1	2	2	3	1	1
w/snow	1	1	<1	0	0	0	0	0	0	0	<1	<1
Typical % relative humidity @1600hrs	64	60	55	52	53	50	46	48	53	61	64	67

Paris

48°44'N, 2°24'E
Elev: 96m (315ft)

It's doubtful that weather is what drives Parisians south in August, given the city's pleasant average high of 24°C/75°F and the rarity of overcast skies (about one day in six). Rain is well distributed through the year, though winter gets long stretches of overcast skies. Paris' famed spring is bright but often unsettled and showery.

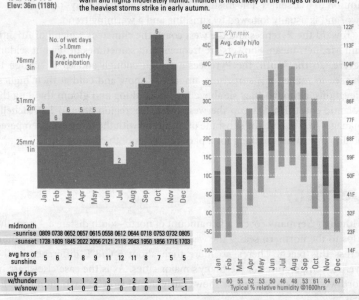

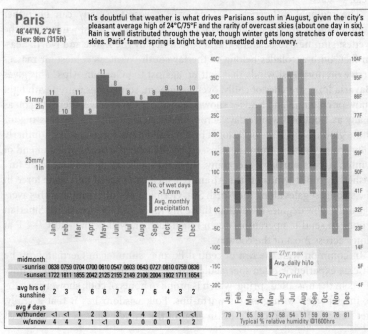

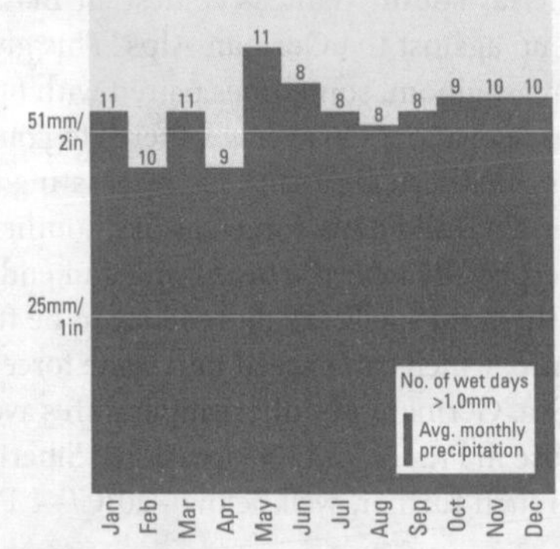

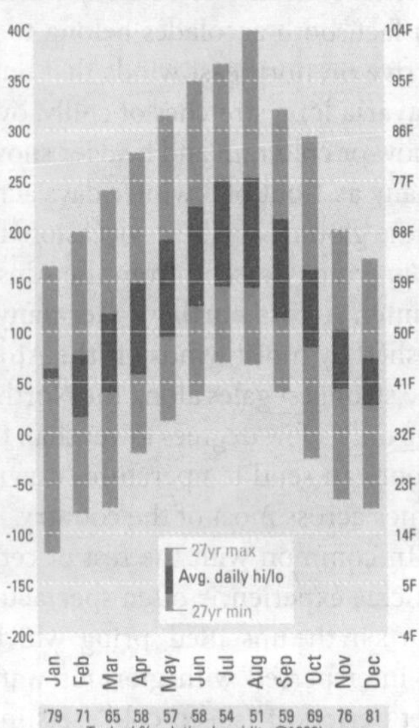

midmonth	Jan	Feb	Mar	Apr	May	Jun	Jul	Aug	Sep	Oct	Nov	Dec
-sunrise	0838	0759	0704	0700	0610	0547	0603	0643	0727	0810	0759	0836
-sunset	1722	1811	1855	2042	2125	2155	2149	2106	2004	1902	1711	1654
avg hrs of sunshine	2	3	4	6	6	7	8	7	6	4	3	2
avg # days w/thunder	<1	<1	1	2	3	3	4	4	2	1	<1	<1
w/snow	4	4	2	1	<1	0	0	0	0	0	1	2
Typical % relative humidity @1600hrs	79	71	65	58	57	58	54	51	59	69	76	81

most frequent and intense from November into April along the coast from **Toulon** to Marseilles. The mistral often arrives abruptly, plays out in a day or two, and is usually followed by sunshine and a warming trend.

Toward the Pyrenees and the **west coast**, the climate is a blend of Atlantic and Mediterranean. Showers are frequent and sometimes heavy but seldom severe. To **the south**, the Pyrenees are high enough and close enough to the sea to make for a surprising amount of snow and cold at their highest elevations, although they sit above the pools of fog and gloom that hug the Atlantic coast. Halfway up the west coast, the region around **La Rochelle** manages to combine the sunshine of **Aquitane** with the more even-tempered quality of Brittany, making for a particularly smooth climatic blend.

Germany

Although **Germany** covers a fair amount of north–south distance – from the Alps to the northern seas – the weather doesn't vary a tremendous amount across this reach. There's no more than about 5°C/8°F difference in the average temperatures of Germany's major cities. As is the case elsewhere in northwest Europe, it rains or snows fairly dependably in each month of the year.

The regional differences that do exist across Germany can be rather counter-intuitive. For instance, you might expect the coldest winters and the wettest summers to be along the northern tier, by **the Baltic** and **North Seas**. In fact, both accolades belong to the **far south**. Winter's coldest air masses arrive on northeast winds that butt up against the **German Alps**. This gives **Bavaria** long stretches of chilly, overcast gloom, sometimes paired with light snow or cold rain and heavier snow at elevation. On average, there's fog on as many as 1 out of 3 winter days across southern Germany. Compensating for these gloomy spells are occasional bright, mild days, forged as dry, southerly *föhn* winds descend from the Alps, especially near the beginning and end of winter. Across **northern Germany**, the mild spells are more likely to be furnished by moist winds off **the Atlantic**, which can exceed hurricane force in the strongest gales along the North Sea. Germany's winter temperatures average only a few degrees lower than those in France, but the occasional Siberian front can send temperatures down much further, well below –20°C/–4°F at times across most of the country.

In common with the rest of central Europe, summer in Germany can be a fickle experience: often spectacular, sometimes miserable. The uncertainty starts in the unsettled spring, which is prone to bouts of cool, showery weather interspersed with glorious warm-ups. Folk wisdom has it that a chilly, wet period of "sheep cold" sets in around the middle of June. As the heart

Berlin

52°23'N, 13°31'E
Elev: 48m (157ft)

Berlin puts a slightly drier, cooler spin on the usual German weather. Arctic and Siberian cold is never far away in winter, although most days are more routinely chilly, with light snow staying on the ground for weeks. Along with the mild summers, May and September are often bright and kind to visitors.

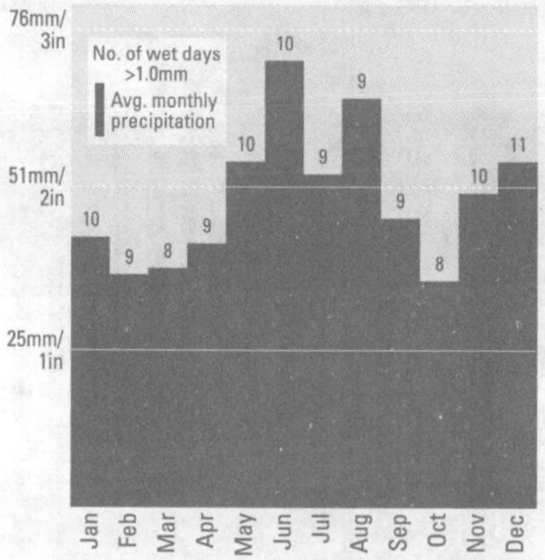

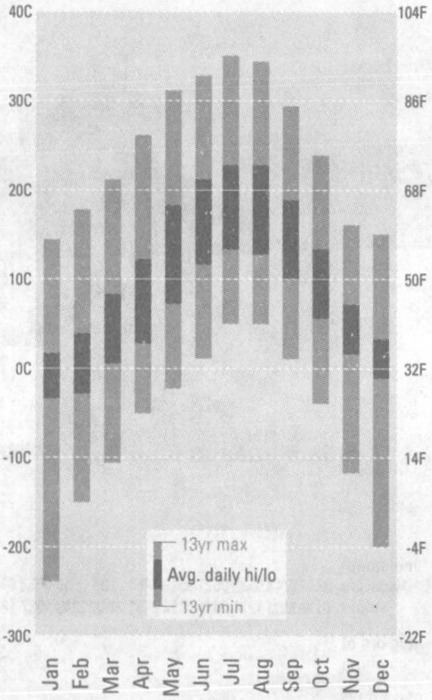

	Jan	Feb	Mar	Apr	May	Jun	Jul	Aug	Sep	Oct	Nov	Dec
midmonth -sunrise	0809	0722	0621	0609	0511	0443	0502	0548	0640	0731	0727	0809
-sunset	1622	1719	1810	2004	2055	2130	2121	2031	1922	1812	1614	1552
avg hrs of sunshine	1	3	4	5	7	7	7	7	5	4	2	1
avg # days w/thunder	<1	<1	1	1	5	7	5	6	3	<1	<1	<1
w/snow	13	11	8	3	<1	0	0	0	0	<1	4	10
Typical % relative humidity @1600hrs	83	76	64	54	50	55	53	54	59	71	81	86

Cologne

50°52'N, 7°10'E
Elev: 99m (325ft)

The Ruhrgebiet is one of the warmer regions in Germany. Late autumn and winter may feel just as cold as anywhere else, though: it rains or snows about every other day and you won't see much sun. Spring and autumn are similar to each other, and it's hard to complain about the warm summer days.

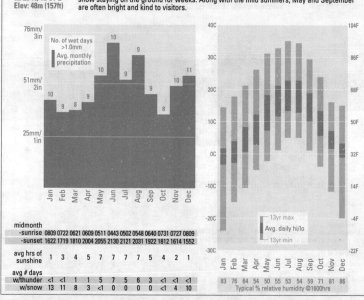

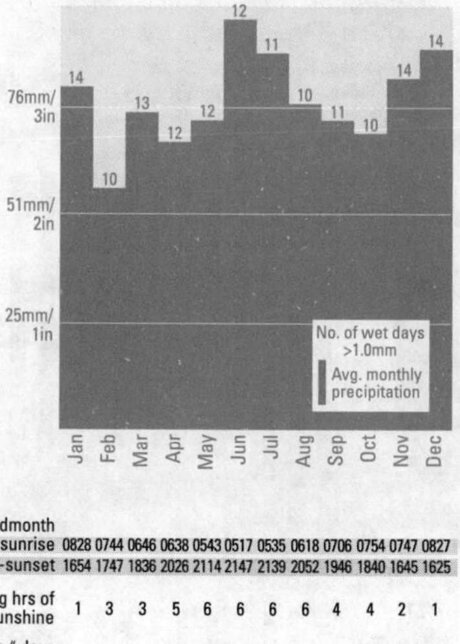

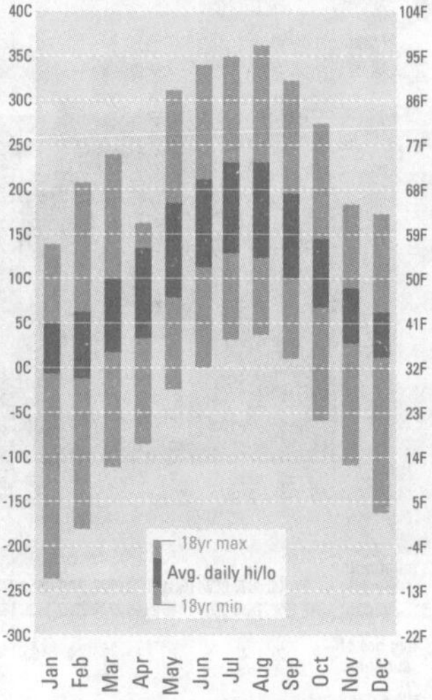

	Jan	Feb	Mar	Apr	May	Jun	Jul	Aug	Sep	Oct	Nov	Dec
midmonth -sunrise	0828	0744	0646	0638	0543	0517	0535	0618	0706	0754	0747	0827
-sunset	1654	1747	1836	2026	2114	2147	2139	2052	1946	1840	1645	1625
avg hrs of sunshine	1	3	3	5	6	6	6	6	4	4	2	1
avg # days w/thunder	1	1	1	2	4	6	5	5	3	1	1	1
w/snow	9	7	5	3	<1	0	0	0	0	<1	3	7
Typical % relative humidity @1600hrs	77	66	62	54	53	56	55	54	62	68	74	79

Hamburg

53°38'N, 10°00'E
Elev: 16m (52ft)

Even though it's the northernmost big city of Germany, Hamburg escapes the worst winter cold, leaning toward damp, sometimes snowy chill and strong wind at times. The cool, often bright spring gets a little less moisture than the cloudier autumn. Summers tend to be cool but can sizzle on rare occasions.

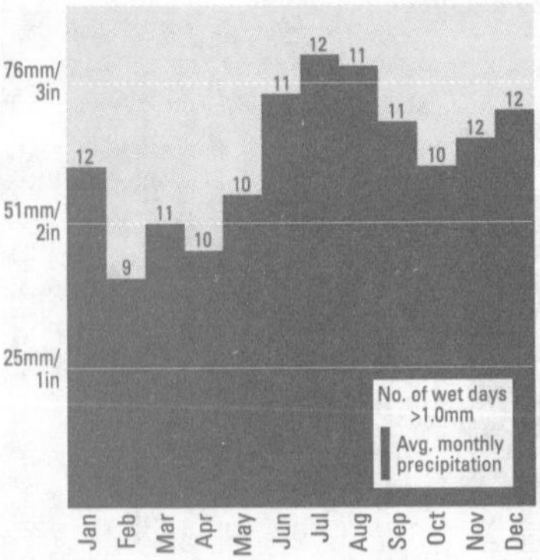

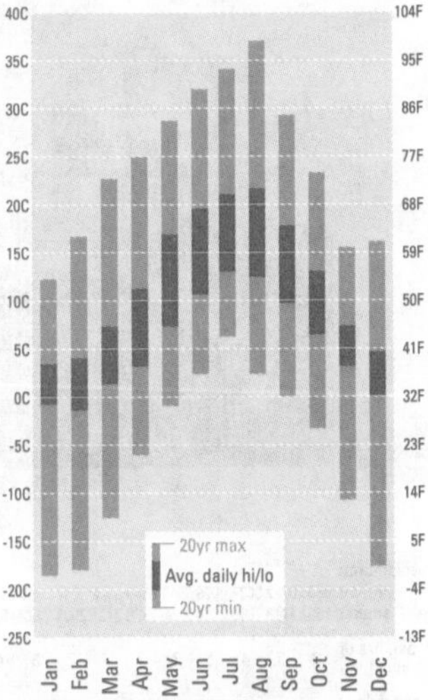

	Jan	Feb	Mar	Apr	May	Jun	Jul	Aug	Sep	Oct	Nov	Dec
midmonth -sunrise	0829	0740	0635	0621	0519	0450	0509	0559	0653	0747	0746	0830
-sunset	1630	1730	1824	2021	2114	2151	2142	2049	1937	1824	1623	1600
avg hrs of sunshine	1	2	3	5	7	7	7	7	5	3	2	1
avg # days w/thunder	1	1	1	2	4	5	5	5	2	1	1	1
w/snow	11	10	7	4	<1	0	0	0	0	<1	3	8
Typical % relative humidity @1600hrs	86	77	69	60	57	60	61	60	67	75	83	87

Munich

48°08'N, 11°42'E
Elev: 529m (1735ft)

Munich is ideally situated for maximum weather variety. Winters can bring long spells of icy cold from Siberia or balmy breezes cascading over the Alps. Oktoberfest takes place during Munich's best climatic window, with little risk of summer's raging thunderstorms or the snow that can fall into April.

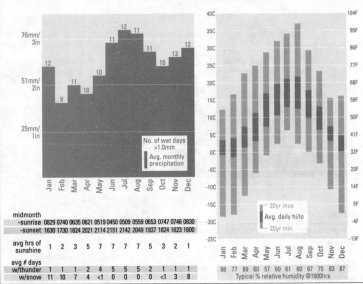

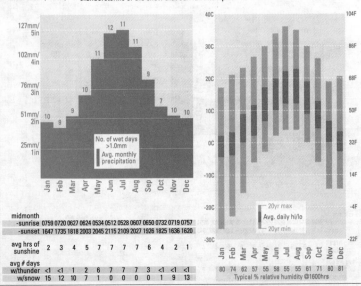

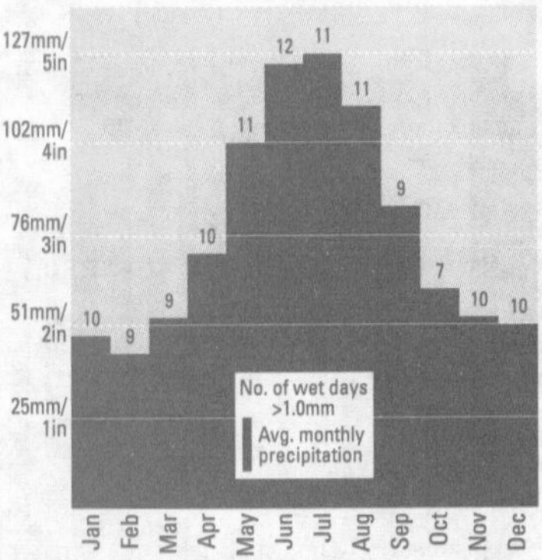

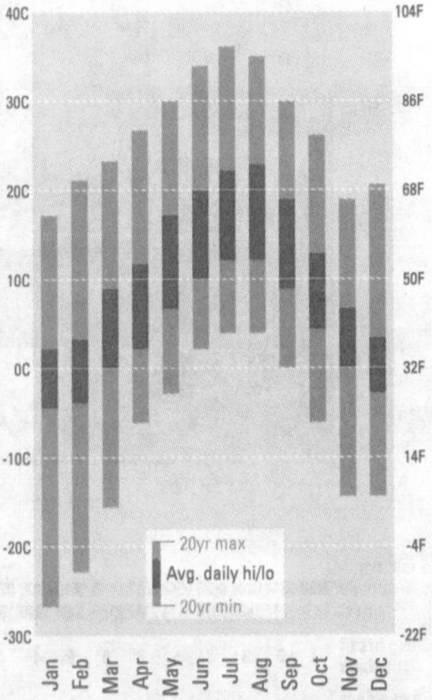

	Jan	Feb	Mar	Apr	May	Jun	Jul	Aug	Sep	Oct	Nov	Dec
midmonth -sunrise	0759	0720	0627	0624	0534	0512	0528	0607	0650	0732	0719	0757
-sunset	1647	1735	1818	2003	2045	2115	2109	2027	1926	1825	1636	1620
avg hrs of sunshine	2	3	4	5	7	7	7	7	6	4	2	1
avg # days w/thunder	<1	<1	1	2	6	7	7	7	3	<1	<1	<1
w/snow	15	12	10	7	1	0	0	0	0	1	9	13
Typical % relative humidity @1600hrs	80	74	62	57	55	58	55	55	61	71	80	81

of Europe heats up with the approach of the summer solstice, cool, damp air can indeed be drawn into Germany from the still-chilly waters of the North and Baltic Seas. By July, warm weather becomes more dependable. In fact, it's sometimes very hot. Readings above 35°C/95°F have occurred at most of the lower elevations, and 30°C/86°F is not uncommon. Germany sees a good amount of thunderstorm action in the summer, with hail especially likely across the southern half. (**Munich** is infamous for its powerful storms; see "Hail", p.74.) The thunder abates quickly in September and October, when morning fog may be the only deterrent to otherwise gorgeous autumn weather. By November, you're likely to find cloudy, chilly conditions setting in most everywhere.

Overall, the north and west slopes of major mountain ranges and hilly regions tend to get more rainfall than the eastern sides. Germany's wettest corner is the **Black Forest** area, which averages more than 1000mm/39in. The east is the driest; parts of the **Elbe Valley** may get only 500mm/20in in a typical year.

Greece

It's hard to summarize the climate in a land with 2000 islands, high mountain peaks, and exposure to air masses that range from sub-tropical to polar. Nearly all of **Greece** is ruled by a Mediterranean regime, which means that precipitation is focused in the cool season. Summers are virtually rain-free: the best chance for a shower is toward the north, but even **Thessaloniki** only gets 1 or 2 a month at best. Since temperatures heat up sharply at lower elevations, the relative humidity isn't the best guide to how sultry the air can feel. A typical **Athens** night in July and August stays above 22°C/72°F. Greece's hottest pockets include the interior lowlands of **Peloponnesia** and **Thessalia**. Along the coast, the mid-day heat is often broken by a light sea breeze, and many spots get the persistent northerly wind in summer known as the *etesian*. Elevation is another cure for the swelter. The average temperature drop with height is sharper in Greece than in many other locations. On a hot afternoon, you can expect it to be as much as 8°C/14°F cooler for every 1000m/3300ft you climb.

Occasional rains start making their presence felt in October, and late autumn can be a thundery time. Even in mid-winter, however, substantial rains are usually punctuated by sunny spells known as **halcyon days**. The heaviest amounts fall on the western slopes of the **Pindus** and **Peloponnes mountains**, often in the form of snow. Cold air has trouble making it through the maze of Greece's mountains and islands. This means that coastal cities in the north are substantially cooler on average than Athens and other south-

Athens
37°54'N, 23°44'E
Elev: 21m (69ft)

Mid-summer in Athens can be a trial: heat, humidity and smog team up on the unwary visitor. May offers more temperate, still mostly dry weather. The same applies to late September and early October, before winter rains (some thundery) pick up. Freezes are rare in mid-winter, and many days are bright.

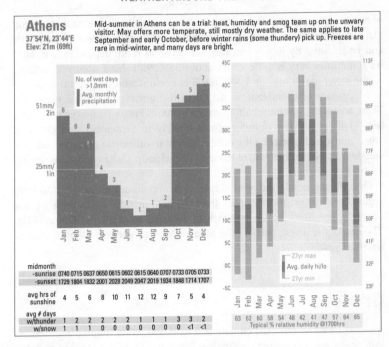

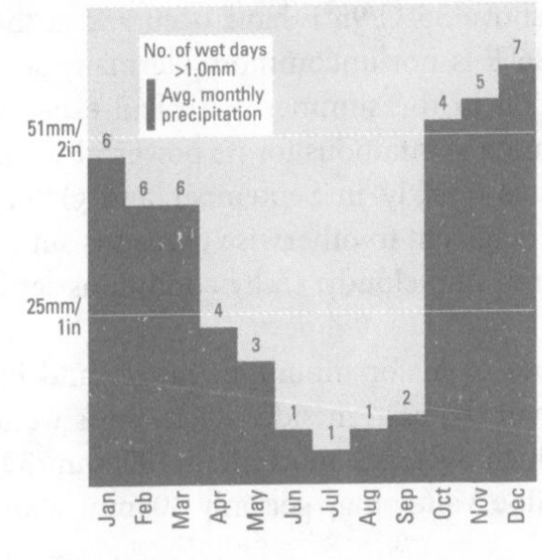

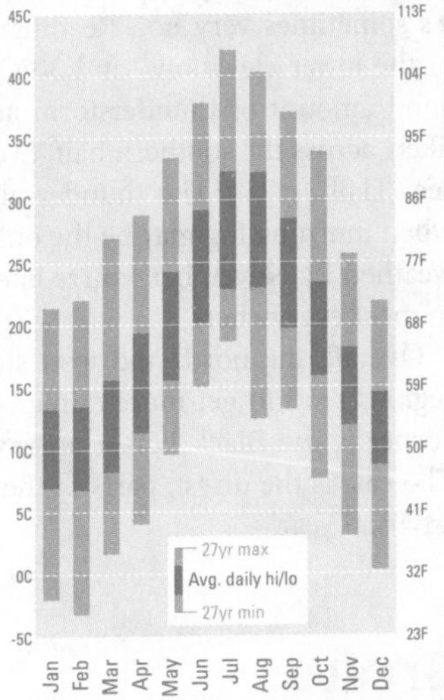

	Jan	Feb	Mar	Apr	May	Jun	Jul	Aug	Sep	Oct	Nov	Dec
midmonth -sunrise	0740	0715	0637	0650	0615	0602	0615	0640	0707	0733	0705	0733
-sunset	1729	1804	1832	2001	2028	2049	2047	2019	1934	1848	1714	1707
avg hrs of sunshine	4	5	6	8	10	11	12	12	9	7	5	4
avg # days w/thunder	1	2	2	2	2	2	1	1	1	3	3	2
w/snow	1	1	1	0	0	0	0	0	0	0	<1	<1
Typical % relative humidity @1700hrs	63	62	60	58	54	48	42	41	47	57	64	65

ern destinations. Hard freezes are rare even on the northern coast, but a round or two of light snow each year may sweep as far south as the shores of Peloponnesia, and frost is frequent in some valleys. The winter rains slacken gradually through spring; it takes until June for Thessaloniki to enter its dry summer regime. Spring is an ideal time to catch the Grecian sun and enjoy the still-green countryside before the temperatures hit their summer peaks. (See "European holiday islands", p.284, for more on Greek island weather.)

Greenland

Most of this semi-independent Danish province is slathered by the Northern Hemisphere's largest sheet of ice, rising more than 3300m/10,000ft high at its centre. By and large, tourists wisely stick to **Greenland**'s coasts, where summers are chilly but bright. During the peak June-to-August travel season, average highs remain above 6°C/40°F along the entire **west coast**, although readings may vary sharply from one harbour or airport to another, in accordance with topography and other local effects. The icy **east coast** is cooler, especially north of **Ammassalik**. South of the Arctic circle, a typical mid-summer afternoon hovers around 10°C/50°F in **Nuuk**, but day-to-day shifts can be dramatic: in July Nuuk has seen freezes as well as 24°C/75°F warmth. Summer readings above 16°C/60°F are not uncommon across the green pastures of the far south. Rains pass across southern Greenland every two or three days on average, usually becoming lighter as you move north. In the winter, coastal visitors can expect lots of dark, cold and snow, although an occasional pulse of warm air may send temperatures just above freezing along parts of the southern coast. Only a few researchers experience the heart of Greenland, where summer temperatures struggle to approach the freezing mark and winter readings can vacillate from –20°C/–4°F to –60°C/–76°F. Clear days tend to be the coldest here, while warm-ups are often laced with clouds and snow-blowing wind. Although it's invariably light, the interior's snowfall does add up: ice cores drawn from central Greenland reveal aspects of climate from over 100,000 years ago.

Hungary

The open fields of the **Great Hungarian Plain** cultivate one of Europe's most vivid four-season climates: visitors from the American Midwest will feel at home. Summers are bathed in sunshine, with frequent afternoon and

Budapest
47°26'N, 19°14'E
Elev: 185m (607ft)

Befitting its central location, Budapest merges aspects of the weather regimes from all directions. A distinctly dry, mild autumn leads into a cloudy, cold winter, with frequent light rain or snow. Springs are variable, and summers are sunny, warm, and thundery. Humidity in the Danube Valley makes it feel fairly sultry.

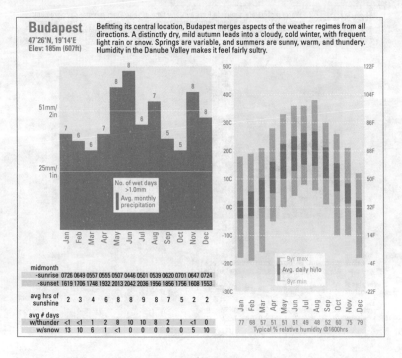

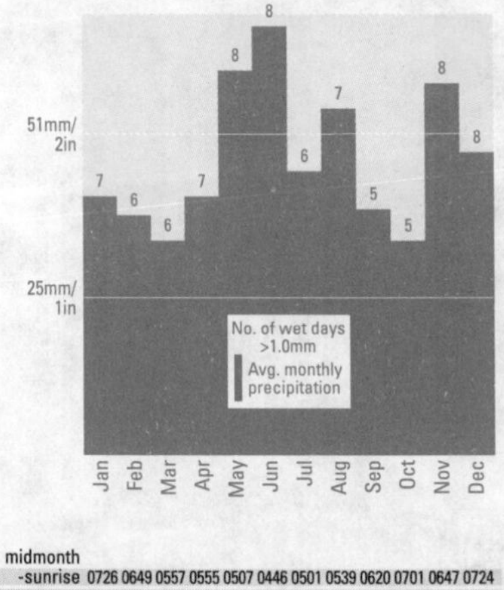

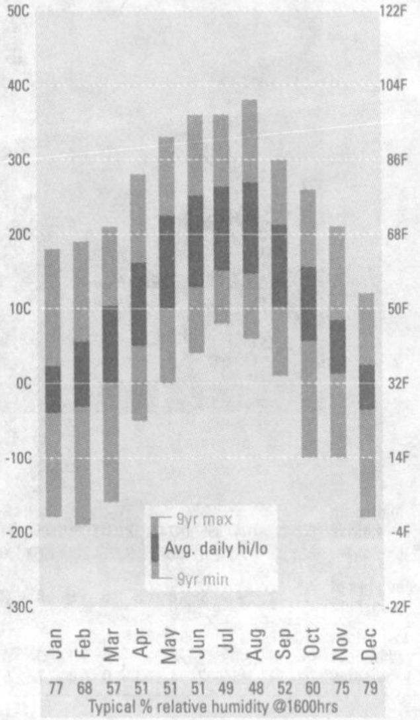

	Jan	Feb	Mar	Apr	May	Jun	Jul	Aug	Sep	Oct	Nov	Dec
midmonth -sunrise	0726	0649	0557	0555	0507	0446	0501	0539	0620	0701	0647	0724
-sunset	1619	1706	1748	1932	2013	2042	2036	1956	1856	1756	1608	1553
avg hrs of sunshine	2	3	4	6	8	8	9	8	7	5	2	2
avg # days w/thunder	<1	<1	1	2	8	10	10	8	2	1	<1	0
w/snow	13	10	6	1	<1	0	0	0	0	0	5	10

evening thunderstorms rolling across the prairie. Especially in the **Danube valley**, there's often enough moisture in the air for it to feel noticeably humid, though not intolerably so. It can get very hot on a few days each year, especially toward the **south** and **east**. The plains dry out quickly in early autumn – a great time for seeing the countryside – before a second peak of rainfall brings a few periods of damp chill from late October through December. Light snows are frequent in January, but Hungary's snow cover isn't as persistent as, say, the Ukraine's. Even in mid-winter, one snowfall often melts before the next arrives. The **Bakony** and other higher points remain a bit colder and may suffer a heavier round of snow on occasion from storms that pull up moisture from **the Adriatic**. By April **the lowlands** are warming fast in the strong Hungarian sun.

Iceland

It may not quite live up to its name, but **Iceland** is indeed a chilly place. **Reykjavik** is the world's northernmost capital (aside from Nuuk, Greenland), and it may also be the one with the coldest summers: in July, the average

Reykjavik

64°08'N, 21°54'W
Elev: 61m (200ft)

Warm jackets are a must in Reykjavik any time of year. Even summer days are cool, and there's usually a wind, sometimes fierce (this city gets over 40 gales a year). Snow and rain often mix during the damp, cold winter. Clouds are prevalent year round, but the sun can poke through even on short winter days.

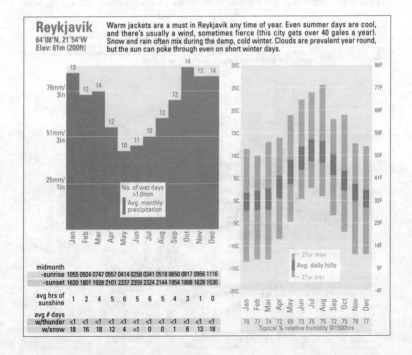

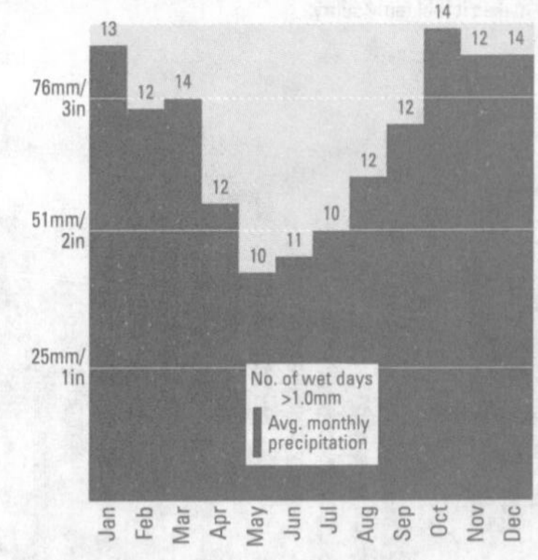

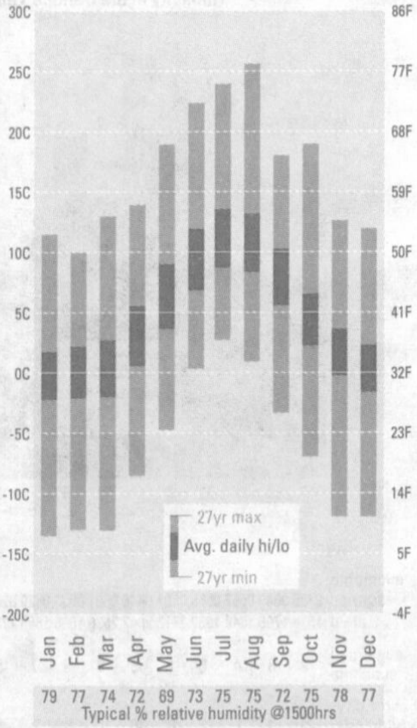

	Jan	Feb	Mar	Apr	May	Jun	Jul	Aug	Sep	Oct	Nov	Dec
midmonth -sunrise	1055	0924	0747	0557	0414	0258	0341	0518	0650	0817	0956	1116
-sunset	1620	1801	1928	2101	2237	2359	2324	2144	1954	1808	1628	1530
avg hrs of sunshine	1	2	4	5	6	5	6	5	4	3	1	0
avg # days w/thunder	<1	<1	<1	<1	<1	<1	<1	<1	<1	<1	<1	<1
w/snow	18	16	18	12	4	<1	0	0	1	6	13	18

highs stay below 14°C/57°F. Warm ocean currents and the most favoured track of North Atlantic storm systems both run close to Iceland. This helps keep the island very windy and fairly cloudy year round; you may find yourself waiting a long while to see clear skies over Iceland. Summer is somewhat drier than winter, with showers or spells of light rain once or twice a week. The **volcanic desert** just north of the **Vatnajökull icecap** gets less than half the precipitation of Reykjavik. None of the island gets very warm: only parts of the northeast rise above 20°C/68°F with any regularity. The mildest, brightest conditions are often on the **north coast**, downstream of the prevailing south winds that climb and descend the central highlands. Temperature changes are driven less by day and night effects than by the regular passage of fronts, so a mild evening may be followed by a chilly, damp day. Travel beyond Reykjavik is a challenge in winter, as snowstorms may close the road that rings the island. Most of Iceland is snow-covered from about November to April; a few winter rains eat away at the snowpack over Reykjavik and points south and east. At any time of year, a period of southeast winds may replace the usually crystalline air with hazier skies, courtesy of the British Isles and the Continent.

Ireland

Ireland is less likely than the UK to surprise visitors with unexpected weather. Positioned further into **the Atlantic**, the Emerald Isle has a more uniform temperature regime than the UK, and over 50 percent more rain as a whole. Cold snaps tend to lose their punch by the time they make it across the sea to Ireland; the chilliest ever observed here is only –19°C/–2°F, and the island's south coast is as mild as Cornwall. Snow covers the ground less than a week a year on average, with more in the **Wicklow mountains** and other higher elevations, and little or none on the Atlantic shores. Summer temperatures nationwide are on the cool side, close to those found in Edinburgh and Glasgow. The rains in mid-summer are about as frequent and heavy as in England and Wales, increasing as you travel to the **west coast**. Thunder seldom interrupts a conversation here; the greatest number of storms – more than ten a year – develop across the northwest corner, often during the winter. As a rule, winter rainfall is generous nationwide. The west coast sees deluges and gales similar to those on Scotland's west coast, with temperatures not quite as chilly. Somewhat sheltered, the **eastern lowlands** are the least damp pocket of Ireland in the winter, similar to England's Thames basin. On average, **Dublin** sees no more wet days than London, though the rain here does tend to be a bit heavier.

Dublin

53°26'N, 6°15'W
Elev: 85m (279ft)

Visitors may be surprised to find a better-than-even chance they'll stay dry on a given day in Dublin, although Co. Wicklow and points west are more prone to dampness. Rainfall varies little from summer to winter, although a few light snows typically fall. Heat waves and cold spells are less likely than in England.

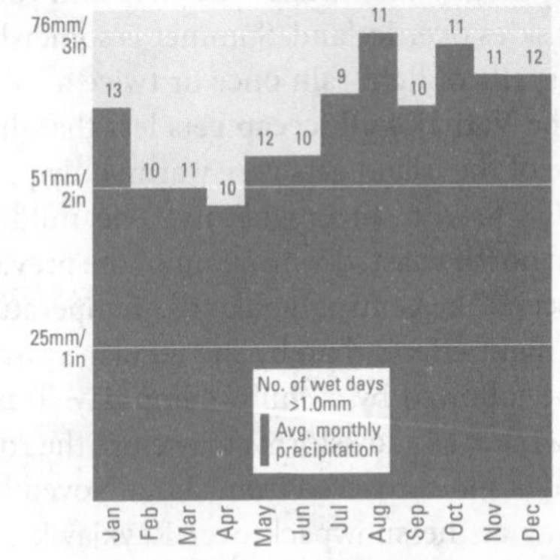

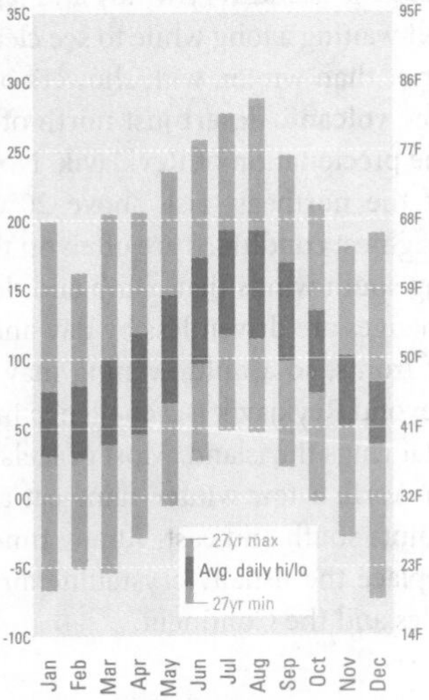

	Jan	Feb	Mar	Apr	May	Jun	Jul	Aug	Sep	Oct	Nov	Dec
midmonth -sunrise	0833	0744	0640	0626	0525	0456	0515	0604	0658	0752	0750	0834
-sunset	1636	1735	1829	2025	2118	2155	2146	2054	1941	1829	1629	1606
avg hrs of sunshine	2	3	4	5	6	6	5	5	5	3	2	2
avg # days w/thunder	<1	<1	<1	<1	1	1	1	1	<1	<1	<1	<1
w/snow	6	5	4	2	<1	0	0	0	0	0	1	2
Typical % relative humidity @1500hrs	78	75	70	67	67	68	68	68	70	74	77	81

Limerick

52°42'N, 8°55'W
Elev: 20m (66ft)

Although Limerick's cool-to-mild temperatures are on par with Dublin's, it rains a bit more often and more heavily here (even more so as you head west). Snow is less frequent here than in the east. Moisture is focused in the winter months, although you could get wet any time of year.

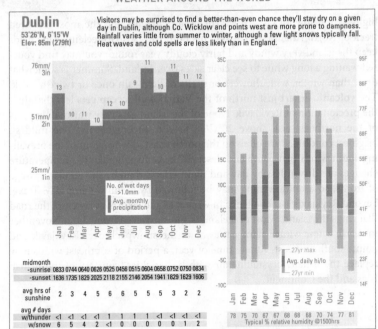

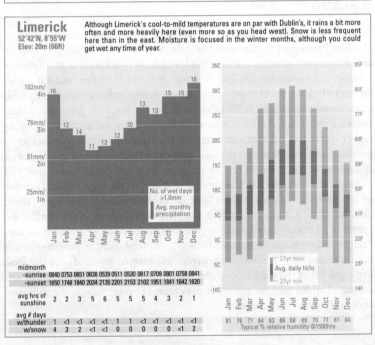

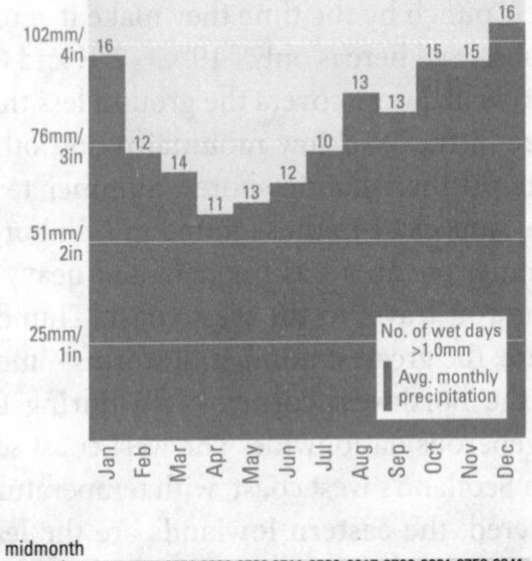

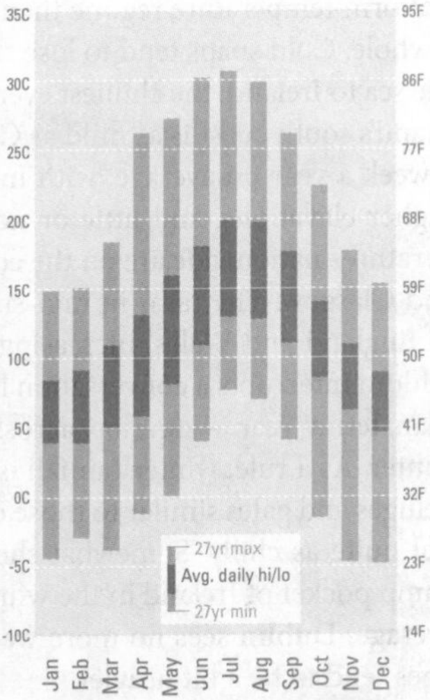

	Jan	Feb	Mar	Apr	May	Jun	Jul	Aug	Sep	Oct	Nov	Dec
midmonth -sunrise	0840	0753	0651	0638	0539	0511	0530	0617	0709	0801	0758	0841
-sunset	1650	1748	1840	2034	2126	2201	2153	2102	1951	1841	1642	1620
avg hrs of sunshine	2	2	3	5	6	5	5	5	4	3	2	1
avg # days w/thunder	1	<1	<1	<1	<1	1	1	<1	<1	<1	<1	<1
w/snow	4	3	2	<1	<1	0	0	0	0	0	<1	2
Typical % relative humidity @1500hrs	81	76	71	64	63	66	68	69	70	77	81	84

Malin Head
55°22'N, 7°20'W
Elev: 25m (82ft)

The weather on this windswept northern outpost is typically Irish (only more so). Frequent rounds of rain, about half as heavy here again as in Dublin, are interspersed with sun. Even in summer, temperatures seldom top 22°C/72°F. Spring is only about half as wet as autumn; both are just a touch cooler than July.

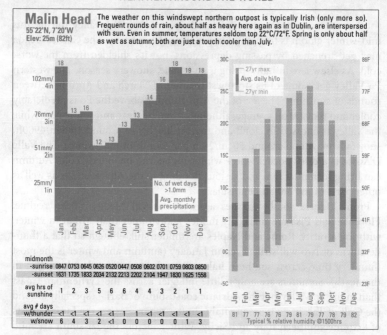

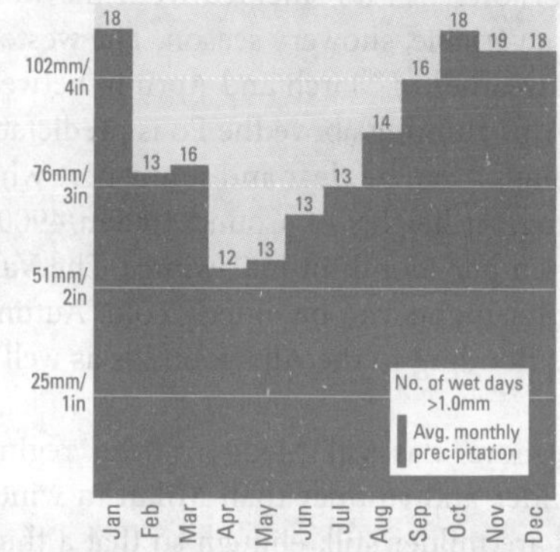

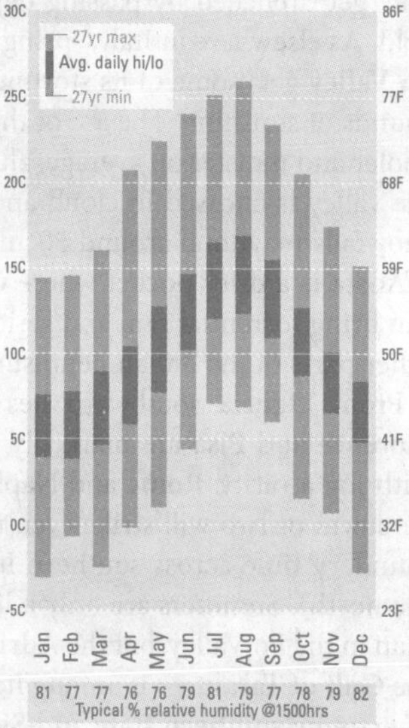

	Jan	Feb	Mar	Apr	May	Jun	Jul	Aug	Sep	Oct	Nov	Dec
midmonth -sunrise	0847	0753	0645	0626	0520	0447	0508	0602	0701	0759	0803	0850
-sunset	1631	1735	1833	2034	2132	2213	2202	2104	1947	1830	1625	1558
avg hrs of sunshine	1	2	3	5	6	6	4	4	3	2	1	1
avg # days w/thunder	<1	<1	<1	<1	<1	1	1	<1	<1	<1	<1	<1
w/snow	6	4	3	2	<1	0	0	0	0	0	1	3

Italy

San Marino | Vatican City

The cradle of the Renaissance was also the location where weather observing as we know it evolved, thanks to the efforts of Torricelli, Galileo and other Italian scientists and inventors (see "Taking the weather's pulse", p.30). There is plenty to observe weatherwise across this varied country. Jutting into the sea, **Italy** is a sitting duck for Mediterranean depressions. These centres of low pressure spin up over the water, especially in the **Gulf of Genoa**. As they drift inland, they bring pulses of rain and thunder that vary in character depending on the time of year and the part of the country. **The Alps** and the **Apennines** produce more wintry regimes at elevation; they also help create a wide array of micro-climates at lower levels by blocking the regular passage of Atlantic fronts, which allows moist air from the south to get into the mix.

The **Po Valley** – running across northern Italy from Turin to Venice – has one of the nation's sharpest and most distinctive climates. Summers are warm, with oppressively muggy periods that build toward thunderous downpours. After a brief, golden respite in September, autumn becomes progressively damper and chillier. From December through February, dense fog and cold, clammy air plague the valley. Visibility can fall below 1km/0.6 miles for days on end. The **eastern Po Valley** is also prone to cold *bora* winds that plunge south from eastern Europe. Rain and snow across the valley during this mid-winter gloom is frequent, but usually not too heavy. The Gulf of Genoa may get drenched by passing depressions, but it's shielded from the worst cold. As elsewhere in Italy, spring is a variable, showery season. The **western Po Valley** gets some of its stormiest weather in March and April in between rounds of sunshine. The arc of the Alps running above the Po is predictably cooler and moister on average, although it can be clear and sunny here while the valley is encased in cloud and fog. At heights of around 1500m/4900ft, snowfalls may total around 50cm/20in per month in mid-winter. The **Valle D'Aosta** is a drier pocket where winter nights can be bitterly cold. Autumn can bring torrential rain and serious flooding to the Alps foothills as well as other parts of the Italian peninsula.

From **Liguria** southward lies a more classical Mediterranean regime. **Florence** and **Pisa** are markedly milder and sunnier than **Milan** in winter, with fog a rarity. **Rome** and **Naples** are milder still, enough so that a thunderstorm or two will strike even in January (autumn and winter is the most thundery time across southern Italy, although it's drier and milder than in the north). Summers are only a little muggier across the **Tyrhennian Coast** than in the Po Valley, but the **Adriatic coast** south of **Bari** – especially around the **Gulf of Taranto** – is among Italy's hotter regions, as are the **Ionian coasts** of extreme **southern Italy** and **Sicily**. Winters across the far south are mild

Florence
43°48'N, 11°12'E
Elev: 44m (144ft)

Although it's equidistant from Milan and Rome, the Florence climate feels more southern than northern. Cold winds dry out after climbing the Apennines, so winter cold tends to be bright and damp spells mild. It's showery into May; summers can be hotter than in the Po Valley. Autumn brings long spells of rain.

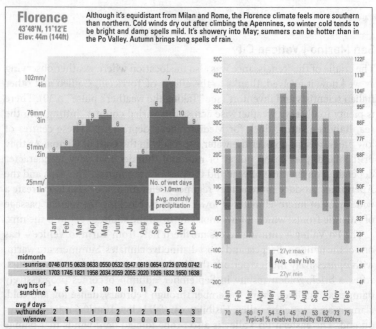

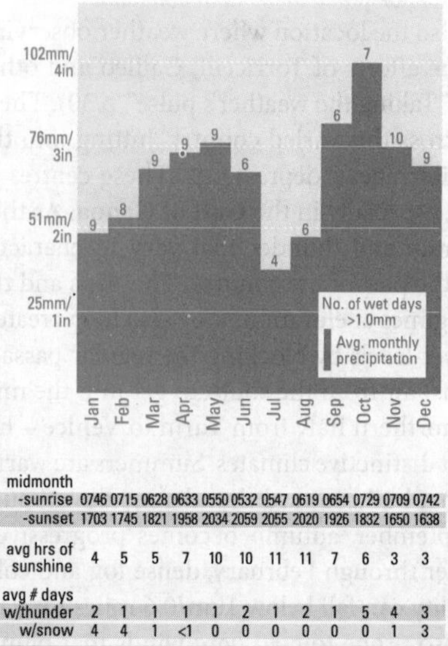

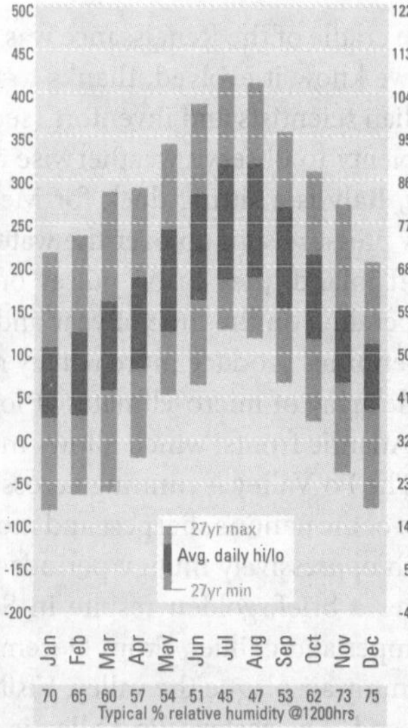

	Jan	Feb	Mar	Apr	May	Jun	Jul	Aug	Sep	Oct	Nov	Dec
midmonth -sunrise	0746	0715	0628	0633	0550	0532	0547	0619	0654	0729	0709	0742
-sunset	1703	1745	1821	1958	2034	2059	2055	2020	1926	1832	1650	1638
avg hrs of sunshine	4	5	5	7	10	10	11	11	7	6	3	3
avg # days w/thunder	2	1	1	1	1	2	1	2	1	5	4	3
w/snow	4	4	1	<1	0	0	0	0	0	0	1	3
Typical % relative humidity @1200hrs	70	65	60	57	54	51	45	47	53	62	73	75

Milan

45°37'N, 8°44'E
Elev: 211m (692ft)

There's no mistaking summer for winter in Milan. The warm season gets plenty of sun with frequent afternoon storms. Fog and cold are entrenched from December into March, with occasional rain or snow. Autumn and spring are the wettest seasons; June and September are the most temperate months.

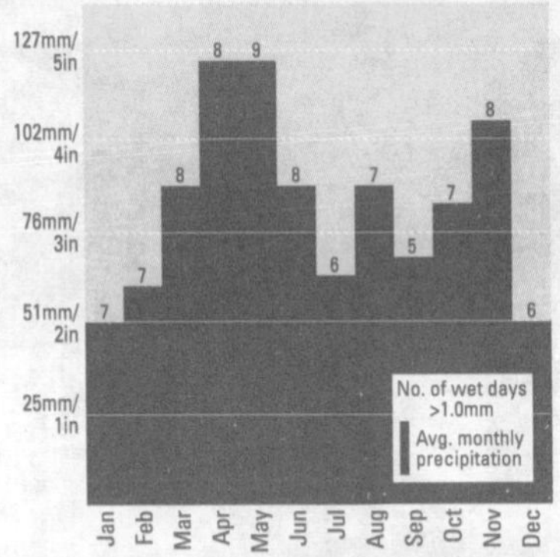

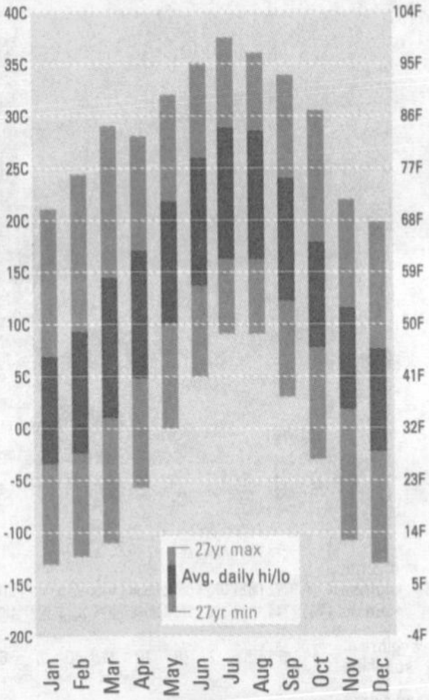

	Jan	Feb	Mar	Apr	May	Jun	Jul	Aug	Sep	Oct	Nov	Dec
midmonth -sunrise	0802	0728	0638	0640	0555	0535	0550	0625	0703	0741	0724	0759
-sunset	1707	1752	1831	2011	2049	2116	2111	2033	1937	1840	1655	1642
avg hrs of sunshine	2	3	5	6	7	8	9	8	6	4	2	2
avg # days w/thunder	<1	0	1	4	8	10	9	10	5	2	<1	<1
w/snow	4	3	2	<1	<1	0	0	0	0	<1	1	3

	Jan	Feb	Mar	Apr	May	Jun	Jul	Aug	Sep	Oct	Nov	Dec
Typical % relative humidity @1600hrs	68	61	55	55	58	57	55	55	59	66	68	68

Naples

40°53'N, 14°18'E
Elev: 88m (289ft)

Similar to Rome, Naples enjoys mild winters with spells of sun between showers. Summers are a touch warmer than in Rome, although readings above 32°C/90°F aren't too common. Expect frequent downpours from October into December. It can thunder any time of year.

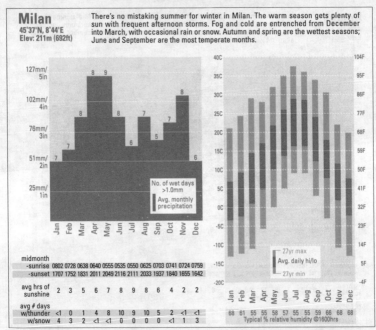

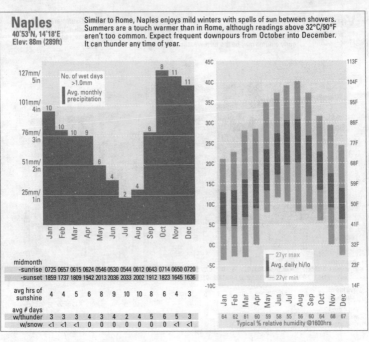

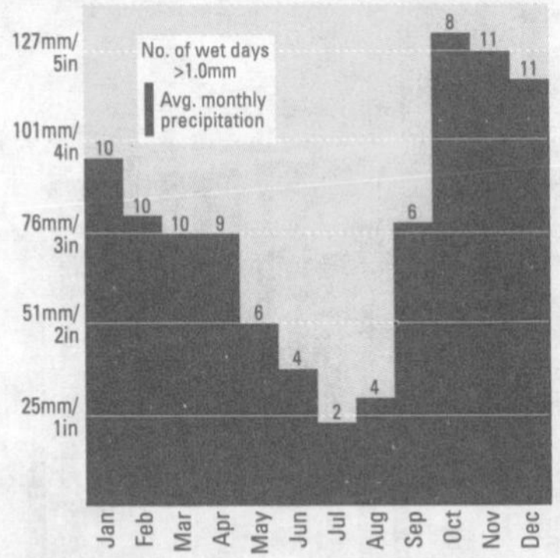

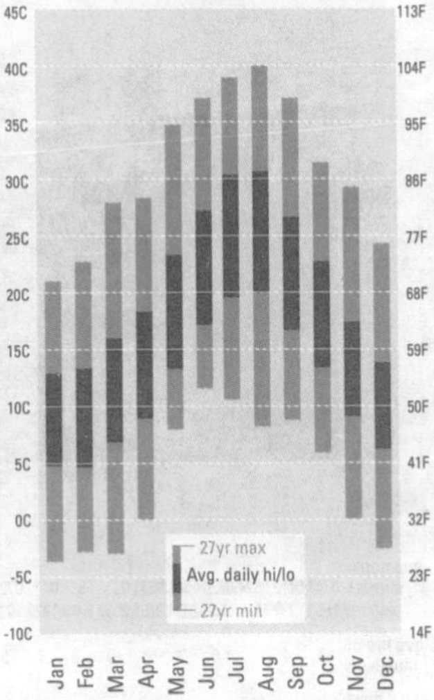

	Jan	Feb	Mar	Apr	May	Jun	Jul	Aug	Sep	Oct	Nov	Dec
midmonth -sunrise	0725	0657	0615	0624	0546	0530	0544	0612	0643	0714	0650	0720
-sunset	1859	1737	1809	1942	2013	2036	2033	2002	1912	1823	1645	1636
avg hrs of sunshine	4	4	5	6	8	9	10	10	8	6	4	3
avg # days w/thunder	3	3	3	4	3	4	2	4	5	6	5	3
w/snow	<1	<1	<1	0	0	0	0	0	0	0	<1	<1

	Jan	Feb	Mar	Apr	May	Jun	Jul	Aug	Sep	Oct	Nov	Dec
Typical % relative humidity @1600hrs	64	62	61	60	59	58	55	56	60	64	68	67

Rome

41°48'N, 12°14'E
Elev: 3m (10ft)

Rome's warm season (May–September) is salubrious indeed. Mid-summer heat is typically cut in the afternoon by the potentino sea breeze; it showers only once every couple of weeks. Rains can be frequent from late October into December. Bright spells alternate with showery periods in winter and spring.

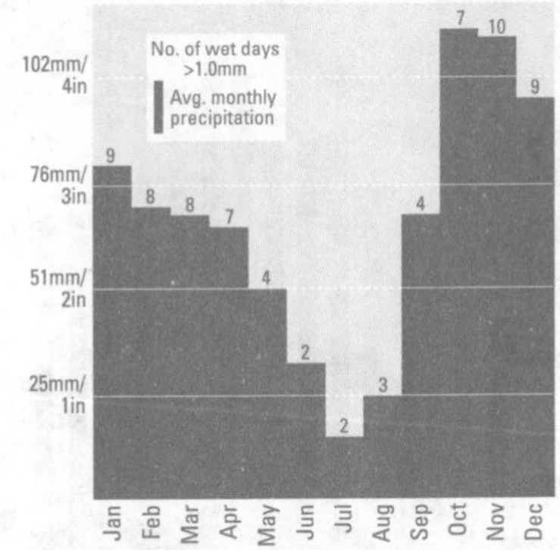

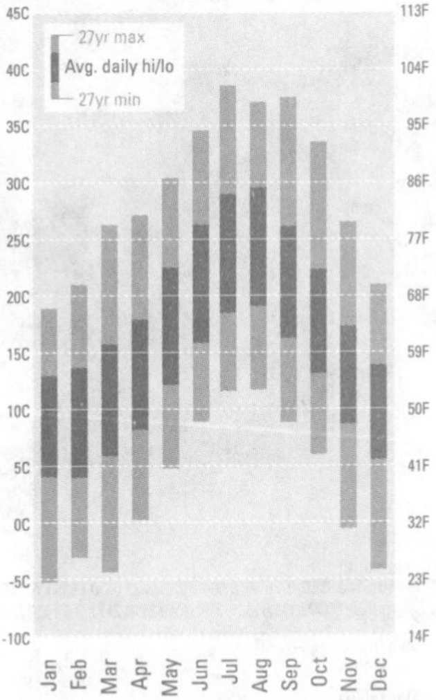

	Jan	Feb	Mar	Apr	May	Jun	Jul	Aug	Sep	Oct	Nov	Dec
midmonth -sunrise	0736	0707	0624	0632	0552	0536	0549	0619	0651	0723	0700	0731
-sunset	1705	1744	1817	1952	2024	2048	2045	2011	1921	1830	1651	1641
avg hrs of sunshine	4	5	5	7	8	10	11	10	8	6	4	4
avg # days w/thunder	3	3	3	4	2	3	2	3	5	6	5	3
w/snow	<1	1	<1	0	0	0	0	0	0	0	<1	<1
Typical % relative humidity @1600hrs	67	66	68	69	68	68	69	68	68	70	70	69

Venice

45°30'N, 12°20'E
Elev: 6m (20ft)

This island city gets harsher weather than visitors might expect. Mid-summers are steamy with heavy evening storms at times. East winds in the autumn may push acqua alta flooding across town, and moisture adds to the winter chill. The shoulder seasons of late spring and early autumn may be the best weather bets.

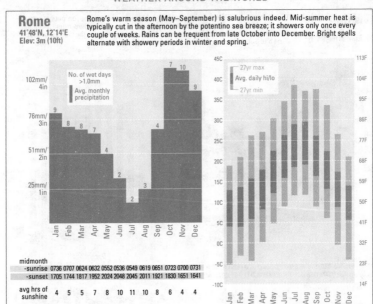

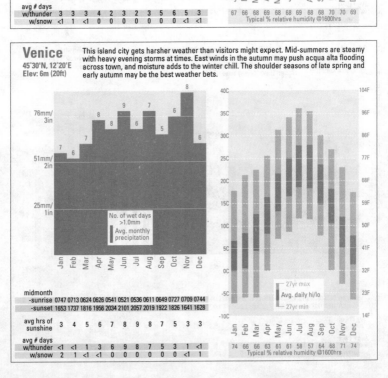

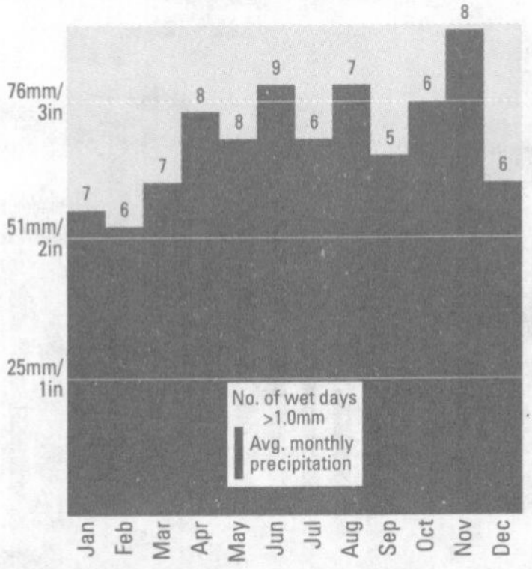

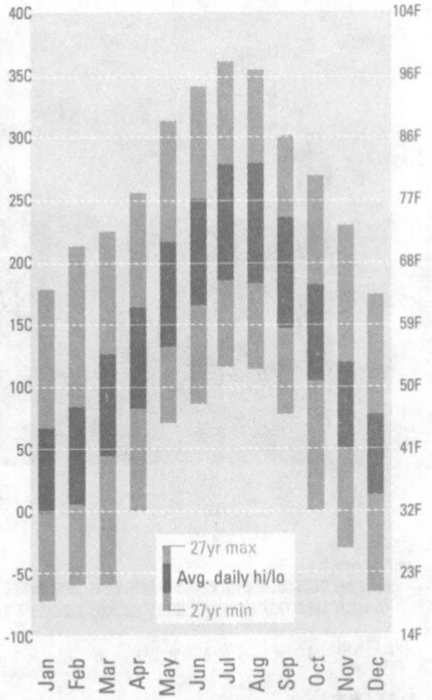

	Jan	Feb	Mar	Apr	May	Jun	Jul	Aug	Sep	Oct	Nov	Dec
midmonth -sunrise	0747	0713	0624	0626	0541	0521	0536	0611	0649	0727	0709	0744
-sunset	1653	1737	1816	1956	2034	2101	2057	2019	1922	1826	1641	1628
avg hrs of sunshine	3	4	5	6	7	8	9	8	7	5	3	3
avg # days w/thunder	<1	<1	1	3	6	9	8	7	5	3	1	<1
w/snow	2	1	<1	<1	0	0	0	0	0	0	<1	1
Typical % relative humidity @1600hrs	74	66	66	63	61	61	58	57	64	68	71	74

but frequently damp and thundery. The highest terrain, including **Mt Etna**, can experience surprisingly heavy snows on occasion. As for the two independent states within Italy's borders, neither **San Marino** nor **Vatican City** are sizeable enough to get weather that differs from the surrounding areas.

Latvia

Estonia

Scandinavians will feel at home with the climate of **Latvia** and its northern neighbour, **Estonia**. Both countries adjoin the **Baltic Sea**, with a west coast wrapping around the **Gulf of Riga**. Proximity to water gives these countries a cool, humid regime year round. Winter temperatures are relatively mild for the latitude – no colder, on average, than in the Ukraine – although the north coast of Estonia gets a few winter gales that add to the chill. Light snows are common across the country, along with periodic thaws (especially near the coast) that prevent the snowpack from getting too deep. Showers and a few steady rains occur about every 2 or 3 days on average in summer. The high latitude ensures plenty of mid-summer sun, however. Autumn rains aren't

Riga
56°58'N, 24°04'E
Elev: 3m (10ft)

Mid-summers are cool but bright in this gulfside city. Expect a shower every third day or so, becoming more frequent into August and September. Winters brings clouds and light snow, but occasional thaws help to keep the snowpack down. Late May and June are sunnier than summer and nearly as mild.

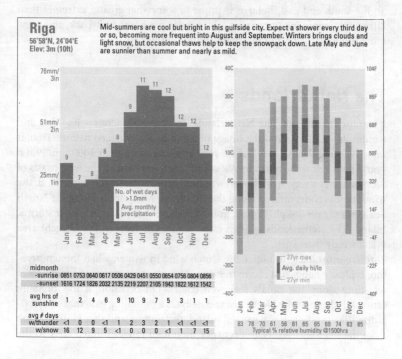

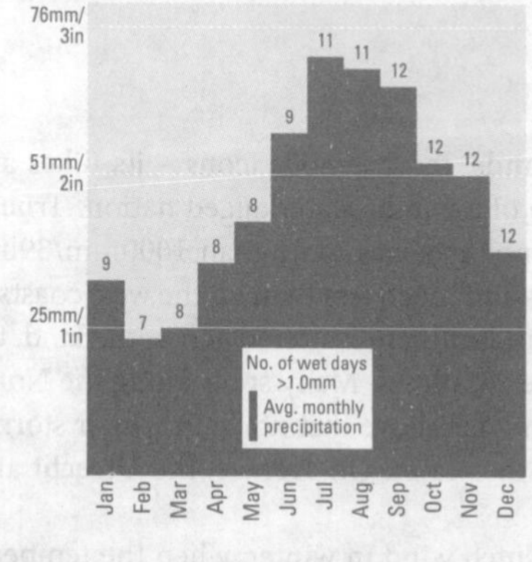

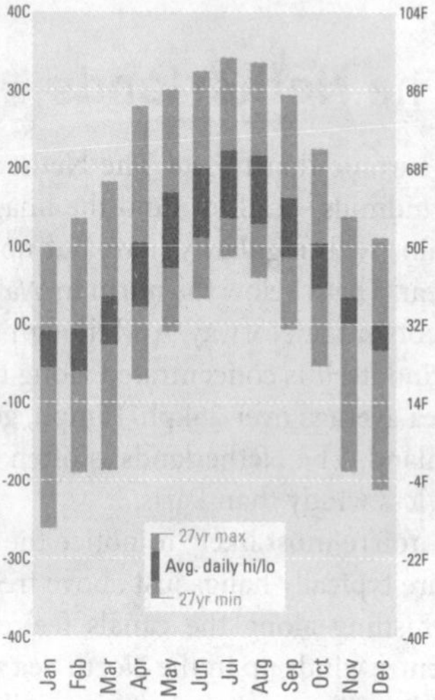

midmonth	Jan	Feb	Mar	Apr	May	Jun	Jul	Aug	Sep	Oct	Nov	Dec
-sunrise	0851	0753	0640	0617	0506	0429	0451	0550	0654	0756	0804	0856
-sunset	1616	1724	1826	2032	2135	2219	2207	2105	1943	1822	1612	1542
avg hrs of sunshine	1	2	4	6	9	10	9	7	5	3	1	1
avg # days w/thunder	<1	0	0	<1	1	2	3	2	1	<1	<1	<1
w/snow	16	12	9	5	<1	0	0	0	<1	1	7	15
Typical % relative humidity @1500hrs	83	78	70	61	56	61	65	65	69	74	83	85

especially heavy, but they become more frequent by September. Off-season visitors might be better off in the sunnier months of May and June.

Lithuania

Belarus

Most of **Lithuania** is only a few hours' drive from the **Baltic coast**, but the inner parts of this country – including the touristed areas of **Vilnius** and **Trakai** – experience more of a dramatic change from summer to winter than their closeness to the sea might suggest. Winters are cold nationwide, though distinctly less so along the Baltic coast, where thaws are more frequent. Points along the **southwest coast** of Lithuania (along with an adjacent fragment of Russia) average more than 2°C/4°F warmer than Vilnius in the winter. Dustings of snow occur every couple of days in winter, with a few bouts of rain along the coast. Spring warms and brightens up quickly, and it can get quite showery and thundery as early as May across parts of **the east**. Summers are mild across the nation, with a few days of heat possible in July and August. It cools down quickly in autumn, and the rains become lighter as well (except along the coast, which can get doused into November). To the south and east, **Belarus** is prone to somewhat greater extremes than Lithuania – more intense winter cold, a more persistent snowpack, and a few more thunderstorms and hot spells in the summer.

The Netherlands

It's ironic that two of **The Netherlands**' most famous icons – its dikes and windmills – call to mind the image of a windy, waterlogged nation. True, it rains with regularity here, but no spot averages more than 1000mm/39in a year. That's below the norm in Wales and much less than on the west coasts of Scotland or Norway. And though windmills are scattered across the land, the wind itself is concentrated along the coastlines. Many spots along the **North Sea** average over 25kph/16mph, going far above that norm in winter storms. Inland, The Netherlands is often rather calm. On average, the **Utrecht** area is less windy than Paris.

You're most likely to notice the Dutch wind in winter, when the temperature typically hangs just above freezing. The moist air makes the stiff breeze whistling along the canals feel even colder. On occasion, a low-pressure centre winds up in the North Sea and delivers a pounding gale that can jeopardize dikes and push waters well inland, reclaiming land that Holland took from the sea long ago. With no major mountain ranges immediately to the

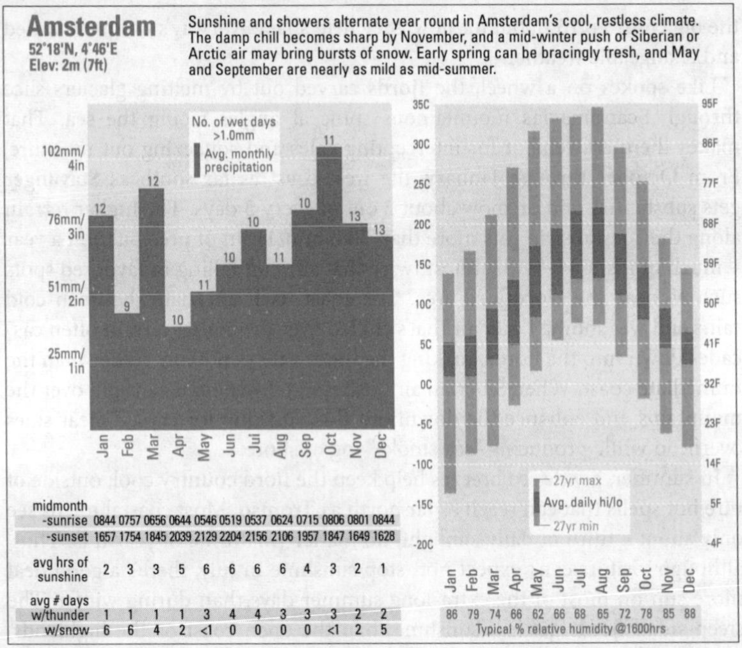

north and east, The Netherlands is also prone to blasts of **Arctic** or **Siberian air** that can last for a week or two, turning the waterways to ice. Cold waves and snow are a bit more frequent and prolonged across the **northeastern reaches** of the country, next to Germany. Breezes off the cool North Sea in summer help protect The Netherlands from many spells of heat. (When it does get hot, though, not even the coast is immune.) Expect some clouds even on a bright, dry day. Clear skies are rare in the Netherlands: in summer, only about 1 in 9 days have clear skies. Your best chance of sun is along the coast and in the **northern islands**. Holland's **southwest coast** (around **The Hague**) is often drenched during the autumn, but on a annual basis, **Hage Veluwe National Park** and environs get the most rain.

Norway

The **Norway** of picture books isn't so much a region as a ribbon, wrapping from southwest to northeast around more than 1600km/1000 miles of coastline. Off the **west coast** lies some of the world's warmest high-latitude water. This part of the **Atlantic basin** never freezes, even during weeks of winter darkness. The ocean currents that bring warmth here from the Gulf Stream and its northern extension might get diverted someday (see "Will Europe get the big chill?", p.154), but for now, they help shape Norway's uniquely varied and changeable weather.

Like spokes on a wheel, the **fjords** carved out by melting glaciers slice through Scandinavia's mountainous spine at angles facing the sea. That makes them convenient for intercepting gales and squeezing out moisture. From October through January, the west coast as far south as **Stavanger** gets substantial rain or snow about 2 out of every 3 days. The higher terrain along the coastal strip gets more than 3000mm/118in of precipitation a year – much of it snow – with almost twice that amount falling in favoured spots such as near the **Nordfjord** area. The coast itself alternates between cold rains and wet snows. Drier air that's chilled over the higher terrain often cascades down into the fjords, making the inner valleys notably colder than the immediate coast. When Siberian air pushes into Sweden, it can spill over the mountains and enhance the downfjord flow, making for crystal-clear skies overhead while producing "sea smoke" fog offshore.

In summer, gentle sea breezes help keep the fjord country cool, outside of rare hot spells that can reach as far north as **Tromso**. Most coastal towns see only about a third of daily sunlight, no matter what time of year it is. Thus, although visitors can't expect non-stop sunshine in July, there's a good deal more sun on most of the extra-long summer days than during winter. The steep-sided fjords get less sunshine than the open coast or the highlands. Showers (mostly light) occur about every other day in mid-summer, and the more substantial autumn rains begin as early as late August. The **Finnmark coast** stays quite chilly even in summer and gets wildly stormy in winter as coastal lows sweep by. Eastern Finnmark is lucky to get more than two weeks of clear skies a year.

On the other side of the mountains, **southeastern Norway** has a more Swedish-flavoured climate. Parts of the **upper Randsfjorden** are almost semi-arid, with the average yearly precipitation as low as 300mm/12in. Snowfall in the **Oslo** area is frequent but may amount to little more than flurries. Some winters are markedly colder or wetter than others, however. Relative mild spells, borne on southwest winds from the Atlantic, are offset by intrusions of bitter Arctic cold. Early spring is especially prone to Siberian air masses; they're in place on about half the days in a typical March and April, often resulting in frigid mornings but bright, sunny afternoons. June and July can be thundery but with more sunshine and warmth than the west coast gets. Occasional strings of bright days can bring readings above 30°C/86°F.

Bergen

60°18'N, 5°13'E
Elev: 50m (164ft)

Any month can bring a deluge in Bergen, where more than half the days of the year see some rain or snow (and seldom just a sprinkle or flurry). Spring brings the best chance of a dry spell, often accompanied by cool north winds. Mid-summer is cloudier here than in most spots east of the mountains.

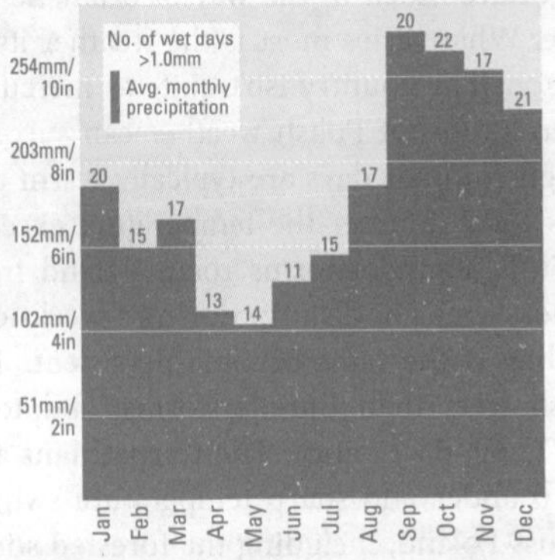

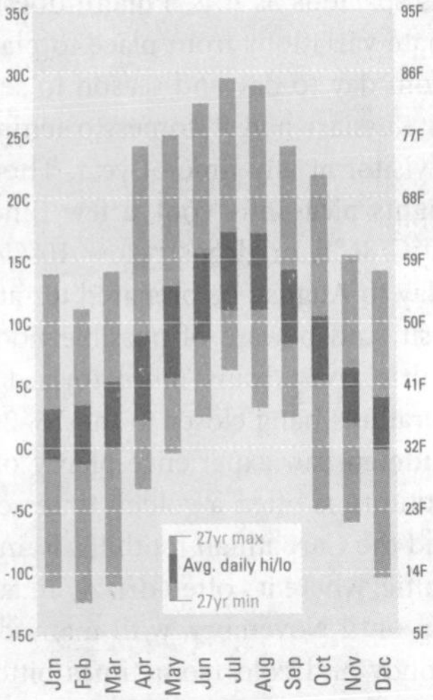

	Jan	Feb	Mar	Apr	May	Jun	Jul	Aug	Sep	Oct	Nov	Dec
midmonth -sunrise	0929	0819	0657	0622	0500	0412	0439	0551	0706	0818	0838	0939
-sunset	1608	1729	1841	2058	2213	2308	2249	2134	2001	1830	1608	1529
avg hrs of sunshine	1	2	3	5	6	6	5	5	3	2	1	0
avg # days w/thunder	2	1	1	<1	1	1	1	1	1	1	2	1
w/snow	16	13	12	7	1	<1	0	0	0	1	7	13

Jan	Feb	Mar	Apr	May	Jun	Jul	Aug	Sep	Oct	Nov	Dec
79	72	68	62	60	63	66	69	70	72	78	81

Typical % relative humidity @1600hrs

Oslo

59°54'N, 10°37'E
Elev: 17m (56ft)

Much like nearby Stockholm, Oslo endures a bitingly cold winter, peppered with frequent snow – usually light. In return, summers are typically glorious. Spring is the driest season, although a bit less sunny here than in points north and east. Winds are usually light around Oslo; the city is prone to mid-winter fog.

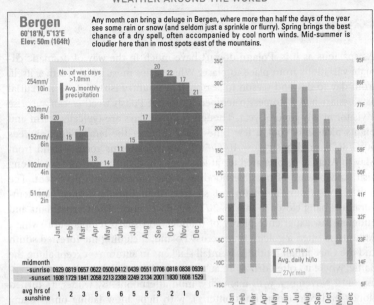

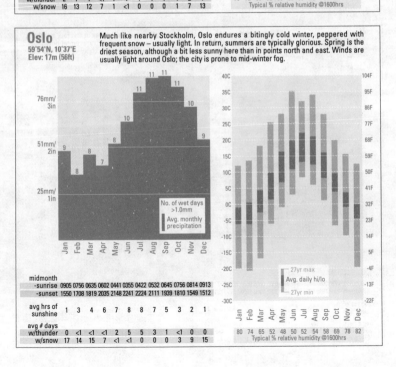

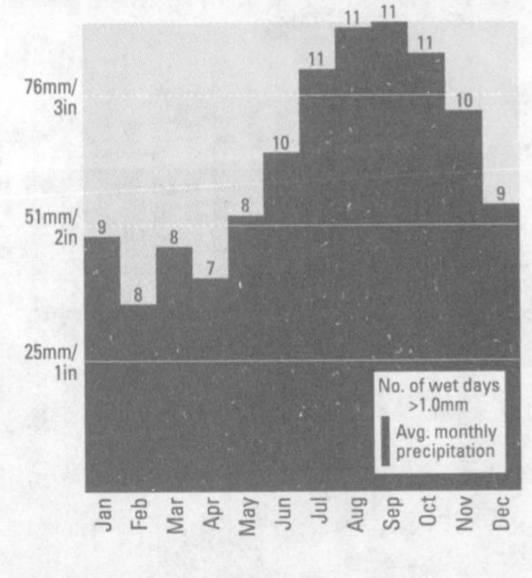

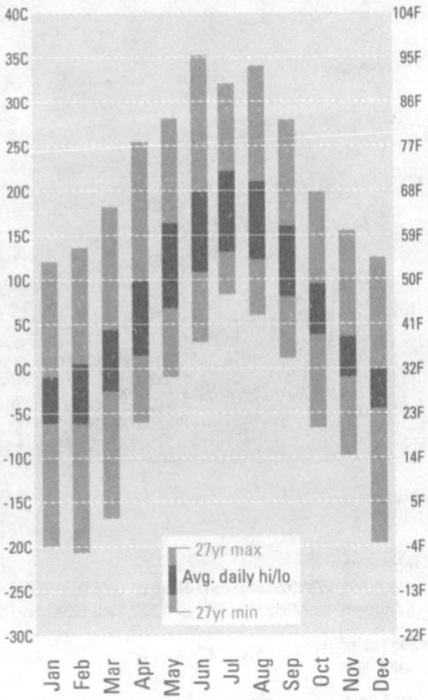

	Jan	Feb	Mar	Apr	May	Jun	Jul	Aug	Sep	Oct	Nov	Dec
midmonth -sunrise	0905	0756	0635	0602	0441	0355	0422	0532	0645	0756	0814	0913
-sunset	1550	1708	1819	2035	2148	2241	2224	2111	1939	1810	1549	1512
avg hrs of sunshine	1	3	4	6	7	8	8	7	5	3	2	1
avg # days w/thunder	0	<1	<1	<1	2	5	5	3	1	<1	0	0
w/snow	17	14	15	7	<1	<1	0	0	0	3	9	15

Jan	Feb	Mar	Apr	May	Jun	Jul	Aug	Sep	Oct	Nov	Dec
80	74	65	52	48	50	52	54	58	69	78	82

Typical % relative humidity @1600hrs

Poland

As spacious as it is, **Poland** doesn't have much in the way of dramatic climate variations from place to place. What varies most is the weather itself, from day to day and season to season. The country isn't quite as mercurial as Russia when it comes to temperatures, but Polish weather can surprise a visitor at any time of year. Though summer days are typically warm and nights pleasantly cool, a few times each summer the temperature exceeds 30°C/86°F, or drops below 10°C/50°F. Thunderstorms roam Poland from May to August: be prepared for at least one good storm during a week-long visit, and beware of possible flooding if the rains become persistent. The **Baltic coast** leans more toward showers than thunderstorms, and temperatures hang closer to the 16–20°C/60–68°F range. The **Carpathians** and **Sudeten** can experience plenty of thunder and sharp temperature swings. Autumn is often a golden time across Poland, including the **forested south** and the Carpathian foothills. Rainfall slackens in autumn – except along the Baltic, where it's often drizzly in autumn – and most nights stay above freezing until November, with many dry, mild afternoons. By winter it's turned cloudy and cold (sometimes bitterly), with light snow falling on as many as half the days. The ground is usually snow-covered through January and

Krakow

50°05'N, 19°48'E
Elev: 237m (777ft)

Like a mini-monsoon, roughly half of Krakow's annual rainfall arrives in just four months – between May and August. Most days are dry, but an afternoon or evening downpour or a steadier dose of rain is seldom too far away. Winters are chilly with occasional rounds of downslope mildness. Autumn is a choice pick.

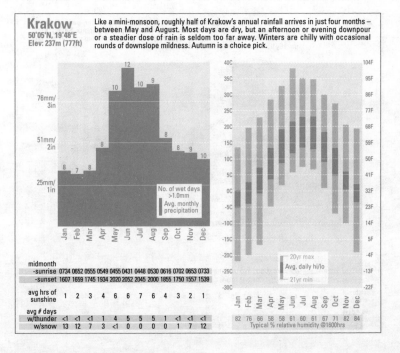

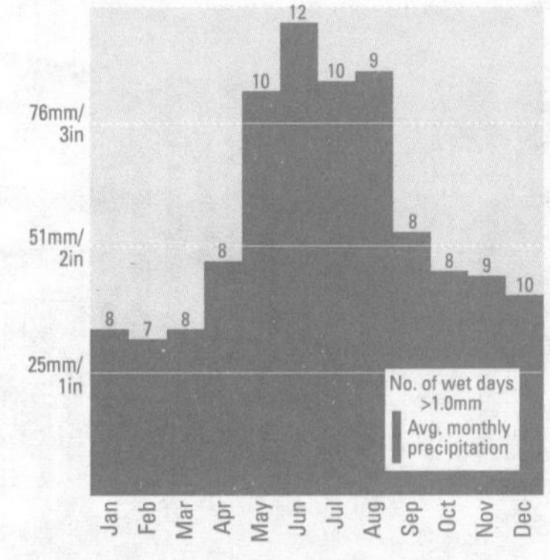

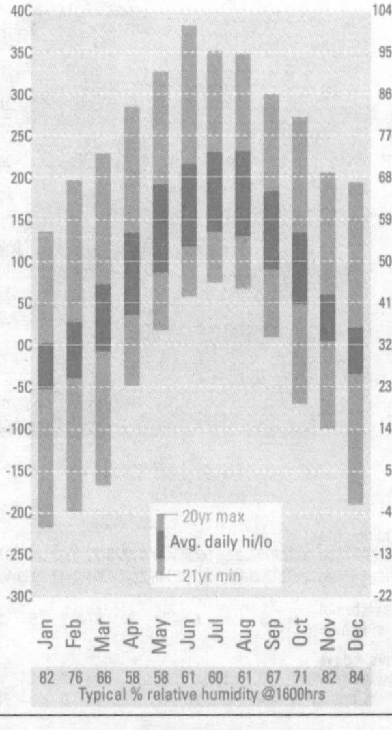

	Jan	Feb	Mar	Apr	May	Jun	Jul	Aug	Sep	Oct	Nov	Dec
midmonth -sunrise	0734	0652	0555	0549	0455	0431	0448	0530	0616	0702	0653	0733
-sunset	1607	1659	1745	1934	2020	2052	2045	2000	1855	1750	1557	1539
avg hrs of sunshine	1	2	3	4	6	6	7	6	4	3	2	1
avg # days w/thunder	<1	<1	<1	1	4	5	5	5	1	<1	<1	<1
w/snow	13	12	7	3	<1	0	0	0	0	1	7	12
Typical % relative humidity @1600hrs	82	76	66	58	58	61	60	61	67	71	82	84

Warsaw

52°10'N, 20°58'E
Elev: 107m (351ft)

Much of the year offers fine weather for seeing Warsaw. Spring and autumn are your best bets for sunny, dry, mild days and cool nights. There's little serious heat in mid-summer. Winters are predictably chilly with frequent light snow. Any time of year can bring a dramatic shift above or below the norms.

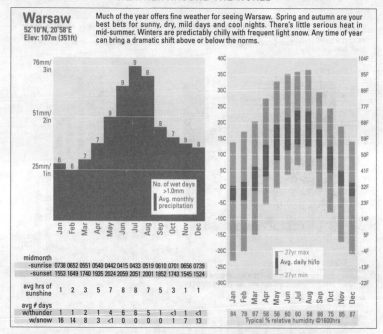

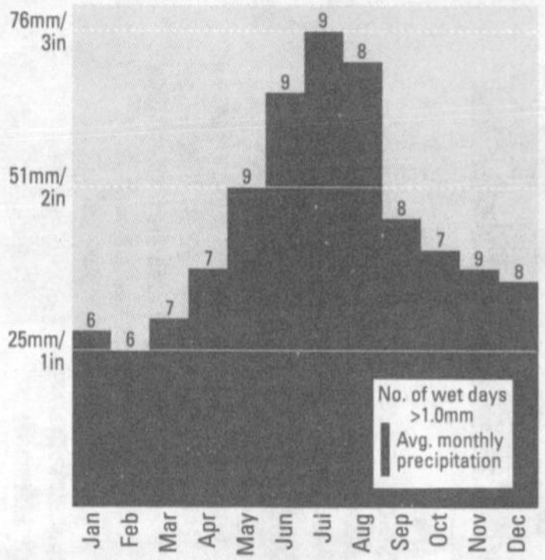

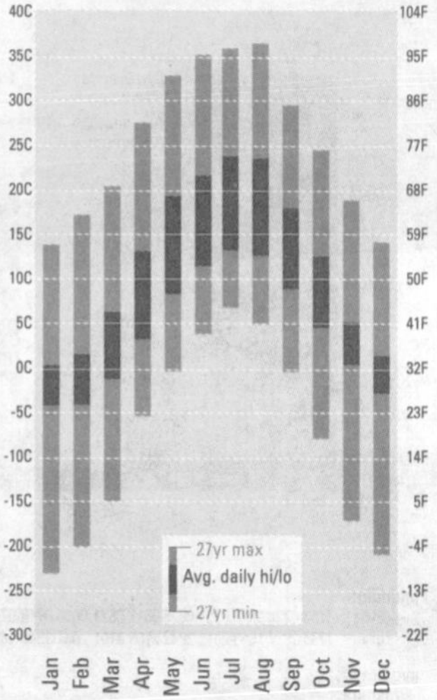

	Jan	Feb	Mar	Apr	May	Jun	Jul	Aug	Sep	Oct	Nov	Dec
midmonth -sunrise	0738	0652	0551	0540	0442	0415	0433	0519	0610	0701	0656	0739
-sunset	1553	1649	1740	1935	2024	2059	2051	2001	1852	1743	1545	1524
avg hrs of sunshine	1	2	3	5	7	8	8	7	5	3	1	1
avg # days w/thunder	1	1	2	1	4	6	6	5	1	<1	1	<1
w/snow	16	14	8	3	<1	0	0	0	0	1	7	13

	Jan	Feb	Mar	Apr	May	Jun	Jul	Aug	Sep	Oct	Nov	Dec
Typical % relative humidity @1600hrs	84	79	67	58	56	60	60	58	66	75	85	87

February across the north and east. Even on the Baltic coast, the precipitation is often snow (but rarely heavy). Some coastal winters are more blustery than others. A few winter days have been known to climb above 10°C/50°F, especially across the south, as dry, *föhn*-type winds flow over the mountains and descend. Like autumn, spring is one of the drier seasons across Poland, with more than a few crystal-clear days, even along the Baltic. Temperature gyrations are at their most extreme in spring; you might encounter shirt-sleeve warmth or below-freezing chill.

Portugal

Flanking the west coast of the Iberian Peninsula, **Portugal** meets and greets Atlantic storm systems throughout the winter months, intercepting much of the rain before it can reach the plain or anywhere else in Spain. The mountains in the **northern half** of Portugal accentuate this effect. Some northern stations get as much winter rain as the west coasts of Ireland or Scotland, with considerable snow at altitude. Further **south** – where storms give more glancing blows to the provinces of **Faro** and **Beja** – it may rain only twice a week during the wet season. Winter temperatures along the coasts are comparable to those around coastal Spain, while the mountains

Lisbon
38°47'N, 9°08'W
Elev: 123m (403ft)

Temperatures in Lisbon are strikingly similar to those in Los Angeles, except for cooler winter days. Hot spells are rare and limited to the summer. The frequent rains from October–April are seldom very heavy, and sunshine usually intervenes before long.

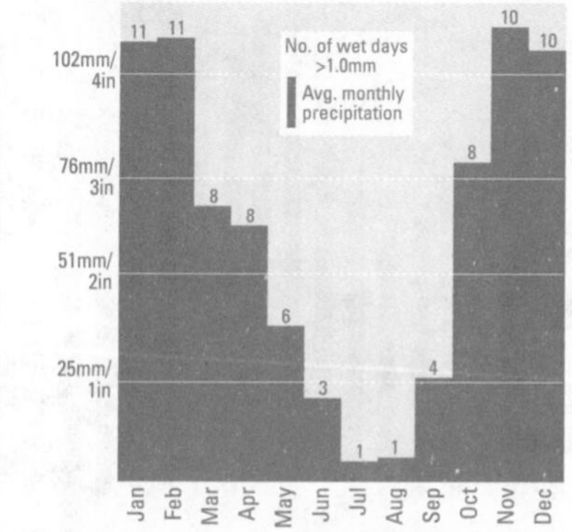

	Jan	Feb	Mar	Apr	May	Jun	Jul	Aug	Sep	Oct	Nov	Dec
midmonth -sunrise	0753	0728	0648	0701	0625	0611	0624	0650	0718	0746	0719	0747
-sunset	1739	1814	1843	2013	2042	2103	2101	2031	1945	1859	1723	1716
avg hrs of sunshine	5	5	7	8	9	10	11	11	9	7	5	5
avg # days w/thunder	1	1	1	2	1	1	1	<1	1	2	2	1
w/snow	<1	0	0	0	0	0	0	0	0	0	0	0

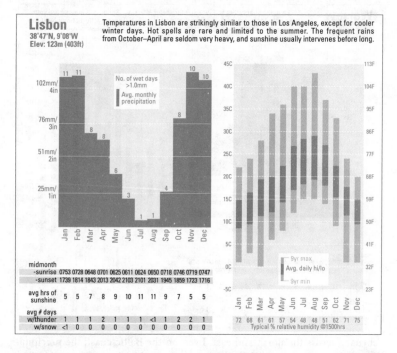

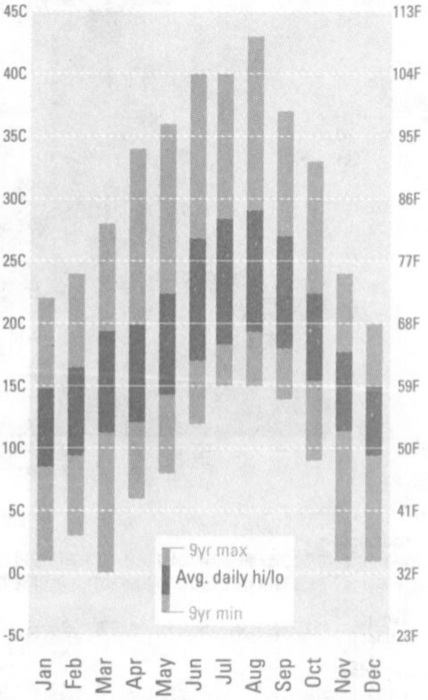

Jan	Feb	Mar	Apr	May	Jun	Jul	Aug	Sep	Oct	Nov	Dec
72	68	61	61	57	54	48	48	51	62	71	75

Typical % relative humidity @1500hrs

are considerably colder. Summers are a treat in much of Portugal, with the Atlantic adding only a bit of humidity. Afternoon highs in July and August average 24–30°C/75–86°F along the coast, cooler in the **higher terrain** and hotter in parts of **the southeast** (a high of 50°C/122°F was once recorded at Riodades). Light showers in summer are limited to **the north**, and even there they strike perhaps once a week or so. Much of the country is as sunny as the Costa del Sol. Spring and autumn are fine times to visit the southern half of Portugal: temperatures are mild, and the rains largely diminish by May and stay infrequent till November.

Romania

Moldova

Much like Bulgaria to its south, **Romania** has a four-season climate modulated by extensive highlands. Except for the highest parts of the **Carpathians**, most of the country will reach 26°C/79°F on a typical mid-summer day. Humidity is highest along the broad **Danube valley**. The warm season is the wetter time across the heart of Romania, but the thunderstorm action peaks early – in May and June. Otherwise, there's not a lot of moisture; even for

Bucharest
44°30'N, 26°8'E
Elev: 91m (298ft)

Bucharest gets a thunderstorm about every third day in June, its wettest month. As summer heats up, showers diminish. Afternoons can be toasty in July and August, with a touch of humidity. Winter cold doesn't begin in earnest until November. Light snow is common during the cloudy winter and into early spring.

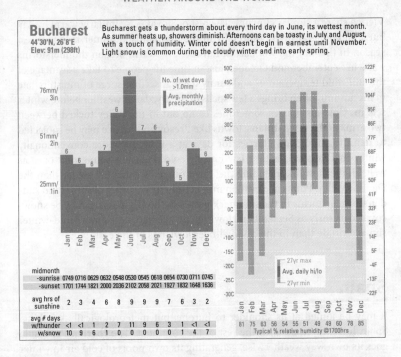

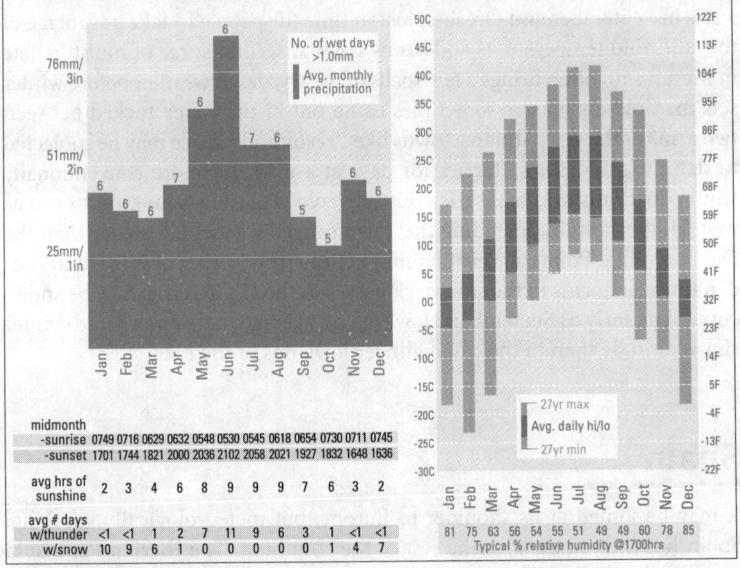

	Jan	Feb	Mar	Apr	May	Jun	Jul	Aug	Sep	Oct	Nov	Dec
midmonth -sunrise	0749	0716	0629	0632	0548	0530	0545	0618	0654	0730	0711	0745
-sunset	1701	1744	1821	2000	2036	2102	2058	2021	1927	1832	1648	1636
avg hrs of sunshine	2	3	4	6	8	9	9	9	7	6	3	2
avg # days w/thunder	<1	<1	1	2	7	11	9	6	3	1	<1	<1
w/snow	10	9	6	1	0	0	0	0	0	1	4	7
Typical % relative humidity @1700hrs	81	75	63	56	54	55	51	49	49	60	78	85

central Europe, Romania is a relatively dry land. **Transylvania** sees a bit more cloud cover, as the prevailing westerly winds ascend the Carpathians; late summer and early autumn are ideal for exploring the hills. Snow falls about every 3 or 4 winter days across most of **the lowlands** of Romania (it's more frequent and prolonged at elevation, extending well into spring). Winter's cold and snow are well entrenched across **northeast Romania** and adjacent **Moldova**, where the northeast *crivat* winds from Russia are strongest and January-to-February thaws are less common than they are further south. Romania's mildest and driest corner is the **Black Sea** coastal strip. Winter winds can feel raw here, even with most days rising above the freezing point, but summers are sunny and temperate, with few days of thunder or intense heat.

Slovakia

Slovakia's climate is split by the mountain ranges at its heart. The **Danube lowland** across the **southwest** is the nation's warm spot. Enough moisture trickles up from the Mediterranean so that many days in July and August feel a touch humid across the plains and on the Danube (where **Bratislava** and

Vienna share a similar climate). The rest of Slovakia is similarly warm by day but a bit cooler at night, especially in **the highlands**. Summer is the wettest season in this not too wet part of Europe. The warm-season thunderstorms wane in September and October, just in time for people to take advantage of the still-mild temperatures and sunny days. A second surge of moisture late in the autumn often brings a few spells of dreary, damp weather before winter sets in. Cold air masses sometimes camp out in the valley tucked between two arms of **the Carpathians**: towns like **Presov** and **Kosice** may be subjected to dank fog and snow flurries for days at a time. It's more conventionally wintry in Bratislava, with a few bitterly cold days, but some mild ones as well. Higher terrain, including the **Tatra range**, is generally colder than the Danube valley. Points at elevation may get a hefty covering of snow (and sudden thunderstorms in the summer); passes in the **High Tetras** may be snow-covered as early as September. However, on average there's more mid-winter sun at altitude than in the clouded-over lowlands of the east.

Spain

It may be joined at its shoulder to Europe, but meteorologically speaking, **Spain** has little to do with the rest of the continent. The **Iberian Peninsula** goes its own way weatherwise, generating its own pools of cold air in winter and dry-roasting on its own in summer. Even if it were as pancake-flat as The Netherlands, Spain would probably have interesting weather, situated as it is between the Atlantic – whose perpetually cool currents run south along the coast of Portugal – and the Mediterranean, which warms and cools with the seasons. Throw in a crisscrossing array of mountain ranges and a central plateau, and you have the makings of Europe's most diverse climate, one that would take a book to describe adequately. Entire seasons can be spanned by crossing from one mountain valley to the next, much less traversing the country.

As in most of Europe, winter brings a fair number of wet days across Spain, particularly along the lush **north coast**, where heavy winter rains have carved fjord-like inlets along the **Catabrica peaks**. Light snow falls on only a handful of days each year across the lower southern plateau, including **Madrid**. The snow is slightly more frequent across the **northern plateau** and far heavier and more persistent across the mountain ranges. The plateau also gets crisp, windy spells that may feel incongruously chilly to a visitor beneath the bright Spanish sun. The worst cold stays clear of the coastlines, which seldom drop to near freezing. The tables are turned in summer: coastal highs tend to average below 30°C/86°F while the interior heats up, especially across the lower elevations of the south (see p.269).

Barcelona

41°17'N, 2°04'E
Elev: 6m (20ft)

June and July – after the spring rains, but before the autumn thunderstorms – are prime months for visiting Barcelona. Winter days tend to be cool, especially when the north tremonte wind blows, but mild spells are frequent and the rain is spotty. Even the wettest months get ample sun.

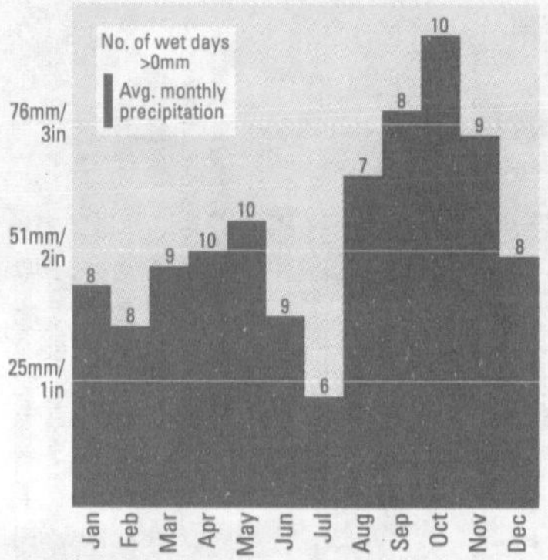

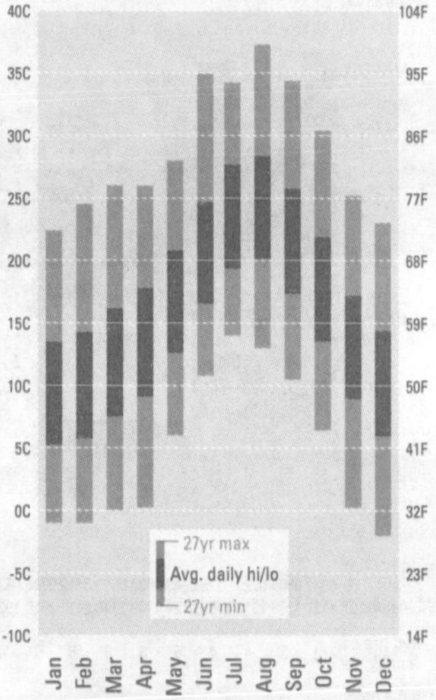

	Jan	Feb	Mar	Apr	May	Jun	Jul	Aug	Sep	Oct	Nov	Dec
midmonth -sunrise	0815	0747	0704	0713	0634	0618	0631	0701	0732	0803	0740	0810
-sunset	1747	1825	1858	2032	2103	2126	2123	2051	2001	1911	1733	1723
avg hrs of sunshine	5	6	6	7	8	9	10	9	7	6	5	5
avg # days w/thunder	1	1	2	2	3	4	3	6	6	5	2	1
w/snow	1	<1	<1	0	0	0	0	0	0	0	0	<1
Typical % relative humidity @1600hrs	65	63	64	66	70	68	66	66	66	66	66	64

Madrid

40°27'N, 3°33'W
Elev: 582m (1909ft)

Madrid is at its mildest and moistest during spring and autumn, with periods of showers and light rain. Spits of thunder are most likely on a warm summer afternoon. Winter can bring sharp chill and light snow. Whatever the season, you'll rarely go more than a couple of days without sunshine.

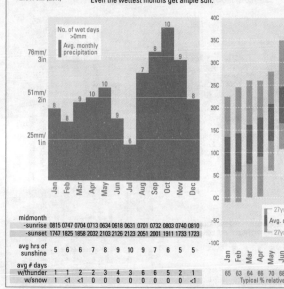

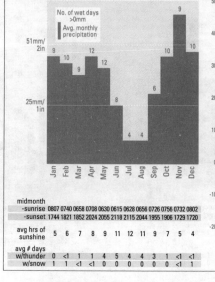

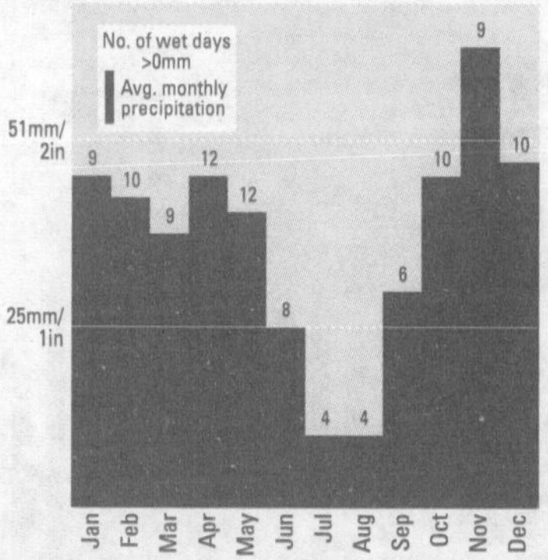

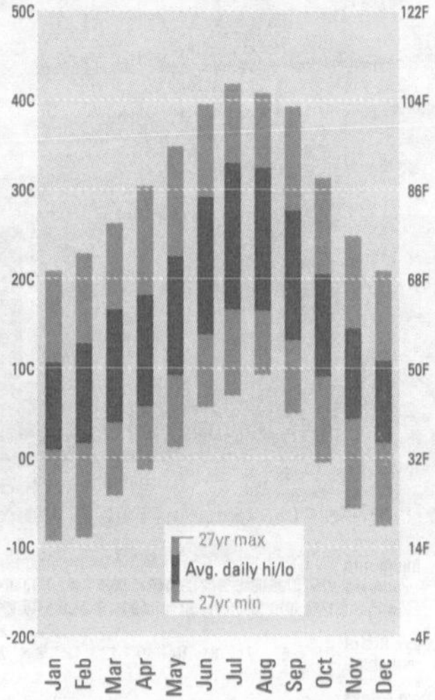

	Jan	Feb	Mar	Apr	May	Jun	Jul	Aug	Sep	Oct	Nov	Dec
midmonth -sunrise	0807	0740	0658	0708	0630	0615	0628	0656	0726	0756	0732	0802
-sunset	1744	1821	1852	2024	2055	2118	2115	2044	1955	1906	1729	1720
avg hrs of sunshine	5	6	7	8	9	11	12	11	9	7	5	4
avg # days w/thunder	0	<1	1	1	4	5	4	4	3	1	<1	<1
w/snow	1	1	<1	<1	0	0	0	0	0	0	<1	1
Typical % relative humidity @1600hrs	62	57	47	48	44	38	30	31	38	51	60	66

San Sebastian

43°21'N, 1°48'W
Elev: 8m (26ft)

The Atlantic's influence is obvious on this coastal town. It rains roughly every other day, and seasonal shifts are gentle: the typical high in January isn't far below the average low in July. Be ready for occasional sharp extremes of heat and cold. Sunshine is limited by Spanish standards.

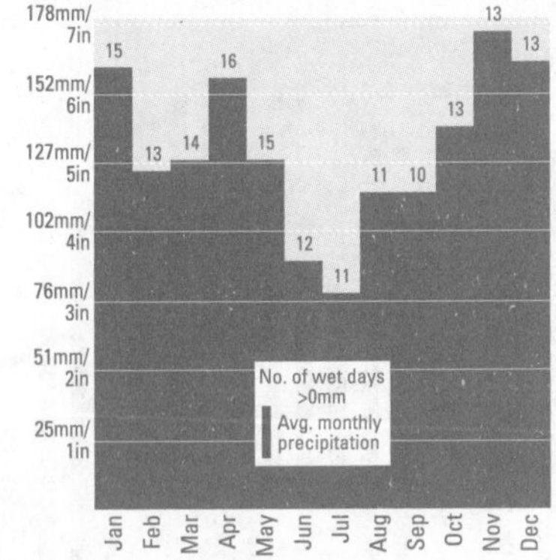

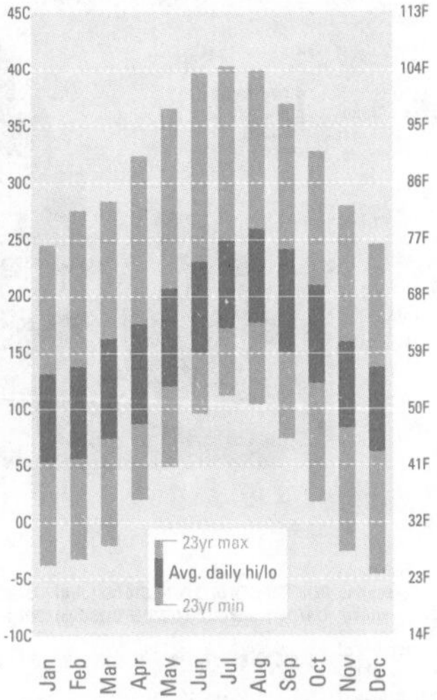

	Jan	Feb	Mar	Apr	May	Jun	Jul	Aug	Sep	Oct	Nov	Dec
midmonth -sunrise	0837	0806	0720	0725	0643	0626	0640	0712	0746	0821	0800	0833
-sunset	1757	1838	1913	2050	2124	2149	2146	2111	2018	1925	1743	1732
avg hrs of sunshine	3	3	4	4	5	6	6	6	5	4	3	3
avg # days w/thunder	1	1	1	1	2	2	2	2	1	1	1	1
w/snow	1	1	<1	0	0	0	0	0	0	0	<1	<1
Typical % relative humidity @1600hrs	70	65	65	66	67	69	69	70	68	67	68	69

Seville

37°25'N, 5°54'W
Elev: 31m (102ft)

Seville is far enough south for a single, protracted wet season peak from October–April. The coolish, partly cloudy weather is punctuated by warm spells that intensify in the variable (but often fine) spring and autumn. Once the rain stops, heat becomes the norm, joined in mid-summer by moderate humidity and intense sun.

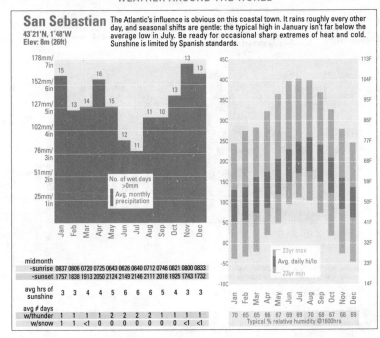

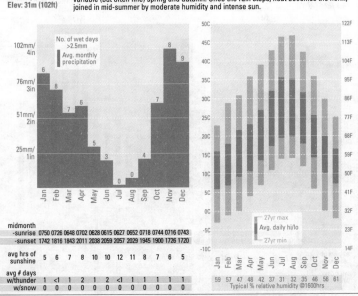

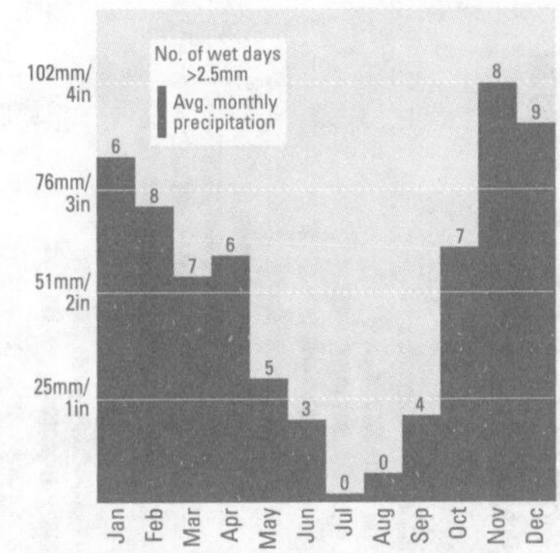

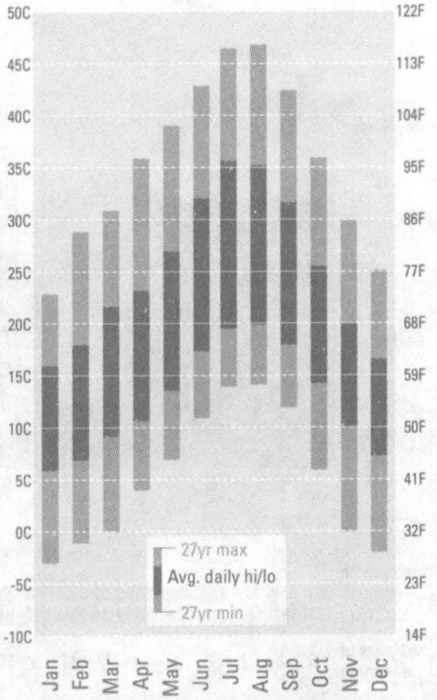

	Jan	Feb	Mar	Apr	May	Jun	Jul	Aug	Sep	Oct	Nov	Dec
midmonth -sunrise	0750	0726	0648	0702	0628	0615	0627	0652	0718	0744	0716	0743
-sunset	1742	1816	1843	2011	2038	2059	2057	2029	1945	1900	1726	1720
avg hrs of sunshine	5	6	7	8	10	10	12	11	8	7	6	5
avg # days w/thunder	1	<1	1	2	1	2	<1	1	1	1	1	1
w/snow	0	0	0	0	0	0	0	0	0	0	0	0
Typical % relative humidity @1600hrs	59	57	47	48	42	37	31	32	35	46	56	61

Unlike the rest of the continent, Spain gets its most widespread moisture in the spring and autumn, as the polar jet stream migrates north and south, respectively. Especially in the spring, the rains tend to be showery, with thunder common in mid-summer, and there's typically plenty of sunshine before and after a warm-season storm. As the rains intensify again in September and October, the **eastern Pyrenees** and **northern Mediterranean provinces** fall prey to heavy downpours that can trigger dangerous flash floods with little or no warning. "Normal" is a slippery term for rainfall in this region, which is frequently either inundated or drought-stricken.

As you travel **south along the coast**, the climate becomes more classically Mediterranean. The **Costa Blanca** and, especially, the aptly named **Costa del Sol** bask in sunshine. Winters are mild and summers are warm, but seldom hot, with only moderate humidity. The rains from autumn through spring are little more than a brief distraction once a week or so, and when it's not raining, it's often crystal clear. The same applies throughout much of **Andalucia**, where there's little or no cloud on 100 to 150 days a year. That much sun is bound to result in some heat, of course. The lowlands of Andalucia often soar above 38°C/100°F – sometimes a good deal hotter – in July and August. The air can be humid as well, albeit rain-free. Escape from the heat is close at hand, however. The **Sierra Nevada** carry snow year round, and the higher Pyrenees (including **Andorra**) tend to stay pleasantly cool as well. The Atlantic-cooled summer air on Spain's north coast is similar to that in Britain.

Sweden

Although **Sweden** extends across nearly as much latitude as Norway, its north–south sweep is on the east side of Scandinavia's mountain corridor rather than the west side. That makes all the difference. With the mountains largely closing the door to Atlantic flow – though it's left ajar in places – Sweden runs distinctly colder and drier on average than its Nordic counterpart.

The **southwest** corner of Sweden, especially **Göteborg** and other spots along the **Kattegat coast**, is really a climatic extension of The Netherlands and Denmark. It's damp and chilly in winter, with frequent light rain intermingled with snowfalls, and showery, but bright, and mild in summer. An east–west transition zone of sorts cuts across Sweden close to the latitude of **Stockholm**. North of this belt, Sweden has a much more continental climate, with summers just as mild as they are to the south, but far colder winters. The **Gulf of Bothnia** typically freezes over for months, which enables bitterly cold air to slide westward from Russia or south from the Arctic

Göteborg

57°47'N, 11°53'E
Elev: 53m (174ft)

Just across the Kattegat from Denmark, Göteborg has a similar marine climate. The worst Siberian cold is rather muted here, with rain sometimes edging out snow (a cold rain, mind you). Summer days are cool even for Scandinavia, and it rains almost twice as hard in autumn as in spring.

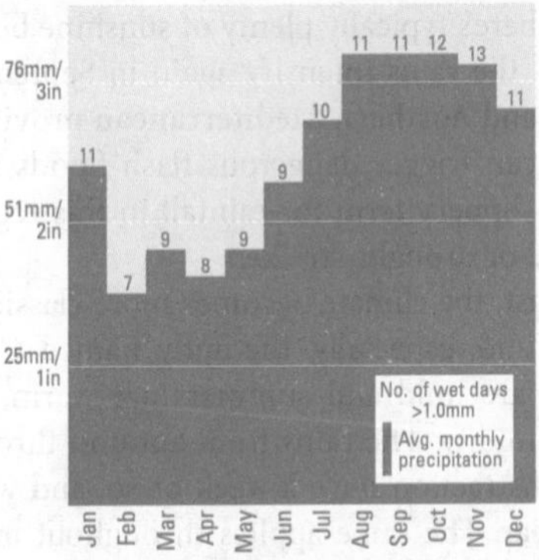

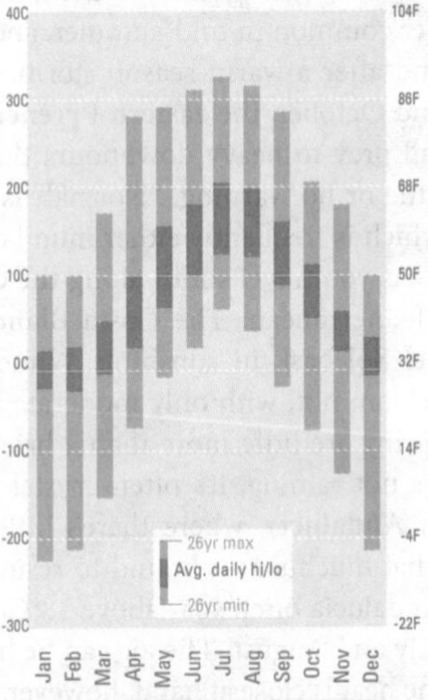

	Jan	Feb	Mar	Apr	May	Jun	Jul	Aug	Sep	Oct	Nov	Dec
midmonth -sunrise	0845	0744	0629	0603	0450	0411	0434	0536	0642	0747	0757	0850
-sunset	1559	1710	1815	2023	2129	2215	2201	2057	1932	1809	1556	1525
avg hrs of sunshine	1	3	4	6	8	9	8	7	5	3	2	1
avg # days w/thunder	0	0	0	0	<1	1	2	2	1	<1	0	0
w/snow	10	6	2	4	0	0	0	0	0	<1	4	9

	Jan	Feb	Mar	Apr	May	Jun	Jul	Aug	Sep	Oct	Nov	Dec
Typical % relative humidity @1600hrs	89	87	76	70	62	62	64	67	72	77	82	87

Stockholm

59°39'N, 17°57'E
Elev: 61m (200ft)

Snow falls in Stockholm about every other day from December through March. It's usually a gentle affair, and thaws can bring the city well above freezing before the next cold wave hits. Spring days are often crystal-clear, but most nights stay nippy till May. Summer days are bright and often quite warm.

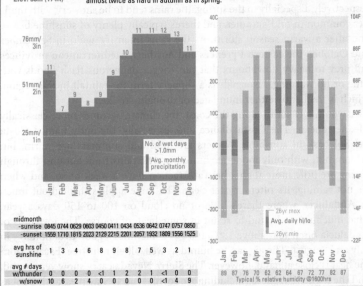

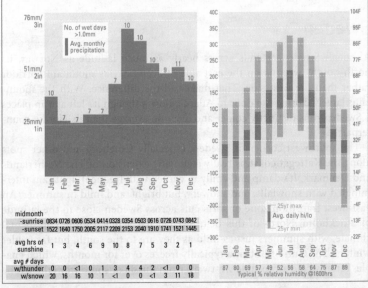

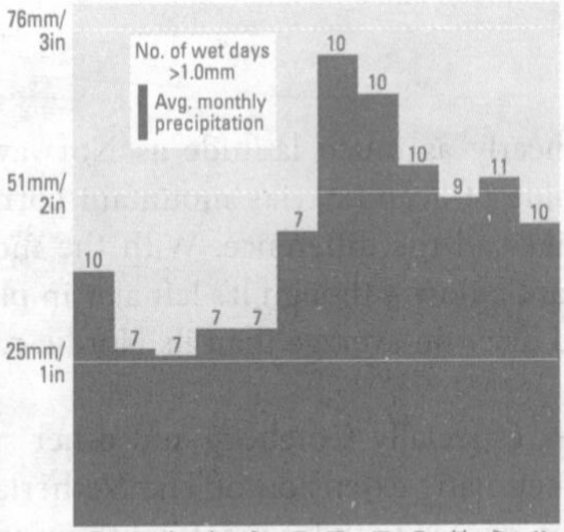

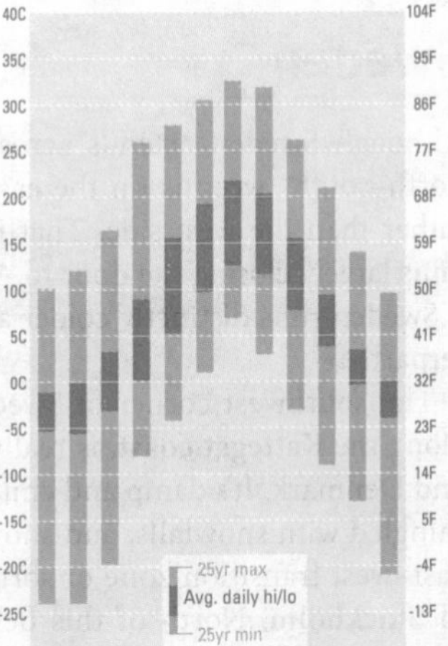

	Jan	Feb	Mar	Apr	May	Jun	Jul	Aug	Sep	Oct	Nov	Dec
midmonth -sunrise	0834	0726	0606	0534	0414	0328	0354	0503	0616	0726	0743	0842
-sunset	1522	1640	1750	2005	2117	2209	2153	2040	1910	1741	1521	1445
avg hrs of sunshine	1	3	4	6	9	10	8	7	5	3	2	1
avg # days w/thunder	0	0	<1	0	1	3	4	4	2	<1	0	0
w/snow	20	16	16	10	1	<1	0	0	<1	3	11	18

	Jan	Feb	Mar	Apr	May	Jun	Jul	Aug	Sep	Oct	Nov	Dec
Typical % relative humidity @1600hrs	87	80	69	57	49	52	56	58	64	75	87	89

Ocean without passing over water. These frigid periods in mid-winter are marked by low clouds and light snow that can last for days. By contrast, as Atlantic winds cross the mountains they often become dry and relatively mild across **central** and **northern Sweden**, and temperatures can push a few degrees above freezing. A few low-pressure centres twirl across the **Baltic Sea** in most winters, bringing a heavier dose of snow to southern and central Sweden.

Spring is the driest, least cloudy period across most of Sweden. As the sun intensifies, spring fever takes over, although nights can still be quite cold and brief showers may scoot across the landscape. By early May, the snow has usually melted across all but the far north and the mountains, and temperatures then take a sharp turn upward. Sweden's mid-summer warmth fuels frequent showers and thunderstorms, some of them bearing decent downpours. These typically form along the mountains in the afternoon and move eastward by evening; thus, the Gulf of Bothnia coast often gets more afternoon sun than the mountains do. The **lowlands near Stockholm** and **Uppsala** tend to be Sweden's warmest region in mid-summer. As you head north from there, increasing clouds and a lower sun angle are partially offset by extra daylight, so temperatures cool down only slightly with latitude. As far north as **Lapland**, the average highs in July exceed 16°C/60°F, and a thunderstorm may interrupt the 24-hour daylight. Nonetheless, mid-summer frost can occur above 1000m/3300ft, and the winter cold here is among the fiercest in Scandinavia. If you're hoping to catch Lapland at its mildest, plan on arriving during the relatively brief window between late June and early August. Because Sweden covers so much north–south distance, there's often unsettled weather somewhere along its length at any given time – when it's beautiful in the south, it may be wretched in the north, or vice versa.

Switzerland

Geography has everything to do with the weather in this very mountainous country, especially across **the Alps** that dominate the country's southern half. The north slopes of the Alps are prone to the same cold waves that put Germany on ice. Meanwhile, south slopes are exposed to Mediterranean flow that's often balmy but laden with moisture. As a result, the northern side of the Alps tends to get steadier rain or snow and less sunshine (a tendency accentuated by their orientation away from the sun). The southern side gets more sun and shorter bursts of more intense rain.

It's no wonder the Alps are a world mecca for cold-season sports. Winter storms that parade across the Mediterranean, or blow in from the Atlantic,

Geneva

46°15'N, 6°08'E
Elev: 416m (1364ft)

Otherwise similar to cities further east, Geneva manages to escape the worst effects of the cold mid-winter bise. It also gets fewer föhn warm-ups. Summers are quite thundery, with temperatures a shade warmer and skies a bit sunnier than in most of Switzerland.

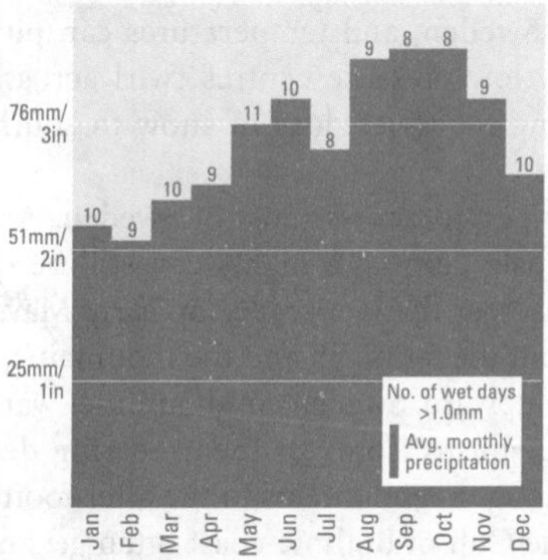

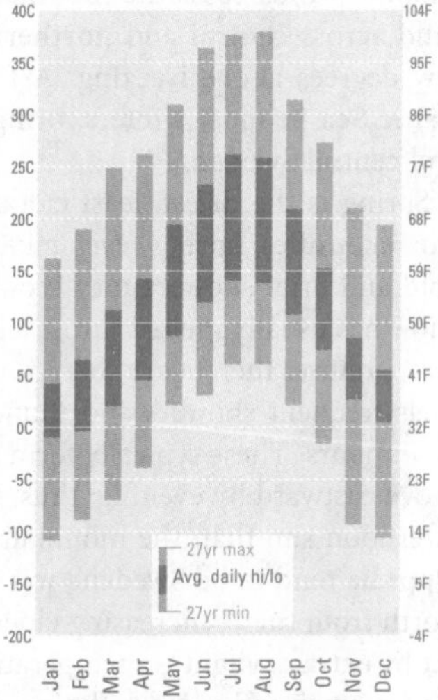

	Jan	Feb	Mar	Apr	May	Jun	Jul	Aug	Sep	Oct	Nov	Dec
midmonth -sunrise	0903	0828	0738	0739	0652	0632	0647	0723	0802	0841	0825	0900
-sunset	1805	1850	1930	2112	2150	2218	2213	2134	2037	1939	1752	1739
avg hrs of sunshine	2	3	4	5	6	7	8	7	6	4	2	1
avg # days w/thunder	1	<1	1	2	5	8	7	7	4	1	<1	1
w/snow	10	7	4	2	<1	0	0	0	0	<1	3	7
Typical % relative humidity @1600hrs	73	66	56	51	51	50	46	47	55	66	71	75

Zürich

47°23'N, 8°34'E
Elev: 569m (1867ft)

Temperatures here are close to those prevailing in Germany, with many winters marked by stretches of raw, overcast cold and light snow. Summers tend to be on the mild side. Zürich gets as much or more summer rainfall as any big city in western Europe, yet there's a good deal of sun as well.

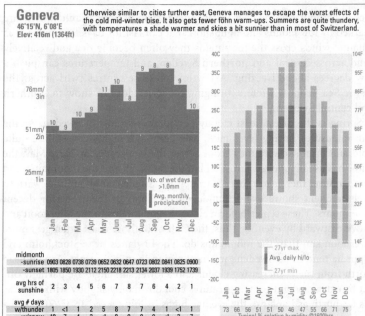

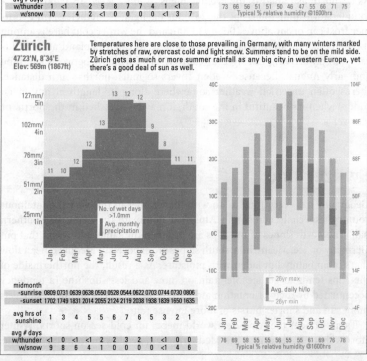

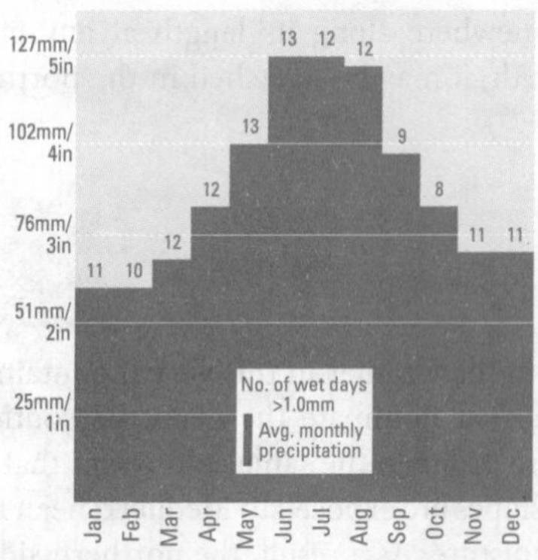

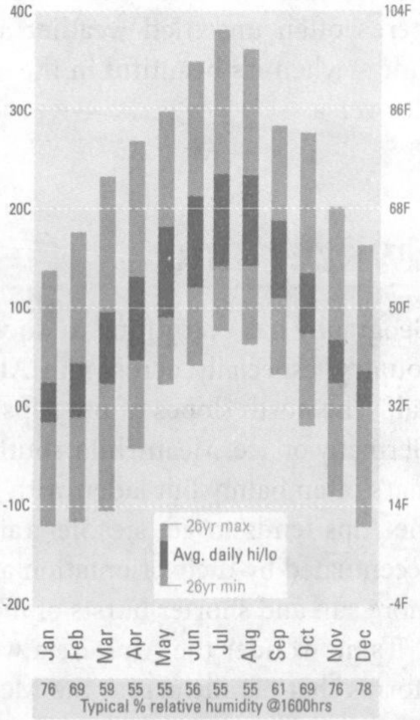

	Jan	Feb	Mar	Apr	May	Jun	Jul	Aug	Sep	Oct	Nov	Dec
midmonth -sunrise	0809	0731	0639	0638	0550	0528	0544	0622	0703	0744	0730	0806
-sunset	1702	1749	1831	2014	2055	2124	2119	2038	1938	1839	1650	1635
avg hrs of sunshine	1	3	4	5	5	6	7	6	5	3	2	1
avg # days w/thunder	<1	0	<1	<1	2	2	3	2	1	<1	0	0
w/snow	9	8	6	4	1	0	0	0	0	<1	4	6
Typical % relative humidity @1600hrs	76	69	59	55	55	56	55	55	61	69	76	78

can drop huge amounts of snow across the higher Alps. However, when it comes to cold, the Swiss champion isn't the Alps at all. **La Brevine** (nicknamed Swiss Siberia) is tucked into one of the many valleys gouged into the **Jura Mountains**, which stretch across northwest Switzerland. Cold air collects with ease in the Jura's steep-sided valleys, which allowed La Brevine to drop to a national record of –42°C/–44°F in January 1987. The same principle applies in somewhat diluted fashion to the much broader **Mittelland**, between the Jura and Alps, that holds the largest Swiss cities. Especially in mid-winter, Siberian air can funnel into this valley on strong northeast *bise* winds and stay for weeks below about 1000–1500m/3300–4900ft. Above this layer of biting cold – which is often accompanied by fog and light snow – skiers may be basking in sunshine and temperatures warmer by 10°C/18°F or more. Smaller valleys within the Alps, protected from the *bise*, may also be sunny and dry, though quite cold at night. Every few days, a round of *föhn* winds will sweep northward, bringing rapid warm-ups and extremely dry conditions. Relative humidities can drop below 10 percent, raising the risk of fire. *Föhn* winds may gust above 140kph/80mph in some north–south valleys immediately along the Alps. The wind usually drops to a more tolerable breeze by the time it reaches the population centres. *Föhn* conditions are a bit more likely in autumn and especially in spring, as warm air makes a run for the continent. The quick switches from showers to sunshine in mid-spring are called "April weather".

As in many other high-altitude regions, summer is thunderstorm season across Switzerland. The Alps and Jura see showers or storms by mid-afternoon almost every day. Most valley cities will get wet every 2 to 4 days on average, with the rains heavier and more frequent as you move toward **Zürich**. Sunshine is plentiful, though – some **central Alpine valleys** tend to stay quite dry – and you may even encounter a bona fide heat wave. If you're planning a hike, you can expect a temperature drop in the neighbourhood of 5°C/9°F per 1000m/3300ft.

Turkey

You can find bitter cold, scorching heat, damp gloom and brilliant sunshine within the borders of **Turkey**'s uncommonly varied climate. The rugged **west** and **south coasts** are firmly Mediterranean. Here, the rains virtually cease from June through September but can be heavy in the cool winter. Along the coastal arc from **Izmir** through **Antalya** to **Adana**, average highs in July and August reach 32°C/90°F, and even January days typically reach 12°C/54°F or better. Inland, the vast **Anatolian plateau** isn't too much warmer than the lowlands of Bulgaria or Hungary to the north, although there's noticeably

Ankara

40°07'N, 32°59'E
Elev: 949m (3113ft)

Despite milder coastlines all around, Ankara and its surrounding plateau can be quite chilly – and smoggy – in winter, with a few light snows or cold rains but sunny days as well. After a thundery May and June, summer is mostly dry with warm (sometimes hot) afternoons and cool nights. Rains return by late October.

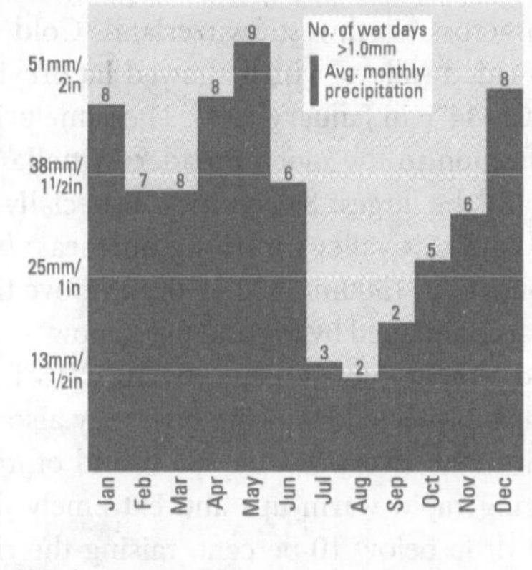

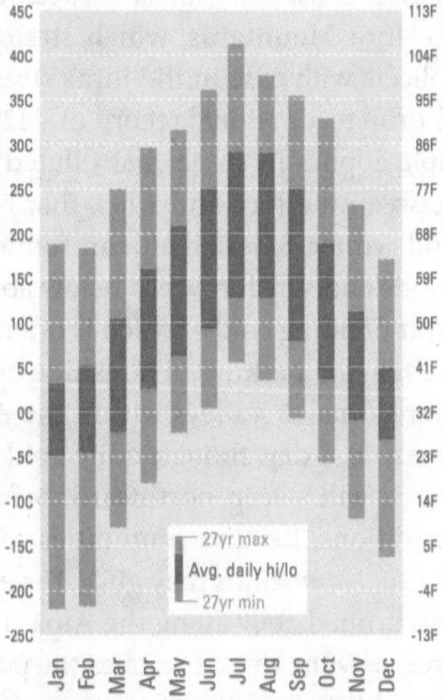

	Jan	Feb	Mar	Apr	May	Jun	Jul	Aug	Sep	Oct	Nov	Dec
midmonth -sunrise	0709	0642	0600	0611	0533	0518	0531	0559	0629	0658	0633	0703
-sunset	1647	1723	1755	1926	1956	2019	2016	1945	1858	1809	1632	1623
avg hrs of sunshine	5	6	7	8	9	11	12	12	10	8	6	4
avg # days w/thunder	<1	1	1	3	10	8	4	3	2	1	1	<1
w/snow	13	10	6	1	<1	0	0	0	0	<1	3	9
Typical % relative humidity @1500hrs	71	62	51	46	44	40	35	33	34	42	54	69

Istanbul

40°58'N, 28°49'E
Elev: 37m (121ft)

Straddling the Aegean and Black Seas, Istanbul has temperature swings from day to night that are typically gentle. Winters are cool, with some wet spells but few downpours; it's never frigid, and many days are bright and mild. Mid-summer is sultry and almost rain-free; May and October are less steamy.

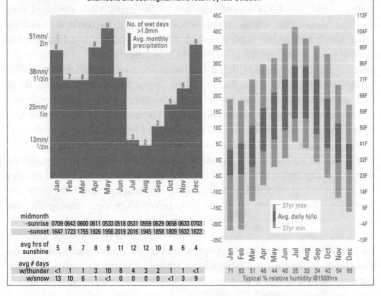

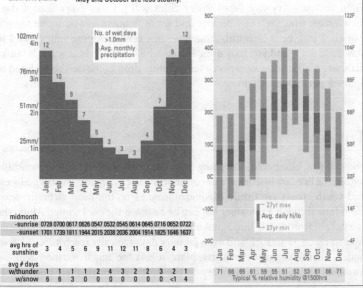

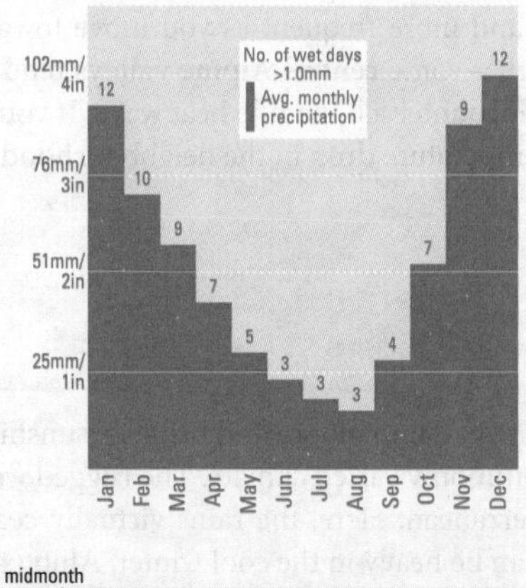

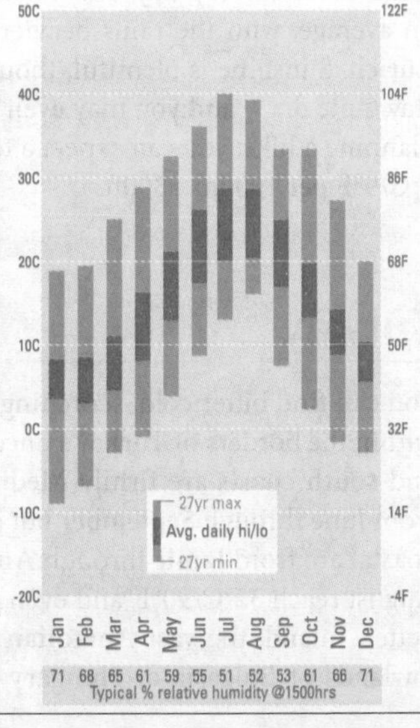

	Jan	Feb	Mar	Apr	May	Jun	Jul	Aug	Sep	Oct	Nov	Dec
midmonth -sunrise	0728	0700	0617	0626	0547	0532	0545	0614	0645	0716	0652	0722
-sunset	1701	1739	1811	1944	2015	2038	2036	2004	1914	1825	1646	1637
avg hrs of sunshine	3	4	5	6	9	11	12	11	8	6	4	3
avg # days w/thunder	1	1	1	1	2	4	3	2	2	3	2	1
w/snow	6	6	3	0	0	0	0	0	0	0	<1	4
Typical % relative humidity @1500hrs	71	68	65	61	59	55	51	52	53	61	66	71

more sunshine here. Late spring is the season of thunderstorms. By July, the plateau is virtually dry until the light rains of October and dustings of snow in winter. The aridity of Anatolia makes big temperature swings possible. In August alone, **Ankara** has seen readings below 4°C/40°F and above 40°C/104°F. The **lowlands** adjoining Syria and Iraq are a few degrees warmer than the plateau: mid-summer highs in **Diyarbakir**, along the **upper Tigris**, often exceed 38°C/100°F.

The **highlands** across the **northeast of Turkey** – where many towns sit above 1500m/4900ft – can be shockingly cold in winter. Both **Kars** and **Erzurum** stay well below freezing on a typical January day. Sunny periods alternate with light snows that pile up slowly but steadily. After the thunderstorms of springtime come and go, northeast Turkey becomes pleasantly warm by day and cool by night during the summer. Yet another distinct climatic regime prevails along the **Black Sea coast**, including the **Bosporus** and **Istanbul**. This is Turkey's cloudiest and most temperate zone, where a substantial rain can fall any time of year. Cold air masses from Europe and Asia pick up moisture and warmth from the Black Sea, so that winters here are dependably on the cool side, with frequent rains (and perhaps a light snow every few weeks) but without the occasional frigid snaps of the interior. Summer showers aren't terribly numerous, and the days here are only a bit warmer than the mild, humid nights. The cool-season rains tend to start earlier in the autumn as you head toward the eastern end of the Black Sea coast.

Ukraine

Like southern Canada, the **Ukraine** experiences summers warm enough to allow grains to thrive despite the high latitude. Even in **Kiev**, toward Ukraine's northern border, high temperatures average 22°C/72°F, or better, for more than three months of the year. Summer storms usually drop enough rain to keep crops healthy (although drought is always a concern), and even the thundery days usually see some sunshine. Both late spring and early autumn tend to be quite sunny as well, with fewer showery days than midsummer. Winter sets in by November, and snow typically covers the ground for months. A few thaws can send temperatures just above freezing, while the worst cold waves can push them well below –20°C/–4°F.

There's more to Ukraine than endless wheat fields. Eastern Europeans have long gone to **Crimea** to escape the worst of their winters. Surrounded by the seldom-frozen **Black Sea**, most of the Crimean peninsula averages more than 5°C/9°F warmer than Kiev in winter (although light snow is still common), and it's a touch warmer than Kiev in summer as well. Just over a row of lime-

Kiev
50°21'N, 30°55'E
Elev: 125m (410ft)

Warmer than Paris in June, and Stockholm in September, Kiev gets plenty of mild weather to compensate for its Eastern European winter. January can bring a cloudy thaw or a bright, bitter day, although clouds and light snow are more typical. One or two thundery downpours may hit during a mid-summer week.

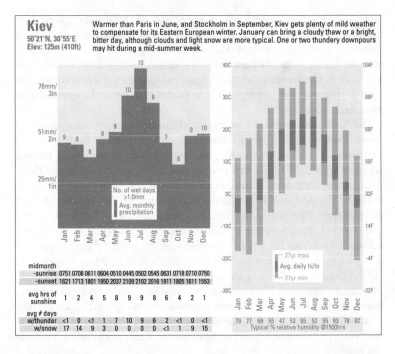

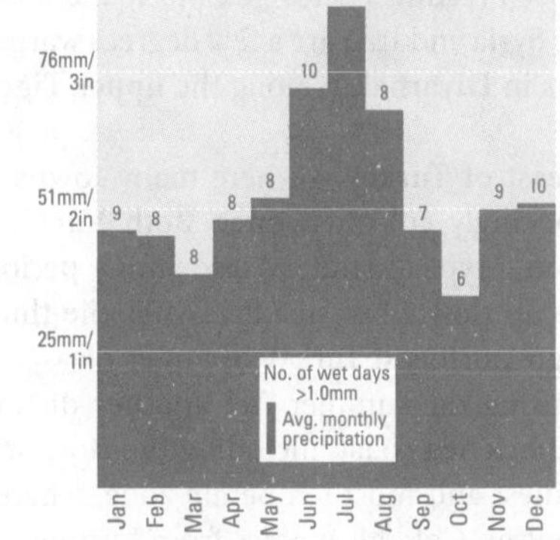

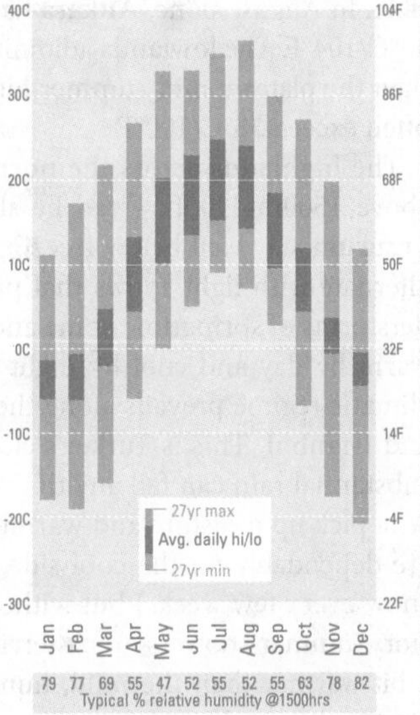

midmonth -sunrise	0751	0708	0611	0604	0510	0445	0502	0545	0631	0718	0710	0750
-sunset	1621	1713	1801	1950	2037	2109	2102	2016	1911	1805	1611	1553
avg hrs of sunshine	1	2	4	5	8	9	9	8	6	4	2	1
avg # days w/thunder	<1	0	<1	1	7	10	9	6	2	<1	0	<1
w/snow	17	14	9	3	0	0	0	0	<1	1	9	15

stone hills, the south Crimea coast, including **Yalta**, is warmer still, with a Mediterranean feel to the climate. It's cool rather than cold in winter (though wetter than the rest of the Crimea), and snow is rare. Mid-summer can be sultry: lows average close to 20°C/68°F and afternoons often top 27°C/80°F. Beyond Crimea, **Odessa** and the surrounding **Black Sea lowlands** are swept by mountain-dried westerly winds passing over **the Carpathians**. Both winter snows and summer thunderstorms tend to be light, and temperatures can soar or plummet quickly – and often – as downslope winds trade off with cold northeasterlies from Russia. The Carpathians are best explored in late summer and early autumn; it can take until May for the mud, slush and flooding of snowmelt season to subside.

United Kingdom

England | Northern Ireland | Scotland | Wales

You're never far from the ocean when you're in the **United Kingdom**. That fact tells you a lot – though hardly everything – about British weather. The climate here feels undeniably moist, yet the rain is by no means constant. Most of **England** gets measurable moisture not quite every other day, often amounting to little more than a good round of drizzle. Many of these rains come and go quickly: some parts of **central England** average little more than an hour a day of precipitation. Atlantic lows tends to make their way across Ireland and **Scotland**, which keeps those regions considerably wetter and breezier than England. **Wales** and the **Lake District** also see a good deal more rain than southeast England. Persistent rain over a few days may cause localized river flooding; on the other hand, it's possible to go a week or longer without any rain at all.

Since Britain's rains tend to come in briefer spells during summer than in winter, the sun pokes out much more regularly then. The brightest spots are often along the coastlines, as clouds build just inland during the day and form offshore at night. Many parts of England get about a dozen thunderstorms each year, mostly on summer evenings and nights. Though mild by US standards, a few of these produce hail and even a weak tornado. Once or twice each summer (more often in some years), high pressure parks over the UK, producing a few days of heat that may feel intense for visitors. No location in the UK had ever reached the 100°F mark until the astounding heat wave of 2003: on 10 August of that year, it hit 38.5°C (101.3°F) in Faversham, Kent, and both Heathrow Airport and the Royal Botanic Gardens in London topped 100°F. Scotland tends to be a few degrees cooler than England – a trend especially noticeable in summer, when 20°C/68°F constitutes a warm day for a town in **the Highlands**. **The Shetlands** and **Outer Hebrides** are even more vulnerable to gusty, chilly conditions in summer, not to mention hurricane-force winds and seemingly interminable rain and cloud in winter.

Although sunny spells do occur even as the days grow darker, the overcasts of late autumn and early winter can seem relentless – and sometimes they are. **London** once suffered an entire month (December 1890) without a single hour of sunshine. As winter sets in, gales frequently buffet the coasts, especially toward the north and west. While mid-winter is seldom bitterly cold, it's consistently chilly. The UK's warmest winter temperatures actually occur along the northeastern sides of Scottish mountains when Atlantic southwesterlies approach. As they flow over the peaks and descend, they can warm to 16°C/60°F or greater.

Winters vary greatly in the extent of snow and below-freezing cold snaps they produce. Snow almost never falls on the western fringes of **Cornwall** and **Pembroke**, while the **Pennines** and the **uplands of Wales** average more than 40 snow days a year and parts of **the Grampians** over 70. Snow may persist near the highest peaks well beyond winter (Britain's last continuous snow patch, near **Ben MacDhui**, melted in September 1933, although snow persists in some summers). Prolonged spells of bright, crisp weather become more likely in February and especially March, as the strong Atlantic flow weakens a bit. A few cold air masses may still slide down from the Arctic or push across from the Continent, bringing late frost. In general, rainfall

Aberdeen

57°12'N, 2°13'W
Elev: 65m (213ft)

Even near the coast, northern Scotland runs chilly year round and dry by Scottish standards. In winter, heavy rain alternates with powdery snow, and temperatures can be bitter or quite mild. Clouds and cool air may infiltrate the long summer days. Be prepared for much harsher conditions in the Grampians.

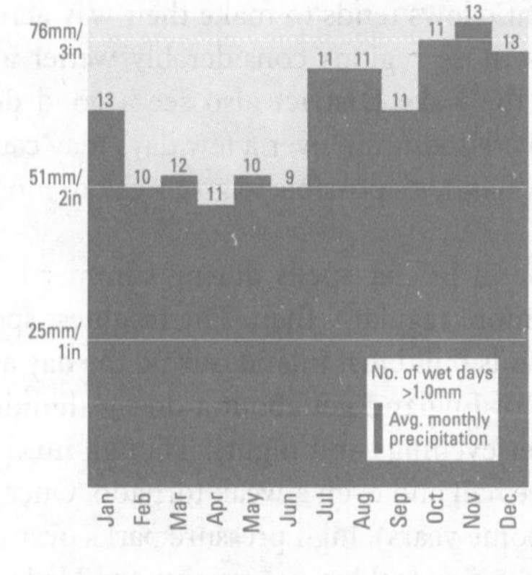

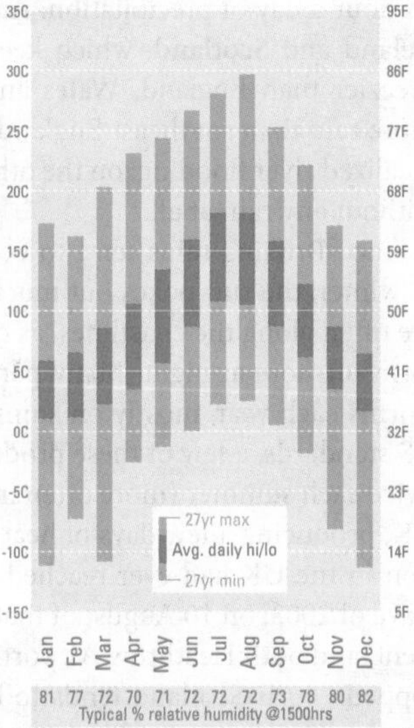

midmonth	Jan	Feb	Mar	Apr	May	Jun	Jul	Aug	Sep	Oct	Nov	Dec
-sunrise	0820	0720	0608	0543	0432	0355	0417	0517	0621	0724	0733	0825
-sunset	1542	1651	1754	2000	2104	2149	2136	2033	1910	1749	1538	1508
avg hrs of sunshine	2	3	4	5	6	6	5	5	4	3	2	1
avg # days w/thunder	<1	0	<1	<1	<1	1	1	1	<1	<1	<1	<1
w/snow	11	9	9	6	1	<1	0	0	0	1	5	8

Belfast

54°36'N, 5°53'W
Elev: 5m (16ft)

Even for the British Isles, Belfast is a wet town. Atlantic storms don't slam into it so much as drag themselves over it, dropping rain or snow on nearly two of three winter days. April into June brings the most sunny, dry days; mid-summer showers and clouds segue to persistent autumn rains.

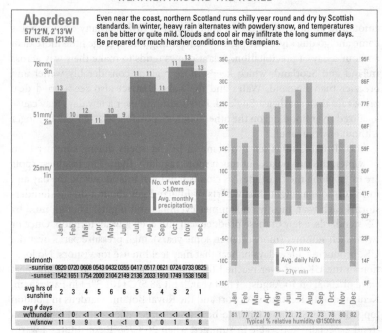

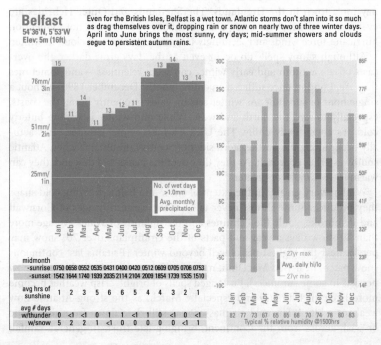

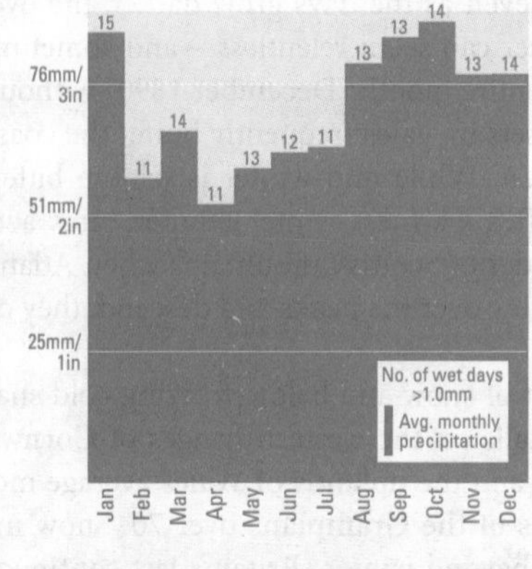

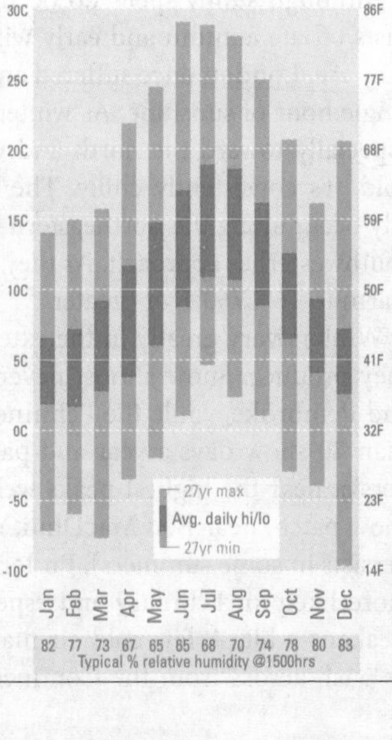

midmonth	Jan	Feb	Mar	Apr	May	Jun	Jul	Aug	Sep	Oct	Nov	Dec
-sunrise	0750	0658	0552	0535	0431	0400	0420	0512	0609	0705	0706	0753
-sunset	1542	1644	1740	1939	2035	2114	2104	2009	1854	1739	1535	1510
avg hrs of sunshine	1	2	3	5	6	6	5	4	4	3	2	1
avg # days w/thunder	0	<1	<1	0	1	1	<1	1	0	<1	0	<1
w/snow	5	2	2	1	<1	0	0	0	0	0	<1	1

Cardiff

51°24'N, 3°21'W
Elev: 67m (220ft)

Cardiff's west-coast location means that wet Atlantic storms arrive here before reaching England. Rainfall amounts are almost twice that of London's, although the number of rainy days isn't very much higher. The nearby Cambrians are soggier still. Temperatures are similar to those across most of southern England.

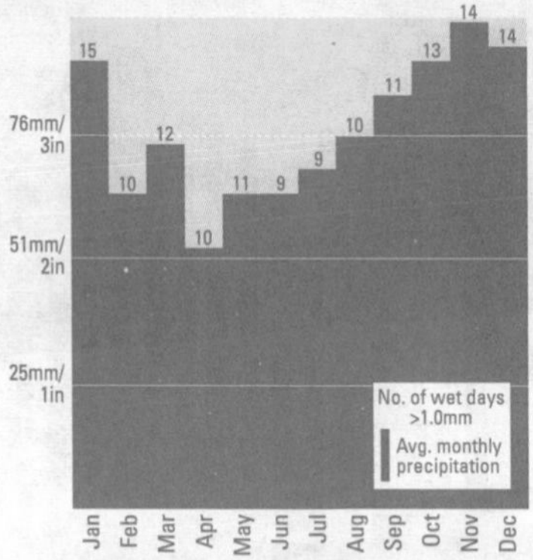

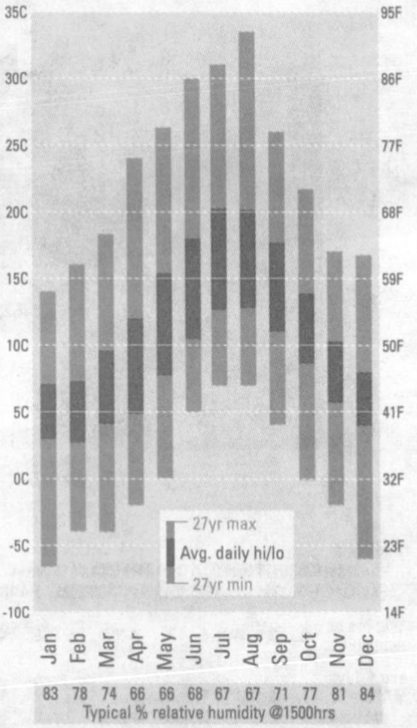

	Jan	Feb	Mar	Apr	May	Jun	Jul	Aug	Sep	Oct	Nov	Dec
midmonth -sunrise	0745	0701	0601	0552	0456	0429	0447	0532	0621	0710	0704	0745
-sunset	1607	1702	1751	1943	2031	2105	2057	2009	1902	1754	1558	1538
avg hrs of sunshine	2	3	4	6	7	7	7	6	5	3	2	3
avg # days w/thunder	<1	<1	<1	<1	1	2	1	1	1	1	1	<1
w/snow	5	5	3	2	<1	0	0	0	0	0	1	2

Edinburgh

55°57'N, 3°21'W
Elev: 41m (135ft)

As with much of central Scotland, Edinburgh's wet and dry spells are tricky to chart. The usual winter rains are peppered with typically light snows, and there's a mid-summer period with frequent showers. Warm days are rare even in the summer; late spring provides long twilights and the best chance of dry weather.

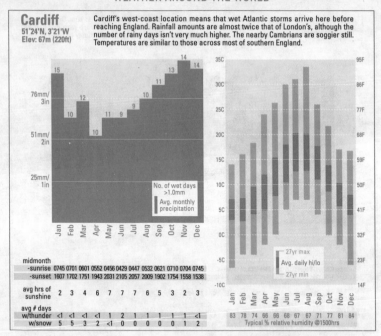

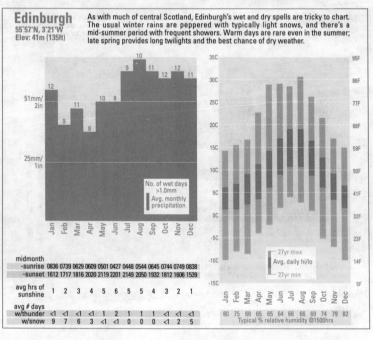

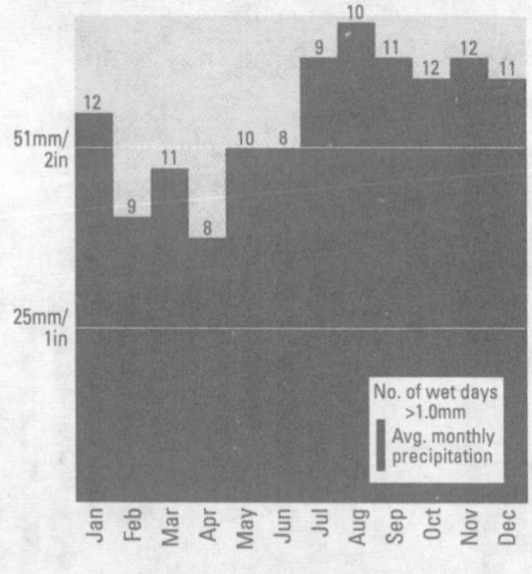

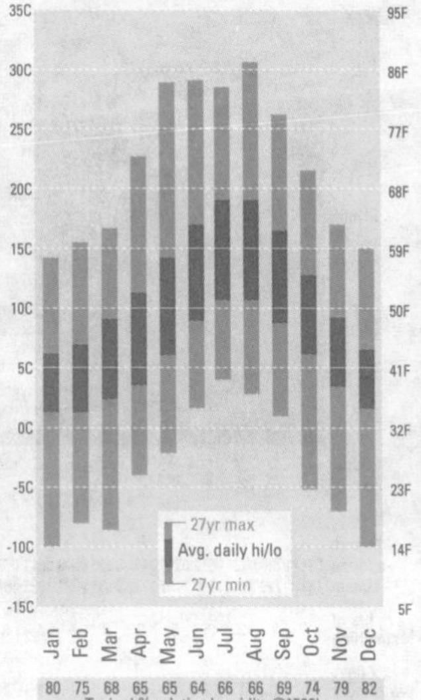

	Jan	Feb	Mar	Apr	May	Jun	Jul	Aug	Sep	Oct	Nov	Dec
midmonth -sunrise	0836	0739	0629	0609	0501	0427	0448	0544	0645	0744	0749	0838
-sunset	1612	1717	1816	2020	2119	2201	2149	2050	1932	1812	1606	1539
avg hrs of sunshine	1	2	3	4	5	6	5	5	4	3	2	1
avg # days w/thunder	<1	<1	<1	<1	1	2	1	1	1	<1	<1	<1
w/snow	9	7	6	3	<1	<1	0	0	0	<1	2	5

London

51°09'N, 0°11'W
Elev: 62m (203ft)

London's much-fabled weather hides variations within its familiar strains. On the whole, it's mild in summer and chilly in winter. Cold and hot spells are rare but can be quite dramatic. To avoid clouds and crowds, visit in May (just as sunny as mid-summer) or September (less bright than May but still mild).

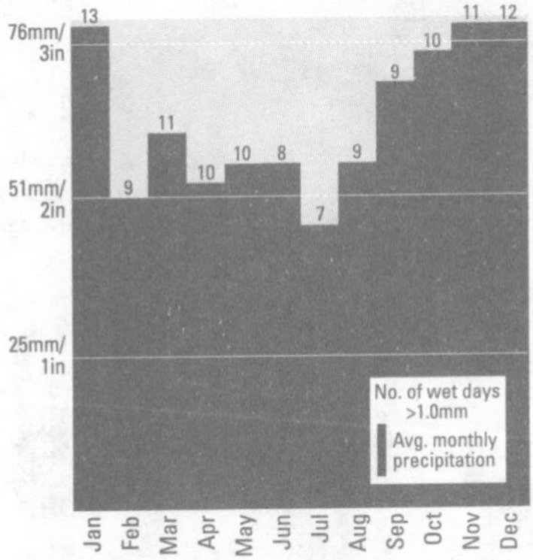

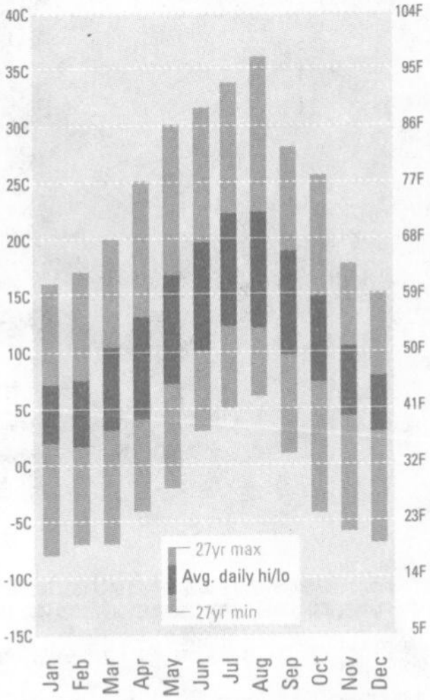

	Jan	Feb	Mar	Apr	May	Jun	Jul	Aug	Sep	Oct	Nov	Dec
midmonth -sunrise	0757	0713	0614	0605	0509	0444	0501	0545	0634	0722	0716	0757
-sunset	1621	1715	1804	1955	2043	2116	2109	2021	1914	1807	1612	1552
avg hrs of sunshine	2	3	4	5	7	7	7	6	5	4	2	2
avg # days w/thunder	1	1	1	1	2	3	2	2	2	1	<1	1
w/snow	6	6	4	2	<1	0	0	0	0	<1	1	4

Jan	Feb	Mar	Apr	May	Jun	Jul	Aug	Sep	Oct	Nov	Dec
80	74	68	60	59	61	59	59	64	72	78	82

Typical % relative humidity @1500hrs

Manchester

53°21'N, 2°16'W
Elev: 78m (256ft)

Although its reputation for rain precedes it, Manchester gets only about ten percent more wet days than London. It does tend to rain more heavily and steadily here than in London in both summer and winter (even more so as you go north). Westerly breezes often sweep in from the nearby Irish Sea.

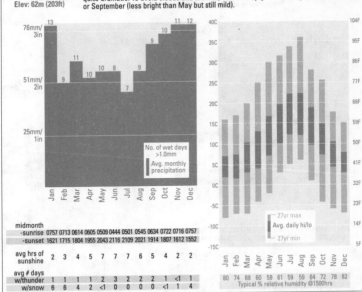

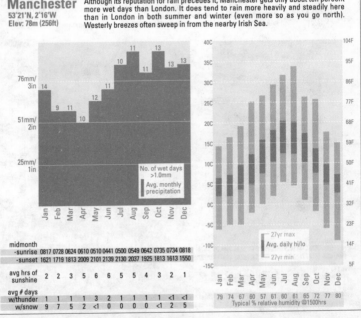

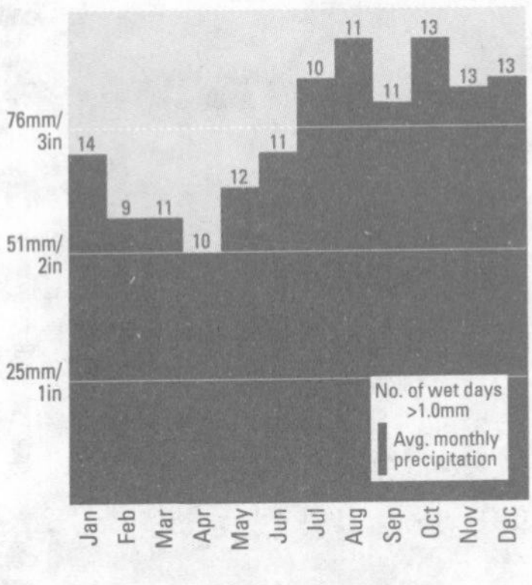

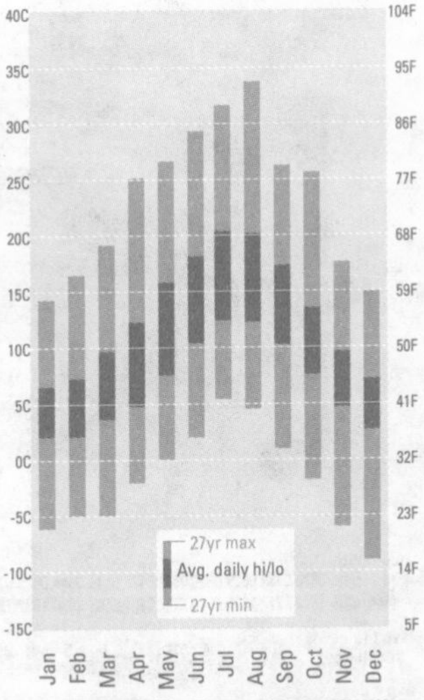

	Jan	Feb	Mar	Apr	May	Jun	Jul	Aug	Sep	Oct	Nov	Dec
midmonth -sunrise	0817	0728	0624	0610	0510	0441	0500	0549	0642	0735	0734	0818
-sunset	1621	1719	1813	2009	2101	2139	2130	2037	1925	1813	1613	1550
avg hrs of sunshine	2	2	3	5	6	6	5	5	4	3	2	1
avg # days w/thunder	1	1	1	1	3	2	1	1	1	1	<1	<1
w/snow	9	7	5	2	<1	0	0	0	0	<1	2	5

Jan	Feb	Mar	Apr	May	Jun	Jul	Aug	Sep	Oct	Nov	Dec
79	74	67	60	57	61	60	61	65	72	77	80

Typical % relative humidity @1500hrs

The power of Britain's peaks

When it comes to snowfall, as well as most every other aspect of British weather, **elevation** makes a huge difference. In few other parts of the world does the vegetation and the feel of the weather change so dramatically with a rise of just 200m/660ft or so. Trees begin to disappear from many mountainous areas above 700m/2300ft. Plants struggle in Britain's higher terrain for at least two reasons. Nightly lows during the winter tend to average just below freezing for many months, which truncates the growing season, and summertime highs stay just short of the subsistence point for many types of vegetation. Walkers and mountain climbers should be aware of the altitude's strong effect on weather and be prepared for conditions that may change even more erratically than expected.

amounts slacken across the UK in the spring. Temperatures are rapidly nearing their summer peaks in May, which gets as much sun as July in many spots (except for the northeast coasts, where clouds are still rolling in from the North Sea).

Western Balkans and vicinity

Albania | Bosnia/Herzegovina | Croatia | Macedonia | Serbia and Montenegro | Slovenia

The **western Balkans** are a laboratory for weather extremes. Conditions can vary in striking fashion across small distances from coast to nearby peak, even from valley to valley. The **Dinaric ridge** runs parallel to the **Adriatic Sea** across **western Serbia** and **north** along the border between **Croatia** and **Bosnia/Herzegovina**. Mediterranean lows slam moisture against the **Dinaric Alps**, especially in the winter, and the result is one of Europe's wettest climates. Some spots along the western slopes get over 4000m/157in of rain and melted snow in a typical year. Coastal towns get less, but many exceed 1000m/39in. However, the sogginess of winter on the Adriatic isn't as unrelenting as it is on the Norway coast. Plenty of in-between days are bright and sunny with mild afternoons, and a gusty *bora* may bring chilly air from eastern Europe. The coast warms up in spring but doesn't really dry out until June. Mid-summer can be blazingly sunny and quite muggy. Inland valleys can scorch – **Mostar**, on the Neretva River, is among Europe's hottest towns – but a light northerly breeze (the *maestro)* helps cool the coast. Expect a shower or storm every week or so until the deluges of October arrive.

The **northern lowlands** of Serbia and Bosnia/Herzegovina, along with the bulk of Croatia, are more akin to central Europe in their climate. Rainfall amounts decrease toward the east, and winters are cloudy even in the drier regions. In mid-winter, *föhn* breezes may flow northward across the Alps and warm the lowlands above 10°C/50°F, or cold air may funnel through **the**

Carpathians toward **Belgrade** on a *kosava* wind. Summer temperatures can soar well above the comfortable norms for several days in a row. With sea breezes blocked by the coastal mountains, the heat can feel stifling until a shower or thunderstorm brings relief.

Although it borders on the Mediterranean (just barely), **Slovenia** has a climate strongly coloured by the hues of central Europe. The **Karst highlands** serve as a barrier that keeps the full effects of the sea limited to the sliver of Slovenian coast fronting the Gulf of Trieste. Although this region is largely sunny and on the mild side, winter rains are especially heavy; when continental air does invade, it's often through a round of howling *bora* winds (here called the *burja)* that can gust over 160kph/100mph. Higher altitudes of the **north** and **west** are mild in summer, but prone to be quite wintry. Early in the summer, the Mediterranean's presence is felt across the country with frequent, thundery downpours every third day or so between spells of fine weather. The storms taper off in September, replaced later in the autumn by off-and-on chilly showers. The average winter temperatures across eastern Slovenia aren't vastly different from those in Hungary, but warm spells visit more readily. It's a bit drier and hotter in summer and colder in winter across **the lowlands** of the **far east**.

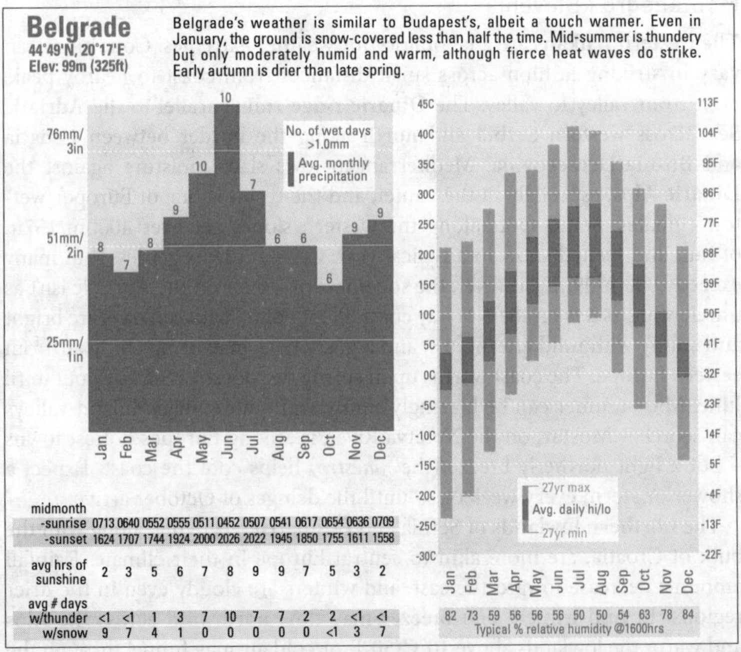

Ljubljana

46°13'N, 14°29'E
Elev: 362m (1187ft)

Bring your umbrella: this is one of the wettest capitals in Europe. No season is free of moisture, although thunder is most likely in summer. On the plus side, much of the rain is showery and brief. Even the wet summer months have plenty of dry days, and there's ample sun outside of mid-winter.

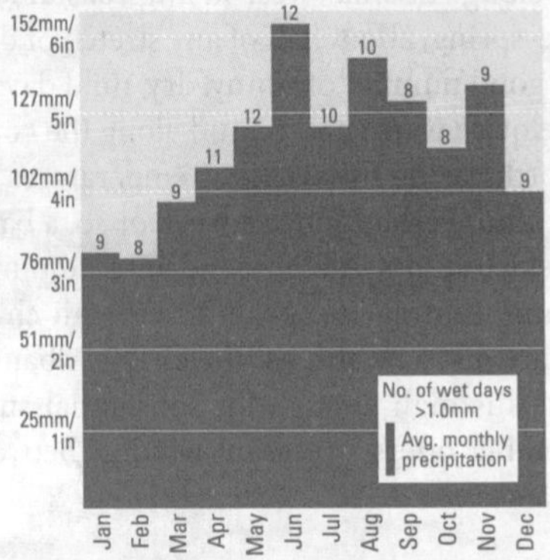

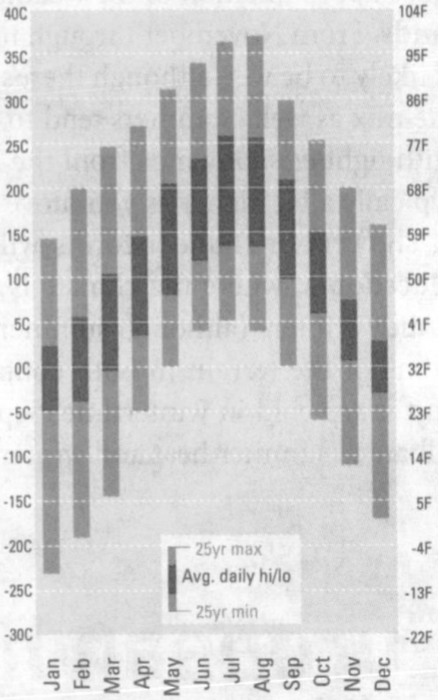

	Jan	Feb	Mar	Apr	May	Jun	Jul	Aug	Sep	Oct	Nov	Dec
midmonth -sunrise	0741	0706	0616	0616	0530	0510	0525	0601	0640	0719	0703	0739
-sunset	1642	1727	1807	1949	2028	2055	2051	2012	1914	1816	1630	1616
avg hrs of sunshine	2	3	4	5	7	7	8	7	5	4	2	1
avg # days w/thunder	<1	<1	1	2	7	11	10	9	5	3	1	1
w/snow	9	9	7	2	<1	0	0	0	0	<1	4	9
Typical % relative humidity @1600hrs	77	65	57	52	52	55	53	53	60	68	78	84

Tirane

41°20'N, 19°47'E
Elev: 73m (241ft)

The rains here aren't necessarily tropical deluges, but they're heavy by European standards from autumn through spring, separated by delightfully sunny periods. Summer is mostly dry and quite warm, with occasional spells of serious heat. The gusty, chilly bora winds of winter are weaker here than further north.

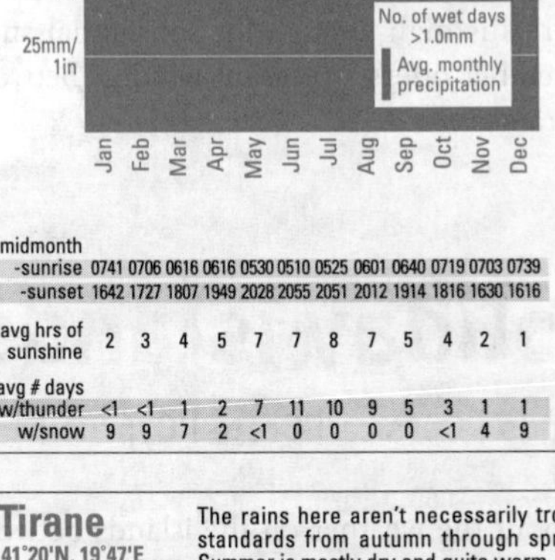

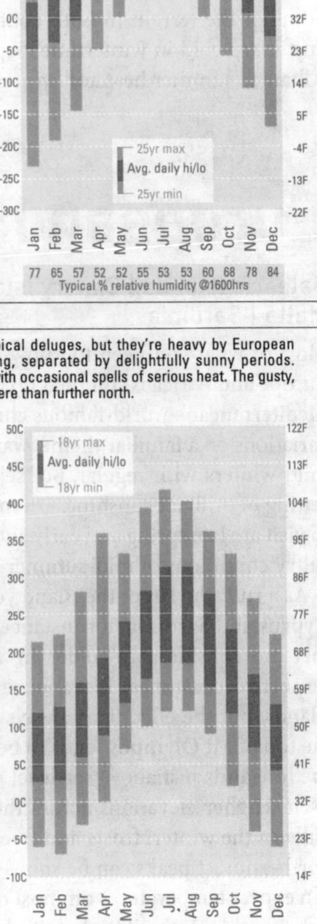

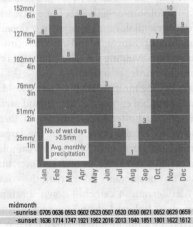

	Jan	Feb	Mar	Apr	May	Jun	Jul	Aug	Sep	Oct	Nov	Dec
midmonth -sunrise	0705	0636	0553	0602	0523	0507	0520	0550	0621	0652	0629	0659
-sunset	1636	1714	1747	1921	1952	2016	2013	1940	1851	1801	1622	1612
avg hrs of sunshine	4	4	5	6	8	10	11	10	8	7	5	4
avg # days w/thunder	n/a	n/a	n/a	n/a	n/a	n/a	n/a	n/a	n/a	n/a	n/a	n/a
w/snow	n/a	n/a	n/a	n/a	n/a	n/a	n/a	n/a	n/a	n/a	n/a	n/a
Typical % afternoon relative humidity	n/a	n/a	n/a	n/a	n/a	n/a	n/a	n/a	n/a	n/a	n/a	n/a

Albania has a classically warm Mediterranean climate. If you double the annual rainfall of Rome but leave the temperatures as they stand, you have a close approximation of the weather along Albania's **west-facing coastal lowlands.** From November through mid-spring, about half of any stretch of days is likely to be wet, although there's a good number of sunny, dry, mild days in the mix as well. Summers tend to be quite warm and humid along the coast, with light *etesian* winds from the north easing the swelter. Temperatures are typically a bit higher as you move inland. Perhaps once a week or so, a batch of showers or thunderstorms will skirt across Albania and into adjoining **Macedonia,** where the climate is more central-European: drier, with chilly, cloudy winters (although summers are just as warm as in western Albania). Some of the terrain in both countries is high enough for substantial snow and sharp cold in wintertime. September offers a pleasant window between Albania's summer heat and winter rains.

European holiday islands

Balearic Islands | Canary Islands | Corsica | Crete | Cyprus | Malta | Sardinia

Most of what you need to know about the weather on the islands between Europe and Africa is summed up in the name of the sea that holds them. The Mediterranean's world-famous climate plays out on all of its islands through variations on a familiar theme: warm-to-hot summers that are bone-dry, and mild winters with regular, but seldom torrential, rains, all cast in a default setting of brilliant sunshine. On many **Mediterranean islands**, the best times to visit are late spring and early autumn, thus avoiding the rains of winter, the sultry conditions of mid-summer and the hordes of tourists.

As a rule, the larger the island you're on, the more weather variety it holds. **Cyprus** and **Sardinia**, for instance, are both big enough to have inland areas that are considerably colder in winter and hotter in summer than their respective coastlines. Elevation also plays a role, especially on the larger islands, but the effect isn't always straightforward. On Cyprus, for example, the top of **Mt Olympos** tends to be pleasantly mild even in mid-summer, but **the lowlands** just above sea level may be considerably hotter than the coast itself. Higher elevations across the Mediterranean can chill down substantially in the winter: frosts and freezes aren't unheard of, and the highest and northernmost peaks can be snow-covered for months. Even at sea level, you can expect temperatures on most of the islands to dip below 10°C/50°F regularly in mid-winter. Both longitude and latitude play a role in island weather: as you'd expect, more **northerly islands** are slightly cooler – especially in winter – and the westernmost islands, particularly the **Balearics**, also tend to be a touch cooler and less humid than the islands near Greece and Turkey, although northerlies across the Aegean tend to ease the worst heat there. Since each island has its own micro-climatic features, the city charts at the end of this section should be taken to represent each city's immediate vicinity and not the entire island.

Overall, the Mediterranean delivers relatively few weather surprises. Every so often a slow-moving depression drifts across the sea, which can bring an unusually long string of damp, unsettled, sometimes thundery weather. Autumn is an especially fertile time for these systems, particularly from October onward. During the spring, depressions spinning off fast-heating North Africa are more likely to bring a different weather headache: the **scirocco**, a hot, dusty wind from the Sahara. Islands closest to the African coast, such as **Malta**, are most likely to get a *scirocco* at its full desiccating power. The sea adds moisture and helps temper the worst of the heat, but its smooth surface also helps the wind to gain strength by the time it reaches southern Europe. From the other direction, summer brings the **etesian** northerly winds, also known as the *maéstro* in the Ionian Sea and the *meltémi* in the Aegean. Although these winds may relieve the heat, they can blow at gale force during the day, especially by the time they reach **Crete**.

Ajaccio

41°55'N, 8°48'E
Elev: 9m (30ft)

This is among the wetter Mediterranean islands in autumn and winter. It rains on the coast and snows at elevation two or three times a week (both can be heavy). Mid-summer is dry and somewhat humid, warm by day and mild at night. Late spring is cooler but less damp than early autumn.

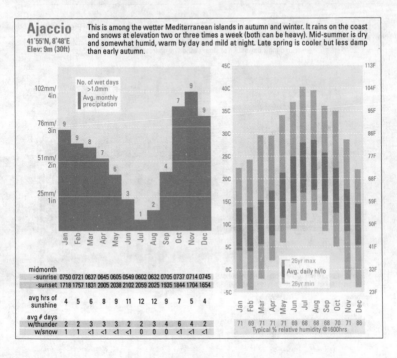

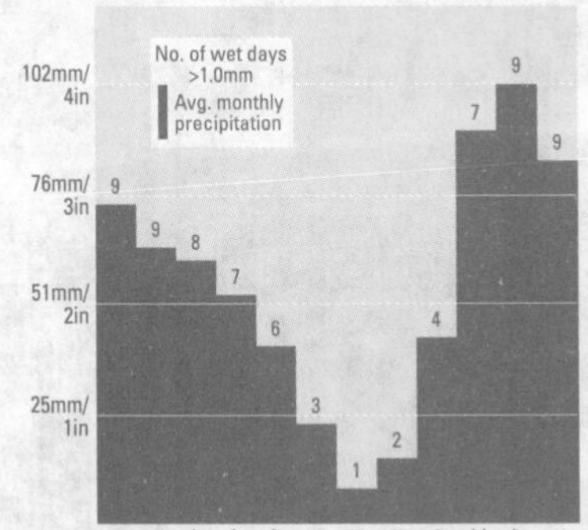

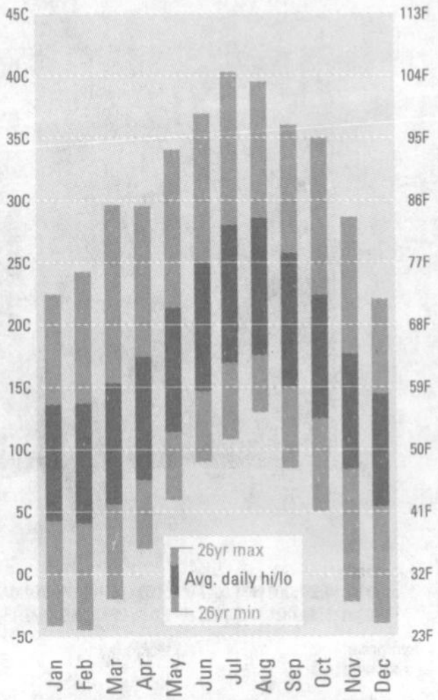

	Jan	Feb	Mar	Apr	May	Jun	Jul	Aug	Sep	Oct	Nov	Dec
midmonth -sunrise	0750	0721	0637	0645	0605	0549	0602	0632	0705	0737	0714	0745
-sunset	1718	1757	1831	2005	2038	2102	2059	2025	1935	1844	1704	1654
avg hrs of sunshine	4	5	6	8	9	11	12	12	9	7	5	4
avg # days w/thunder	2	2	3	3	3	2	2	3	4	6	4	2
w/snow	1	1	<1	<1	<1	<1	0	0	0	<1	<1	<1

	Jan	Feb	Mar	Apr	May	Jun	Jul	Aug	Sep	Oct	Nov	Dec
Typical % relative humidity @1600hrs	71	69	71	71	71	69	68	68	68	70	71	86

Cagliari

39°15'N, 9°03'E
Elev: 1m (3ft)

Facing southward and ringed by hills, Cagliari is shielded from cool northerly winds. This keeps it warmer and drier than Sardinia in winter, though often sultry in summer. Blasts of scirocco wind from the Sahara can bring toasty weather in April or scorching heat in July. Any month may bring thunder.

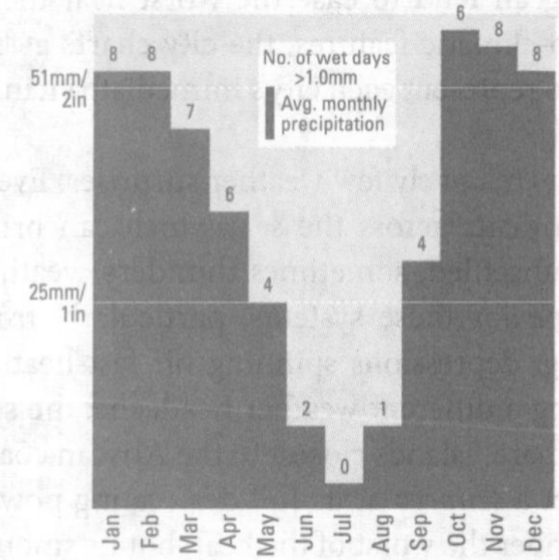

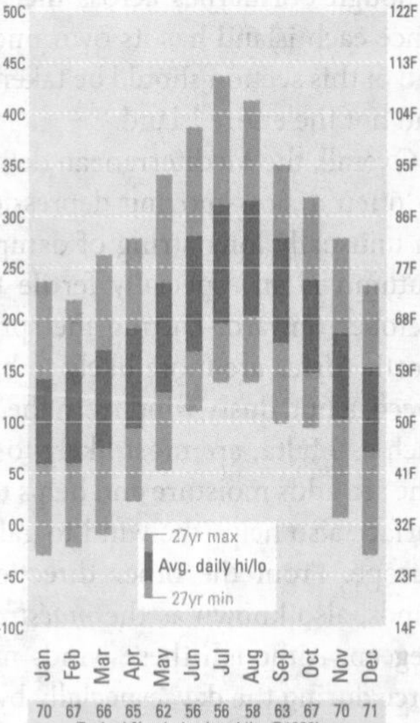

	Jan	Feb	Mar	Apr	May	Jun	Jul	Aug	Sep	Oct	Nov	Dec
midmonth -sunrise	0742	0716	0636	0647	0611	0557	0610	0636	0705	0733	0707	0736
-sunset	1725	1800	1830	2001	2030	2052	2050	2020	1933	1845	1709	1702
avg hrs of sunshine	4	5	6	7	9	10	11	10	8	6	5	4
avg # days w/thunder	2	3	2	3	2	3	2	3	5	5	3	2
w/snow	<1	<1	<1	0	0	0	0	0	0	0	0	<1
Typical % relative humidity @1600hrs	70	67	66	65	62	56	56	58	63	67	70	71

Iraklion

35°20'N, 25°11'E
Elev: 39m (128ft)

Stationed on the north coast of Crete, Iraklion avoids the extremes of Greek climate. Winter days are mild and summer afternoons warm. A frequent summer sea breeze can become an irritating meltémi gale. A scirocco from Africa, especially in spring, can bring a day or two of strong wind and oppressive heat.

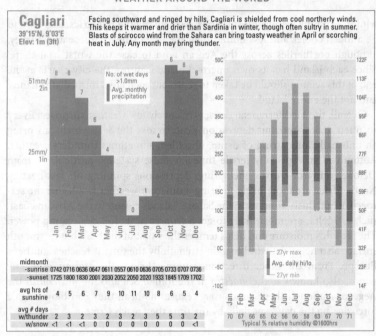

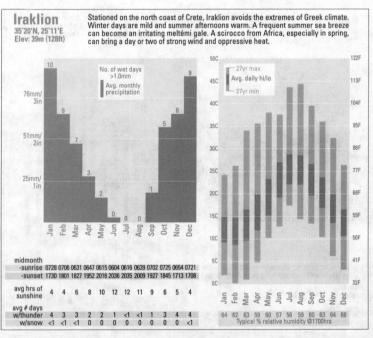

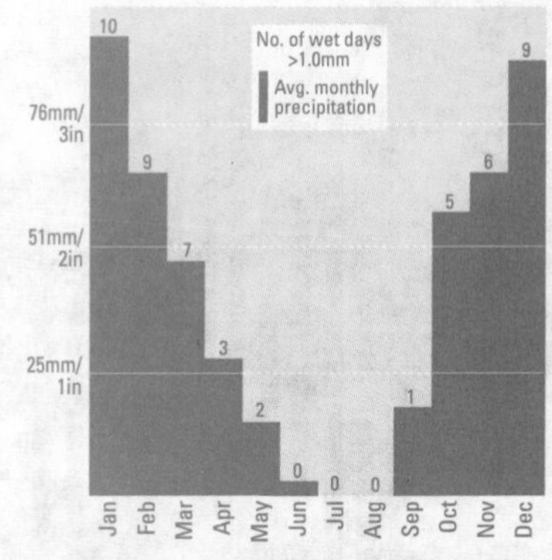

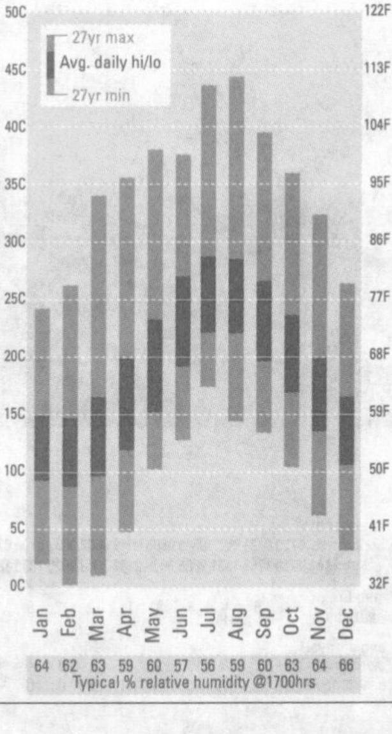

	Jan	Feb	Mar	Apr	May	Jun	Jul	Aug	Sep	Oct	Nov	Dec
midmonth -sunrise	0728	0706	0631	0647	0615	0604	0616	0639	0702	0725	0654	0721
-sunset	1730	1801	1827	1952	2016	2036	2035	2009	1927	1845	1713	1708
avg hrs of sunshine	4	4	6	8	10	12	12	11	9	6	5	4
avg # days w/thunder	4	3	3	2	2	1	<1	<1	1	3	4	4
w/snow	<1	<1	<1	0	0	0	0	0	0	0	0	<1
Typical % relative humidity @1700hrs	64	62	63	59	60	57	56	59	60	63	64	66

Nicosia

34°53'N, 33°38'E
Elev: 2m (7ft)

Nicosia's inland location boosts its temperature range. Winter nights are quite cool, even on the coast, and summer days can scorch, often topping 30°C/86°F inland. Winter brings occasional bouts of rain, with a good deal of thunder. The high Troodos get a regular snowpack. Nicosia sees flurries every year or two.

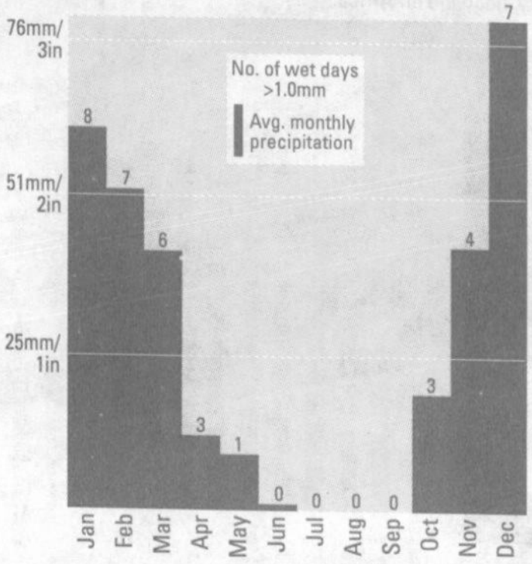

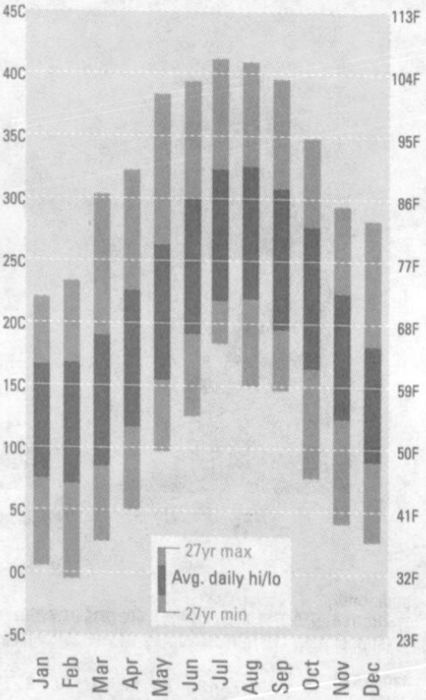

	Jan	Feb	Mar	Apr	May	Jun	Jul	Aug	Sep	Oct	Nov	Dec
midmonth -sunrise	0653	0632	0557	0614	0542	0531	0543	0605	0628	0651	0619	0646
-sunset	1657	1728	1753	1917	1942	2001	2000	1934	1853	1811	1640	1635
avg hrs of sunshine	6	7	7	9	11	12	13	12	11	9	7	6
avg # days w/thunder	5	4	5	2	2	1	<1	<1	<1	4	5	6
w/snow	0	0	0	0	0	0	0	0	0	0	0	0
Typical % relative humidity @1700hrs	67	64	64	59	57	56	57	58	55	58	61	67

Palma

39°33'N, 2°44'E
Elev: 8m (26ft)

Except for early autumn, when Mediterranean lows can provide a stretch of wet days, the Balearic Islands rarely get much rain of substance. Mid-winter showers are just enough to break up periods of sunshine (although nights are often nippy into May). It's virtually dry from late May through August.

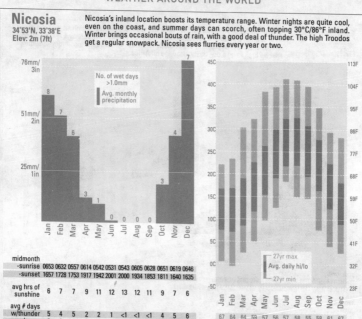

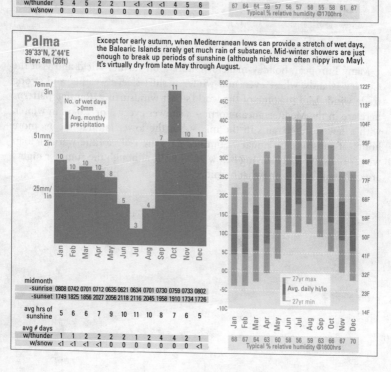

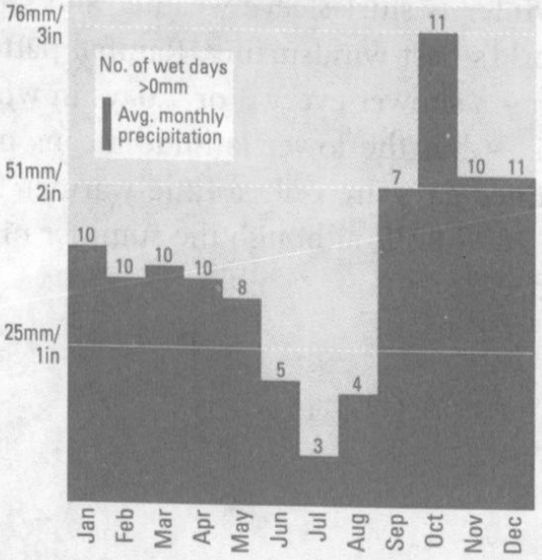

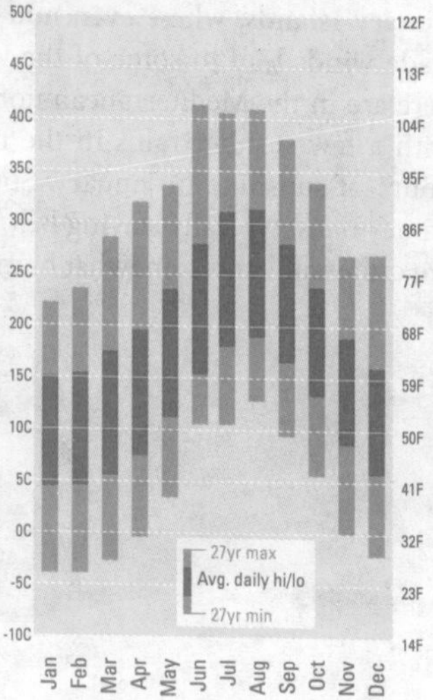

	Jan	Feb	Mar	Apr	May	Jun	Jul	Aug	Sep	Oct	Nov	Dec
midmonth -sunrise	0808	0742	0701	0712	0635	0621	0634	0701	0730	0759	0733	0802
-sunset	1749	1825	1856	2027	2056	2118	2116	2045	1958	1910	1734	1726
avg hrs of sunshine	5	6	6	7	9	10	11	10	8	7	6	5
avg # days w/thunder	1	1	2	2	2	2	1	2	4	4	2	1
w/snow	<1	<1	<1	<1	0	0	0	0	0	0	0	<1
Typical % relative humidity @1600hrs	68	67	64	63	60	58	56	59	63	66	67	70

Tenerife

28°03'N, 16°34'W
Elev: 72m (236ft)

You're almost guaranteed mild weather across the Canaries, which sit well out of the path of polar air masses yet within a cool Atlantic current. Winter rains fall in sporadic spells, and May–September are virtually rain-free. The moderate humidity of summer may be broken by a dry, dusty, hot wind off Africa.

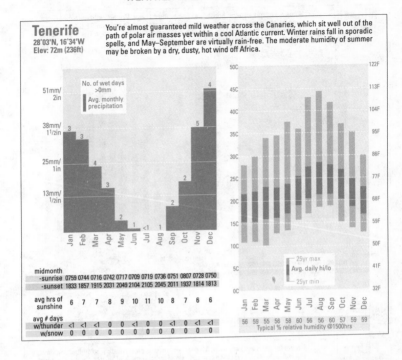

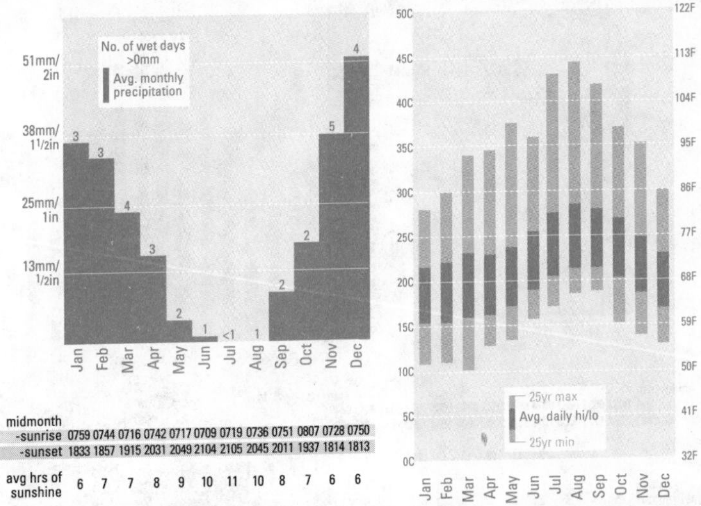

	Jan	Feb	Mar	Apr	May	Jun	Jul	Aug	Sep	Oct	Nov	Dec
midmonth -sunrise	0759	0744	0716	0742	0717	0709	0719	0736	0751	0807	0728	0750
-sunset	1833	1857	1915	2031	2049	2104	2105	2045	2011	1937	1814	1813
avg hrs of sunshine	6	7	7	8	9	10	11	10	8	7	6	6
avg # days w/thunder	<1	<1	<1	0	0	<1	0	0	<1	0	<1	<1
w/snow	0	0	0	0	0	0	0	0	0	0	0	0

	Jan	Feb	Mar	Apr	May	Jun	Jul	Aug	Sep	Oct	Nov	Dec
Typical % relative humidity @1500hrs	56	59	55	56	58	60	56	56	60	57	59	59

Many European holiday-makers venture a bit further afield and visit the **Canary Islands**, where even mid-winter is shirt-sleeve weather and strong trade winds lead to some of the world's best windsurfing. Rainfall patterns here are in the Mediterranean norm – a shower every 3 or 4 days in winter, with a few steadier rains in the mix – but the lower latitude means more hours of sunshine in January. Summer days are only a little warmer than winter ones, typically staying below 28°C/82°F, although the summer nights are quite mild and somewhat humid.

Middle East

A crossroads of culture and faith for millennia, the **Middle East** is also a place where giant weather regimes rub shoulders. Cold high pressure centred in Siberia pushes its way into the region each winter, which can be a surprisingly chilly season in the heart of the Middle East. Some of **Iran**'s mountains remain snow-capped year round. Wet winter storms from the Mediterranean help moisten the coastal strips and mountains of **Lebanon** and **Israel**; the storms then dry up across the great inland deserts of the **Arabian peninsula**. Some of these depressions become revitalized near the **Persian Gulf**, delivering quick shots of rain to the lower elevations and snow to the mountains. When a depression lingers, normally arid deserts and valleys may experience local flooding.

Clear skies and blazing heat rule most of the Middle East during the summer, although dust and haze often cut down on visibility. Only the southern corner of Arabia manages an occasional thunderstorm. Almost every location beneath the highest mountains reaches at least 30°C/86°F on a summer afternoon, and many spots soar well above 40°C/104°F. Perhaps the most notable regional differences in summer are in humidity. Interior deserts can feel as dry as the dust that coats them, while towns near the **Red Sea** and Persian Gulf can be incredibly humid. No wonder the coastal resorts are most popular in winter, when the humidity goes down and the heat is balmy rather than blistering.

Spring and autumn aren't considered true seasons in the Middle East. Like neighbouring Asia – whose monsoons help drive Middle Eastern climate – the region oscillates between summer and winter modes. The shifts between these take only two or three weeks to play out. Typically, the transitions occur in late spring and mid-autumn, but they may be a bit early or late in any given year, so the averages convey a more lengthy transition than actually tends to occur. April and May – sandwiched between the winter rains and summer heat – are good times to see the Mediterranean countries at their greenest.

Iran

Ringed by high ridges, **Iran** has a mountain-and-plateau climate that belies its location next to two major bodies of water. The cold Asian high dominates in winter, but as it waxes and wanes, temperatures across parts of the interior may shoot above 16°C/60°F, or drop below –18°C/0°F. Mediterranean depressions arrive about once a week on average from January to March, often bearing a respectable amount of rain or snow. However, most of the

Tehran
35°41'N, 51°12'E
Elev: 1191m (3906ft)

Shielded by mountains from the Caspian Sea, Tehran features a dry climate that's chilly and variably cloudy in winter, hot and dry in summer. Fog settles in every three or four days in mid-winter, when rain or snow is infrequent but occasionally heavy. Gusty west winds can strike year round, sometimes reaching gale force in spring and autumn.

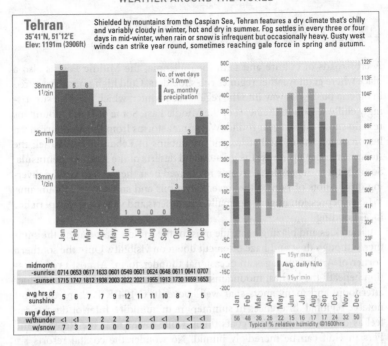

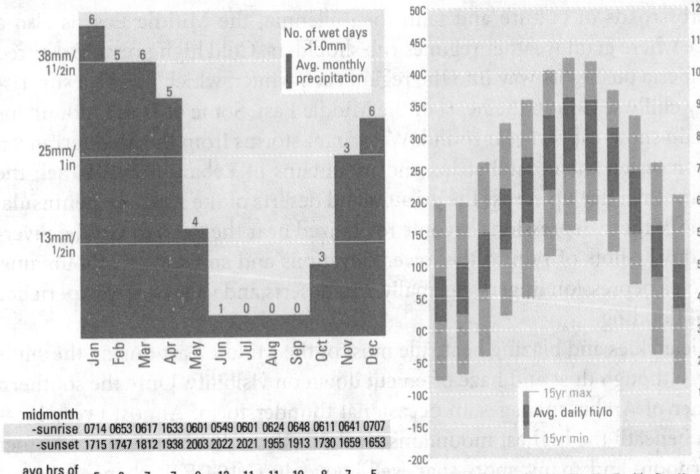

	Jan	Feb	Mar	Apr	May	Jun	Jul	Aug	Sep	Oct	Nov	Dec
midmonth -sunrise	0714	0653	0617	1633	0601	0549	0601	0624	0648	0611	0641	0707
-sunset	1715	1747	1812	1938	2003	2022	2021	1955	1913	1730	1659	1653
avg hrs of sunshine	5	6	7	7	9	12	11	11	10	8	7	5
avg # days w/thunder	<1	<1	1	2	2	2	1	<1	<1	1	<1	<1
w/snow	7	3	2	0	0	0	0	0	0	0	<1	2
Typical % relative humidity @1600hrs	56	48	36	26	22	15	16	17	17	24	32	50

country averages less than 250mm/10in of annual precipitation, with higher amounts toward the west. When the country's northern peaks intercept Asian fronts, the **Caspian coast** – especially the western part – can get very gloomy. Some spots average more than 100 wet days and 2000mm/79in of precipitation each year. The dry summer features steady northwest winds and temperatures on the plateau (including **Tehran**) that rank as the hottest in the world for their elevation. Humidity levels are modest except on the **Persian Gulf coast**, which is just as oppressively hot and humid as Kuwait or Qatar. The **lowlands** of the far southwest (next to Iraq) are hotter still.

Israel

For its size, **Israel** boasts unusually varied weather. Like its neighbours to the north, it has a Mediterranean coastal zone that gets regular winter rains. Annual amounts trend downward as you move south from the **Haifa area** (which gets around 630mm/25in, with a couple of soggy days every winter week) toward the **Egyptian border** (which sees less than 200mm/8in). Hardly any rain falls along the abbreviated **Gulf of Aqubu** coast, Israel's warmest corner. Here, afternoons typically warm to above 20°C/68°F in winter, and

Jerusalem

31°52'N, 35°13'E
Elev: 759m (2490ft)

Despite Jerusalem's inland location, its cool winter is just as damp as coastal Tel Aviv's. The occasional snow is usually light (though not always), and temperatures seldom dip much below freezing. A few hot, windy spells punctuate the fine spring and autumn weather; summer is warm and bone-dry.

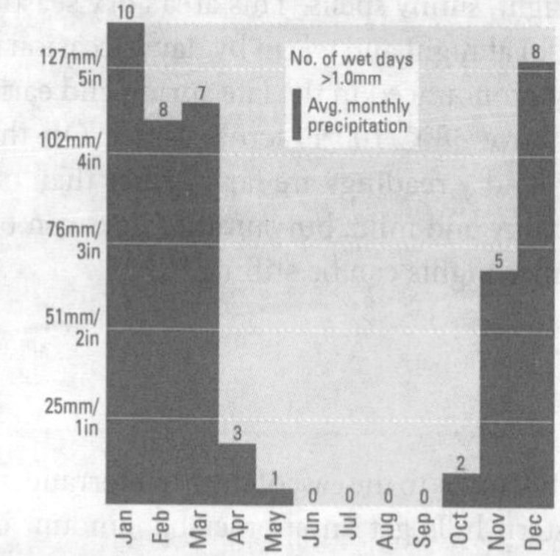

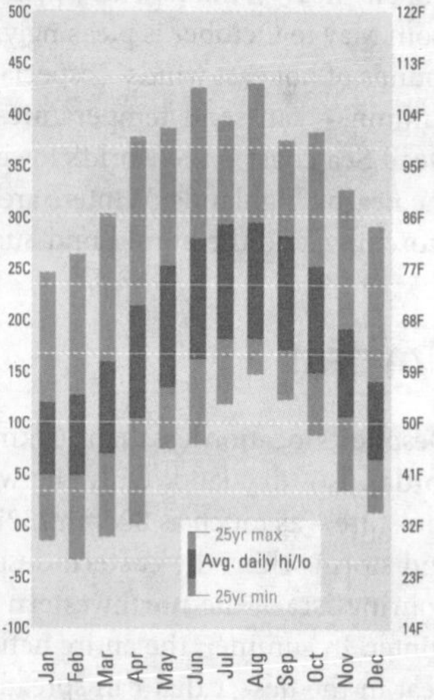

	Jan	Feb	Mar	Apr	May	Jun	Jul	Aug	Sep	Oct	Nov	Dec
midmonth -sunrise	0640	0622	0550	0611	0542	0533	0544	0604	0623	0542	0607	0632
-sunset	1657	1725	1747	1908	1929	1946	1946	1923	1846	1707	1640	1637
avg hrs of sunshine	6	9	7	9	11	13	12	12	10	9	8	6
avg # days w/thunder	1	2	1	1	1	<1	0	<1	<1	3	2	2
w/snow	2	3	1	<1	0	0	0	0	0	0	0	1

	Jan	Feb	Mar	Apr	May	Jun	Jul	Aug	Sep	Oct	Nov	Dec
Typical % relative humidity @1700hrs	71	67	66	53	48	47	51	55	57	61	66	69

Tel Aviv

32°00'N, 34°54'E
Elev: 49m (161ft)

The Mediterannean helps insulate Tel Aviv from severe winter chill. Roughly every other day sees a shower or storm, with sun in between. Mid-spring is often gorgeous but variable – spells of dry heat can top 32°C/90°F. Summers are warm and humid, with ample sunshine and sea breezes. Morning fog is common even in the rainfree summer.

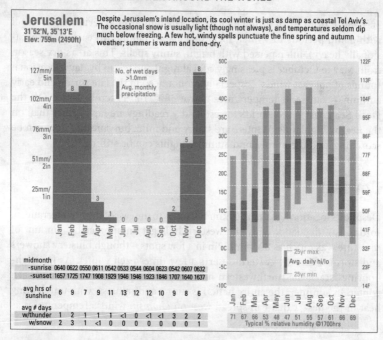

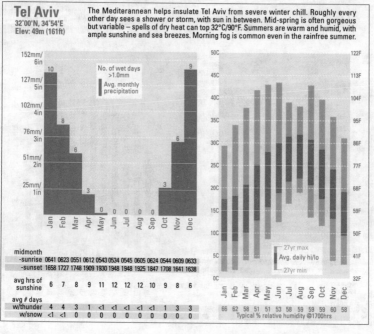

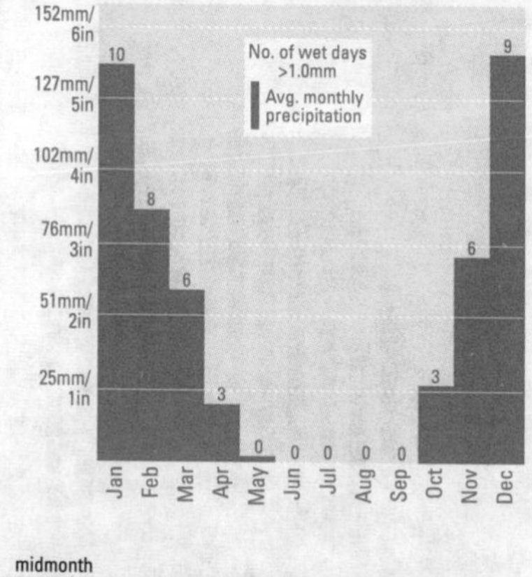

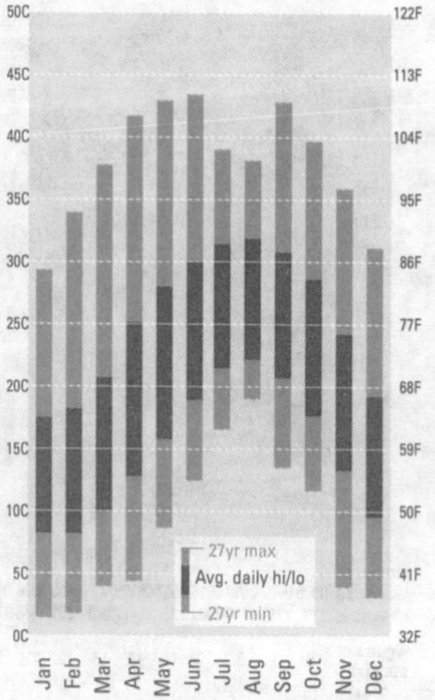

	Jan	Feb	Mar	Apr	May	Jun	Jul	Aug	Sep	Oct	Nov	Dec
midmonth -sunrise	0641	0623	0551	0612	0543	0534	0545	0605	0624	0544	0609	0633
-sunset	1658	1727	1748	1909	1930	1948	1948	1925	1847	1708	1641	1638
avg hrs of sunshine	6	7	8	9	11	12	12	12	10	9	8	6
avg # days w/thunder	4	4	3	1	<1	<1	<1	<1	<1	1	3	3
w/snow	<1	<1	0	0	0	0	0	0	0	0	0	0

	Jan	Feb	Mar	Apr	May	Jun	Jul	Aug	Sep	Oct	Nov	Dec
Typical % relative humidity @1700hrs	66	62	58	51	51	53	58	59	59	59	60	58

30°C/86°F from April through October. The hills around **Jerusalem** are cool with damp periods in winter, including a snowfall or two each year (occasionally heavy), interspersed with bright, sunny spells. This area's dry season from May to October is pleasingly cool at night and warm by day. Occasional rounds of hot east winds – especially pronounced in the late spring and early autumn – can send temperatures above 38°C/100°F across Israel. On the **Dead Sea coast** – the world's lowest land – readings are far warmer than on the nearby highlands. Winters are sunny and mild, but summer days can be scorching, and the humid mid-summer nights can be stifling.

Jordan

Despite its location less than 160km/100 miles to the east of the Mediterranean, **Jordan** is a dry land. Only the western hills get an appreciable amount of moisture – as much as 800mm/32in in a few spots – though blustery showers and storms skim the **eastern deserts** a few times each winter. As in neighbouring Israel, the **northwestern hills** may pick up a snowfall or two each winter. In summer, the entire nation is sunny and rainless, with unrelenting heat in the desert that can spread west on *khamsin* winds. Temperatures as

Al Aqabah
29°38'N, 35°01'E
Elev: 51m (167ft)

Temperatures on this coastal resort gradually rise from mild mid-winter levels to become seriously sultry by July. Rains are infrequent (especially during the parched summer) and usually quite light. Evenings can drop below 8°C/40°F anytime between November and April, though it's usually milder than that.

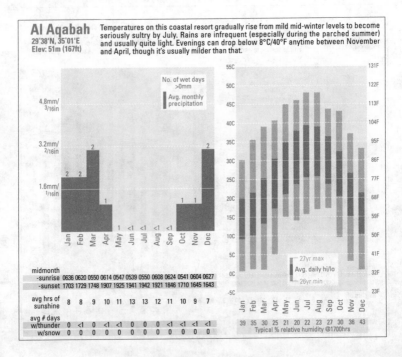

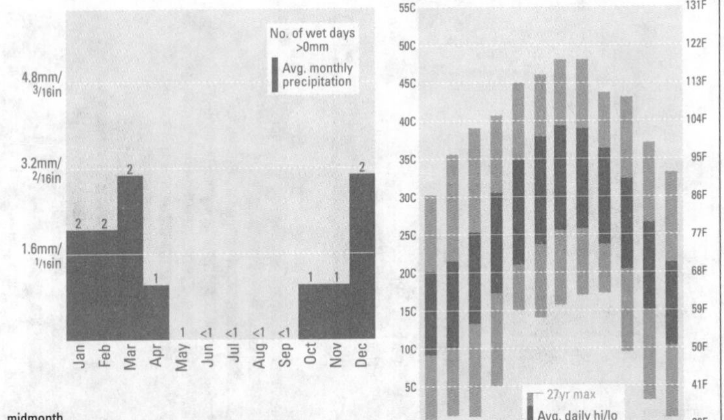

	Jan	Feb	Mar	Apr	May	Jun	Jul	Aug	Sep	Oct	Nov	Dec
midmonth -sunrise	0636	0620	0550	0614	0547	0539	0550	0608	0624	0541	0604	0627
-sunset	1703	1729	1748	1907	1925	1941	1942	1921	1846	1710	1645	1643
avg hrs of sunshine	8	8	9	10	11	13	13	12	11	10	9	7
avg # days w/thunder	0	<1	0	<1	<1	0	0	0	<1	<1	<1	<1
w/snow	0	0	0	0	0	0	0	0	0	0	0	0

	Jan	Feb	Mar	Apr	May	Jun	Jul	Aug	Sep	Oct	Nov	Dec
Typical % relative humidity @1700hrs	39	35	30	25	21	20	22	23	27	30	36	43

Amman

31°59'N, 35°59'E
Elev: 773m (2535ft)

Although Jordan's capital gets almost as many cloudy, wet days as Jerusalem, it has about half as much rain. Fog is frequent in winter, but thunder is rare. Summer days are seldom extremely hot, and humidities are fairly low, with nights often cool and breezy. Dusty days are most common in spring.

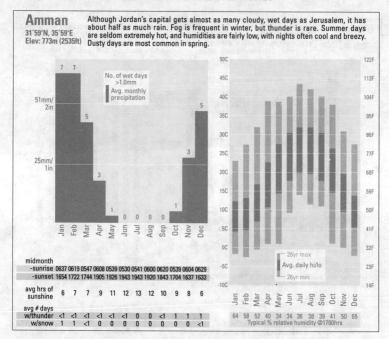

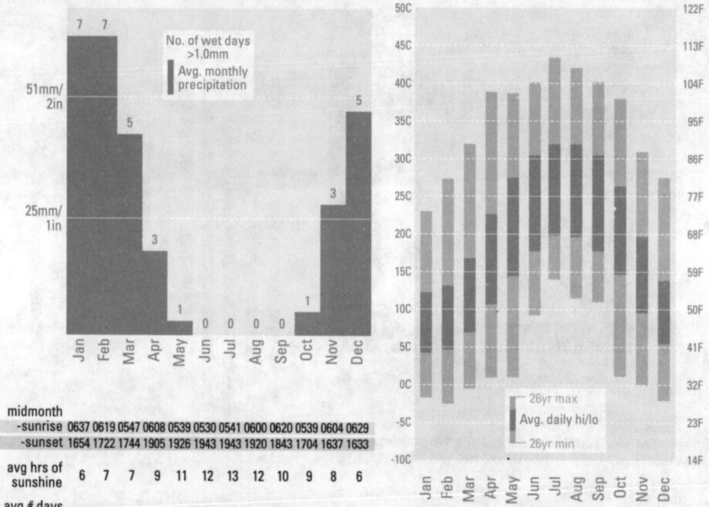

	Jan	Feb	Mar	Apr	May	Jun	Jul	Aug	Sep	Oct	Nov	Dec
midmonth -sunrise	0637	0619	0547	0608	0539	0530	0541	0600	0620	0539	0604	0629
-sunset	1654	1722	1744	1905	1926	1943	1943	1920	1843	1704	1637	1633
avg hrs of sunshine	6	7	7	9	11	12	13	12	10	9	8	6
avg # days w/thunder	<1	<1	<1	<1	<1	<1	0	0	<1	1	1	1
w/snow	1	1	<1	0	0	0	0	0	0	0	0	<1
Typical % relative humidity @1700hrs	64	58	52	40	34	34	36	38	39	41	50	65

well as sultriness increase as you descend toward the **Red Sea coast**, more than 350m/1150ft below sea level.

Along the coastline, and in the Great Rift Valley that runs northward from the Red Sea and southward to **Al Aqabah**, summertime can be brutal. Low temperatures regularly hang above 26C/79F with high humidity to boot, and the afternoon readings, which normally top 35C/95F with ease, have been known to soar above 45C/113F in both July and August. The valley can sizzle even in mid-winter, although more often than not it's pleasantly cool.

Lebanon

Thanks in large part to its formidable mountains – some of which rise to 2500m/8200ft – **Lebanon** is one of the most well-watered spots in the Middle East. The **mountains** get enough snow for skiing in winter and can offer temperatures as low as the freezing point in summer. Even the **coastal hills** get 5 to 10 days of snow each year, and typically more than 1000mm/39in of annual precipitation. The coast itself is moderated by the Mediterranean and protected from Asian cold, producing an enviable climate (apart from the

Beirut
33°49'N, 35°29'E
Elev: 19m (62ft)

The wettest major Middle Eastern city, Beirut gets almost half as much rain again as London. Over 500mm/20 inches can fall in January and February, when nearly half the days get rain interspersed with sunshine. Showers slacken by spring, when khamsin wind and heat can strike. Summer and early autumn are dependably warm, dry, and sunny (but humid).

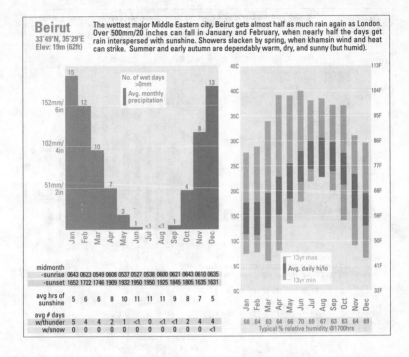

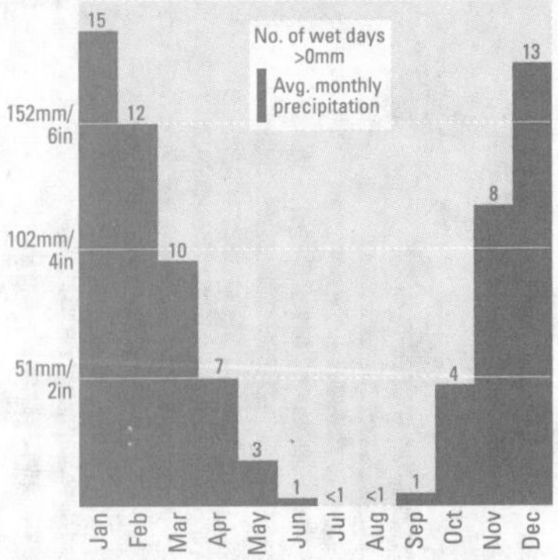

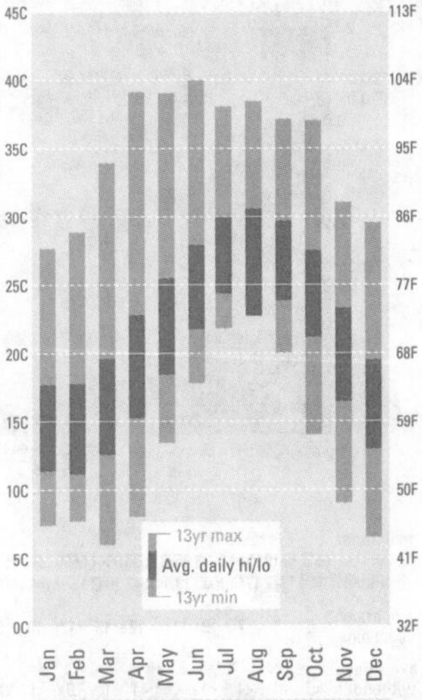

	Jan	Feb	Mar	Apr	May	Jun	Jul	Aug	Sep	Oct	Nov	Dec
midmonth -sunrise	0643	0623	0549	0608	0537	0527	0538	0600	0621	0643	0610	0635
-sunset	1652	1722	1746	1909	1932	1950	1950	1925	1845	1805	1635	1631
avg hrs of sunshine	5	6	6	8	10	11	11	11	9	8	7	5
avg # days w/thunder	5	4	4	2	1	<1	0	<1	<1	2	4	4
w/snow	0	0	0	0	0	0	0	0	0	0	0	<1

	Jan	Feb	Mar	Apr	May	Jun	Jul	Aug	Sep	Oct	Nov	Dec
Typical % relative humidity @1700hrs	68	64	63	64	66	70	69	67	63	63	64	69

winter dampness). It virtually never freezes at the coast, where summer days are fairly humid but rarely exceed 32°C/90°F.

Oman

Wrapped around the eastern end of the Arabian peninsula, **Oman** is influenced by both the **Arabian Sea** and **Persian Gulf**. The nation's most populated area, along the **Gulf of Oman coast**, offers truly world-class sultriness. From May through September, the coastal strip is more hot and humid than Miami ever gets. Nights regularly hang above 30°C/86°F. The often-dusty *shamal* winds of the Persian Gulf normally diminish as they reach Oman. From November to March, the coast sees ample sunshine and plenty of warmth, with only a stray shower or storm every week or two. The lower Arabian Sea coast around **Salalah** gets some cloud and rainfall during the summer on southwest monsoonal winds. Perhaps once a decade, a **tropical cyclone** threatens Oman from the Arabian Sea. As in Yemen, the higher terrain is more moist than the lowlands.

Muscat

23°35'N, 58°17'E
Elev: 15m (49ft)

Year-round summer prevails in this coastal city. Afternoon humidities in July and August are remarkably high, given the torrid heat. A northeast sea breeze eases things somewhat. Offshore winds prevail in the much more pleasant winter, when it rains lightly every few days. Dust and haze are common from March to September.

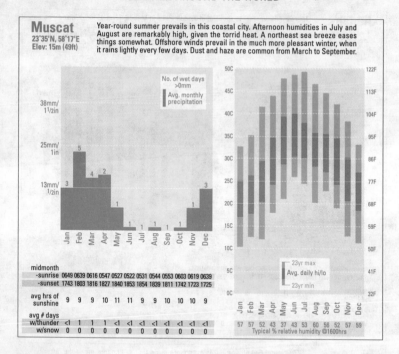

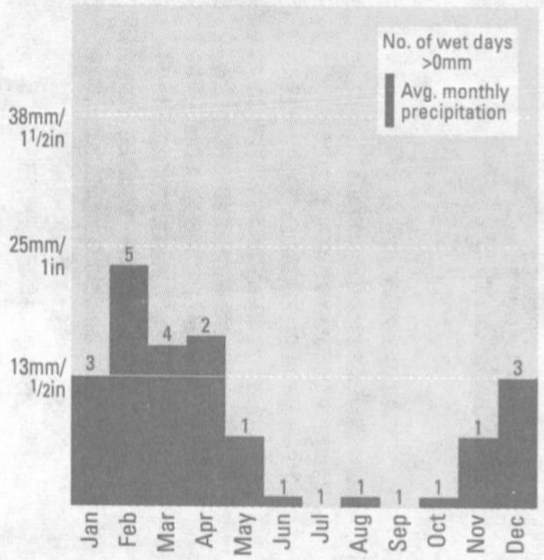

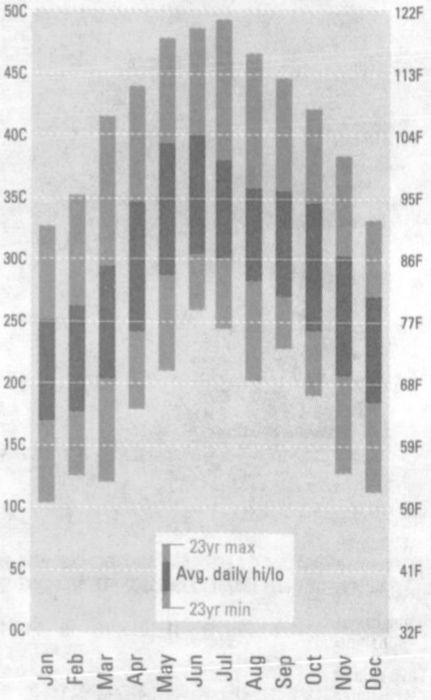

	Jan	Feb	Mar	Apr	May	Jun	Jul	Aug	Sep	Oct	Nov	Dec
midmonth -sunrise	0649	0639	0616	0547	0527	0522	0531	0544	0553	0603	0619	0639
-sunset	1743	1803	1816	1827	1840	1853	1854	1839	1811	1742	1723	1725
avg hrs of sunshine	9	9	9	10	11	11	9	9	10	10	10	9
avg # days w/thunder	<1	1	1	1	<1	<1	<1	<1	<1	<1	<1	<1
w/snow	0	0	0	0	0	0	0	0	0	0	0	0

	Jan	Feb	Mar	Apr	May	Jun	Jul	Aug	Sep	Oct	Nov	Dec
Typical % relative humidity @1600hrs	57	57	52	43	37	43	53	60	56	52	57	59

Saudi Arabia

Bahrain | Iraq | Kuwait | Qatar | United Arab Emirates

With the world's largest expanse of sand at its heart, **Saudi Arabia** lives up to its reputation for dust and heat – at least between May and October – although the summer mode may arrive and depart a little ahead or behind schedule. The **Arabian peninsula** slopes eastward from a substantial rise near the west coast. West of that ridge, destinations near the **Red Sea**, such as **Mecca** and **Jiddah**, experience sultry summer humidity to accompany the heat. A few thunderstorms bring relief in the far southwest, next to Yemen. Along the **Persian Gulf coast** – which includes the nations of **Kuwait**, **Bahrain**, **Qatar**, and the **United Arab Emirates** and the southeast delta of **Iraq** – the extremely muggy summers (fed by moisture evaporating from the warmest sea on Earth) are accompanied by the *shamal*. This steady northwest wind is funnelled down the Gulf between the Azores high and Asian monsoonal low. It's a nearly constant breeze in June and July, although it blows more fiercely in brief, dusty spells toward the beginning and end of the hot season. The great **Empty Quarter**, hot as it is, escapes the *shamal* and the humidity, which makes the summer air slightly more bearable. Once the winter regime sets in, weak lows scoot across north Arabia every few days. They can bring strong

Jiddah

21°40'N, 39°09'E
Elev: 12m (39ft)

There's little weather to speak of in Jiddah – and in Mecca, 72km/45 miles inland – other than heat and humidity, both intensified from April–November. Sunshine is near-constant and winds are usually light, although dust may blow a few days each month. Only a couple of thunderstorms strike each winter; when it does rain, floods occur easily.

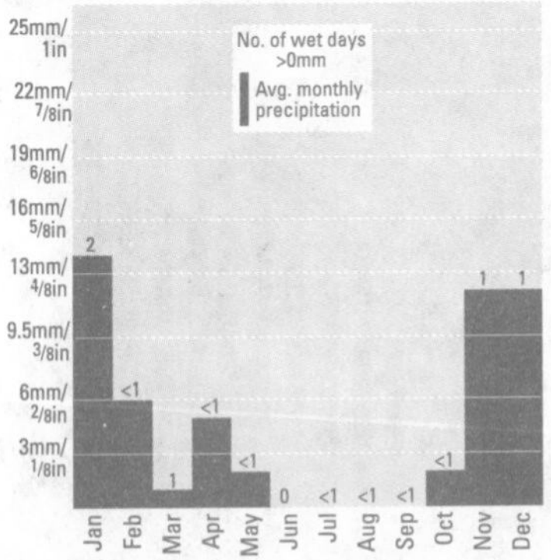

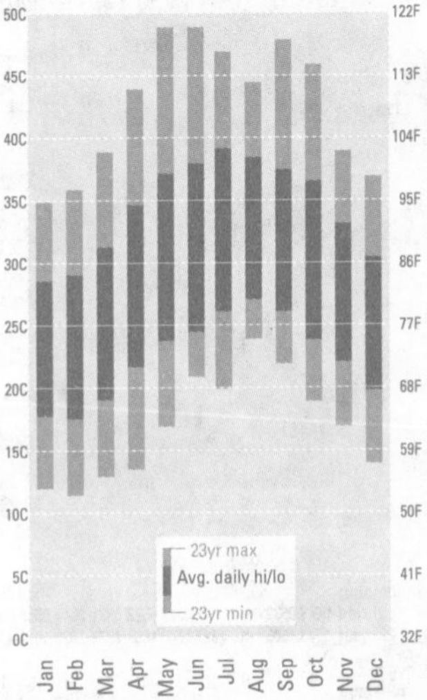

	Jan	Feb	Mar	Apr	May	Jun	Jul	Aug	Sep	Oct	Nov	Dec
midmonth -sunrise	0704	0655	0632	0604	0545	0540	0549	0601	0610	0619	0635	0654
-sunset	1801	1821	1833	1843	1855	1907	1909	1854	1827	1759	1741	1743
avg hrs of sunshine	7	8	9	9	11	11	10	10	10	10	8	7
avg # days w/thunder	1	<1	<1	1	<1	<1	<1	<1	<1	<1	1	1
w/snow	0	0	0	0	0	0	0	0	0	0	0	0
Typical % relative humidity @1500hrs	46	44	43	42	43	43	38	43	53	50	49	47

Riyadh

24°43'N, 46°43'E
Elev: 612m (2007ft)

The typical Riyadh summer is even more torrid than in Phoenix, minus the latter's humidity and thunder. What little rainfall there is begins around December and peaks in March and April (when about every fourth day gets a bit of moisture) before abruptly ending. Winters are cool, but frost is quite rare.

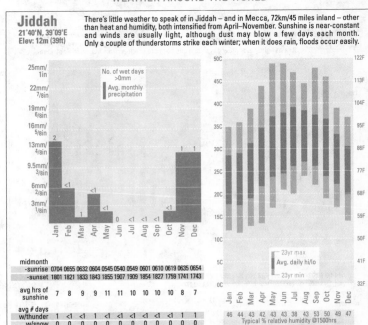

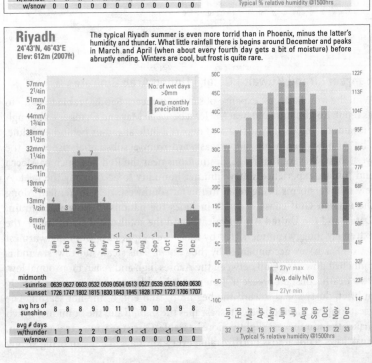

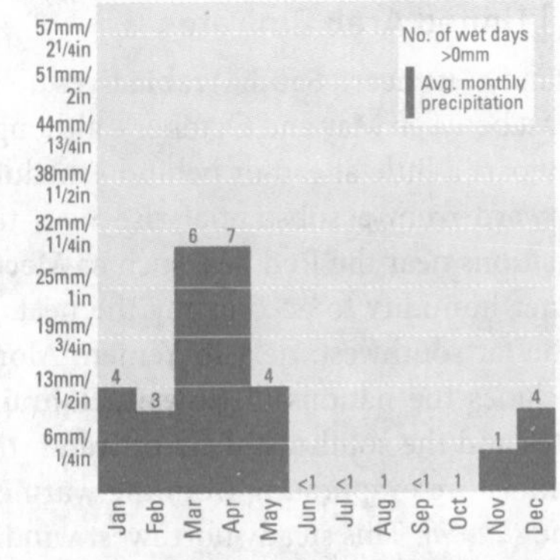

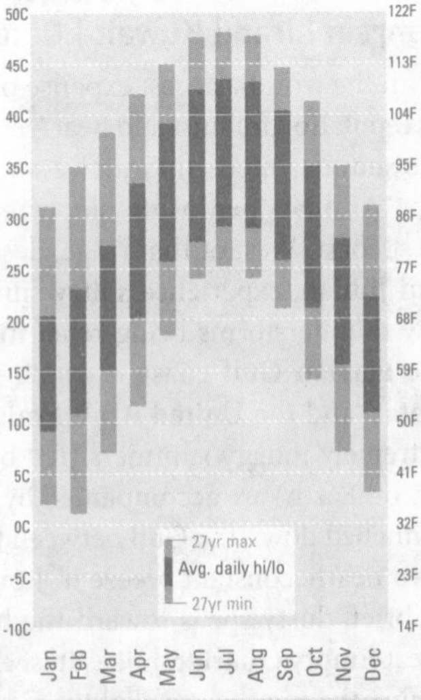

	Jan	Feb	Mar	Apr	May	Jun	Jul	Aug	Sep	Oct	Nov	Dec
midmonth -sunrise	0639	0627	0603	0532	0509	0504	0513	0527	0539	0551	0609	0630
-sunset	1726	1747	1802	1815	1830	1843	1845	1828	1757	1727	1706	1707
avg hrs of sunshine	8	8	8	9	10	11	10	10	10	10	9	8
avg # days w/thunder	1	1	2	2	1	<1	<1	<1	0	<1	<1	1
w/snow	0	0	0	0	0	0	0	0	0	0	0	0
Typical % relative humidity @1500hrs	32	27	24	19	13	8	8	8	9	13	22	33

winds and a round of thunder, especially toward spring. Across the Persian Gulf, a passing front may bring a brief winter-style *shamal*. Winter temperatures and humidities along the Persian Gulf are roughly similar to those in Los Angeles or San Diego, but with fewer rainy spells. The desert tends to be a bit warmer and even drier than the Persian Gulf coast, although nights after cool fronts can dip close to freezing. Iraq's desert is a few degrees cooler than the peninsula in winter but just as scorching in summer. The **Red Sea coast** stays warm and humid through the year.

Syria

Syria epitomizes Middle Eastern climate, with a Mediterranean wet-winter regime on the coast and a desert that spans most of the country east of a narrow strip of coastal mountains. **Damascus** is part of the Fertile Crescent, a transition zone near the Lebanon border that gets just enough moisture to keep the desert at arm's length. Winters are chilly in the mountains and adjacent plateau, with a bout of rain once or twice a week and a day or two of snow possible in Damascus. Snows are more heavy and frequent at high elevations. The rains stop from June through August, which are hot by day

Damascus

33°25'N, 36°31'E
Elev: 611m (2004ft)

Damascus sits just inland from the mountains that drain Mediterranean storms of their moisture over Lebanon. This gives the world's oldest continuously inhabited city a desert-flavoured climate. Rain or snow (seldom heavy) falls about once per winter week. Dusty, hot khamsins blow in from the east several times a year.

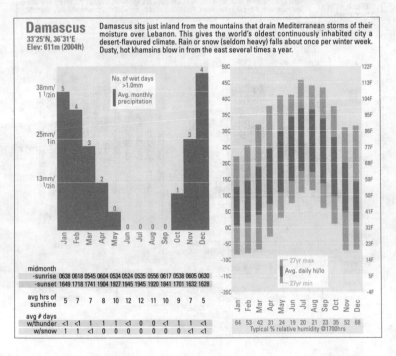

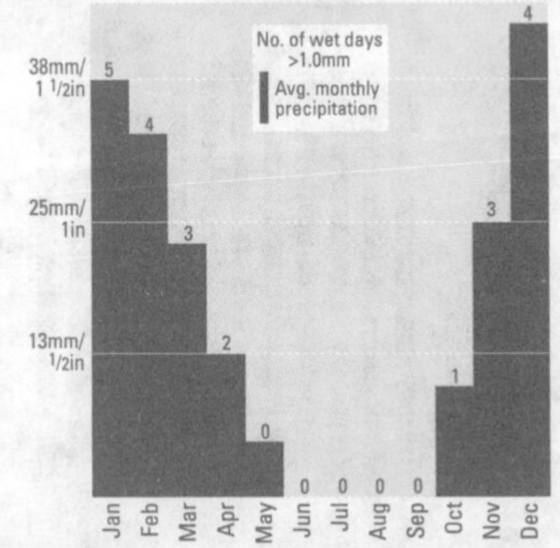

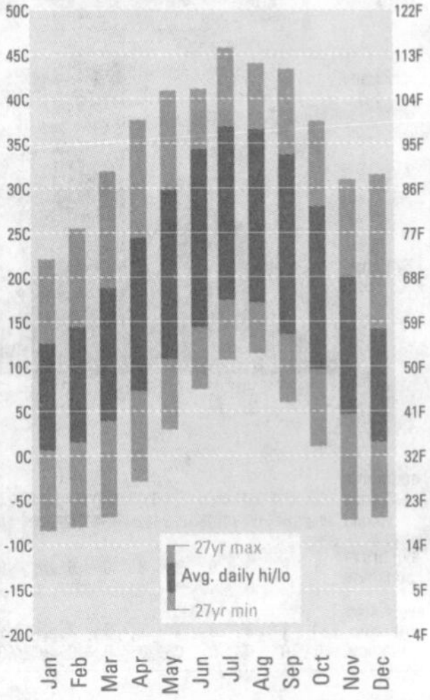

	Jan	Feb	Mar	Apr	May	Jun	Jul	Aug	Sep	Oct	Nov	Dec
midmonth -sunrise	0638	0618	0545	0604	0534	0524	0535	0556	0617	0538	0605	0630
-sunset	1649	1718	1741	1904	1927	1945	1945	1920	1841	1701	1632	1628
avg hrs of sunshine	5	7	7	8	10	12	12	11	10	9	7	5
avg # days w/thunder	<1	<1	1	1	1	<1	0	0	<1	1	1	<1
w/snow	1	1	<1	0	0	0	0	0	0	0	<1	<1

	Jan	Feb	Mar	Apr	May	Jun	Jul	Aug	Sep	Oct	Nov	Dec
Typical % relative humidity @1700hrs	64	53	42	31	24	19	20	21	23	35	52	68

but relatively cool at night – except across the deserts, where even the nights sizzle in midsummer. Winter in the deserts can bring a few days of light rain and even an occasional dusting of snow.

Yemen

Don't expect a rainforest here, but the climate of **Yemen** does have its moist side. Every summer, a powerful low-level jet stream arcs from Somalia toward India. As the strong flow approaches Yemen's soaring **western highlands**, it can trigger substantial rains and thunderstorms. The high terrain picks up other moisture through the rest of the year – snow even falls at the highest peaks – with the most favoured locales experiencing as much as 500mm/20in of precipitation. Summer showers may fall along the **Gulf of Aden** coast, but the rest of the nation, especially the interior desert, is more conventionally arid. Conditions are hot year-round at lower elevations, with perpetual humidity on the coasts. Altitude allows the **high plateau**, surrounded by mountains topping 3500m/11,500ft, to stay amazingly mild.

Sana'a

15°31'N, 44°11'E
Elev: 2190m (7183ft)

Unlike the sizzling deserts to the north and east, Sana'a enjoys a salubrious climate with little extreme heat or cold. A few thundery days are sprinkled through the spring and late summer, with a break in between. Rainfall is usually quite light, although the July–August showers cut down on sunshine.

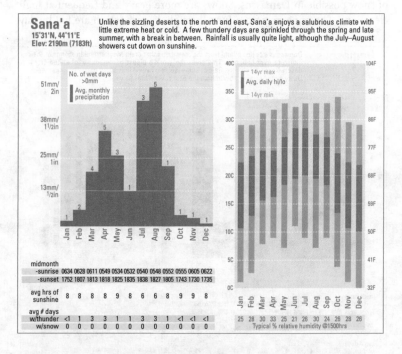

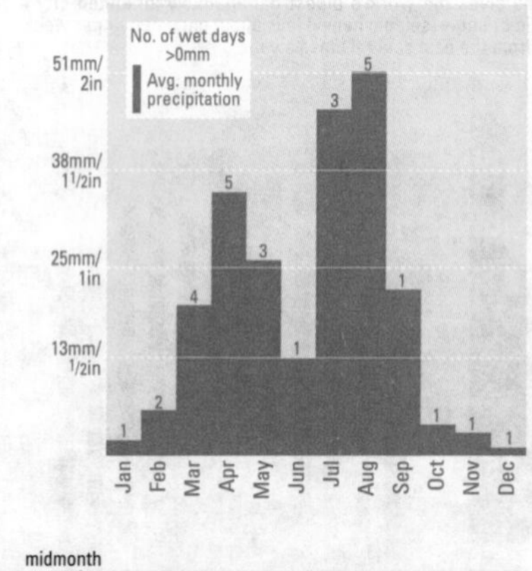

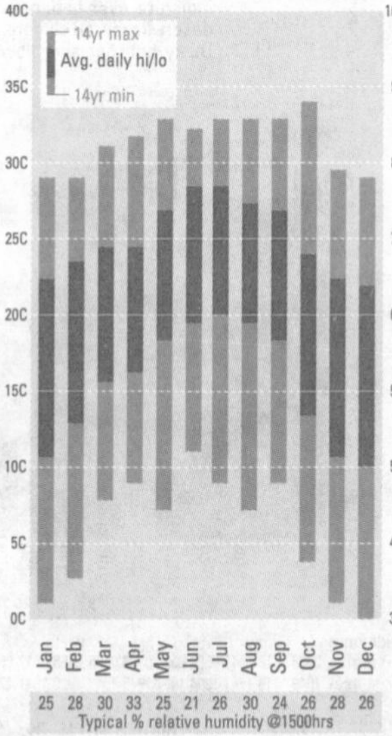

	Jan	Feb	Mar	Apr	May	Jun	Jul	Aug	Sep	Oct	Nov	Dec
Typical % relative humidity @1500hrs	25	28	30	33	25	21	26	30	24	26	28	26

	Jan	Feb	Mar	Apr	May	Jun	Jul	Aug	Sep	Oct	Nov	Dec
midmonth -sunrise	0634	0628	0611	0549	0534	0532	0540	0548	0552	0555	0605	0622
-sunset	1752	1807	1813	1818	1825	1835	1838	1827	1805	1743	1730	1735
avg hrs of sunshine	8	8	8	8	9	8	6	6	8	9	9	8
avg # days w/thunder	<1	1	3	3	1	1	3	3	1	<1	<1	<1
w/snow	0	0	0	0	0	0	0	0	0	0	0	0

Africa

More than any other continent, **Africa** is swept by the back-and-forth migration of meteorology's equator. Africans watch the globe-straddling **intertropical convergence zone (ITCZ)** as intently as a day trader watches stocks. As the ITCZ sloshes north and south over its annual cycle, it brings much of Africa its only period of annual rainfall. The dry-season/wet-season rhythm flavours almost every aspect of life across the heart of the continent. In Botswana, the word for rain, *pula*, is also the nation's motto, the name of its currency and an everyday greeting.

The rainfall contrasts across this continent are stunning. Parts of the **Sahara** see less than 5mm/0.2in of rain a year – barely enough to cover the bottom of a glass – while Réunion Island, off **Madagascar**, holds the world record for the most rain observed in a 24-hour period, a mind-boggling 1825mm/72 inches. In general, the wet and dry seasons and locations are fairly well-defined across Africa. Thus, it's possible for a visitor to target, say, the lushness after weeks of rain have ended, or the tail end of a dry period, when more places may be accessible than after the rains begin.

Nowhere is the contrast between dry and wet more acute than through the middle of West Africa. The parched Sahara and the lush **Guinea coast** run side by side along an east-to-west band more than 1600km/1000 miles long. This stretch includes the semi-arid, drought-prone **Sahel**, whose name derives from an Arabic term for border. During the northern spring, the ITCZ slowly migrates into the Sahel and toward the Sahara. The northernmost penetration of the ITCZ varies from year to year, and the timing and length of its visit can make the difference between life or death for herders and farmers who work this unforgiving land. From July through September, dramatic bands of thunderstorms sweep west along the ITCZ in waves that sometimes form Atlantic hurricanes. By August, the ITCZ is typically so far inland that the Guinea coast sees a relative lull in rainfall before the "little

The howling harmattan

When the rains stop across the **Sahel**, it's time for the **harmattan** ("north wind" in the Hausa language). This gusty, dusty northeast flow is a less-chilly version of the persistent northerlies that accompany Asia's winter monsoon. Sometimes gentle and sometimes fierce, the *harmattan* dominates much of western Africa from November to March, sometimes penetrating all the way to the Guinea coast. At their worst, *harmattan* winds can suffuse the Sahel and points south with a thick, irritating haze that cuts visibility to a few hundred metres or less, especially in the morning. The wind-driven grit can cause respiratory problems, hamper travel and generally ruin one's day. A 1973 plane crash in Kano, Nigeria, that killed 176 people was attributed to poor visibility in a severe *harmattan*.

rains" return. The ITCZ's push south is faster than its movement north, thus making the little rains briefer than the main wet season.

Further north is a kingdom of sand, sun and heat. The Sahara reigns as the world's largest desert, but the air isn't completely dry. Even in summer, there's more moisture per volume of air than you might find in the middle of a Siberian snowstorm. However, the Sahara's temperatures are so scorchingly high that the relative humidities drop toward the single digits, and what paltry rain does manage to develop often evaporates as virga – "ghost rain" – before it reaches the ground. Thanks to the low humidity, Sahara nights can be surprisingly cool, especially in December and January. Winter is also the season for the **harmattan** (see box, p.299), a northeast wind that dominates the Sahel.

On the narrow south end of Africa, near 30°S, is a much smaller counterpart to the Sahara, the **Kalahari Desert**. Between these dry zones is the dampest part of mainland Africa, the forested highlands and valleys of the **Congo Basin**. The Great Rift Valley separates the Congo's year-round rain from the often tortuous wets and drys of eastern Africa. The average rainfall from **Tanzania** to **Ethiopia** is sufficient for farming, but this obscures a marked year-to-year variability – due in part to **El Niño** and **La Niña** – that leaves the area vulnerable to wrenching drought.

Both the north and south ends of Africa extend into the mid-latitudes, where you'll find some of the continent's most varied weather. The **Atlas Mountains** accentuate the divide between the Sahara and the much cooler Mediterranean climate of northern **Morocco** and **Algeria**. Fierce spring sandstorms blow off the Sahara, and the tail ends of winter storms crossing Europe and Asia can bring blustery coolness and a touch of rain to the coast. At the continent's other end, the **Drakensbergs** help produce a smorgasbord of micro-climates across South Africa.

Botswana

For a country that's half desert, **Botswana** has a surprisingly agreeable climate. Sunshine is ample year round, from the driest reaches of the south to the wildlife-studded swamps of the **Okavango Delta**. The stretch from the Delta to **Francistown** is the wettest part of the country, but even during the wet season (October–March) thunderstorms arrive only about every fourth day on average, and they seldom last long; in fact, drought is a recurring problem. Humidity is high but not insufferable in and near the Delta, where fog softens some winter mornings. Conditions are much drier to the south. The mostly rainless winter months see temperatures dip below 10°C/50°F at night throughout Botswana. Light freezes can occur, but the bright afternoons often

Francistown

21°13'S, 27°30'E
Elev: 1000m (3280ft)

In this transition zone between soggy Mozambique and the dry Kalahari, summer thundershowers are sporadic, though sometimes heavy. It's dry from May to September, with sunny, mild days and cool nights. Frost becomes more frequent at higher altitudes.

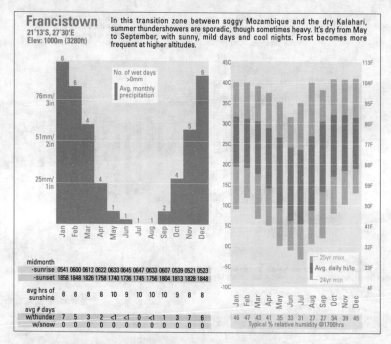

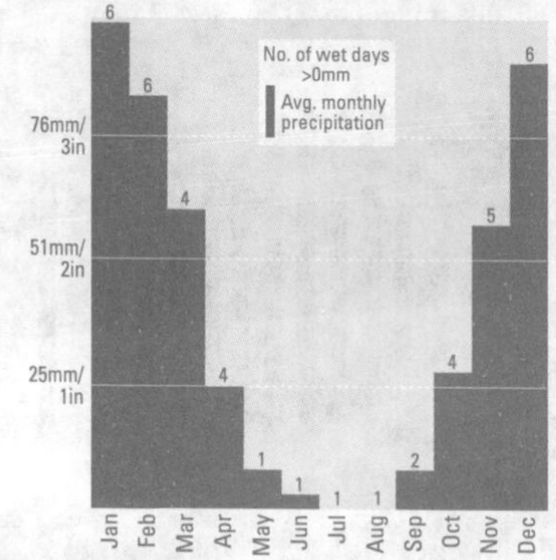

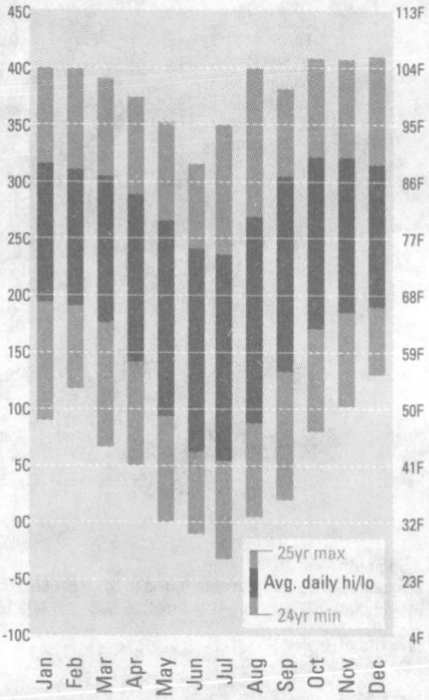

	Jan	Feb	Mar	Apr	May	Jun	Jul	Aug	Sep	Oct	Nov	Dec
midmonth -sunrise	0541	0600	0612	0622	0633	0645	0647	0633	0607	0539	0521	0523
-sunset	1858	1848	1826	1758	1740	1736	1745	1756	1804	1813	1828	1848
avg hrs of sunshine	8	8	8	8	10	9	10	10	10	9	8	8
avg # days w/thunder	7	5	3	2	<1	<1	0	<1	1	3	7	6
w/snow	0	0	0	0	0	0	0	0	0	0	0	0
Typical % relative humidity @1700hrs	46	47	43	41	35	33	31	27	27	34	39	45

top 21°C/70°F. Although the relatively small scope of the Kalahari limits the extent of dust storms, winter squalls can whip across the pans of the northeast and bring a stinging round of windblown salt and dust.

Cameroon

Congo | Gabon | Principe | Såo Tomé

This small, spectacular country lies at the crossroads of several African weather regimes. Few places can out-drench **Cameroon**'s coastal delta, where moist southwesterlies are forced upward by Cameroon Mountain and neighbouring peaks to produce one of the world's wettest climates. **Debundscha**'s annual average is over 1000mm/390in, and the coast is soaked from March to November. Further east, the rains are slightly less heavy, and they taper off around July for a few weeks in between the ITCZ's two passages. Much of the rain arrives in brief, but intense, waves, followed by fresh easterly winds and sunny skies. The narrow strip of Cameroon's **Nord** laps into the **Sahel**. Here, more spotty rains are focused from June to September, after which the climate segues from sultry and wet to scorching and dry. Nights here do provide some relief in December and January, with average lows dropping below

Yaoundé
3°49'N, 11°31'E
Elev: 753m (2470ft)

Even while torrents are falling in July and August on the coast just 160km/100 miles to the west, Yaoundé sees much lighter rain, although it's the cloudiest time of the year. Of the two wet seasons, March–June is more prolonged but less intense than September–October. Days are warm and nights mild year round.

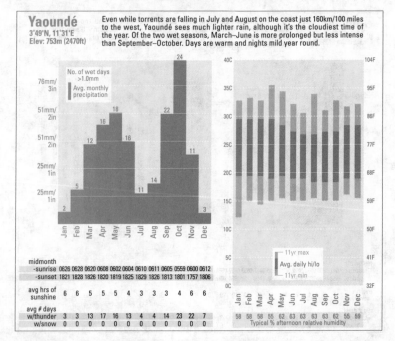

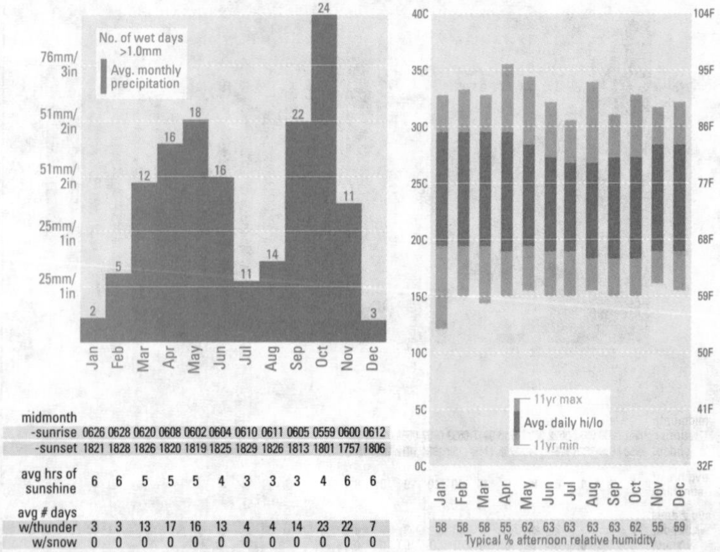

	Jan	Feb	Mar	Apr	May	Jun	Jul	Aug	Sep	Oct	Nov	Dec
midmonth -sunrise	0626	0628	0620	0608	0602	0604	0610	0611	0605	0559	0600	0612
-sunset	1821	1828	1826	1820	1819	1825	1829	1826	1813	1801	1757	1806
avg hrs of sunshine	6	6	5	5	5	4	3	3	3	4	6	6
avg # days w/thunder	3	3	13	17	16	13	4	4	14	23	22	7
w/snow	0	0	0	0	0	0	0	0	0	0	0	0
Typical % afternoon relative humidity	58	58	58	55	62	63	63	63	63	62	55	59

16°C/60°F before the mercury soars back above 32°C/90°F by day. Freezing temperatures are routine above 3000m/10,000ft on Cameroon's peaks.

Although nearby **Gabon** straddles the equator (as does aptly named **Equatorial Guinea**), Gabon's climate reflects more of a Southern Hemisphere influence, as does that of the adjacent **Congo**. Temperatures peak between February and April, and the long October-to-May wet season is punctuated by a relative lull in February. July and August are the mildest and driest months, with the trend accentuated as you head south along the coast from **Cape Lopez**, a stretch where waters are chilled by the Benguela Current. Even during the dry season, low clouds often shroud the sky, with the sun shining only about a third of the time. Humidity is high year-round. The volcanic islands of **Såo Tomé** and **Principe** pull off an unusual combination: perpetually sultry air (and torrential rains on the southwest slopes) above cool Benguela water.

Democratic Republic of the Congo

Central African Republic

At the heart of equatorial Africa, the **Democratic Republic of the Congo** (formerly Zaïre) has one of the more dependable climatic regimes of the

Bangui

4°24'N, 18°31'E
Elev: 366m (1200ft)

A touch of intense Sahel-style spring heat shows up in Bangui, where highs typically exceed 33°C/91°F from January to April. Otherwise, temperatures stay tropically warm and humid year round. Thunderstorms are heavy and frequent from March to November. Winter is mostly dry and sunny.

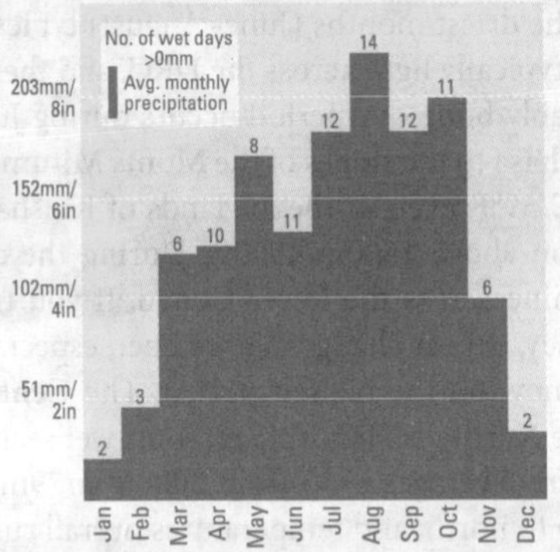

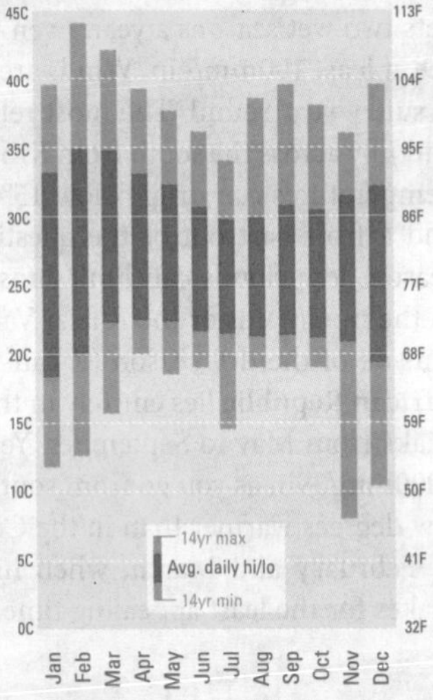

	Jan	Feb	Mar	Apr	May	Jun	Jul	Aug	Sep	Oct	Nov	Dec
midmonth -sunrise	0558	0601	0552	0540	0533	0535	0541	0543	0537	0531	0533	0545
-sunset	1752	1800	1758	1752	1752	1758	1802	1758	1745	1732	1728	1737
avg hrs of sunshine	7	7	6	7	6	6	5	4	5	6	6	6
avg # days w/thunder	4	8	13	14	15	15	15	17	16	21	14	5
w/snow	0	0	0	0	0	0	0	0	0	0	0	0
Typical % relative humidity @1600hrs	40	39	44	57	60	67	67	68	68	68	64	52

Kinshasa

4°23'S, 15°26'E
Elev: 312m (1023ft)

Kinshasha only sees rain about every other day during the October–May wet season, and it often falls in the night or early morning. Both heat and humidity peak during the wet season. Nights get a bit cooler and less sultry during the dry stretch from June–August, although the Congo Basin stays very humid.

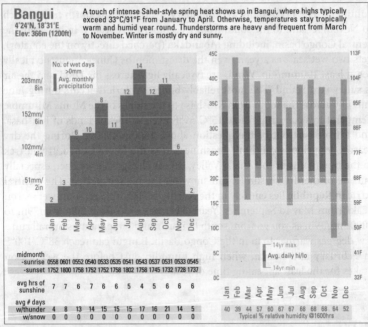

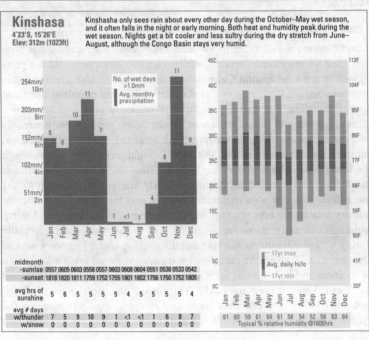

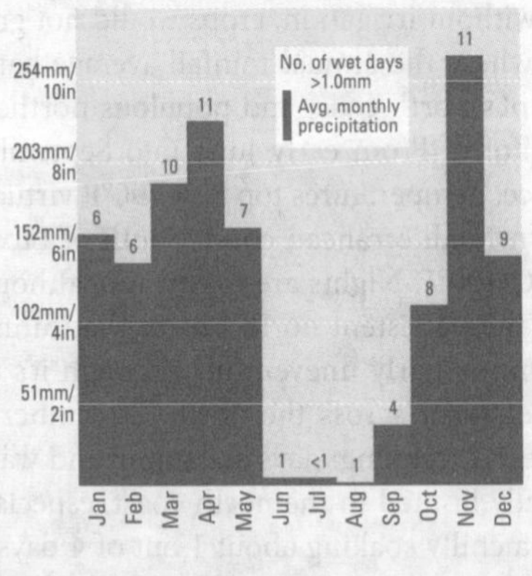

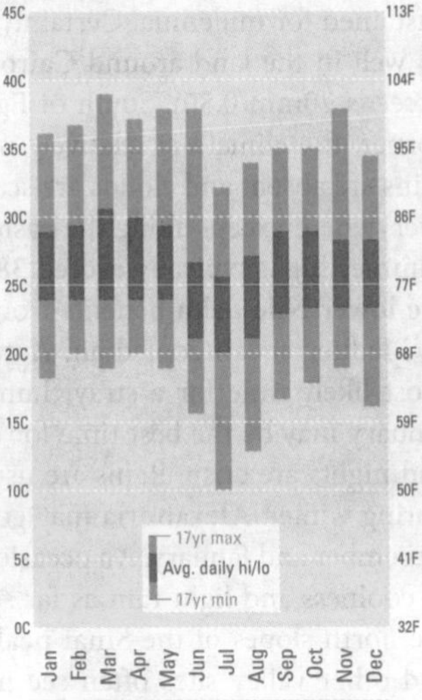

	Jan	Feb	Mar	Apr	May	Jun	Jul	Aug	Sep	Oct	Nov	Dec
midmonth -sunrise	0557	0605	0603	0558	0557	0603	0608	0604	0551	0538	0533	0542
-sunset	1818	1820	1811	1759	1752	1755	1801	1802	1756	1750	1752	1805
avg hrs of sunshine	5	6	5	5	5	5	4	5	5	5	5	4
avg # days w/thunder	7	5	9	10	9	1	<1	<1	1	6	8	7
w/snow	0	0	0	0	0	0	0	0	0	0	0	0
Typical % relative humidity @1600hrs	61	60	59	61	64	61	58	54	52	58	63	64

continent. The ITCZ pulls wet, thundery weather to the north from March to November and to the south from December to February. In between, the central Congo basin, including **Mbandaka** (perched directly on the equator), gets two wet seasons a year. Even the driest months (June–August) typically see at least 100mm/4in. Winds are typically light across the DRC, and the air is sultry year round. The most reliably bone-dry period occurs during June and July across the south from **Kinshasa** to the slopes of the **Monts Mitumba**. Temperatures can drop below 15°C/59°F, even in the lowlands of Kinshasa, and a frost isn't out of the question above 1500m/4900ft. During the dry season, low clouds can limit sunshine across the lower Congo. If you trek to the mountains or the Ruzizi Valley, expect changeable weather, especially outside of the dry season (it can snow on the highest peaks). The **Central African Republic** lies entirely in the Northern Hemisphere, so its wet season peaks from May to September. Yearly averages drop from 2000mm/79in to 1000mm/39in as you go from south to north, and temperatures overall run a few degrees warmer than in the Congo Basin. **Bangui** can reach 38°C/100°F in February and March, when high humidity just before the wet season makes for the least appealing time of year.

Egypt

Libya | Sudan

Adaptation is the key to life in the **Nile Valley**, where agriculture has been sustained for millennia. Certainly, without irrigation, crops would not grow so well in the land around **Cairo**, where the annual rainfall average barely exceeds 20mm/0.8in. South of **Egypt**'s north coast and populous northeast corner, the climate is relatively uniform. From early June into September, rains are absent and clouds are scarce. Temperatures top 32°C/90°F virtually everywhere except along the eastern Mediterranean coast. South of **Luxor**, summer highs typically exceed 38°C/100°F. Nights are toasty, too, although the **lower Nile delta** benefits from the persistent north breeze that funnels in Mediterranean-cooled air. Autumn is fairly uneventful, although it's the most likely time for a stray thunderstorm across the north. December or January may be the best time for desert trekking: days are sunny and warm and nights are crisp. Rains are usually limited to the north coast, especially during winter. **Alexandria** may get a chilly soaking about 1 out of 4 days in December and January. An occasional maritime storm can bring a day or two of coolness and light rain as far south as Cairo, and snow may briefly coat the north slopes of the **Sinai peaks**. Throughout the cooler months, Cairo and other valley sites often see night and morning fog and smog. As the Sahara heats up in spring, Mediterranean storms are drawn southward into

Alexandria

31°12'N, 29°57'E
Elev: 7m (23ft)

The Mediterranean climate regime barely reaches this coastal city. Only a few winter days are damp, sometimes accompanied by a raw, chilly wind (don't expect freezes, though). Most summer days are less hot but more humid than in Cairo. Autumn is a choice time, with less risk of *khamsin* heat than in spring.

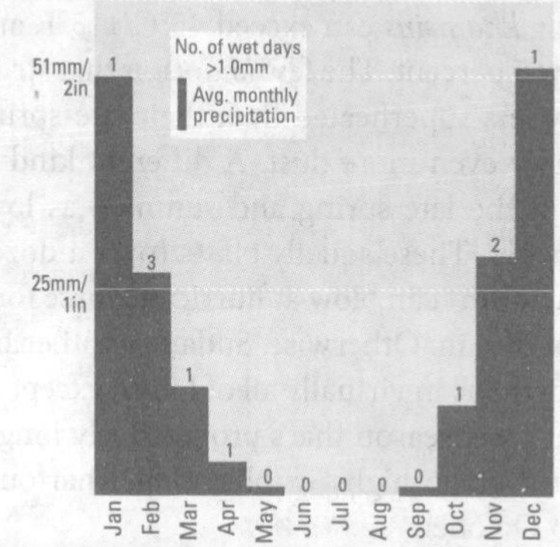

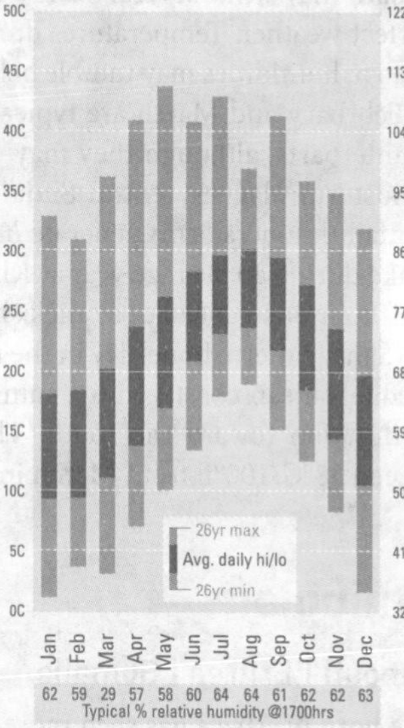

	Jan	Feb	Mar	Apr	May	Jun	Jul	Aug	Sep	Oct	Nov	Dec
midmonth -sunrise	0659	0642	0611	0533	0605	0556	0606	0626	0644	0603	0627	0651
-sunset	1720	1747	1808	1828	1949	2006	2006	1943	1906	1729	1702	1659
avg hrs of sunshine	6	7	8	9	10	12	12	11	10	9	7	6
avg # days w/thunder	2	1	1	<1	<1	<1	0	<1	<1	<1	1	2
w/snow	0	0	0	0	0	0	0	0	0	0	0	0

Cairo

30°08'N, 31°24'E
Elev: 74m (243ft)

One of the driest, sunniest cities on the planet, Cairo gets little more than 20mm/0.8 inches of rain annually. However, occasional winter depressions can deliver clouds and some light rain to Cairo (more in Alexandria). Winters tend to be cool to mild and summers hot. A few *khamsins* bring dust, wind, and heat each spring.

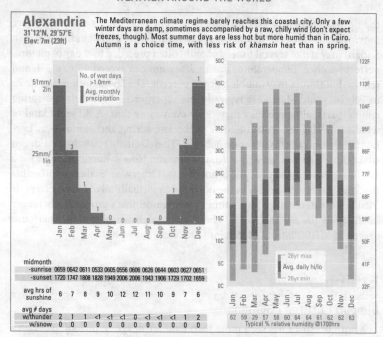

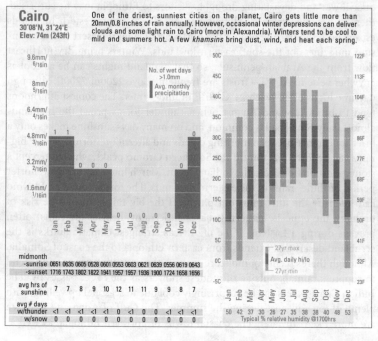

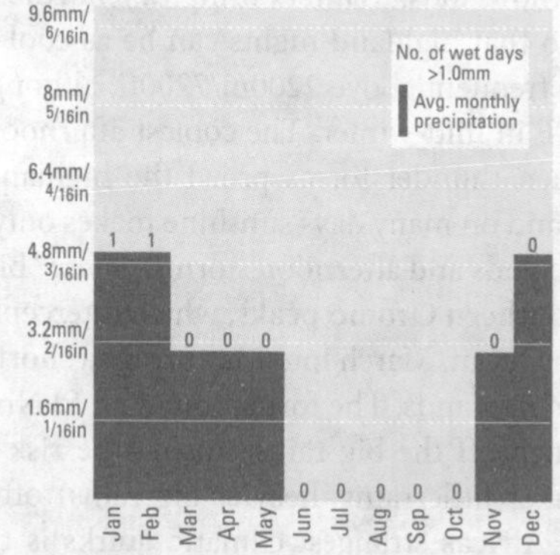

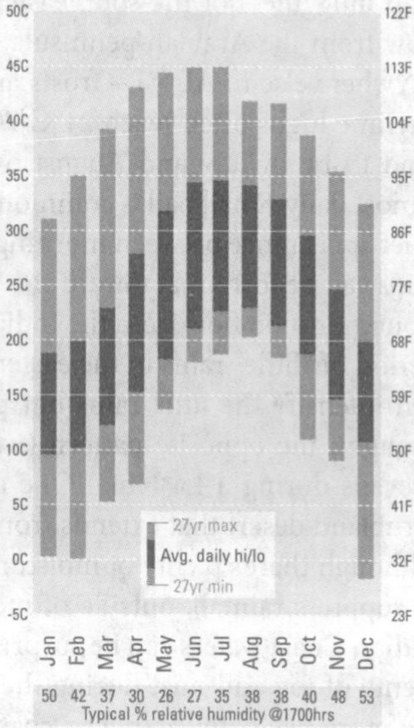

	Jan	Feb	Mar	Apr	May	Jun	Jul	Aug	Sep	Oct	Nov	Dec
midmonth -sunrise	0651	0635	0605	0528	0601	0553	0603	0621	0639	0556	0619	0643
-sunset	1716	1743	1802	1822	1941	1957	1957	1936	1900	1724	1658	1656
avg hrs of sunshine	7	7	8	9	10	12	11	11	9	9	8	7
avg # days w/thunder	<1	<1	<1	<1	<1	0	<1	0	0	<1	<1	<1
w/snow	0	0	0	0	0	0	0	0	0	0	0	0

the desert, where they kick up little rain – but great clouds of dust – before moving across the Nile valley. These *khamsin* storms (*ghibili* in **Libya,** *irifi* in **Sudan**) may strike several times a month, often preceded by a day of picture-perfect weather. Temperatures during *khamsins* can exceed 45°C/113°F, and relative humidities may tumble below 5 percent. The few *khamsins* that strike in February and March are typically less superheated than their late-spring counterparts, although they may carry even more dust. A different kind of windstorm plagues **central Sudan** in the late spring and summer, as low-rainfall thunderstorms generate *haboobs*. These squally blasts (over a dozen strike during an average year at **Khartoum**) can blow at hurricane force for a half-hour or so ahead of a quick shot of rain. Otherwise, Sudan's weather fits the Sahel pattern: bone-dry in the north (as in virtually all of Libya, except its Mediterranean coast), with a summer wet season that's progressively longer in duration toward the south. The average highs in blistering Khartoum exceed 38°C/100°F from March into October.

Ethiopia

Djibouti | Eritrea | Somalia

The horn of Africa has seen more than its share of agony, including decades of civil war and devastating drought. Visitors to **Ethiopia** may be surprised, then, at how agreeable its climate can be. The nation's western highlands escape both the sizzling heat of the northeast's **Danakil Desert** and the non-stop mugginess of the southeastern lowlands. Winters bring dry northeast flow from the Arabian peninsula, so that highland nights can be as cool as anywhere else in Africa – frosts are frequent above 2200m/7200ft, although daytime highs often reach 21°C/70°F in mid-winter. The coolest afternoons tend to be in July and August, when thunderstorms prowl the highlands almost daily. Small hail is common, and on many days sunshine makes only a brief appearance between morning clouds and afternoon storms. These "big" rains can be torrential toward the **Southern Oromo peaks**, which intercept a stout monsoon flow from the Indian Ocean. March into May brings a shorter period of "little" rains to the eastern highlands. The formation of an **El Niño** can intensify the little rains but pinch off the big rains, raising the risk of drought; the opposite pattern (weaker little rains, heavier big rains) often prevails during a **La Niña**. One of Africa's strangest climatic quirks is the scrubland desert that extends from eastern Ethiopia to the coast of **Somalia**. Although there's plenty of moisture around, the prevailing upper winds tend to suppress rainfall, outside of modest showers as the ITCZ passes by. The **Indian Coast** waters can be surprisingly cool due to seasonal upwelling, with plenty of fog and mist during the summer. The **Gulf of Aden** and **Red Sea** are more like bathwater. The coastal towns of northern Ethiopia, as well as

Addis Ababa

8°59'N, 38°48'E
Elev: 2355m (7724ft)

The average temperatures in this high-altitude city are close to 16°C/60°F year round, with a wider daily range during the dry, sunny winter. Showers occur in spring, and thunderstorms strike on many afternoons in July and August. When El Niño takes shape, the summer rains typically diminish but spring rains intensify.

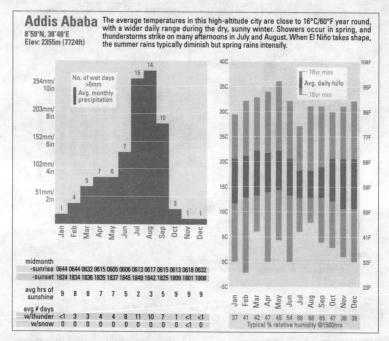

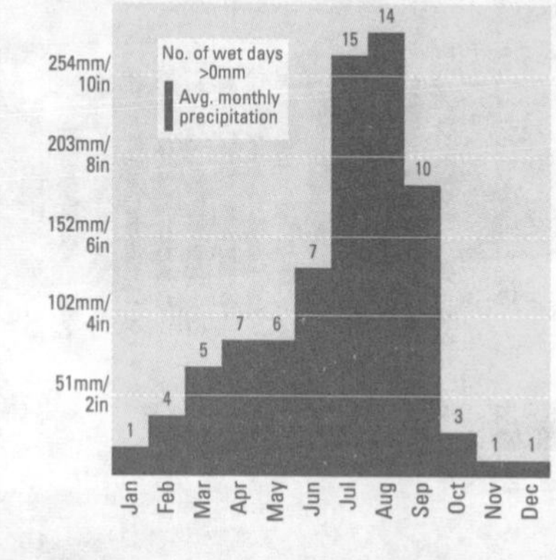

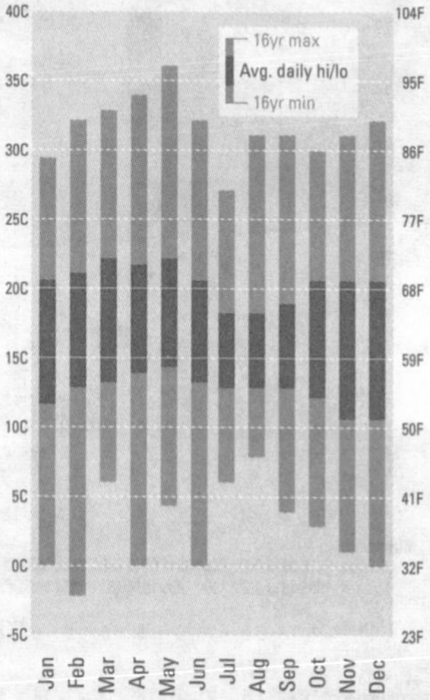

	Jan	Feb	Mar	Apr	May	Jun	Jul	Aug	Sep	Oct	Nov	Dec
midmonth -sunrise	0644	0644	0632	0615	0605	0606	0613	0617	0615	0613	0618	0632
-sunset	1824	1834	1836	1835	1837	1845	1849	1842	1825	1809	1801	1808
avg hrs of sunshine	9	8	8	7	7	5	2	3	5	9	9	9
avg # days w/thunder	<1	3	3	4	4	8	11	10	7	1	<1	<1
w/snow	0	0	0	0	0	0	0	0	0	0	<1	0
Typical % relative humidity @1500hrs	37	41	42	47	45	54	68	68	65	47	38	38

those of **Djibouti** and **Eritrea**, tend to simmer year round. A few rainy spells in winter are induced by a narrow zone of converging winds along the coast. Otherwise, imagine Saharan heat combined with tropical humidity and you get the picture.

Ghana

Benin | Ivory Coast | Liberia | Togo

Ghana's climate is the most distinctive among the small nations clustered along the **Guinea coast** of West Africa. The land is distinctive, too: huge tracts of virgin rainforest have disappeared, with more savanna than you'd expect so far south. Yet even without the hand of humanity, the ecosystem of Ghana, along with nearby **Togo** and **Benin**, wouldn't be the same as that of their neighbours. The Guinea coast angles from southwest to northeast from Cape Three Points to Lomé. The prevailing southwest winds diverge as they parallel the coast, and cool ocean waters rise to the surface in the summer. Thus, **Accra**'s rainfall averages less than 750mm/30in a year – August is as dry and mild as January – and a stiff sea breeze year round helps keep temperatures a touch cooler than in nearby cities like Lagos or Abidjan. The

Accra
5°36'N, 0°10'W
Elev: 69m (226ft)

A cool offshore current helps keep Accra drier than much of the Guinea coast, especially in July and August, as towering storms build just inland. Even in the wet spring and autumn, the majority of days see no rain. Year-round humidity is accompanied by a stout sea breeze. Harmattan haze may reach the coast in mid-winter.

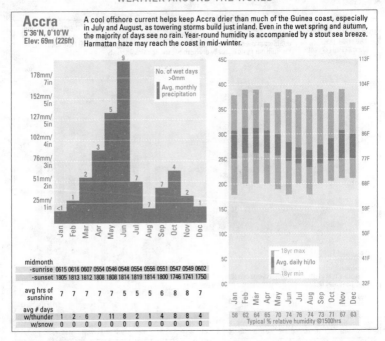

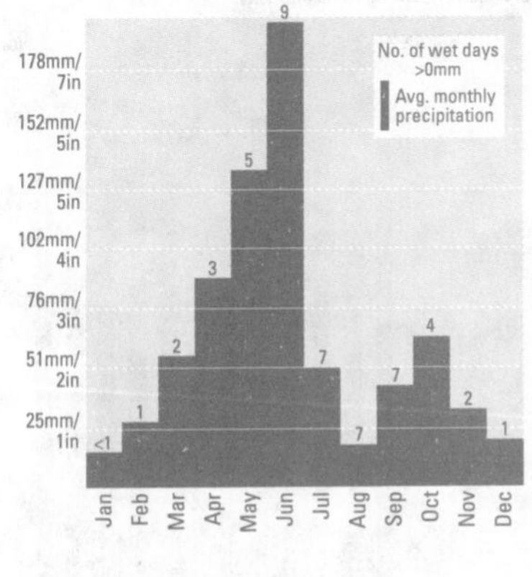

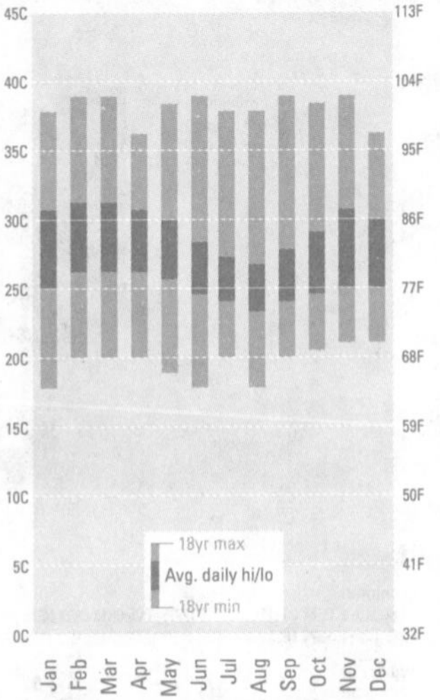

	Jan	Feb	Mar	Apr	May	Jun	Jul	Aug	Sep	Oct	Nov	Dec
midmonth -sunrise	0615	0616	0607	0554	0546	0548	0554	0556	0551	0547	0549	0602
-sunset	1805	1813	1812	1808	1808	1814	1819	1814	1800	1746	1741	1750
avg hrs of sunshine	7	7	7	7	7	5	5	5	6	8	8	7
avg # days w/thunder	1	2	6	7	11	8	2	1	4	8	8	4
w/snow	0	0	0	0	0	0	0	0	0	0	0	0

	Jan	Feb	Mar	Apr	May	Jun	Jul	Aug	Sep	Oct	Nov	Dec
Typical % relative humidity @1500hrs	58	62	64	65	70	74	76	74	73	71	67	63

rains are almost twice as heavy only an hour's drive inland, and this more-standard regime covers most of Ghana, the **Ivory Coast** and **Liberia**, with southern Togo and Benin getting just a bit less rain. The ITCZ brings a single wet season in summer to the north, but spring and autumn rains across the south. From late autumn through winter, visibility-slashing *harmattan* dust storms may interfere with travel and make sightseeing difficult. They're most frequent across the northern parts of the Guinea-coast nations; they can extend as far as the coast on a few days each winter, especially in Togo and Benin, although each year varies greatly.

Kenya

Uganda

Generalizing about Kenyan weather is a dangerous practice. Given the highlands of the west, the semi-arid lowlands, the moist coast and the volcanic peaks of Kirinyaga and – just across the border – Kilimanjaro, distinct climate zones abound. Even climatology is of limited use, since a hallmark of Kenyan rainfall is its year-to-year variability. Much of this equatorial nation gets two wet seasons associated with the northward (March–May) and southward (October–December) progression of the ITCZ. During an **El Niño**, the

Mombasa

4°02'S, 39°37'E
Elev: 55m (180ft)

Any time of year is beach weather in Mombasa, with warmth and sea breezes the norm. Temperatures are hottest and nights sultriest during the January–March dry season. Showers are scattered through the rest of the year, with the heaviest downpours occuring in April and (especially) May.

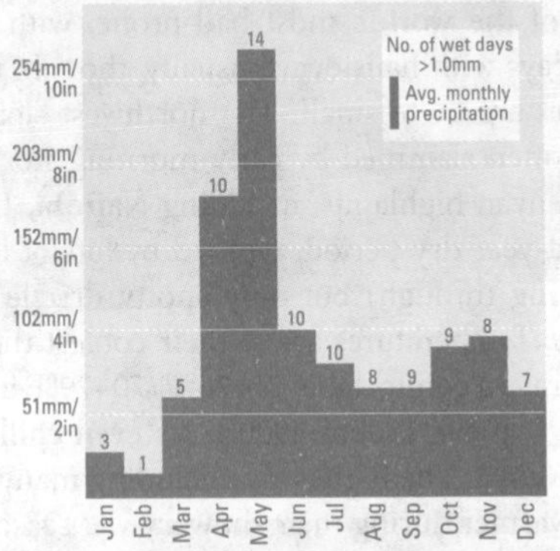

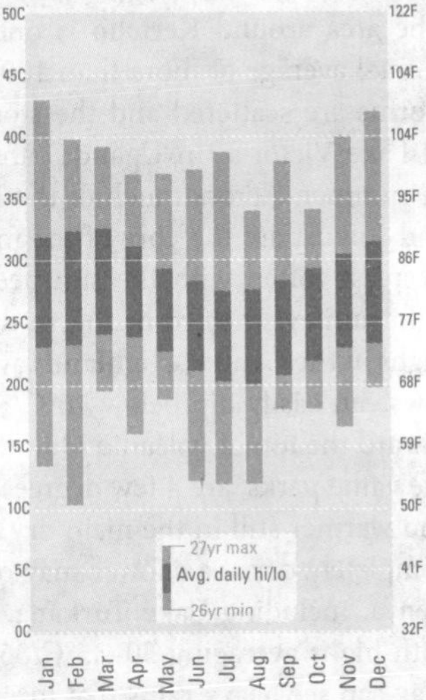

	Jan	Feb	Mar	Apr	May	Jun	Jul	Aug	Sep	Oct	Nov	Dec
Typical % relative humidity @1500hrs	61	59	60	66	70	67	67	66	65	66	67	66
midmonth -sunrise	0621	0629	0627	0621	0620	0625	0630	0627	0614	0602	0557	0606
-sunset	1841	1843	1834	1822	1816	1819	1825	1825	1819	1813	1815	1827
avg hrs of sunshine	9	9	9	8	7	7	7	8	8	9	9	8
avg # days w/thunder	1	1	2	3	1	<1	<1	0	<1	<1	1	2
w/snow	0	0	0	0	0	0	0	0	0	0	0	0

Nairobi

1°19'S, 36°55'E
Elev: 1624m (6327ft)

Temperatures are spring-like year round in this elevated city. The average is near 20°C/68°F, with little serious heat or cold. There are two short, well-defined wet seasons from April–May and November–December, when sunshine is interspersed with showers or thunder. July and August are a bit drier than January–March, but much cloudier.

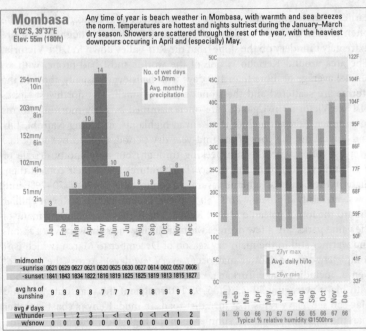

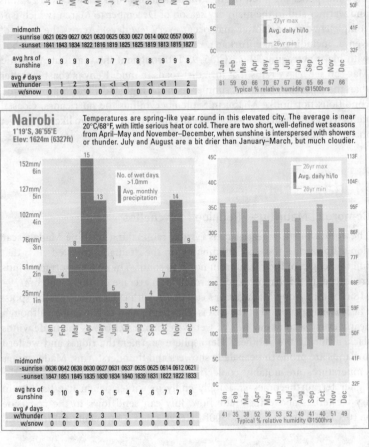

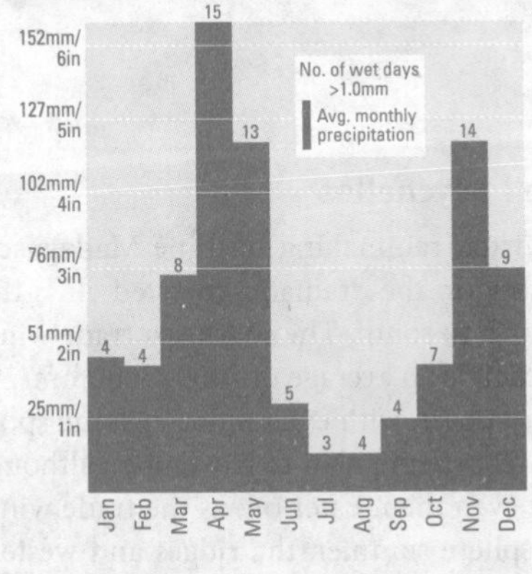

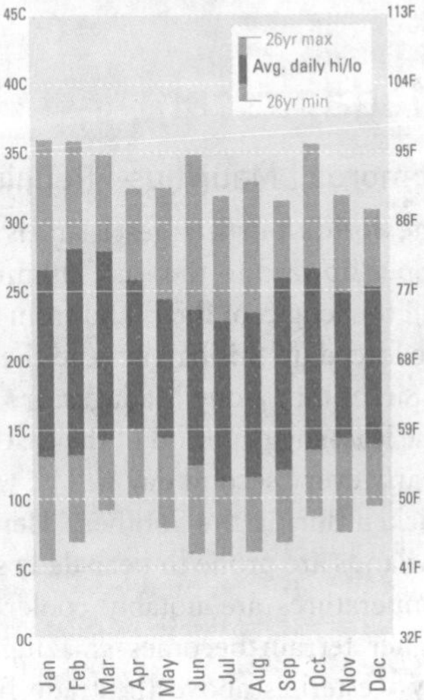

	Jan	Feb	Mar	Apr	May	Jun	Jul	Aug	Sep	Oct	Nov	Dec
Typical % relative humidity @1500hrs	41	35	38	52	56	53	52	49	41	40	51	49
midmonth -sunrise	0636	0642	0638	0630	0627	0631	0637	0635	0625	0614	0612	0621
-sunset	1847	1851	1845	1835	1830	1834	1840	1839	1831	1822	1822	1833
avg hrs of sunshine	9	10	9	7	6	5	4	4	6	7	7	8
avg # days w/thunder	1	2	2	5	3	1	1	1	1	1	2	1
w/snow	0	0	0	0	0	0	0	0	0	0	0	0

earlier rainy season may be paltry but the latter one enhanced; the opposite often applies during **La Niña**. Afternoons from June through August are extremely thundery on the slope from the **Rift Valley** down to **Lake Victoria**. The area around **Kericho** is one of the world's most hail-prone, with an annual average of more than 100 days with hailstorms; usually, though, the storms are scattered and the stones are quite small. The northwest slopes of Lake Victoria (in **Uganda**) are often slammed by early-morning storms in summer. Across much of the Kenyan highlands, including **Nairobi**, July and August are the core of the mid-year dry period, typified by lots of low overcast (often with the sun peeking through) but only spotty drizzle or light rain every third or fourth day. Temperatures are at their coolest then: highs in the **Central Highlands** average a comfortable 21–26°C/70–79°F, and lows can easily dip below 10°C/50°F above 1800m/5900ft. It's even chillier toward the top of volcanic **Mount Kenya**. The highlands, including many of the game parks, are a few degrees warmer during their showery wet seasons and warmer still in the main dry season of December to March (which isn't completely dry – a shower may occur every few days). Across the north of Kenya, including **Lake Turkana**, conditions are much more arid and hot, with highs averaging 30–35°C/86–95°F throughout the year, but April and May can see heavy rains that turn dirt roads to mud. **Kenya's coast** is wetter than the adjacent plains but sunnier than the highlands, with warm, humid conditions not unlike Miami in September, minus the hurricanes. Coastal rainfall peaks sharply from April to early June, when it's more likely to be wet than dry on any given day.

Madagascar

Comoros | Mauritius | Réunion | Seychelles

The world's fourth largest island is a virtual rainmaking machine. **Madagascar** slopes upward to the east, culminating in the dramatic, forested cliffs that run the length of the island from north to south. The southeast trade winds are forced up this barrier, and the result is an average of 2000–3000mm/79–118in of rain along Madagascar's east coast, with even more in some spots. The least soggy period in **the east** is from September to November, although nearly every second day will be wet even then. Later on, as the trade winds slacken during the Southern Hemisphere summer, the ridges and western lowlands are prone to near-daily showers and thunderstorms. Madagascar's temperatures are notably cooler than on most tropical islands, and the higher terrain becomes amazingly chilly in winter, when it often sits in dry westerlies above the trade flow. July can send lows to the cold side of 10°C/50°F above 1200m/3900ft. Heavy rains – especially from tropical

Antananarivo

18°48'S, 47°29'E
Elev: 1276m (4185ft)

In the midst of Madagascar's central plateau, Antananarivo gets most of its rain from November–March, with heavy summer showers (many thundery) almost every other day. Temperatures are strikingly cool for the latitude during the plateau's mostly dry winter. Some nights dip below 10°C/50°F in Antananarivo, even chillier at altitude.

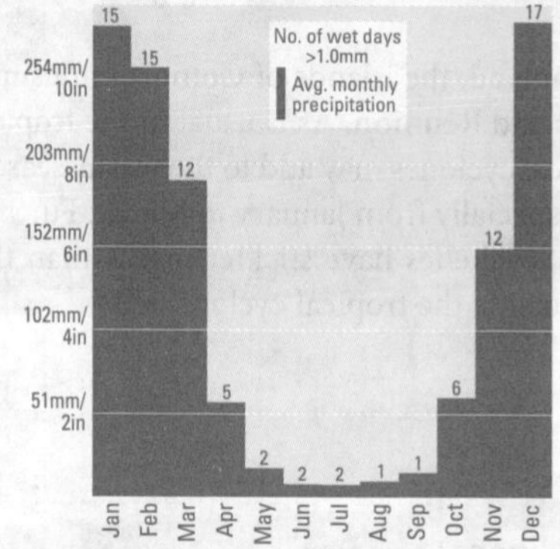

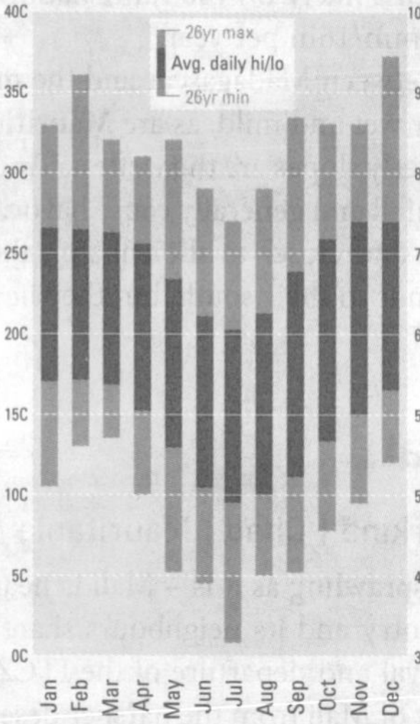

	Jan	Feb	Mar	Apr	May	Jun	Jul	Aug	Sep	Oct	Nov	Dec
midmonth -sunrise	0525	0543	0553	0600	0609	0620	0623	0611	0546	0521	0505	0507
-sunset	1834	1825	1805	1740	1723	1721	1729	1739	1745	1751	1805	1823
avg hrs of sunshine	7	6	6	7	7	7	7	8	8	8	8	6
avg # days w/thunder	8	8	9	3	1	0	0	<1	1	6	11	12
w/snow	0	0	0	0	<1	0	0	0	0	0	0	0

	Jan	Feb	Mar	Apr	May	Jun	Jul	Aug	Sep	Oct	Nov	Dec
Typical % relative humidity @1500hrs	60	62	60	56	53	54	52	50	43	44	50	58

Plaisance

20°26'S, 57°40'E
Elev: 57m (187ft)

Drenching rains often interrupt the sunshine from January through May. Thunder isn't frequent, but tropical cyclones are a risk. It rains less often at Port Louis and along the northwest coast of Mauritius, especially during the dry season. Humidities are high year round, so the cooler May–November span is more comfortable.

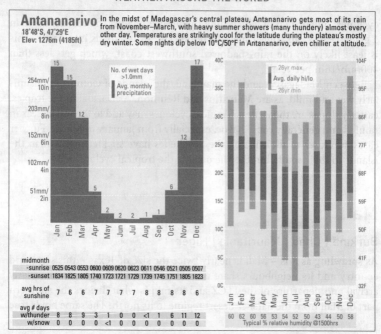

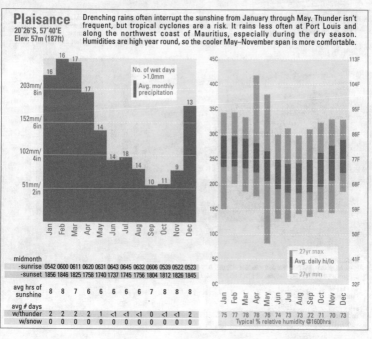

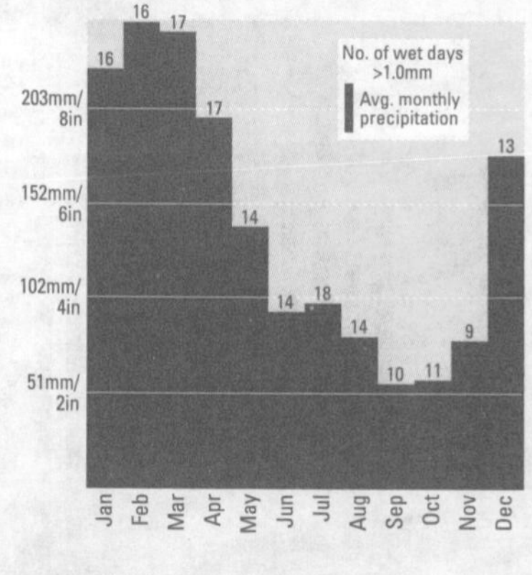

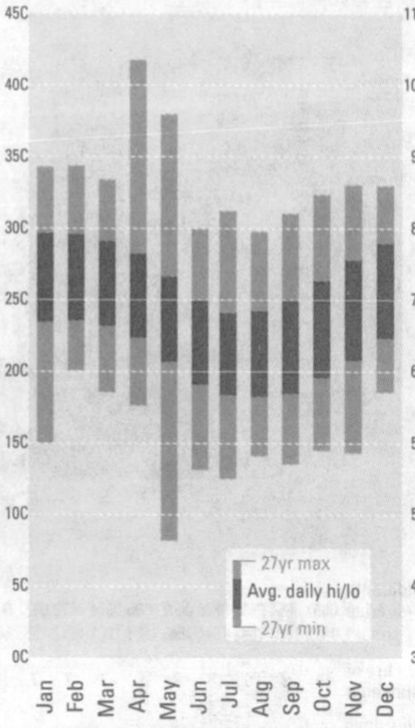

	Jan	Feb	Mar	Apr	May	Jun	Jul	Aug	Sep	Oct	Nov	Dec
midmonth -sunrise	0542	0600	0611	0620	0631	0643	0645	0632	0606	0539	0522	0523
-sunset	1856	1846	1825	1758	1740	1737	1745	1756	1804	1812	1826	1845
avg hrs of sunshine	8	8	7	6	6	6	6	6	7	8	8	8
avg # days w/thunder	2	2	2	2	1	<1	<1	<1	0	<1	<1	2
w/snow	0	0	0	0	0	0	0	0	0	0	0	0

	Jan	Feb	Mar	Apr	May	Jun	Jul	Aug	Sep	Oct	Nov	Dec
Typical % relative humidity @1600hrs	75	77	78	78	76	74	73	73	72	71	70	73

cyclones, one or two of which fling themselves against the eastern shore in a given year – can trigger massive floods and throw roads into chaos. This is least likely on the rain-shadowed **southwest coast**, which gets less than 400mm/16in per year.

Between Madagascar and the mainland, the islands of **Comoros** are similarly wet and mild, as are **Mauritius** and **Réunion**. As is usual in the tropics, easterly slopes are the wettest. Tropical cyclones may add to the warm-season rainfall and generally cause havoc, especially from January to March. Further east and closer to the equator, the **Seychelles** have stickier nights than the islands to their south, but they lie outside the tropical cyclone belt.

Mali

Burkina | Chad | Mauritania | Niger

As sprawling as it is – Mali is nearly twice the size of Texas – this semi-arid country and its neighbours share a common climatic language. The yearly arrival and departure of the ITCZ and its rain is what divides the southern part of Mali from the harsher desert regime to its north. The same holds true

Bamako

12°32'N, 7°57'W
Elev: 381m (1250ft)

In classic Sahel fashion, Bamako endures blazing heat from March to June, as humidity rises and thunderstorms becomes more frequent. Intense mid-summer squall lines sweep through from the east every two or three days. November through March is bone dry, with spells of strong harmattan wind and blowing dust that can last for days.

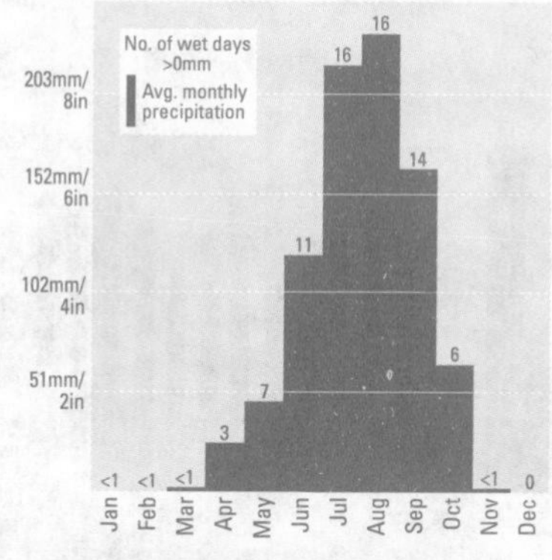

	Jan	Feb	Mar	Apr	May	Jun	Jul	Aug	Sep	Oct	Nov	Dec
midmonth -sunrise	0657	0654	0639	0620	0607	0606	0614	0620	0621	0622	0630	0645
-sunset	1825	1838	1842	1844	1849	1858	1901	1852	1833	1813	1803	1809
avg hrs of sunshine	9	9	9	8	8	8	7	7	7	8	9	9
avg # days w/thunder	<1	0	<1	3	8	11	14	15	12	7	1	<1
w/snow	0	0	0	0	0	0	0	0	0	0	0	0

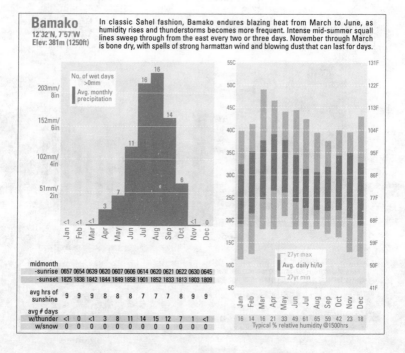

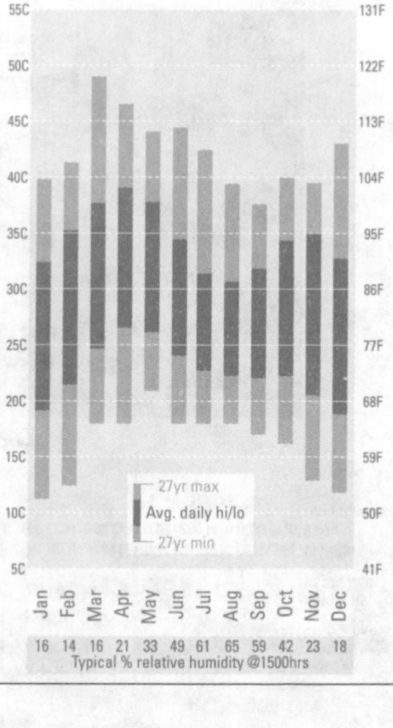

for **Niger** and **Chad** to the east and most of **Mauritania** to the west. The slow onset of the **monsoon** begins in the spring across the southern Sahel with a tantalizing hint of rain in the midst of oppressively hot and increasingly muggy weather (highs often top 38°C/100°F). The north is less humid but even more blazingly hot. By mid-summer, the tongue of moist air marking the ITCZ has pushed up to roughly 20°N. During July and August, locations south of 13°N – including southern Mali and **Burkina** – feel much more like the humid tropics, with rains every day or two and slightly more tolerable heat than in the spring. The belt from around 13° to 18°N remains scorching, but has just enough moisture for spotty showers and thunderstorms, which might total over 150mm/6in in a good summer or almost nothing in a bad year (like those of the 1970s and 1980s that ravaged the Sahel with drought). Most of the summer rains across the Sahel are delivered by easterly waves and squall lines. These great bands of storms originate after easterly flow is funnelled through the mountains of Chad and encounters the humid south-westerlies south of the ITCZ. Each line of storms brings a quick, dramatic bout of gale-force wind and heavy rain, followed by a pleasant cooldown for a few hours. The wet season ends quickly across the Sahel, and December and January can be a treat: rain-free, cool at night, and warm by day. However, if the prevailing northeast winds intensify beyond about 55kph/34mph, they can generate enough dust to produce winter's down side – the irritating **harmattan** (see box, p.299). *Harmattan* dust sometimes extends all the way to coastal Mauritania, where sea breezes normally take a slight edge off the year-round heat.

Morocco

Algeria

A Spaniard need only cross the Strait of Gibraltar to arrive in **Morocco**, so it naturally follows that the **north coast** of this mountainous nation shares a similar Mediterranean climate, with bright, warm summer days and a cool winter laced with rainy spells every two or three days. The story's much the same inland to the **Atlas Mountain foothills** and along the **Atlantic coastline** down past **Essaouira** to **Cap Rhir**. If anything, the Atlantic coast is even more temperate than the Mediterranean: the winter rains are lighter, and it's somewhat milder in winter and cooler in summer, when sea fog can sweep in. The big caveat is the influence of the **Sahara**, lurking just over the Atlas range. When Saharan air pours over the mountains on a southerly wind, the weather quickly turns dry and hot – very hot, at times. In July, when the average highs barely top 26°C/79°F along the coast, a dust-laden *chergui* wind can push temperatures to above 43°C/109°F. By contrast, the **High Atlas** can be as wintry as any other high peaks, with howling winds and snow that can

Casablanca

33°34'N, 7°04'W
Elev: 62m (203ft)

Winter or summer, Casablanca's average temperatures are quite moderate, although hot or cold excursions aren't unknown. A light rain (with snow on high peaks) is typical about every third day from November to March. Other than a touch of morning fog, summer is sunny, with temperatures hotter further inland.

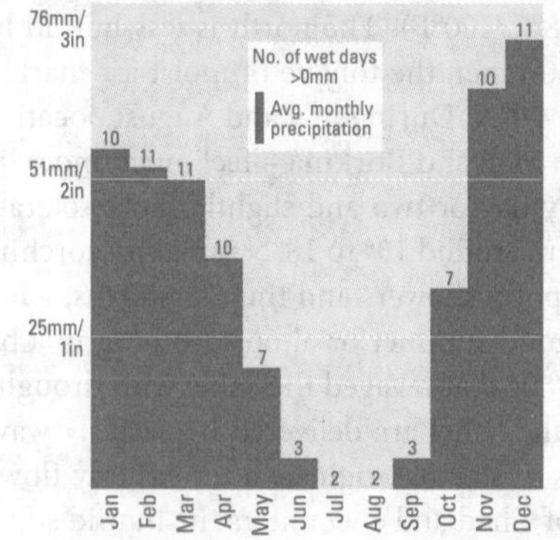

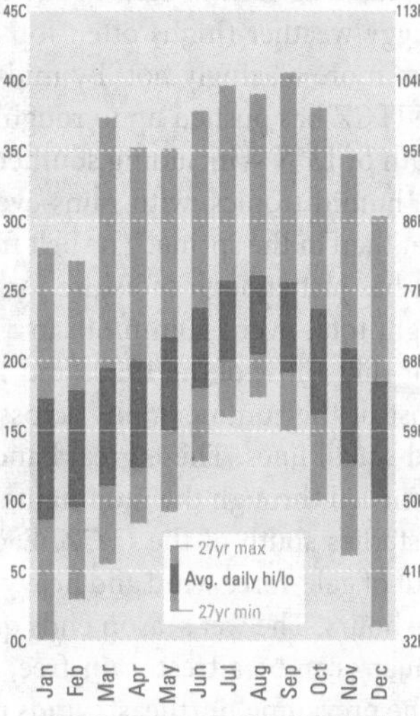

	Jan	Feb	Mar	Apr	May	Jun	Jul	Aug	Sep	Oct	Nov	Dec
Typical % relative humidity @1600hrs	71	72	71	72	71	74	74	74	73	71	73	73

midmonth -sunrise	0636	0616	0543	0502	0431	0421	0432	0454	0515	0536	0603	0628
-sunset	1646	1716	1739	1802	1825	1843	1843	1818	1739	1658	1629	1625
avg hrs of sunshine	6	7	8	9	9	10	10	9	9	8	6	6
avg # days w/thunder	1	1	1	1	1	1	<1	1	1	1	1	1
w/snow	0	0	<1	0	0	0	0	0	0	0	0	0

Marrakesh

31°37'N, 8°02'W
Elev: 466m (1528ft)

Marrakesh is far enough from the coast to make its annual rainfall half that of Casablanca's, although spring can be showery. Mid-summer brings intense (even dangerous) heat and humidity, with a dry thunderstorm once or twice a month. Conditions may be quite different at altitude; snow can pummel the High Atlas in winter.

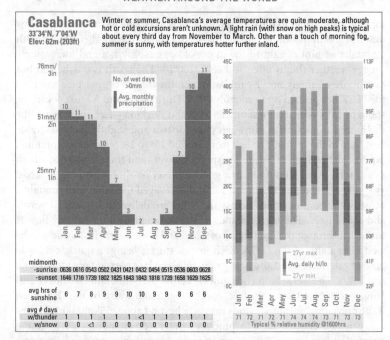

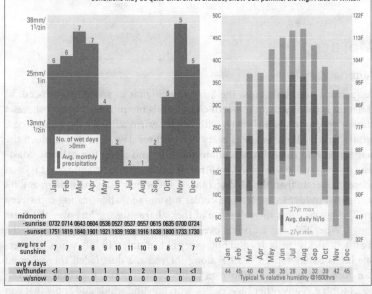

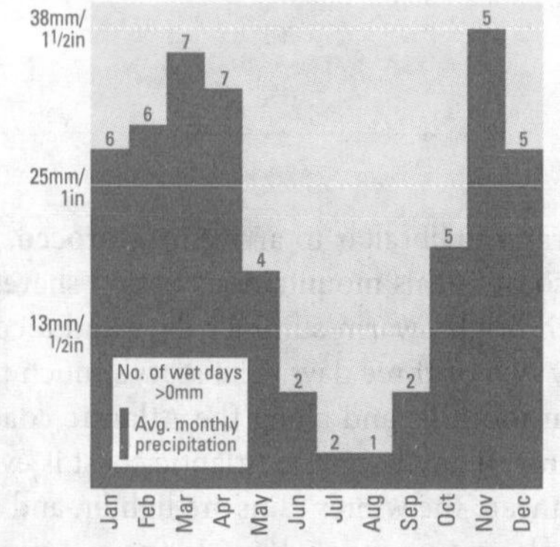

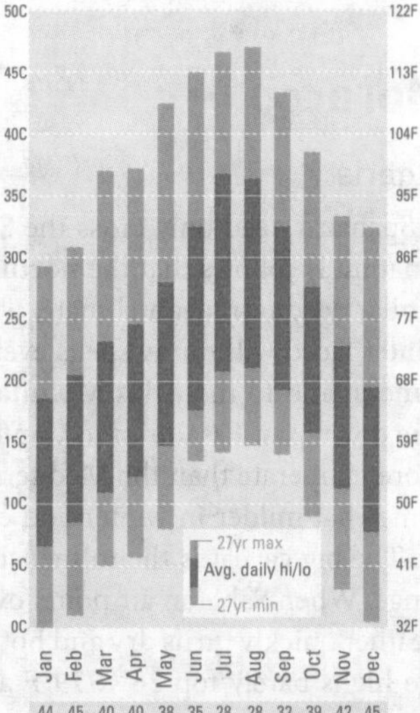

	Jan	Feb	Mar	Apr	May	Jun	Jul	Aug	Sep	Oct	Nov	Dec
Typical % relative humidity @1600hrs	44	45	40	40	38	35	28	28	32	39	42	45

midmonth -sunrise	0732	0714	0643	0604	0536	0527	0537	0557	0615	0635	0700	0724
-sunset	1751	1819	1840	1901	1921	1939	1938	1916	1838	1800	1733	1730
avg hrs of sunshine	7	7	8	8	9	10	11	10	9	8	7	7
avg # days w/thunder	<1	1	1	1	1	1	1	2	1	1	1	<1
w/snow	0	0	0	0	0	0	0	0	0	0	0	0

block passage for days. The Moroccan regime holds true in **Algeria**, except, of course, that there's far more land on the hot side of the Atlas.

Namibia

Angola

If it's dry land you crave, try **Namibia**. Featuring not one, but two types, of desert, temperatures here are still remarkably comfortable. Most of the nation is a high, sunny plateau, which keeps summer readings somewhat in check. Winter mornings are crisp, but the afternoons typically warm to well above 16°C/60°F. A stray winter shower may reach the far north, but otherwise, Namibia's only noteworthy rains fall in summer, when thundershowers strike every few days along the escarpment, seldom bringing heavy downpours. On the seaward side of the escarpment is a classic example of that global rarity – a **coastal desert**. The cold upwelling in the Benguela current helps keep the air stable year round. Some parts of the desolate coast are fog-shrouded more than half the year; they actually get more water from the fog itself than from their scant rainfall. Even the **Namib dunes** as far as 100km/60 miles inland are fogged over on as many as sixty days each year, when the fog may

Windhoek
22°34'S, 17°06'E
Elev: 1725m (5658ft)

Unlike the coastal desert to its west, Windhoek normally gets a few doses of summer rain. Temperatures may rise well above the mild norm or dip to chilly levels, but humidity is seldom high. The parched winter gets sunny days and cool nights, and the mostly dry spring is just as warm and bright as summer.

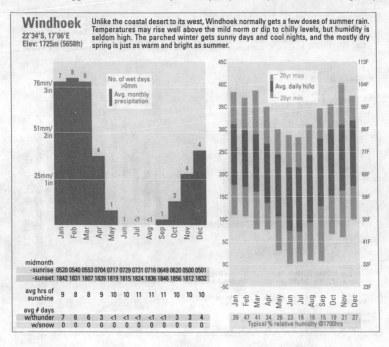

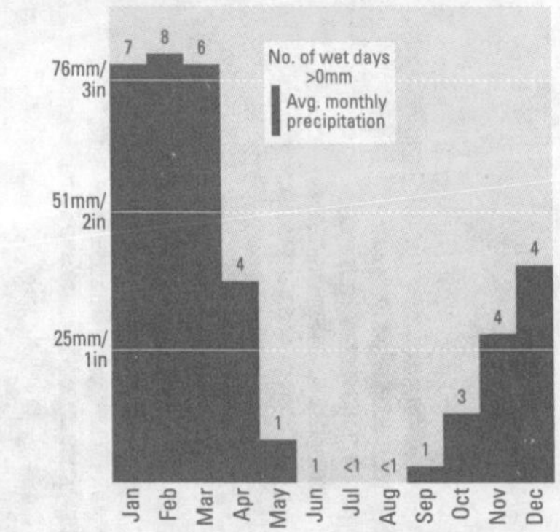

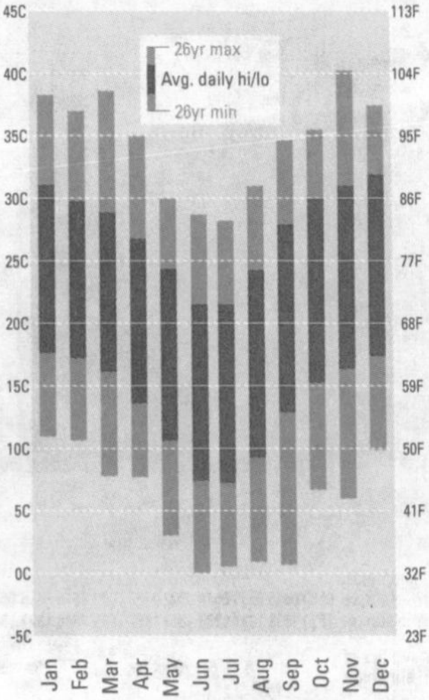

	Jan	Feb	Mar	Apr	May	Jun	Jul	Aug	Sep	Oct	Nov	Dec
midmonth -sunrise	0520	0540	0553	0704	0717	0729	0731	0716	0649	0620	0500	0501
-sunset	1842	1831	1807	1839	1819	1815	1824	1836	1846	1856	1812	1832
avg hrs of sunshine	9	8	8	9	10	10	11	11	11	10	10	10
avg # days w/thunder	7	6	6	3	<1	<1	<1	<1	<1	3	3	4
w/snow	0	0	0	0	0	0	0	0	0	0	0	0
Typical % relative humidity @1700hrs	39	47	41	34	26	23	19	16	15	19	21	27

produce *motreën* (drizzle). Coastal temperatures seldom get below 8°C/46°F or above 25°C/77°F range, except when hot downslope *berg* winds arrive from the interior. The coastal desert extends into southern **Angola**; rainfall becomes more prevalent as you continue north and east. The interior of Angola gets a fair number of thundershowers, largely in the summer.

Nigeria

Africa's most populous nation, **Nigeria** has a climate regime similar to that of the Guinea-coast nations to its west, as the ITCZ brings wet weather northward from March to May and southward from October to December. Only the extreme north is semi-arid in classic Sahel style, while parts of the **lower Niger delta** are drenched with more than 3000mm/118in of rain each year. The **Lagos** area is the least sodden stretch of coastline, though its humidity – and the heat radiating from urban sprawl – help maintain a sticky atmosphere year round. The immediate coast experiences a major dry period in winter with heaviest rains in late spring and early summer, followed by a brief dry spell around August and a second pulse of rain into October. **Harmattan** dust storms are a perennial winter problem, especially toward the north.

Kano

12°02'N, 8°31'E
Elev: 476m (1563ft)

Temperature swings are far greater here than on the Nigerian coast. The dry winter features coolish nights, warm days, and spells of harmattan wind and dust. Serious heat sets in by March, and wilting humidity builds in the spring. Intense thunderstorms are frequent in mid-summer. Daytime heat returns in autumn, but evenings are cooler.

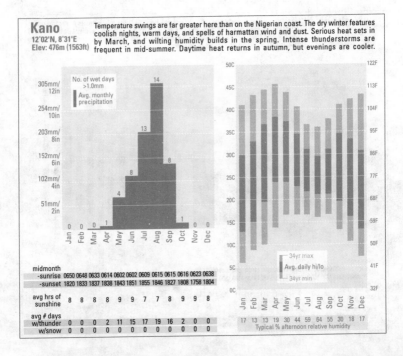

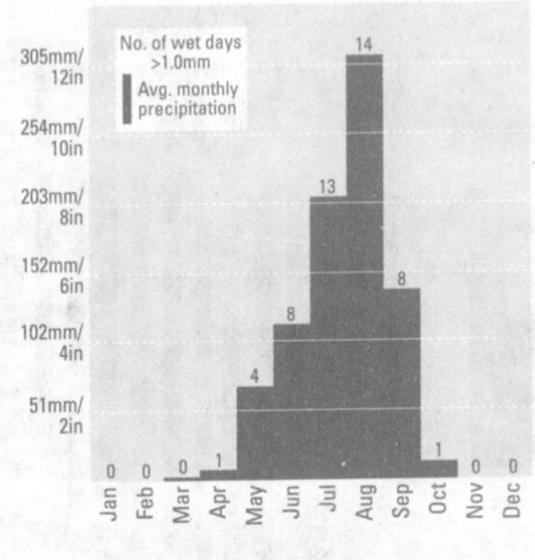

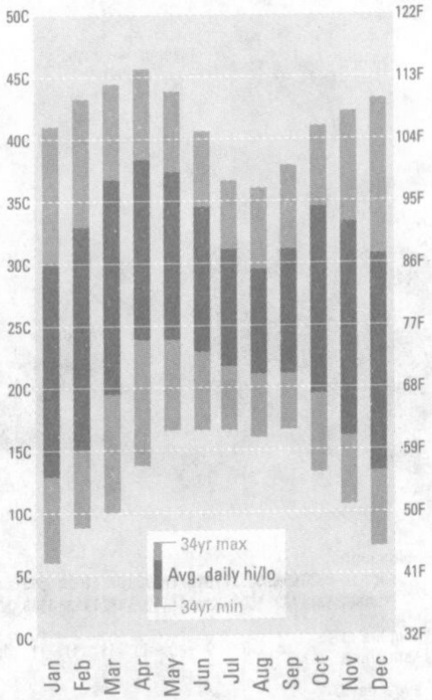

	Jan	Feb	Mar	Apr	May	Jun	Jul	Aug	Sep	Oct	Nov	Dec
midmonth -sunrise	0650	0648	0633	0614	0602	0602	0609	0615	0615	0616	0623	0638
-sunset	1820	1833	1837	1838	1843	1851	1855	1846	1827	1808	1758	1804
avg hrs of sunshine	8	8	8	8	9	9	7	7	8	9	9	8
avg # days w/thunder	0	0	0	2	11	15	17	19	16	2	0	0
w/snow	0	0	0	0	0	0	0	0	0	0	0	0
Typical % afternoon relative humidity	17	13	13	19	30	44	59	64	55	30	18	17

Lagos
6°35'N, 3°20'E
Elev: 38m (125ft)

Temperatures in Lagos are nothing if not dependable – it's usually 26–32°C/79–90°F around the clock any time of year. Humidity jumps from high to oppressive during the long wet season (May–October, with a short break in August), when there is lots of cloud and frequent drenchings, heavier toward the Niger Delta and lighter inland.

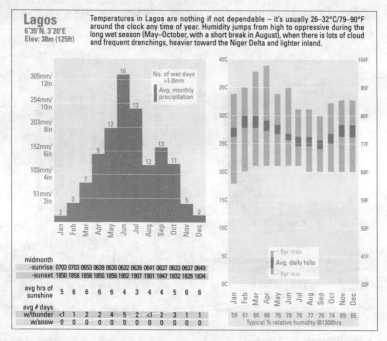

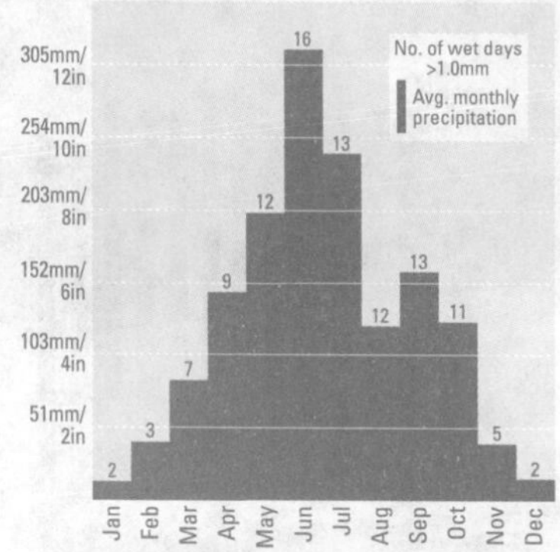

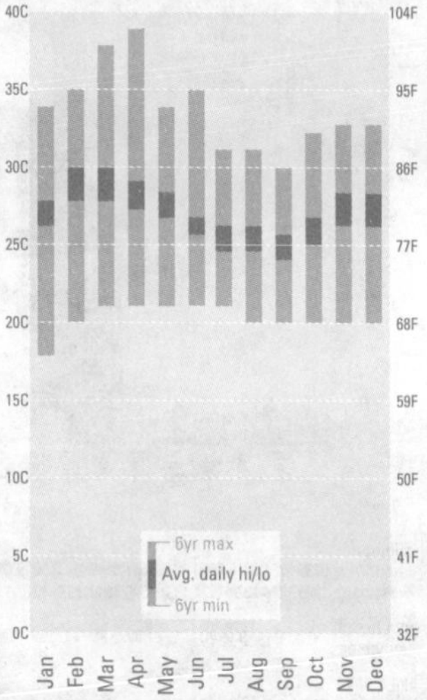

	Jan	Feb	Mar	Apr	May	Jun	Jul	Aug	Sep	Oct	Nov	Dec
midmonth -sunrise	0703	0703	0653	0639	0630	0632	0639	0641	0637	0633	0637	0649
-sunset	1850	1858	1858	1855	1856	1902	1907	1901	1847	1832	1826	1834
avg hrs of sunshine	5	6	6	6	6	4	3	4	4	5	6	6
avg # days w/thunder	<1	1	2	2	4	5	2	<1	2	3	1	1
w/snow	0	0	0	0	0	0	0	0	0	0	0	0

	Jan	Feb	Mar	Apr	May	Jun	Jul	Aug	Sep	Oct	Nov	Dec
Typical % relative humidity @1300hrs	59	61	66	68	76	78	78	77	76	74	69	65

Senegal

Cape Verde | The Gambia | Guinea | Guinea-Bissau | Sierra Leone

This is the end-point for the east-to-west march of weather systems across the Sahel. **Senegal**, as well as **The Gambia** (which it surrounds) is where easterly waves enter the Atlantic after dousing the region with intense rain and thunder. Outside of the June-to-October visit of the ITCZ and its easterly waves, Senegal is virtually rainless. Weather is hot year round, except on the northern coast, which juts into a cool Atlantic current. Africa's westernmost city, **Dakar**, is practically air-conditioned in the dry winter, averaging below 27°C/81°F by day, while much of the interior exceeds 32°C/90°F. More than three-quarters of Dakar's rain falls in August and September, although even then, most days remain dry. The Sahara influence extends to the **Cape Verde** islands 800km/500 miles offshore, which get virtually all their rain from August into early October. Further south, it's a far wetter story, with lots of cloud to boot. Average rainfall tops 1500mm/59in across nearly all of **Guinea** and **Guinea-Bissau**. The coastal reaches of Guinea and **Sierra Leone** average twice that amount – all during a single wet season, albeit a longer one as you move south. In **Freetown** (Sierra Leone), only winter could be considered dry, and the heat and humidity never really abate. The north Guinea high-

Dakar
14°44'N, 17°30'W
Elev: 24m (79ft)

It's easy to avoid most of Dakar's bad weather: stay away from late July to early October, when thundery tropical waves pass overhead every few days en route to potential new roles as Atlantic hurricanes. Otherwise, Dakar is almost rain-free and nicely temperate. Hot, dusty breezes occasionally arrive from the interior, especially in winter.

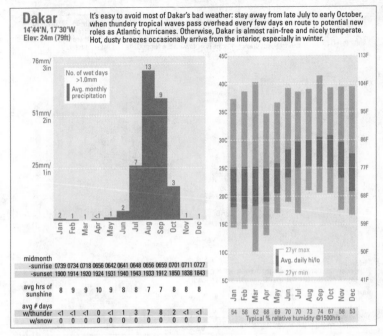

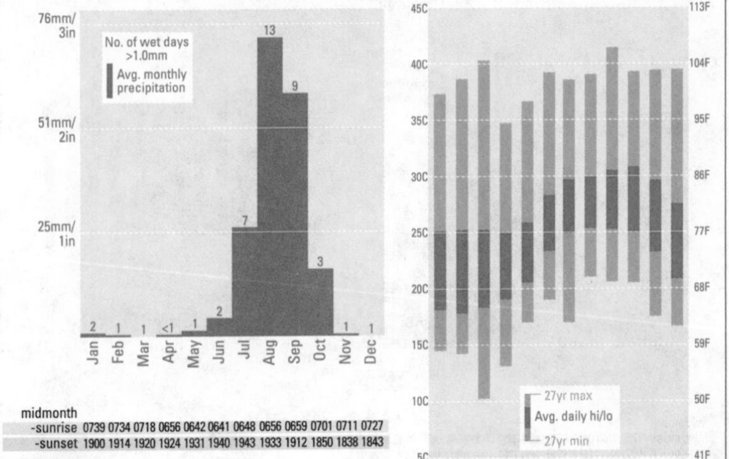

	Jan	Feb	Mar	Apr	May	Jun	Jul	Aug	Sep	Oct	Nov	Dec
midmonth -sunrise	0739	0734	0718	0656	0642	0641	0648	0656	0659	0701	0711	0727
-sunset	1900	1914	1920	1924	1931	1940	1943	1933	1912	1850	1838	1843
avg hrs of sunshine	8	9	9	10	9	8	8	7	7	8	8	8
avg # days w/thunder	<1	<1	<1	0	<1	1	3	7	8	2	<1	<1
w/snow	0	0	0	0	0	0	0	0	0	0	0	0
Typical % relative humidity @1500hrs	54	58	62	68	69	70	70	73	74	67	58	53

lands adjoining the Sahel are a bit less rainy and a touch cooler. Hail may accompany summer storms there.

South Africa

Lesotho | Swaziland

Hail, tornadoes, snowstorms: at first glance the climate of **South Africa** seems to resemble that of the US Midwest. In reality, the extremes here are more localized and not terribly common. Sunshine and moderate warmth are actually the most prevalent features of South African climate.

Jutting into the Atlantic, the southwest corner of the **Western Cape coast** – with **Cape Town** at its centre – has a Mediterranean regime. Its summers are dry and warm (a few notches cooler than, say, Italy or Tunisia) and the cool winters bring periods of rain and drizzle every few days, often during the night or early morning. Along the rest of the coast, from **Port Elizabeth** up to **Natal**, summer is the wetter time, with showers and thunderstorms about every third day on average from October to March. Temperatures warm as you go north, although beachgoers from Miami or Mykonos may find it chilly. Even in January, its warmest month, the **KwaZulu-Natal coast** often stays below 28°C/82°F by day, although its nights are balmy. The

Cape Town

33°59'S, 18°36'E
Elev: 42m (138ft)

In Cape Town's Mediterranean-type climate, winters are cool and variable; rainfall occurs about every third day. Summers are sunny, with gusty winds at exposed coastal spots. Only a few days exceed 32°C/90°F, usually thanks to a warm Bergwind. Watch for the "tablecloth" cloud that shrouds Table Mountain, most often in early summer.

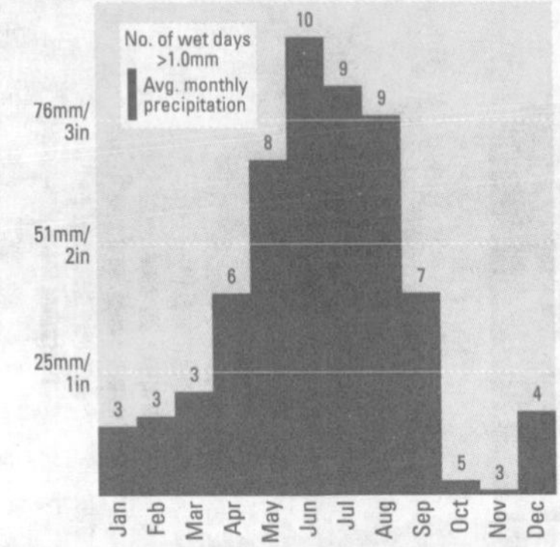

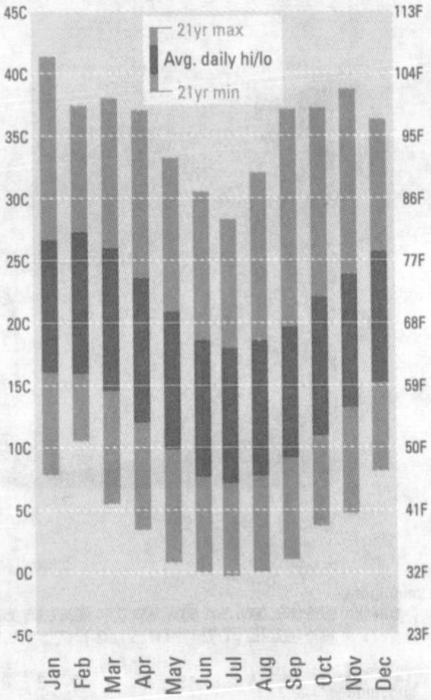

	Jan	Feb	Mar	Apr	May	Jun	Jul	Aug	Sep	Oct	Nov	Dec
midmonth -sunrise	0550	0620	0645	0708	0731	0749	0749	0725	0645	0604	0534	0529
-sunset	2000	1938	1904	1823	1753	1743	1754	1816	1837	1859	1927	1953
avg hrs of sunshine	11	11	9	8	7	6	6	7	7	9	10	11
avg # days w/thunder	<1	<1	1	1	1	<1	<1	<1	<1	<1	<1	<1
w/snow	0	0	0	0	0	0	<1	<1	0	0	0	0
Typical % relative humidity @1700hrs	60	60	63	67	69	70	70	67	65	63	61	60

Durban

29°58'S, 30°57'E
Elev: 8m (26ft)

Summers in Durban are warm and humid enough to feel sub-tropical. It rains about every third day, but Highveld-type thunderstorms are rare. Showers are less frequent in winter, when there's more temperature variation. The averages are mild yet offshore winds can bring a burst of dry heat or a cold snap.

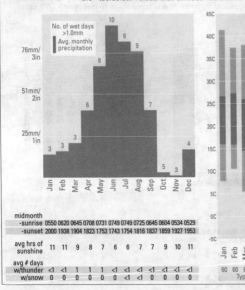

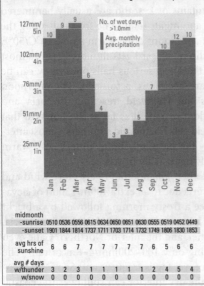

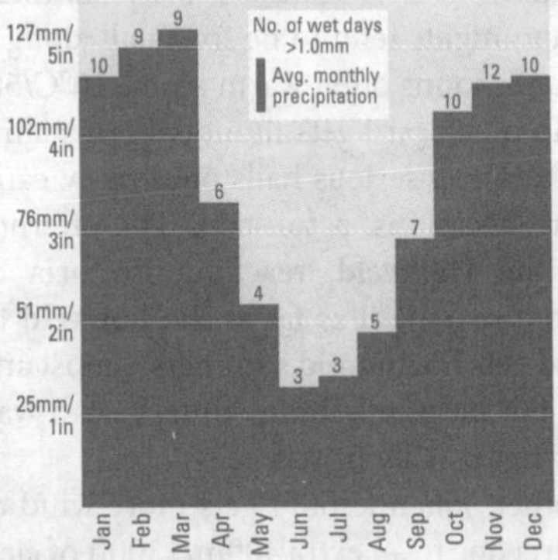

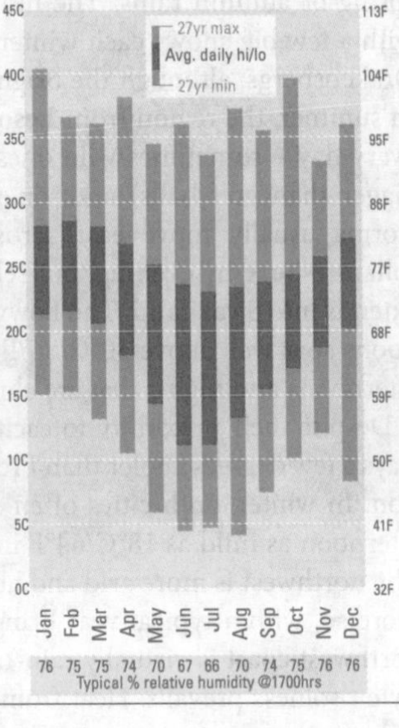

	Jan	Feb	Mar	Apr	May	Jun	Jul	Aug	Sep	Oct	Nov	Dec
midmonth -sunrise	0510	0536	0556	0615	0634	0650	0651	0630	0555	0519	0452	0449
-sunset	1901	1844	1814	1737	1711	1703	1714	1732	1749	1806	1830	1853
avg hrs of sunshine	6	6	7	7	7	7	7	7	6	5	6	6
avg # days w/thunder	3	2	3	1	1	1	1	1	2	4	5	4
w/snow	0	0	0	0	0	0	0	0	0	0	0	0
Typical % relative humidity @1700hrs	76	75	75	74	70	67	66	70	74	75	76	76

Johannesburg

26°08'S, 28°14'E
Elev: 1700m (5576ft)

Hail-bearing storms occur in and near Johannesburg on dozens of summer evenings each year, with flash flooding also a risk. Otherwise, summer weather is quite affable. Winters are mostly dry, with frost on some mornings but mild, bright afternoons. In nearby Pretoria, temperatures average about 3°C/5°F higher.

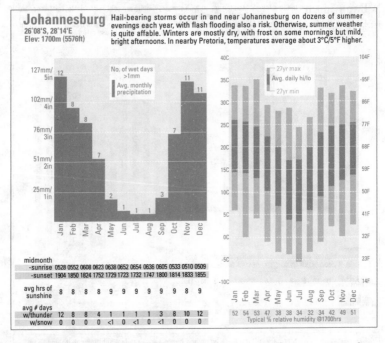

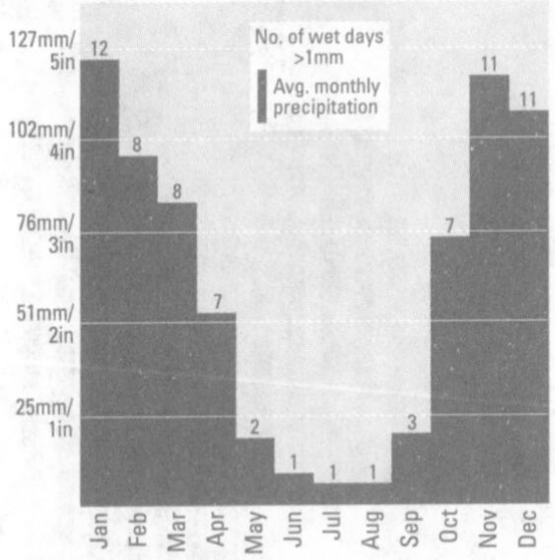

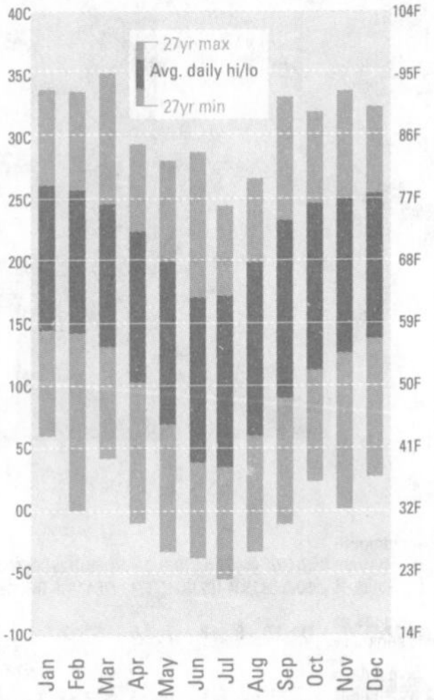

	Jan	Feb	Mar	Apr	May	Jun	Jul	Aug	Sep	Oct	Nov	Dec
midmonth -sunrise	0528	0552	0608	0623	0638	0652	0654	0636	0605	0533	0510	0509
-sunset	1904	1850	1824	1752	1729	1723	1732	1747	1800	1814	1833	1855
avg hrs of sunshine	8	8	8	8	9	9	9	9	9	9	8	9
avg # days w/thunder	12	8	8	4	1	1	1	1	3	8	10	12
w/snow	0	0	0	0	<1	0	<1	0	<1	0	0	0

Jan	Feb	Mar	Apr	May	Jun	Jul	Aug	Sep	Oct	Nov	Dec
52	54	53	47	38	38	34	32	34	42	49	51

Typical % relative humidity @1700hrs

Great Karoo and **Drakensberg mountains** separate the wetter coast from the quite arid interior. On occasion KwaZulu-Natal gets a heavy round of spring or autumn rains. The mountains of **Lesotho** are usually hammered with a few big snows each winter, and nights tend to be frosty all along the Drakensbergs, although the bright afternoons often warm above 10°C/50°F. In summer, the region from **Lesotho** northward gets thunderstorms almost every day – sometimes wild ones that drop serious hailstones (they can be bigger than golf balls) and, on rare occasions, a tornado. The afternoon storms usually move east across the **Highveld**, reaching **Pretoria** and **Johannesburg** by evening, but seldom making it as far as the **Lowveld** that extends into **Swaziland**. The Lowveld gets hot, humid summers – most afternoons soar well above 30°C/86°F – but pleasantly mild winters, with warm afternoons and nights that are chilly but usually frost-free.

Despite their proximity to each other, Johannesburg gets more wind and stays a few degrees cooler than Pretoria due to an extra 400m/1300ft of elevation. In winter, both cities often see a frosty morning followed by a bright afternoon as mild as 18°C/64°F in Johannesburg and 22°C/72°F in Pretoria. The northwest is more arid and hotter, and the interior **Northern Cape** even more so, with a typical year bringing only 100–200mm/4–8in of rain. The northwest coast is virtually rain-free, yet fog is frequent and San Francisco-style coolness prevails. Heat from the dry interior can spill over the coastal ranges through gusty downslope *Bergwinds* that can send readings above 30°C/86°F at the shore almost any time of year.

Tanzania

Burundi | Rwanda

Like Kenya to its north, **Tanzania** is blessed with delightful temperatures – largely as a result of altitude – but cursed by erratic rainfall that plays havoc with food supplies. The average rainfall does tend to run slightly higher than Kenya's; in both countries, summer drought can result when an El Niño is brewing. Most of the lowlands and the high plateau, including the **Serengeti**, average at least 500mm/20in of rain a year, mainly in the form of thundery showers that dot the land from November to May. Much more dependable summer rains fall where the prevailing southeast trades blow against mountains, especially around 2000–3000m/6600–9800ft in altitude. The south flank of **Kilimanjaro** sees over 3000mm/118in of precipitation a year. Summer thundershowers are also frequent and sometimes heavy along the east shores of **lakes Victoria** and **Tanganyika** (and north into **Burundi** and **Rwanda**), as the afternoon sea breeze bumps into the trade winds. Here, however, as in most of Tanzania, there's usually a good deal of sunshine.

Dar es Salaam
6°52'S, 39°12'E
Elev: 55m (180ft)

This coastal city averages a few days of rain each month. Rain is heaviest from March–May, often in the form of brief but intense showers and storms. Just offshore, Zanzibar and Pemba average twice as much rain per year. Heat and humidity are most intense prior to the wet season. The sea breeze is usually vigorous.

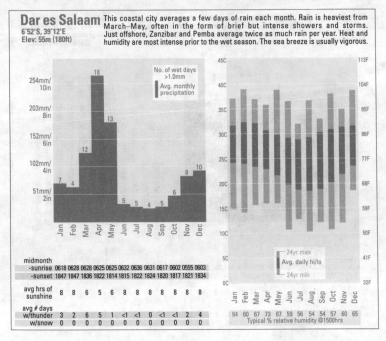

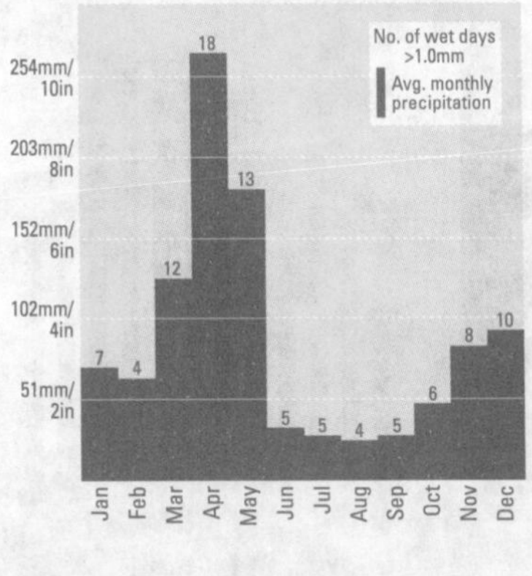

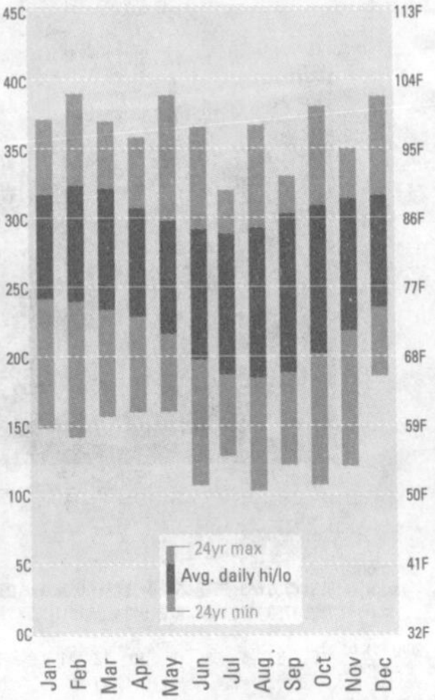

	Jan	Feb	Mar	Apr	May	Jun	Jul	Aug	Sep	Oct	Nov	Dec
Typical % relative humidity @1500hrs	64	60	67	73	67	59	56	54	54	57	60	65

	Jan	Feb	Mar	Apr	May	Jun	Jul	Aug	Sep	Oct	Nov	Dec
midmonth -sunrise	0618	0628	0628	0625	0625	0632	0636	0631	0617	0602	0555	0603
-sunset	1847	1847	1836	1822	1814	1815	1822	1824	1820	1817	1821	1834
avg hrs of sunshine	8	8	6	5	6	8	8	8	8	8	8	8
avg # days w/thunder	3	2	6	5	1	<1	<1	0	<1	<1	2	4
w/snow	0	0	0	0	0	0	0	0	0	0	0	0

Temperatures across the country seldom vary much at all from the mild-to-warm norms. Outside of the high country, almost every place experiences highs year round in the 25–31°C/77–88°F range. Readings are on the cooler end of this spectrum during the June-to-August dry season, but humidities stay fairly high and the risk of frost is low, except in the mountains and toward the southernmost highlands. On very rare occasions, a tropical cyclone moving toward Mozambique can bring gales and heavy rain to the **Tanzanian coast**, which runs warmer and wetter than the interior. Just offshore, the islands of **Zanzibar** and **Pemba** receive up to twice as much rain as the coastal city of **Dar es Salaam**, much of it early in the day.

Tunisia

This tiny but diverse nation echoes the landscape and climate of Algeria in miniature: a mountain barrier separates a scenic coast from a parched interior. The east-facing shore is more vulnerable than Morocco's to cool, rainy winter storms settling down from Europe and heat from the Sahara moving north. Apart from the occasional blast-furnace days induced by Saharan winds, the stretch from **Tunis** to **Sfax** enjoys weather similar to nearby Sicily. The

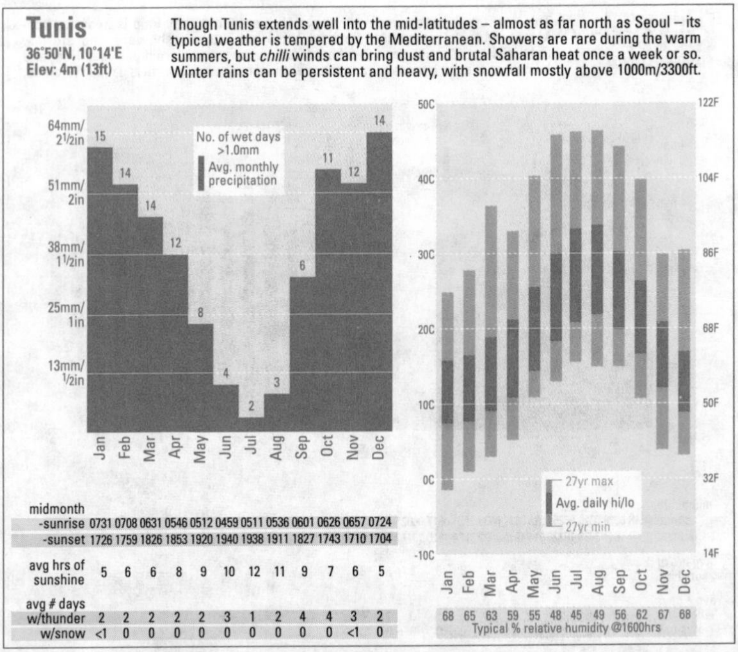

mountains of **central Tunisia** boast more weather variety than most African locales. Winters can bring serious cold (snow may coat the highest peaks), but summer afternoons can reach 38°C/100°F over large areas. When Saharan and European air masses collide during spring and autumn, a desert depression may spin up and bring a sharp round of wind and showers across Tunisia – although just as often, the rain evaporates and the storm delivers only dust.

Zimbabwe

Malawi | Mozambique | Zambia

Much like the Sahel, but without the blazing temperatures, **Zimbabwe** and neighbouring **Zambia** feel the pulse of the ITCZ each year. Its southward push brings widespread showers, thunderstorms and clouds from December into March. The rains are most frequent (almost daily) and the clouds most persistent across Zambia, where the ITCZ is enhanced by intrusions of "Zaïre air" (or "Congo air") – a tongue of moist Atlantic flow that pushes in from the northwest. The savanna of Zimbabwe get wet-season thundershowers every couple of days. Longer spells of *guti* weather – drizzle and fog – accompany a push of cooler air from the southeast every week or so.

Harare

17°55'S, 31°08'E
Elev: 1503m (4930ft)

Harare adds a bit of seasonality to the high-altitude regime found nearer the equator in places like Nairobi. Winds are light and days are mild year round, with lots of sun even during the wet season (November–March). Nights cool down in winter, with frost possible in spots. Mid-spring brings the warmest afternoons.

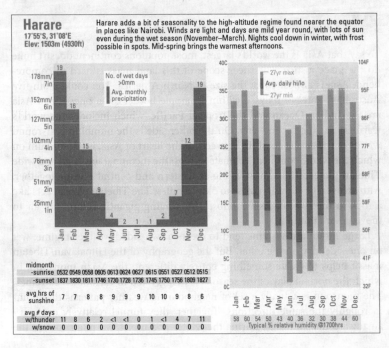

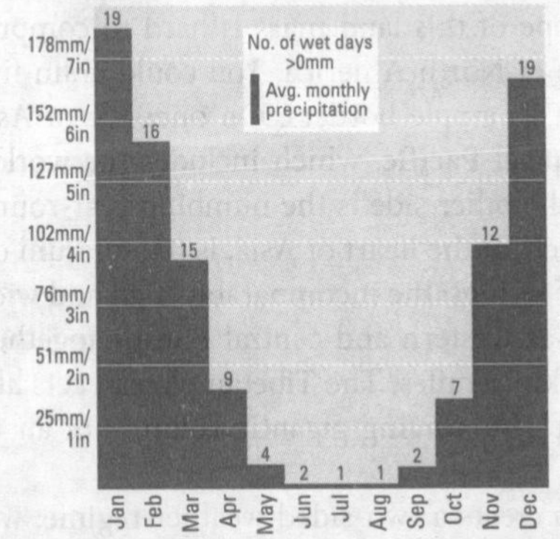

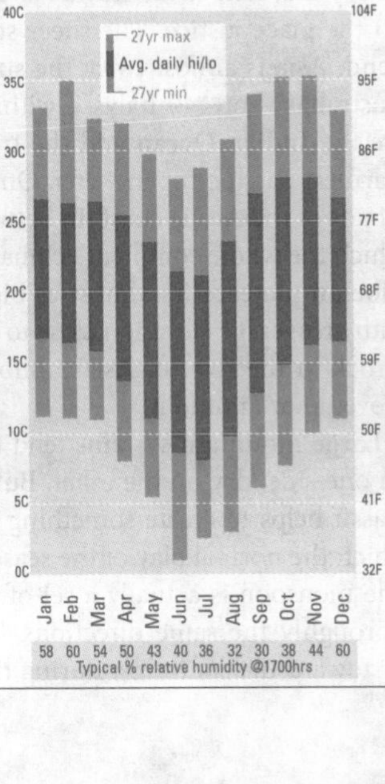

	Jan	Feb	Mar	Apr	May	Jun	Jul	Aug	Sep	Oct	Nov	Dec
midmonth -sunrise	0532	0549	0558	0605	0613	0624	0627	0615	0551	0527	0512	0515
-sunset	1837	1830	1811	1746	1730	1728	1736	1745	1750	1756	1809	1827
avg hrs of sunshine	7	7	8	8	9	9	9	10	10	9	8	6
avg # days w/thunder	11	8	6	2	<1	<1	0	1	<1	4	7	11
w/snow	0	0	0	0	0	0	0	0	0	0	0	0

Daytime temperatures tend to be pleasantly warm year round across both countries and in adjacent **Malawi**. Dry-season highs approach, but seldom top, 90°F/32°C, except toward southwestern Zambia and the hot **Limpopo Valley** of extreme southeast Zimbabwe. Rain is essentially absent from June through September, except in Zimbabwe's eastern highlands, where an occasional winter *guti* may strike. Since most of the land is a plateau above 1000m/3300ft, the winter nights become strikingly cool for the tropics. They usually bottom out around 4–7°C/39–45°F, but hang near 10°C/50°F in western Zambia. Despite the chill, freezes are quite rare on the plateau. The tropical latitude reasserts itself as you drop from the highlands into low-lying **Mozambique**. The swampy southland is warm and muggy in the wet season and mild with occasional showers in winter. As you move north, the coastal rains get heavier and the air stickier. Inland are Mozambique's semi-arid hot spots: the lower Limpopo Valley and the **Tete region**, sandwiched between Zimbabwe and Malawi. Highs run close to 35°C/95°F in October and November before the rains arrive. Although they rarely make landfall here, tropical cyclones can trigger severe flooding from December to April across Mozambique and Malawi.

Asia

To say that **Asia** is the world's largest, most populous continent doesn't quite do the place justice. The sheer scope of this land mass is hard to comprehend. Asia is almost twice the size of North America. You could cram five Australias – or 140 Italys – within its ample borders. On one side of Asia are the **Indian Ocean** and the tropical **Pacific**, which includes the world's warmest patch of open water. On the other side is the numbing year-round winter of the Arctic coast. In between, at the heart of Asia, is the fulcrum on which the whole continent's climate swings: the incomparably high and wide **Tibetan plateau**. It's almost as big as western and central Europe together, with an average elevation close to 5km/3 miles. The Tibetan plateau acts like a rock in the river of Asia's atmosphere, forcing gigantic volumes of air to move up or around it.

Large mountain systems tend to create a two-sided weather regime: wet on one side, dry on the other. But the geography of the Himalayan/Tibetan massif helps to create something entirely different: the **Asian monsoon**, in which the normal play of the seasons is exaggerated to cartoonish extremes. The monsoon is actually a set of neighbouring weather patterns that point in roughly the same directions. Together, they funnel relatively cool, dry air toward the southwest during the winter and moist, warm air toward the northeast in the summer. The most famous monsoon, of course, belongs to **India**, where it virtually defines a way of life for hundreds of millions of people. **China**'s monsoon, which also touches hundreds of millions, is no less vivid in its shaping of the seasons. Even the empty reaches of Siberia are affected by the intense seasonality of this atmospheric seesaw.

There's a plentiful variety of other weather elements at work across Asia. **Deserts** – from tropical to polar – are tucked in between mountain ranges or strewn across the heart of the continent. Asian temperature extremes can be mind-boggling. Parts of India can exceed 45°C/113°F during the tedious pre-monsoon days, and temperatures each winter plummet below –50°C/–58°F in Siberia. The warm waters of the western Pacific offer up the world's biggest crop of **tropical cyclones** (called typhoons in this region), many of which bring devastation to southern and eastern Asia.

In spite of all this meteorological drama, the Asian climate has its plus sides. The very reliability of the monsoon supports agriculture and settlement across a great sweep from India to Japan, where over two billion people live – more than 30 percent of the world's population. The monsoon also gives travellers a better-than-usual sense of what to expect at certain times of year. You needn't bother to bring an umbrella to Calcutta in December, but you'd be rash not to carry one in July.

China

Mongolia | North Korea | South Korea | Taiwan | Tibet

Weather may be one reason why the philosophy of yin and yang developed in China. This ancient culture evolved in a climate rife with dualism. For starters, there's the monsoon. Less publicized than India's, this annual cycle of rain and drought is a fundamental part of life across the densely populated eastern half of the country. Although China faces the Pacific Ocean, it's actually the moisture from the further-away **Indian Ocean** that sweeps up from southeast Asia late each spring to kick off the monsoon. Places that barely see a drop of rain or a flake of snow in January are awash by August. **Beijing** gets almost a hundred times more precipitation in early summer than in early winter. **Severe floods** can strike China's colossal river valleys, especially the **Yangtze**, where more than three million people died from flooding and associated famine and illness in 1931 alone. By late September, the monsoon is being hounded southward by cool northerly winds that strengthen through the autumn and keep much of China dry and cold throughout the winter.

Just as dramatic as the monsoon is the yearly **oscillation in temperatures**, especially in the eastern heart of China. No other populous part of the world gets this intense a combination of bitter winters and sweltering, humid summers. At least China's temperature cycle is more predictable than most. Seasons tend to arrive close to schedule, and day-to-day temperature variations are subdued compared to those in North America. The pulses of cold air that cascade from Siberia onto the east China lowlands each winter can push into Southeast Asia and beyond. At 22°N, **Hong Kong** is one of the few coastal cities anywhere in the tropics that's experienced air close to the freezing mark. Despite all this chill, the intense aridity of Chinese winter means that the eastern cities avoid the persistent snowdrifts of Siberia, although what snow does fall in far northeast China can last for a long time. (Winter isn't as bad as it used to be across northern China – most recent winters have seen temperatures markedly above normal for long stretches.) If January and February are surprisingly cold for China's low latitude, the summers are just as hot and moist as you might expect. Perhaps the ideal time to visit eastern China is October, which can offer gorgeously sunny, crisp weather.

To the **west**, China's more remote reaches offer a wholly different weather experience. The stark **Gobi Desert**, straddling the border of China and **Mongolia**, sits next to the **Tarim Basin**, former home of a giant glacial lake. Both the desert and basin get dry-roasted in the summer. Further north, the climate is more akin to Siberia's, with dry, bone-chilling winters and showery, coolish summers. The vast plateau of **Tibet**, sitting 5km/3 miles above its neighbours, is a meteorological world apart, much of it still unmapped.

East

Although all of this part of China is controlled by the annual monsoon, it's a large enough region to allow for plenty of differences. Winter is on the dry side everywhere, except for a few dustings of snow in the north and light rains once or twice a week further south, especially along the coast. Gusty Siberian cold waves that roll south every few days make some of their biggest inroads up the **Chang Jiang (Yangtze)** valley, chilling cities from **Wuhan** to **Changsha**. The **Nan Ling Mountains** keep the worst cold from reaching the southernmost provinces. Another relative mild spot is the **Sichuan Basin**, including **Chengdu** and **Chongqing**. In the spring, a few intense dust storms usually sweep from **Mongolia** across north and northeast China, stinging the eyes and obscuring visibility. Otherwise, the fast-warming weather – still on the dry side – is quite appealing. The leading edge of the summer monsoon rains typically hits south China by mid-May, then moves north in fits and starts. In June the rains often stall across the Chang Jiang Valley, forming a famous deluge-producing boundary known as the *Mei-yu* front that can linger for several weeks. The summer monsoon normally hits **Beijing** by late July and the far northeast plains in early August. This region's brief rainy season ends quickly as the monsoon retreats by late August. The mountains of the peninsula holding **North Korea** and **South Korea**, surrounded by water on both sides, are nearly as cold as northeast China in winter but more sultry in summer. To the west, the Sichuan Basin gets most of its monsoon rain at night, although its days tend to stay cloudy as well.

The rains often abate for a few weeks in late July and August across the core of east China, between the Chang Jiang Valley and the Nan Ling Mountains. It's then that heat and humidity build to insufferable levels. In such cities as Chongqing, Wuhan and **Nanjing**, nights may stay above 25°C/79°F, and many afternoons soar past 35°C/95°F, with no sea breeze to relieve the tropical torpor. Relief may come only with localized thunderstorms or with a **typhoon** moving inland. Typhoons are most likely to strike the **southeast coast** from July to September. The worst one in at least 50 years, Supertyphoon Saomai, killed more than 400 people in August 2006 when it stormed ashore south of **Wenzhou** packing winds of 150kph (150mph). Even though many recurve back toward the Pacific well before reaching China, about four or five typhoons make landfall in a typical year, and every few years one makes it as far as North Korea. They deserve respect, so be sure to heed warnings if you're near the coast (and watch for potential flooding along its track if you're further inland). Autumn sweeps quickly from north to south, with freshening breezes and rapidly cooling temperatures but plenty of sun. As the monsoon moisture retreats, it makes a last stand across **southwest China**, where multiday rains often fall across **Sichuan**, **Yunnan** and **Guizhou** provinces. The highlands of Yunnan, sitting above the tropical air masses of summer and the shallow cold blasts of winter, enjoy a "springtime all year" climate – exemplified in **Kunming** – that's considered one of China's most agreeable.

Taiwan

Positioned 200km/120 miles off the Chinese mainland, **Taiwan** gets a few spells of wintertime rain, as the monsoon flow rolls over the **East China Sea** and encounters the island's steep slopes. The northeasternmost town, **Chilung**, once suffered 55 consecutive days of winter rain. South of **Tungshih**, the winters are less prone to rain, but milder and more humid than in the coolish north. Summer is sweltering and wet across the whole island, opening with the *Mei-yu* downpours in May and concluding in August and September with sometimes-fierce typhoons that can produce torrents as they encounter the mountainous terrain. On 10–11 September 1963, **Paishih** received 1248mm/49.1in – one of the largest 24-hour rainfalls in Northern Hemisphere weather annals.

Northwest and Mongolia

Shielded from monsoon moisture by the **Tibetan plateau** and other highlands, this is one of Asia's driest pockets. The big divide is the **Tien Shen** range, which produces wildly varied weather across a small area. The basin's overall climate is on par with Beijing's, minus the summer rains – only about a day per month sees moisture. North of the Tien Shen, the **Junggar basin** gets nearly as warm as Beijing in summer but can drop below –30°C/–22°F in winter. Frequent light snows fall on foothill towns like **Ürümqi**, with the heavier snowfall at elevation. The highest peaks maintain a snowpack through the summer. Oddly enough, China's summer hot spot is nearby, in a small basin sitting below sea level just southeast of the Tien Shen, where **Turpan** has reached 49°C/120°F.

Conditions in the **Gobi** of **southern Mongolia** can be brutal. **Ulan Bator** struggles to reach –18°C/0°F on a typical January day, and epic dust storms roar across the south each spring. However, Mongolia's highlands are just far enough east to benefit from monsoon-related showers, and the mid-summer sun helps make temperatures quite pleasant.

Tibet

High-elevation weather usually affects only a sliver of land along a mountain range, but on the enormous **Tibetan plateau** it spreads from horizon to horizon. Many visitors wind up in the more temperate southern valleys, with summer readings much like you might find in central Scandinavia. Those venturing into the less-touristed reaches above 5000m/16,400ft may

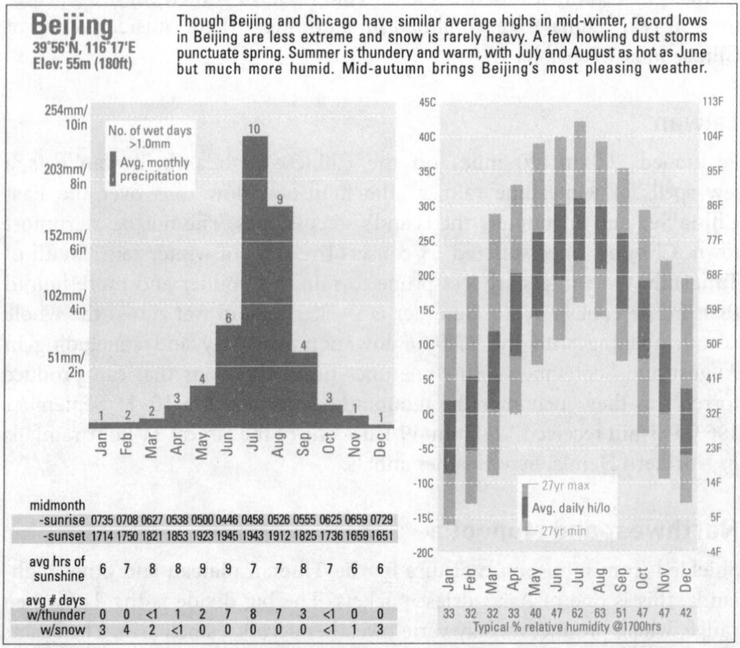

Hong Kong

22°20'N, 114°11'E
Elev: 24m (79ft)

At 22°N, it's amazing the Hong Kong winter is so cool. Think of San Francisco in summer – breezy, sunny, long-sleeve weather, never bitterly cold – with rain about once a week. Rainfall is heavy from May–September, and typhoons are a real risk in summer and early autumn. Transition months are mild and variably wet.

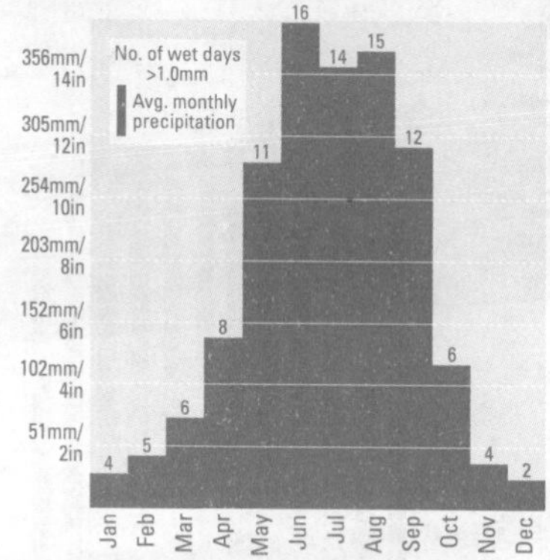

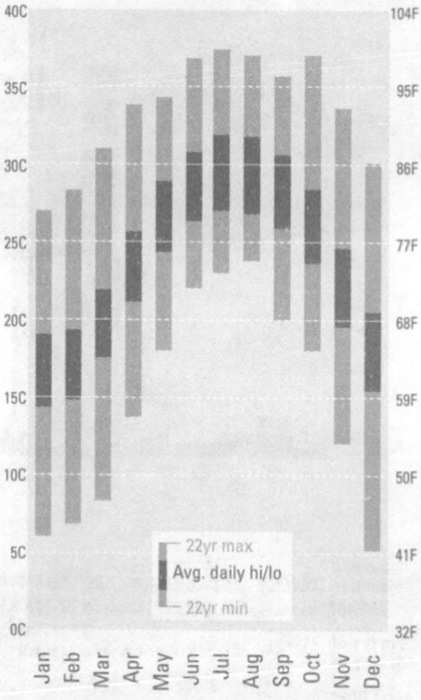

	Jan	Feb	Mar	Apr	May	Jun	Jul	Aug	Sep	Oct	Nov	Dec
midmonth -sunrise	0705	0655	0633	0604	0544	0539	0548	0600	0610	0619	0635	0655
-sunset	1800	1820	1832	1843	1856	1908	1910	1855	1827	1759	1740	1742
avg hrs of sunshine	5	4	3	4	5	5	7	7	6	6	6	6
avg # days w/thunder	0	1	2	5	6	6	7	8	6	1	<1	0
w/snow	0	0	0	0	0	0	0	0	0	0	0	0
Typical % relative humidity @1700hrs	64	71	76	76	76	75	71	73	70	65	61	60

Kunming

25°01'N, 102°41'E
Elev: 1892m (6206ft)

Year-round spring is Kunming's claim to fame. Seasonal averages are nearly as mild as Sydney's, moderated by Indian Ocean flow and altitude. Frequent summer showers occur in the late night and mid-afternoon. Morning frosts and afternoon sun are typical of winter, with a day or two of light snow each year.

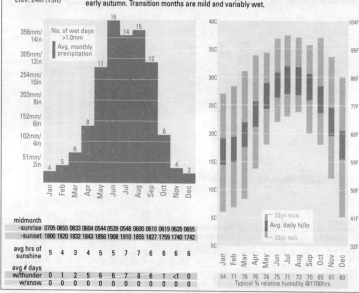

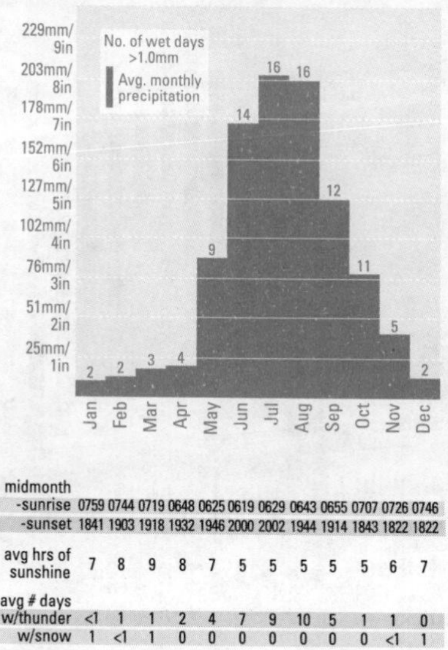

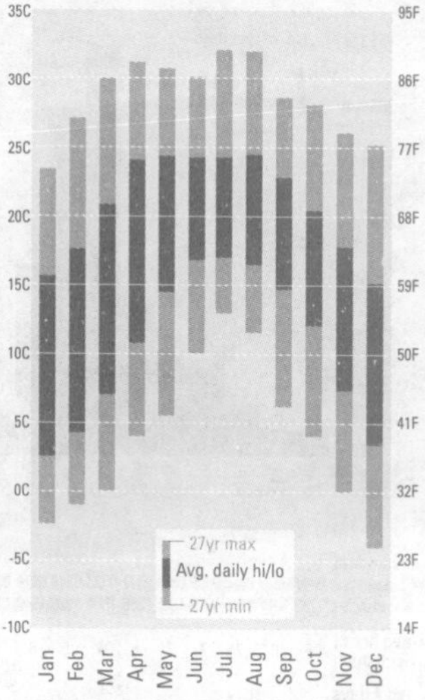

	Jan	Feb	Mar	Apr	May	Jun	Jul	Aug	Sep	Oct	Nov	Dec
midmonth -sunrise	0759	0744	0719	0648	0625	0619	0629	0643	0655	0707	0726	0746
-sunset	1841	1903	1918	1932	1946	2000	2002	1944	1914	1843	1822	1822
avg hrs of sunshine	7	8	9	8	7	5	5	5	5	5	6	7
avg # days w/thunder	<1	1	1	2	4	7	9	10	5	1	1	0
w/snow	1	<1	1	0	0	0	0	0	0	0	<1	1
Typical % relative humidity @1700hrs	41	39	37	39	52	68	72	68	67	62	56	48

Lhasa

29°43'N, 91°02'E
Elev: 3650m (11,976ft)

Tibet's largest city is tucked into one of the plateau's mildest valleys. Nighttime thunder is particularly common during the mild summer, with clouds parting by mid-day and another storm arriving by late afternoon. Winter days are sunny and crisp, with a few days' worth of snow each year, usually early or late in the season.

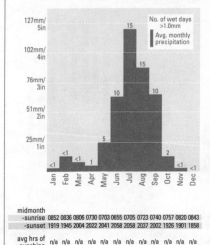

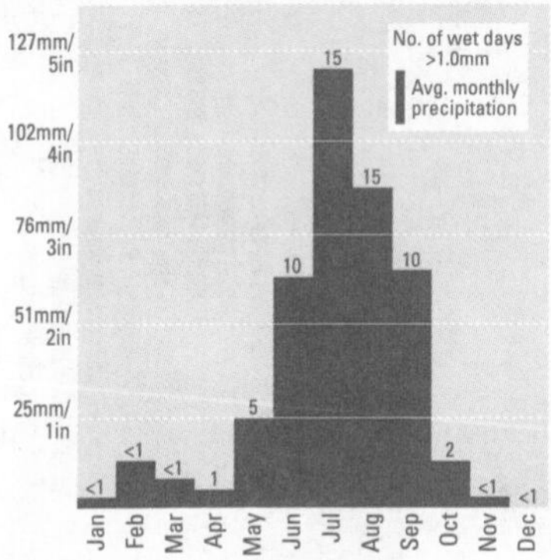

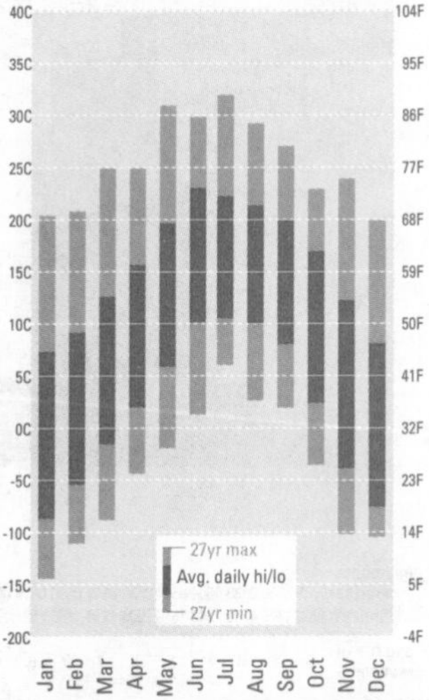

	Jan	Feb	Mar	Apr	May	Jun	Jul	Aug	Sep	Oct	Nov	Dec
midmonth -sunrise	0852	0836	0806	0730	0703	0655	0705	0723	0740	0757	0820	0843
-sunset	1919	1945	2004	2022	2041	2058	2058	2037	2002	1926	1901	1858
avg hrs of sunshine	n/a	n/a	n/a	n/a	n/a	n/a	n/a	n/a	n/a	n/a	n/a	n/a
avg # days w/thunder	0	0	<1	2	7	16	21	20	11	2	0	<1
w/snow	n/a	n/a	n/a	n/a	n/a	n/a	n/a	n/a	n/a	n/a	n/a	n/a
Typical % relative humidity @1200hrs	17	17	22	28	34	43	54	56	51	35	30	23

Seoul

37°33'N, 126°48'E
Elev: 18m (59ft)

A shade cooler and wetter than Beijing, Seoul has a similar monsoon-driven climate, albeit a bit more varied. Frequent light snows alternate with sunshine in mid-winter. July and August are sultry, fairly cloudy, and showery. Spring and autumn days are often bright and delightful. A late-summer typhoon makes it this far northwest once every few years.

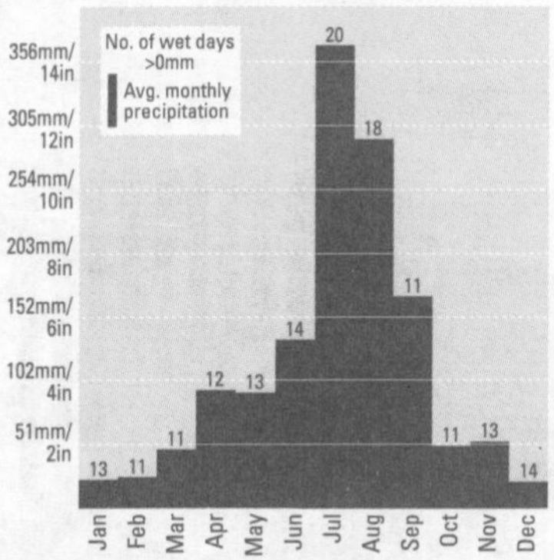

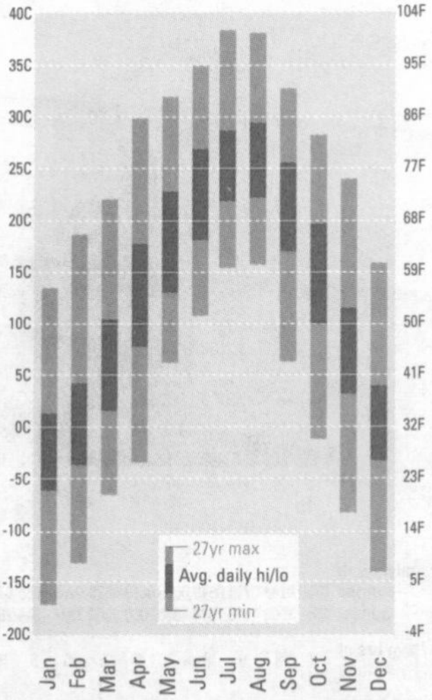

	Jan	Feb	Mar	Apr	May	Jun	Jul	Aug	Sep	Oct	Nov	Dec
midmonth -sunrise	0747	0723	0645	0559	0524	0511	0523	0548	0614	0640	0712	0740
-sunset	1738	1811	1839	1908	1935	1955	1954	1926	1842	1757	1722	1715
avg hrs of sunshine	5	6	7	7	7	6	4	5	6	7	5	5
avg # days w/thunder	<1	<1	<1	1	1	1	3	2	1	2	1	<1
w/snow	9	7	3	<1	0	0	0	0	0	<1	2	8
Typical % relative humidity @1500hrs	54	53	50	49	55	63	71	67	59	51	54	58

Shanghai

31°10'N, 121°26'E
Elev: 7m (23ft)

Shanghai's agreeable spring is followed by several weeks of Mei-yu rain and cloud in June. July and August are sunnier but hot, humid, and showery. Typhoons to the south can bring torrents, although gales are rare. October provides the choicest weather. Winter is cloudy and chilly, especially in January, with several snows (usually light).

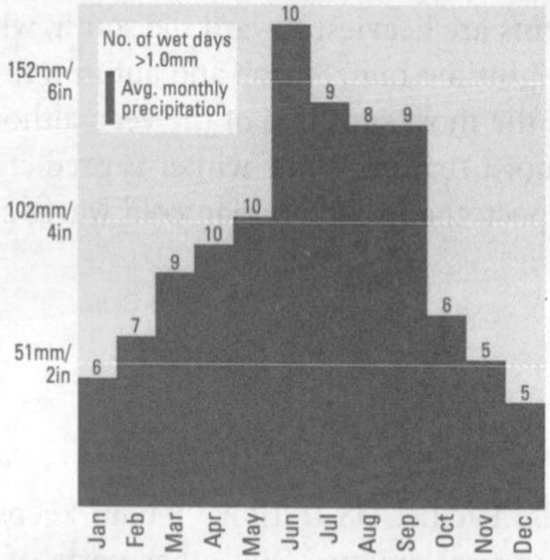

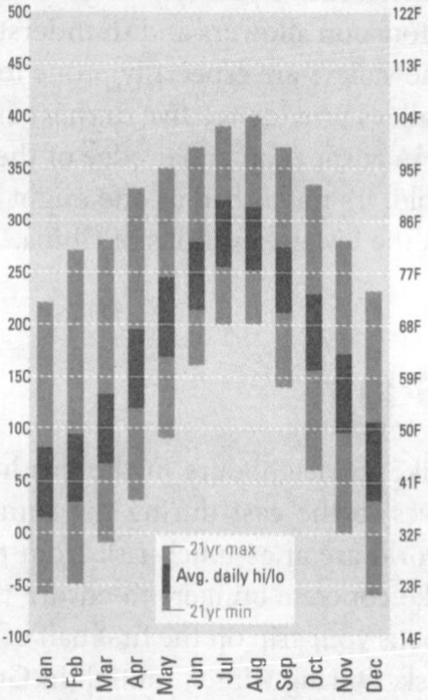

	Jan	Feb	Mar	Apr	May	Jun	Jul	Aug	Sep	Oct	Nov	Dec
midmonth -sunrise	0653	0636	0605	0527	0459	0450	0500	0520	0538	0557	0621	0645
-sunset	1714	1741	1802	1822	1843	1859	1900	1838	1801	1723	1656	1653
avg hrs of sunshine	4	4	5	5	6	5	7	8	5	5	5	5
avg # days w/thunder	<1	<1	1	1	1	2	5	4	2	<1	<1	<1
w/snow	4	2	1	0	0	0	0	0	0	0	0	1
Typical % relative humidity @1700hrs	65	64	68	67	67	73	74	73	72	66	64	61

Urumqi

43°47'N, 87°37'E
Elev: 919m (3014ft)

Siberian air pooling north of the Tien Shan makes this area as cold as high Tibet in winter. Urumqi seldom rises above freezing from December through February, as sunshine alternates with brief fogs and light snow. Summer is thundery and mild (though scorching south of the Tien Shan), but rain is usually light. Spring and autumn are brief transitions.

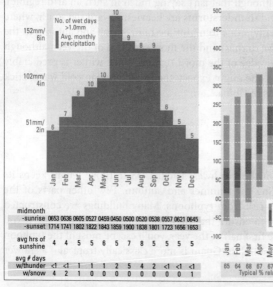

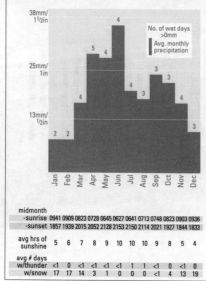

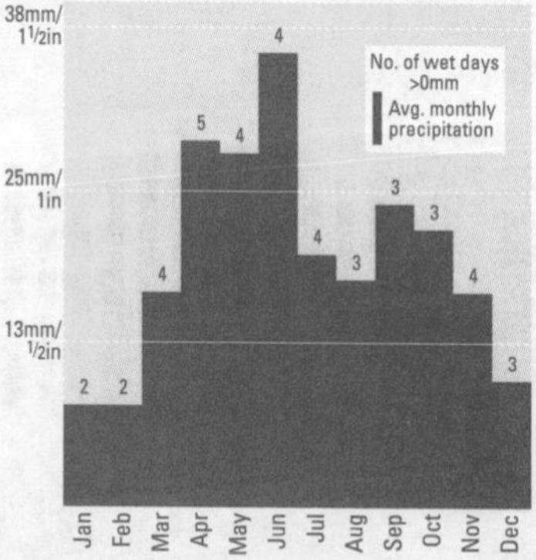

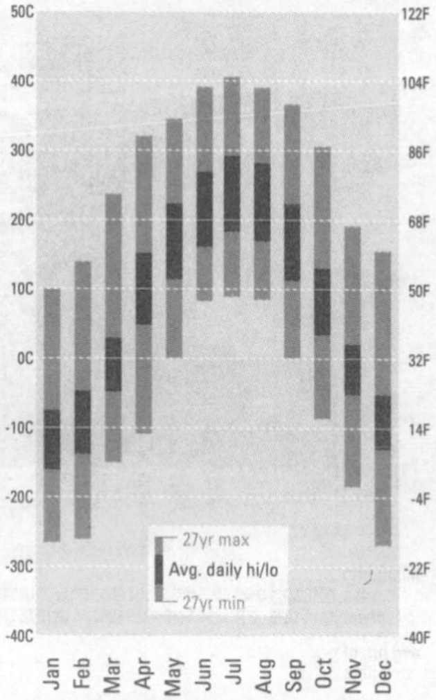

	Jan	Feb	Mar	Apr	May	Jun	Jul	Aug	Sep	Oct	Nov	Dec
midmonth -sunrise	0941	0909	0823	0728	0645	0627	0641	0713	0748	0823	0903	0936
-sunset	1857	1939	2015	2052	2128	2153	2150	2114	2021	1927	1844	1833
avg hrs of sunshine	5	6	7	8	9	10	10	10	9	8	5	4
avg # days w/thunder	<1	0	<1	<1	<1	<1	1	1	<1	0	<1	0
w/snow	17	17	14	3	1	0	0	0	<1	4	13	19
Typical % relative humidity @1700hrs	74	74	63	37	30	31	30	27	31	45	67	76

encounter July days that stay below 10°C/50°F and nights that plummet to freezing. The highest reaches may see more snow in summer than any other time of year (although that isn't saying much in such an arid regime). **Monsoon showers** and **thunderstorms** are heaviest toward the south, where the valleys are especially prone to nighttime rain. Spring and autumn in the valleys provide relative dryness and the most sunshine of the year, although it may snow on either edge of the moist season. While winter is predictably cold, it's not as bad as one might expect; the worst Siberian cold waves stick to the lower elevations of China.

Guam

Like its neighbours in the northwest Pacific, US territory **Guam** keeps its eyes to the **east** during the summer and autumn. Few other parts of the world are at as much risk from **typhoons**. Many buildings are constructed like concrete bunkers to ensure that they last more than a few years. If you focus your trip on the first half of the year, you'll escape most of the typhoon risk. During winter, rainfall in Guam is more classically tropical – a shower

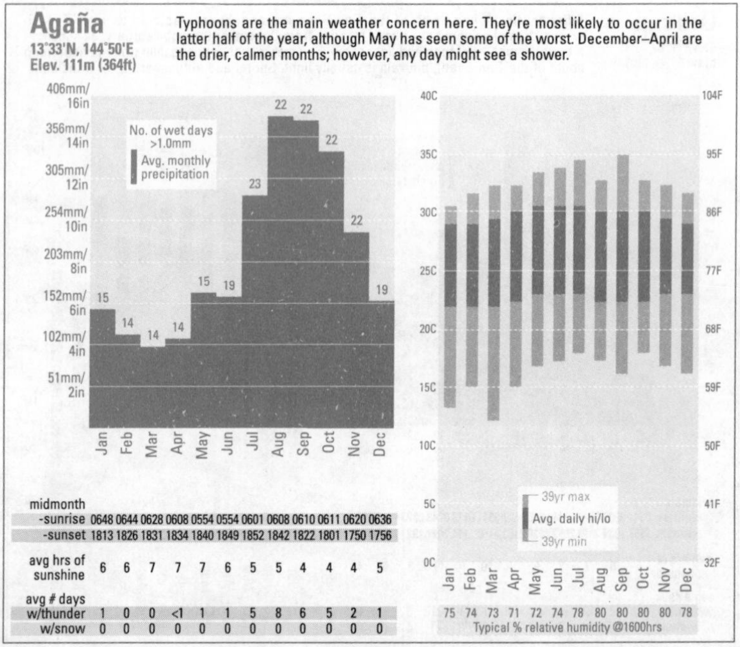

every couple of days – and temperatures are balmy, a couple of degrees lower than in the summer.

India

Afghanistan | Bangladesh | Pakistan

As troublesome as it may be, the **monsoon** is a member of the family here. India simply wouldn't be the same without it. It's a dramatic sight, to be sure: after days of torrid heat and humidity, with only an occasional storm at best, the skies suddenly darken, the southwest winds kick in, and before long the skies open up. And open up they do: on 26 July 2005, Mumbai set a new national record for 24-hour rainfall with an astounding 942mm (37.1in).

The dates on the map below were first calculated in 1940, and they're still good ballpark figures for monsoon arrival. In about half of all years the monsoon is at least three days early or late, but rarely by more than a week. The rains end quickly across the northwest half of India between mid-September and mid-October, but they lag across the south until December in what's known as the **northeast (winter) monsoon**.

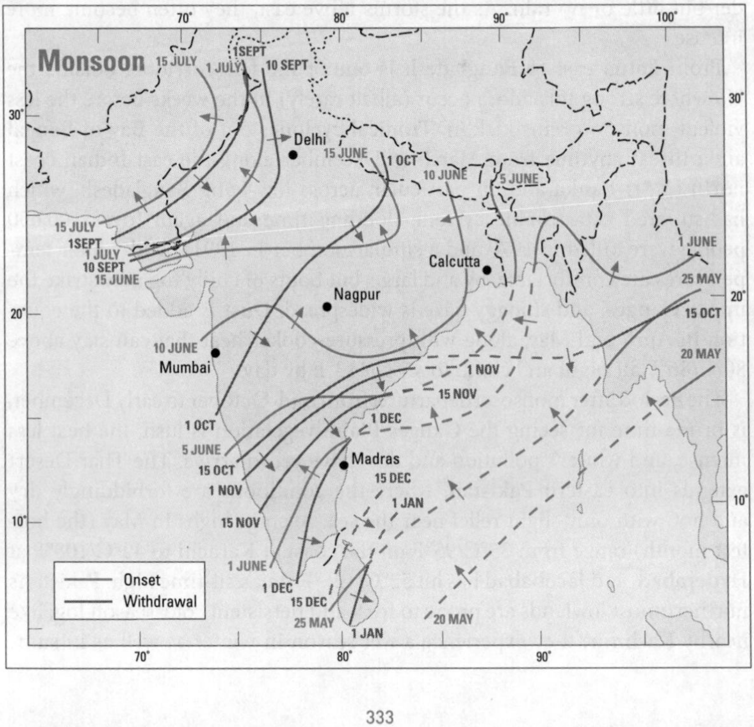

It doesn't rain constantly during the monsoon. About once each summer the rains shift north to the Indian/Nepalese foothills for a week or two and leave the rest of the subcontinent dry. The most likely time for such a "**break monsoon**" is the middle of August. Even when the monsoon is healthy, squalls come and go, and there's enough sunshine in between so that the air will feel hot and humid to any visitor just arriving, especially toward the south. In fact, India south of the Himalayas is almost never bitingly cold. Even **New Delhi** is hard-pressed to muster a frost during the dry winter, when sunny and hazy skies are the rule across the north. When cold waves do strike, they can claim many lives due to the scarcity of cold-weather shelter and clothing.

Northwest and Ganges Plain

One of the world's great weather puzzles is why India's rainfall varies so much from northwest to northeast. Parts of the **Thar Desert** average under 5cm/2 in, a year, whereas **Cherrapunji** experienced the rainiest twelve months ever recorded on Earth in 1860–61 with 26,470mm/1041in. Whatever the reason for the disparity, visitors should plan on progressively wetter conditions towards the east. In the spring, low-pressure centres help whip up *Aandhi* – blinding dust storms across the Thar that may be accompanied by thunder but little or no rain. As the storms move east, they often become more intense.

From **Patna** east to **Bangladesh** is one of the few corridors outside the US where strong tornadoes occur (albeit rarely) in the weeks before the less violent monsoon rains kick in. **Tropical cyclones** out of the **Bay of Bengal** are a threat anytime from March to December along the east Indian coast north of **Sri Lanka** and, in particular, across low-lying Bangladesh, which has suffered catastrophic cyclone flooding time and again (over 100,000 people were killed in 1970 and a similar number in 1991). Mid-winter temperatures are comfortable, by and large, but bouts of chilly **fog** may strike the **upper Ganges**, and smoggy haze is widespread. **Dust** is added to the equation in April and May, along with pressure-cooker heat that can stay above 30°C/86°F all night and climb to 45°C/113°F by day.

The period after monsoon departure, from mid-October to early December, is prime time for seeing the **Ganges Plain**: vegetation is lush, the heat less intense, and winter's pollution and dust have yet to arrive. The Thar Desert extends into eastern **Pakistan**, where the conditions are forbiddingly dry and hot, with only slight relief near the sea. Average highs in May (the hottest month) range from 35°C/95°F on the coast at **Karachi** to 42°C/108°F at **Hyderabad**, and **Jacobabad** has hit 52°C/126°F, Asia's all-time high. Pakistan's northernmost lowlands are prone to frost and persistent cold-season fog; like nearby **Kashmir**, they experience a wet season in winter as well as in sum-

Bangalore

12°58'N, 77°35'E
Elev: 921m (3021ft)

Bangalore's altitude gives it a more temperate climate than other Indian cities. The monsoon lingers into November, and even mid-winter brings a spit of rain every few days, but there's lots of sun as well. The pre-monsoon spring is thundery but not too hot, seldom above 35°C/95°F.

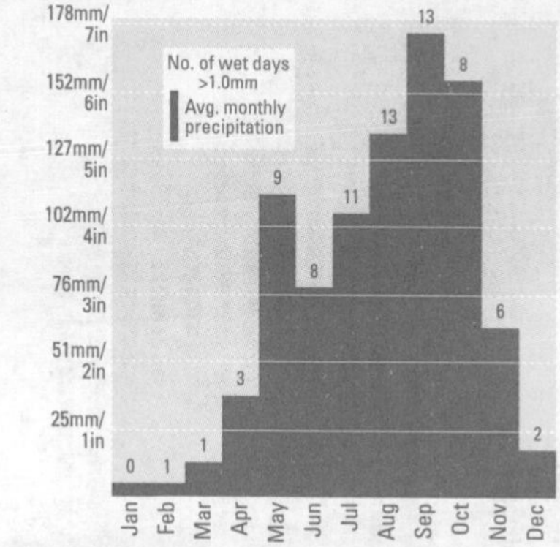

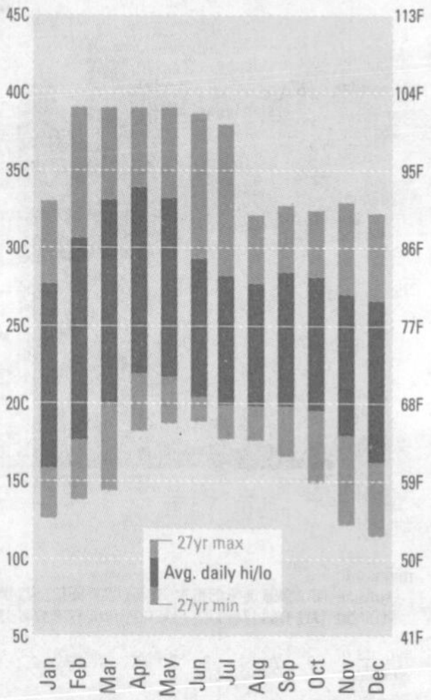

	Jan	Feb	Mar	Apr	May	Jun	Jul	Aug	Sep	Oct	Nov	Dec
midmonth -sunrise	0646	0642	0627	0607	0554	0554	0601	0607	0609	0610	0618	0634
-sunset	1812	1826	1830	1832	1838	1847	1850	1841	1821	1801	1750	1756
avg hrs of sunshine	8	9	9	9	8	5	4	4	5	6	6	7
avg # days w/thunder	<1	<1	1	4	7	3	1	1	4	3	<1	<1
w/snow	0	0	0	0	0	0	0	0	0	0	0	0
Typical % relative humidity @1700hrs	40	33	29	34	45	62	65	69	65	63	61	54

Bikaner

28°03'N, 73°12'E
Elev: 229m (750ft)

By the time the Indian monsoon reaches Bikaner – usually in July – it's a rather limp version of itself, bringing only a few days of heavy rain, although afternoons do cool from their blazing springtime levels. Late autumn and winter are great times to explore the Thar region: intense heat, freezes, and rain are all uncommon.

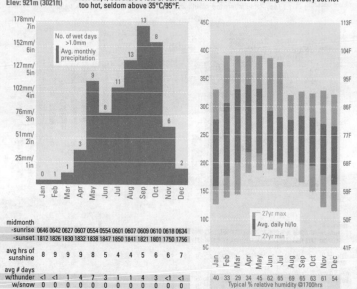

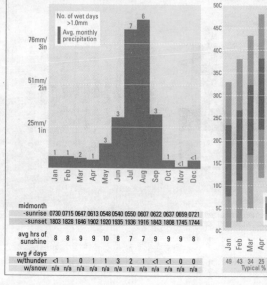

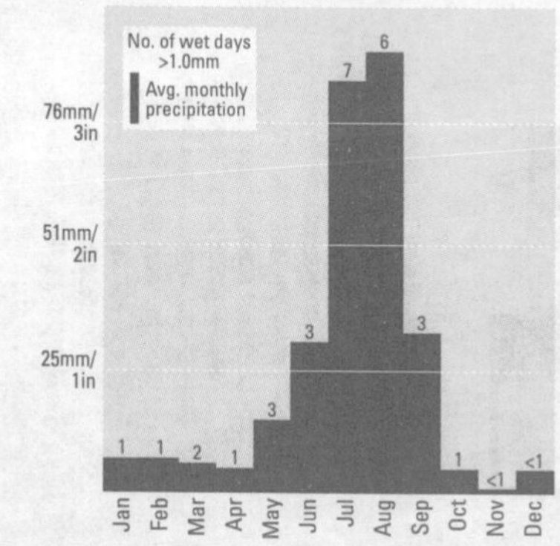

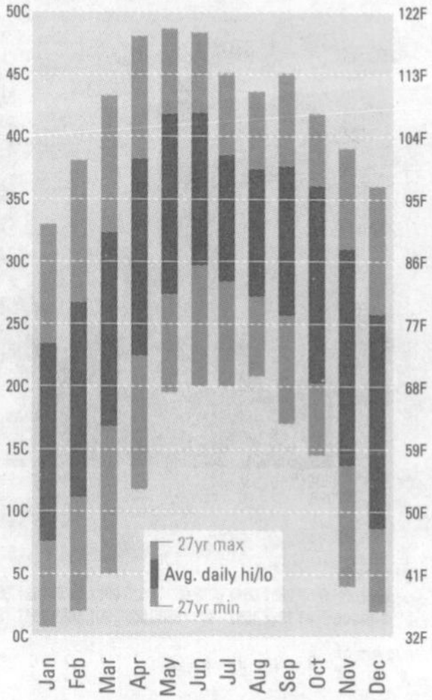

	Jan	Feb	Mar	Apr	May	Jun	Jul	Aug	Sep	Oct	Nov	Dec
midmonth -sunrise	0730	0715	0647	0613	0548	0540	0550	0607	0622	0637	0659	0721
-sunset	1803	1828	1846	1902	1920	1935	1936	1916	1843	1808	1745	1744
avg hrs of sunshine	8	8	9	9	10	8	7	7	9	9	9	8
avg # days w/thunder	<1	1	0	1	1	3	2	1	<1	<1	0	0
w/snow	n/a	n/a	n/a	n/a	n/a	n/a	n/a	n/a	n/a	n/a	n/a	n/a
Typical % relative humidity @1200hrs	49	43	34	25	27	39	58	61	52	36	40	48

Calcutta

22°39'N, 85°27'E
Elev: 6m (20ft)

Warm days are practically a given here, even during the dry winter. December–February offers plenty of sun, although it's often hazy. Stifling pre-monsoon heat follows, broken at times by an intense thunderstorm. Tropical cyclones are a serious threat from May–November; the mouth of the Ganges sees the world's deadliest landfalls.

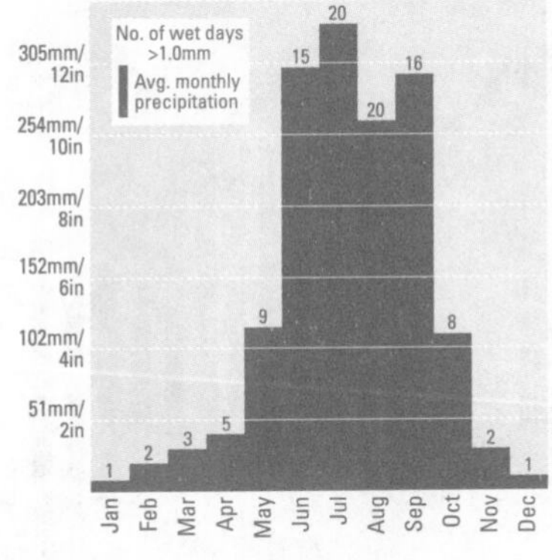

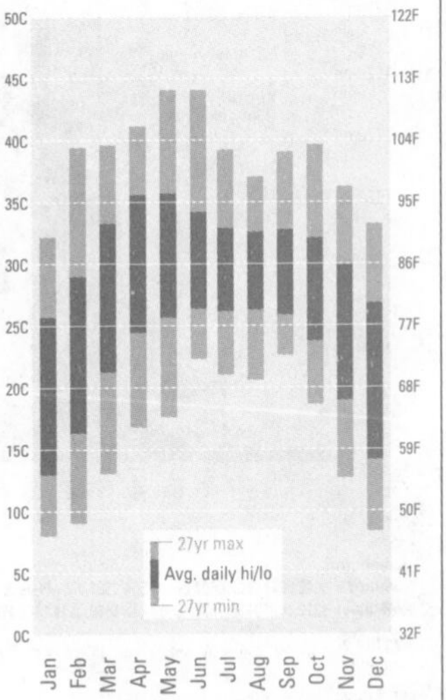

	Jan	Feb	Mar	Apr	May	Jun	Jul	Aug	Sep	Oct	Nov	Dec
midmonth -sunrise	0619	0608	0546	0516	0456	0451	0500	0513	0523	0533	0549	0608
-sunset	1712	1733	1745	1756	1809	1822	1824	1808	1740	1711	1653	1654
avg hrs of sunshine	7	7	7	8	7	4	3	3	4	6	6	7
avg # days w/thunder	<1	2	3	6	8	9	8	9	10	5	<1	<1
w/snow	0	0	0	0	0	0	0	0	0	0	0	0

Jan	Feb	Mar	Apr	May	Jun	Jul	Aug	Sep	Oct	Nov	Dec
44	43	41	50	59	71	78	79	76	62	49	45

Typical % relative humidity @1500hrs

Colombo

6°49'N, 79°53'E
Elev: 5m (16ft)

The best time to see Colombo without rain is in mid-winter, though even then it can pour once a week or so. It's wettest as the monsoon heads north in April–May and returns in October–November. Intense tropical cyclones aren't a major threat here. Head for the hills to escape the constant sultriness.

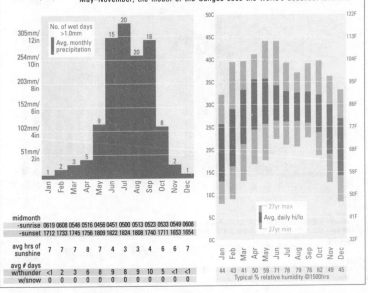

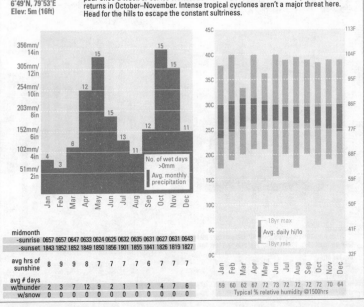

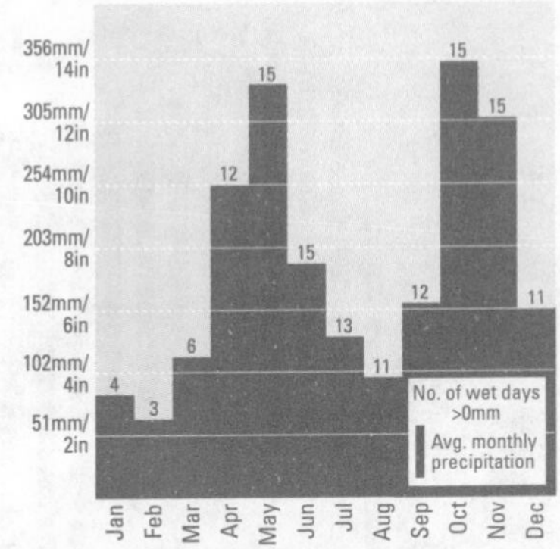

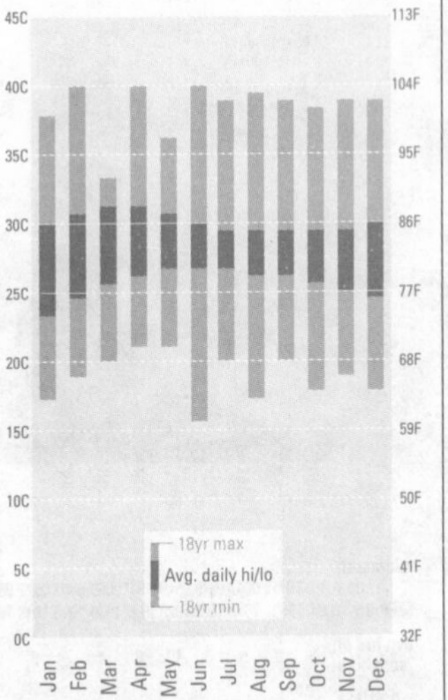

	Jan	Feb	Mar	Apr	May	Jun	Jul	Aug	Sep	Oct	Nov	Dec
midmonth -sunrise	0657	0657	0647	0633	0624	0625	0632	0635	0631	0627	0631	0643
-sunset	1843	1852	1852	1849	1850	1856	1901	1855	1841	1826	1819	1827
avg hrs of sunshine	8	9	9	8	7	7	7	7	6	7	7	7
avg # days w/thunder	2	3	7	12	9	2	1	1	2	4	7	6
w/snow	0	0	0	0	0	0	0	0	0	0	0	0

Jan	Feb	Mar	Apr	May	Jun	Jul	Aug	Sep	Oct	Nov	Dec
59	60	62	67	72	73	72	72	72	72	70	64

Typical % relative humidity @1500hrs

Delhi

28°35'N, 77°12'E
Elev: 216m (708ft)

Late autumn and late winter are ideal times to see Delhi. Many mid-winter afternoons are sunny and mild, although cold air can produce dense fog. Pre-monsoon heat gets serious by April, along with wind and dust. Although the monsoon caps the most intense temperatures, the air stays sultry into September.

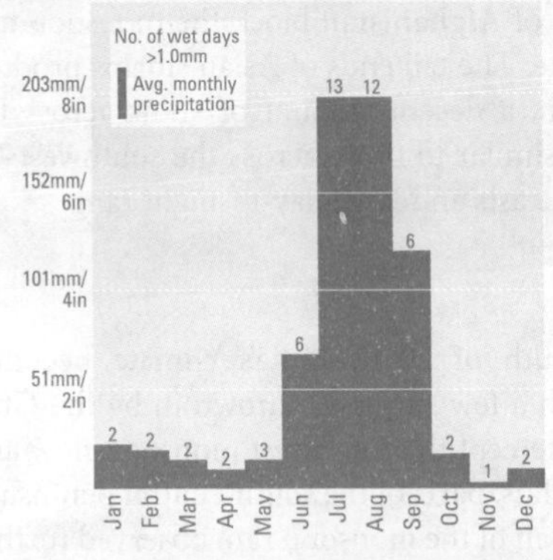

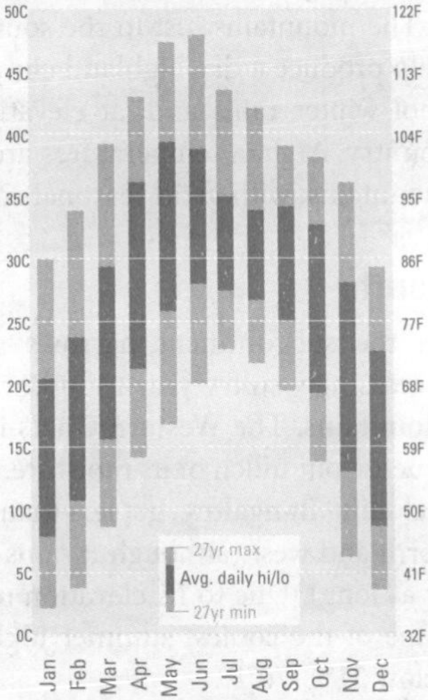

	Jan	Feb	Mar	Apr	May	Jun	Jul	Aug	Sep	Oct	Nov	Dec
Typical % relative humidity @1500hrs	41	35	30	21	24	36	61	64	51	33	31	38

midmonth	Jan	Feb	Mar	Apr	May	Jun	Jul	Aug	Sep	Oct	Nov	Dec
-sunrise	0715	0700	0631	0556	0531	0523	0533	0550	0606	0622	0643	0706
-sunset	1746	1811	1829	1847	1905	1920	1921	1901	1827	1752	1728	1726
avg hrs of sunshine	7	8	8	9	8	7	5	6	7	9	8	7
avg # days w/thunder	1	2	3	2	4	5	6	5	2	1	<1	1
w/snow	0	0	0	0	0	0	0	0	0	0	0	0

Mumbai

19°07'N, 72°51'E
Elev: 14m (46ft)

In classic Indian fashion, Mumbai is dry and warm in winter, broiling in spring, and moist and hot in summer. Winter nights are somewhat warmer and the pre-monsoon heat not quite as intense as in east India. Average July rainfall is roughly twice that of Calcutta's. Epic downpours – some topping 500mm/20 inches a day – often paralyze transport.

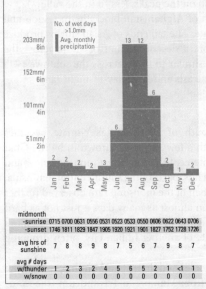

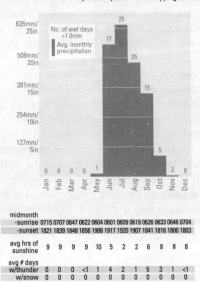

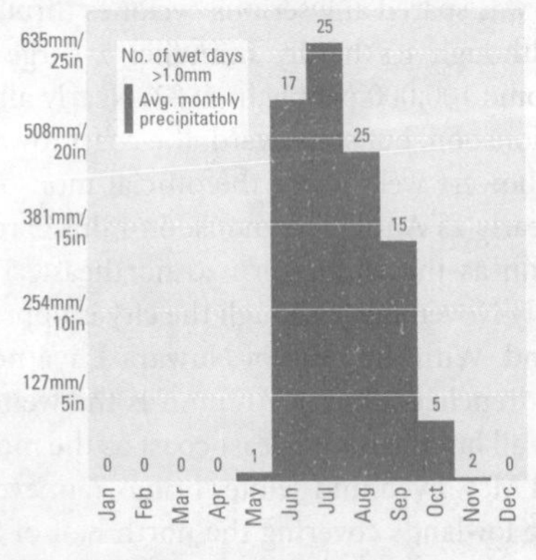

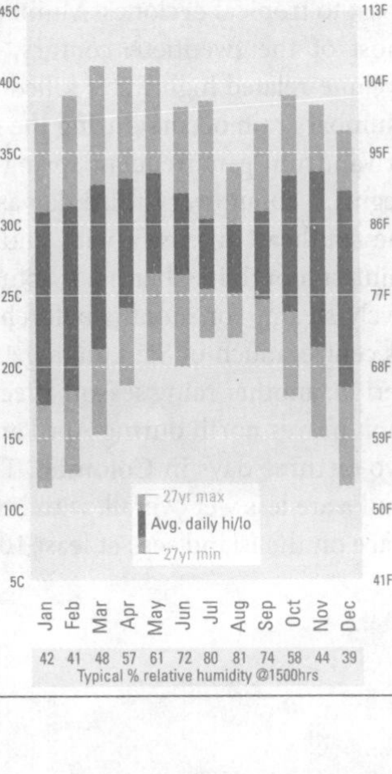

	Jan	Feb	Mar	Apr	May	Jun	Jul	Aug	Sep	Oct	Nov	Dec
Typical % relative humidity @1500hrs	42	41	48	57	61	72	80	81	74	58	44	39

midmonth	Jan	Feb	Mar	Apr	May	Jun	Jul	Aug	Sep	Oct	Nov	Dec
-sunrise	0715	0707	0647	0622	0604	0601	0609	0619	0626	0633	0646	0704
-sunset	1821	1839	1848	1856	1906	1917	1920	1907	1841	1816	1800	1803
avg hrs of sunshine	9	9	9	9	10	5	2	2	6	8	8	8
avg # days w/thunder	0	0	0	<1	1	4	2	1	5	3	1	<1
w/snow	0	0	0	0	0	0	0	0	0	0	0	0

mer. The high mountains of Kashmir and elsewhere across the far northwest get a later start to the monsoon than does **Nepal**, but they have a better chance of **winter snowfall** (more so on the peaks, less so in the valleys).

The mountains just to the south of **Afghanistan** block the monsoon and help produce a dry highland climate. The tail ends of Asian storms produce cool winter rains and, at elevation, a decent amount of snow across the country. Afghan temperatures are similar to those across the southwest US mountains, with vivid seasonal contrasts and large day-to-night ranges.

South

As the subcontinent narrows south of 20°N, India's climate becomes increasingly sultry year round, with a few surprises thrown in by the **Ghat Mountains**. The **Western Ghats** intercept the southwest monsoon flow and squeeze out much of its moisture. Thus, parts of the south-central peninsula, including **Bangalore**, get less than half of the monsoon rain observed further north and west (although it rains on almost as many days – just not as hard or as long). Due to its elevation of close to 1000m/3300ft, Bangalore enjoys some of the coolest summer nights of any big Indian city, often dropping below 21°C/70°F.

Further north, **Pune** is notably cooler and drier than **Mumbai**, just on the other side of the Ghats. The coast from Mumbai to **Trivandrum** (the monsoon's traditional entrance point) is among the wettest places in all of India. Much of the Western Ghats is cloaked in **fog** and relentless **rain** through the summer. Interestingly, this strip is much less prone than the **Bay of Bengal** coast to tropical cyclones. Mumbai was spared any serious cyclones through most of the twentieth century, although it's hardly immune: a surge of cyclone-related high water killed some 100,000 people in 1882. Nearly all of Mumbai's rain occurs during the monsoon, but southward, the coast toward Trivandrum gets frequent spring showers well before the official monsoon begins – about every third day as early as April. The monsoon rains across the southeast intensify with autumn as the winds turn to northeast. The rainiest month in **Chennai** is actually November, although the city is oppressively steamy for months beforehand. With the peak of **Nuwara Eliya** near its centre, much of **Sri Lanka** gets drenched all year. Autumn is the wettest period; another rainy season affects all but the island's east coast as the monsoon moves north during April and May. Even mid-winter brings rain every two or three days in **Colombo**. The lowlands covering the north half of Sri Lanka are less wet overall, with little rain in mid-summer, but virtually every place on the island sees at least 1000mm/39in a year.

Indonesia

Papua New Guinea | Solomon Islands

What **Indonesia** lacks in longitude, it makes up for in its vast east–west spread, extending over 4800km/3000 miles (and even further if you include **Papua New Guinea**). The **volcanic peaks** of Indonesia ensure that the main temperature differences are tied to elevation. As in Malaysia, anticipate a drop of roughly 6°C/11°F for every 1000m/3300ft, with very little difference between the months of the year. Avoiding the rainy season is a more challenging task – and you may want to avoid it, since it brings most of Indonesia more than 1500mm/59in of water. The southwestern chain of islands, including **Bali** and **Timor**, are drier than their northern counterparts. In general, destinations along south and west slopes – the west coast of **Sumatra**, for example – get their heaviest rains from May to October. North-facing locales – such as the bulk of Sumatra, **northern Java** and **Sulawesi** – are wettest from November to April. The same holds true as you move east and south into Papua New Guinea and the **Solomon Islands**, where the Australian monsoon dominates. The soaring peaks on Papua New Guinea can be incredibly cold, with snow clinging to the highest ones. Some Indonesian islands tend more toward a two-peak equatorial pattern, with wet spells from March to

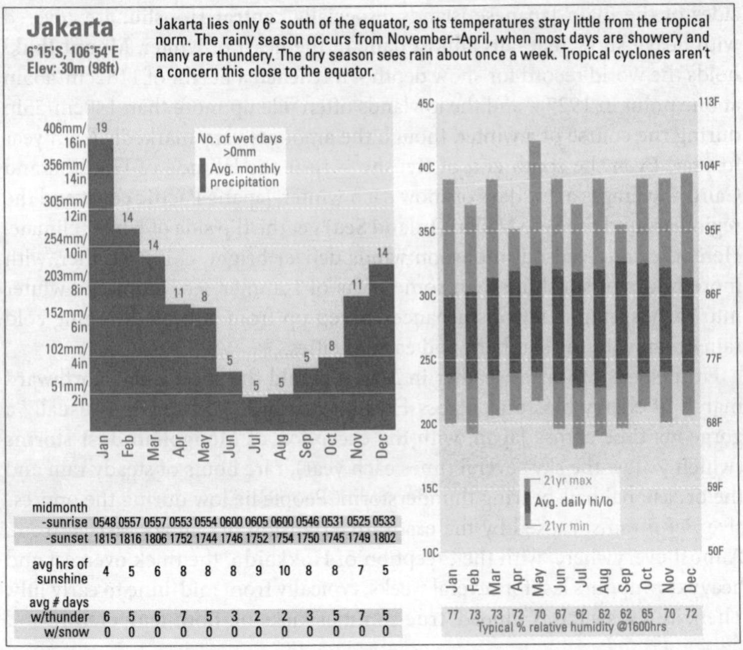

May and near the year's end, plus ample rain even during the dry periods. There are infinite variations to these guidelines, as the array of islands in and near Indonesia can produce small-scale blocking of monsoon flow in either direction. It's worth checking with a reliable local source on the island(s) you plan to visit. **El Niño** years tend toward drought across Indonesia during the November-to-April season, exacerbating the risk of forest fire and smoke.

Japan

The hard-core seasonality of Chinese climate gets moderated on its way to the Japanese archipelago. This chain of over 4000 islands runs from the tropics to the northern mid-latitudes, so there is plenty of north-to-south contrast. However, the broad strokes are similar to China's: wintertime cold and sweaty summer heat are interspersed with distinctly rainy transitions. Although the seasons here are as sharply defined as anywhere else, **Japan**'s maritime location and its **rugged topography** provide a stimulating array of micro-climates and day-to-day weather change.

In winter, as the cold northwest winds of the Asian monsoon pick up moisture across the **Sea of Japan**, they lead to one of the world's great climate contrasts on either side of the mountains that bifurcate Honshu. The western sides of the main Japanese islands, especially **central Honshu**, are covered with heavy snow that falls almost daily in the heart of winter. **Mount Ibuki** holds the world record for snow depth – it reached a height of 1182cm/465in at one point in 1927 – and the lowlands often pile up more than 140cm/55in during the course of a winter, though the amounts vary markedly from year to year. Even the south end of **Kyushu** – near the latitude of Houston and Cairo – averages a few days of snow each winter. Japan's **Pacific coast** and the region around the **Seto Naikai (Inland Sea)** get the flip side of winter climate. Here, the squeezed-out monsoon winds deliver bright, chilly weather, with more hours of sunshine than some parts of summer see. From late winter into early spring, a few disturbances sweep up from Taiwan, bringing cold rain or snow to the southern and eastern cities.

Fitful spells of warm weather in March herald the spectacular northward march of cherry blossoms across Honshu in April. Mid-spring is usually a gorgeous time across Japan, with the exception of **Mongolian dust storms** (which yellow the sky several times each year), rare bouts of steady rain and the occasional hail-bearing thunderstorm. People lie low during the oppressive *Bei-u* rains, caused by the eastward extension of China's *Mei-yu* front. Almost everywhere, with the exception of **Hokkaido**, the thick overcast and heavy downpours last for several weeks, typically from mid-June to early July. Then, as if a switch is flipped, true summer kicks in. Both temperature and

Osaka

34°47'N, 135°27'E
Elev: 15m (49ft)

The Seto Naikai (inland sea) helps make Osaka a notch warmer than Tokyo in summer, as reliable afternoon sea breezes give way to calm but sultry nights. Typhoons are a risk in September and October, although on average more rain results from the multi-day Bei-yu soakers of June. Frequent light snows alternate with mild, sunny days in winter.

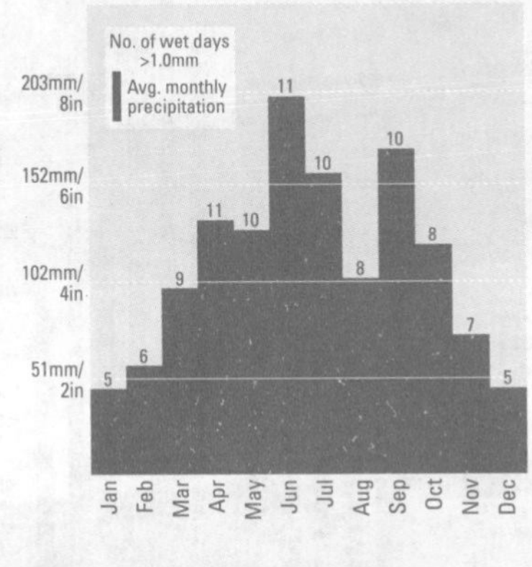

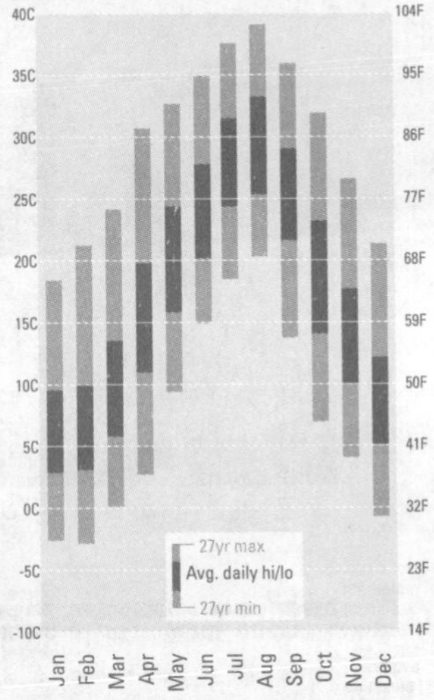

	Jan	Feb	Mar	Apr	May	Jun	Jul	Aug	Sep	Oct	Nov	Dec
midmonth -sunrise	0705	0645	0610	0527	0456	0444	0456	0518	0541	0603	0632	0658
-sunset	1710	1741	1805	1830	1854	1913	1912	1847	1806	1724	1654	1648
avg hrs of sunshine	5	5	5	6	6	5	6	7	5	5	5	5
avg # days w/thunder	1	<1	<1	1	1	2	5	9	3	1	1	<1
w/snow	11	11	4	0	0	0	0	0	0	0	0	2

	Jan	Feb	Mar	Apr	May	Jun	Jul	Aug	Sep	Oct	Nov	Dec
Typical % relative humidity @1700hrs	54	54	54	51	55	59	66	60	63	57	58	56

Sapporo

43°07'N, 141°23'E
Elev: 11m (36ft)

Winter-weather fiends will adore Sapporo. It snows almost every day for months, often heavily. More than 100cm/39 inches can fall with ease. Summers are mild on average but can vary dramatically from day to day and year to year. Late-summer rains become winter snows with little break; the driest, mildest period is from May into July.

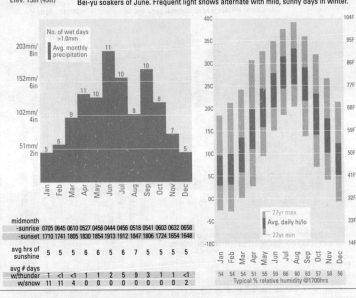

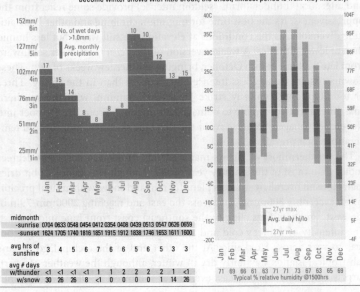

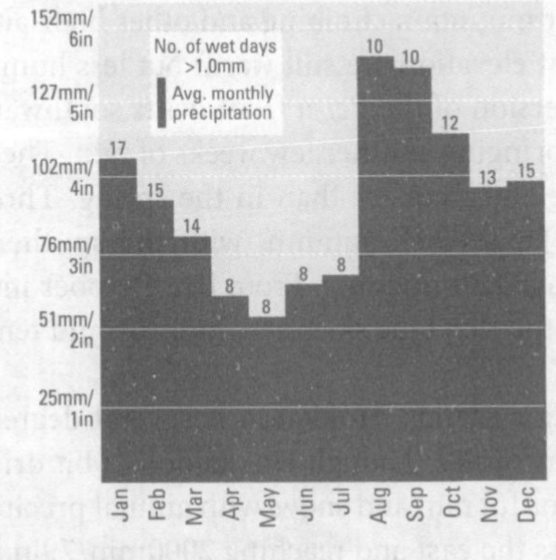

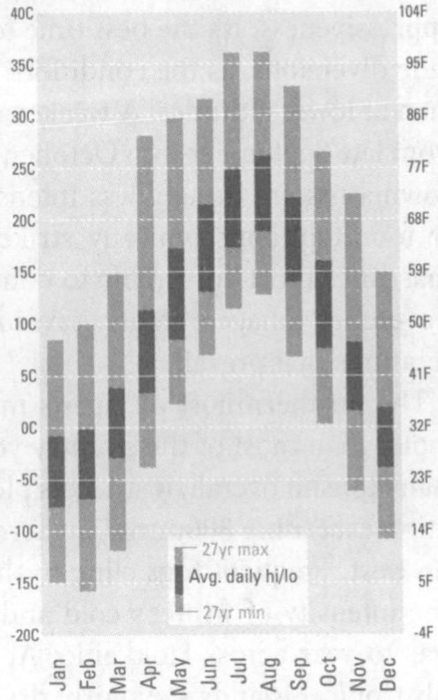

	Jan	Feb	Mar	Apr	May	Jun	Jul	Aug	Sep	Oct	Nov	Dec
midmonth -sunrise	0704	0633	0548	0454	0412	0354	0408	0439	0513	0547	0626	0659
-sunset	1624	1705	1740	1816	1851	1915	1912	1838	1746	1653	1611	1600
avg hrs of sunshine	3	4	5	6	7	6	6	5	6	5	3	3
avg # days w/thunder	<1	<1	<1	<1	1	1	2	2	2	2	1	<1
w/snow	30	26	26	8	<1	0	0	0	0	1	14	26

	Jan	Feb	Mar	Apr	May	Jun	Jul	Aug	Sep	Oct	Nov	Dec
Typical % relative humidity @1500hrs	71	69	66	61	63	71	71	73	67	63	65	69

Tokyo

35°33'N, 139°47'E
Elev: 8m (26ft)

July and August are bright and humid in Tokyo, with a cooling shower or thunderstorm every few days. Summer is flanked by wet periods in late May–June (strings of soggy days) and September–October (a typhoon hits every two or three years). Late autumn is often breezy. Winter is crisp and sunny, with rains increasing by March.

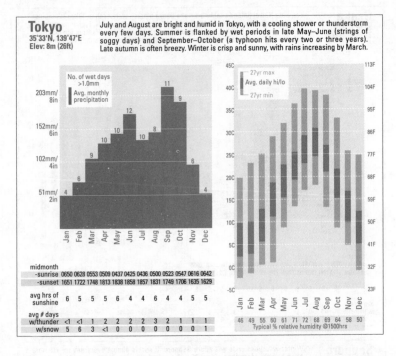

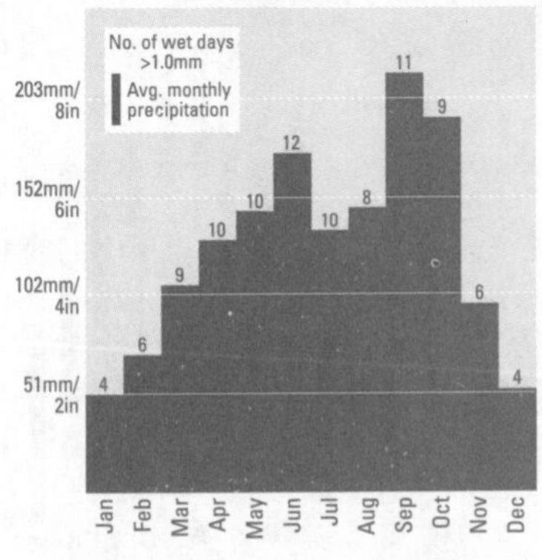

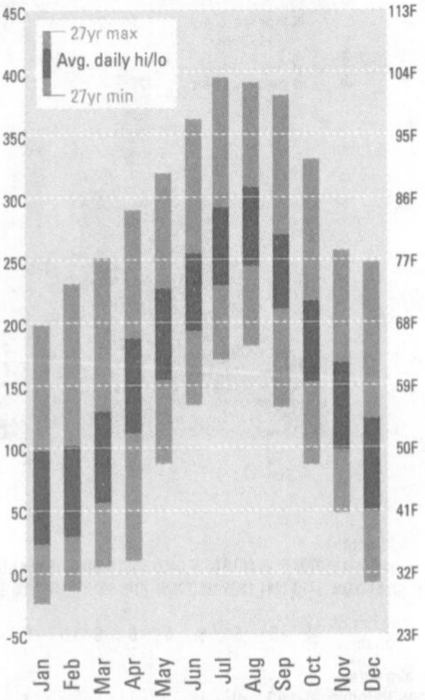

	Jan	Feb	Mar	Apr	May	Jun	Jul	Aug	Sep	Oct	Nov	Dec
midmonth -sunrise	0650	0628	0553	0509	0437	0425	0436	0500	0523	0547	0616	0642
-sunset	1651	1722	1748	1813	1838	1858	1857	1831	1749	1706	1635	1629
avg hrs of sunshine	6	5	5	5	6	4	4	6	4	4	5	5
avg # days w/thunder	<1	<1	1	2	2	2	2	3	2	1	1	1
w/snow	5	6	3	<1	0	0	0	0	0	0	0	1

	Jan	Feb	Mar	Apr	May	Jun	Jul	Aug	Sep	Oct	Nov	Dec
Typical % relative humidity @1500hrs	46	49	55	60	61	71	72	68	69	64	58	50

humidity soar to tropical levels and stay there into September. Sea breezes and afternoon thunderstorms (seldom intense) provide some relief from the oppressiveness. It's the best time for mountain climbing and other high-altitude diversions, as the conditions at elevation are still warm but less humid than at lower altitudes. A weaker version of the *Bei-u* front drifts southward from late September into October, bringing another few weeks of rain. These downpours are usually less intense and frequent than in the spring. Three or four **typhoons** typically strike Japan each autumn, with the **southern coastlines** most vulnerable to damage and flooding. From late October into November, foliage watchers savour the rich blue skies and cool-to-mild temperatures that prevail.

The northernmost of Japan's major islands, **Hokkaido** is several degrees cooler than most of the country year round. Though Hokkaido is a bit drier than Honshu overall, it still gets plentiful rain and snow, with annual precipitation exceeding 800mm/32in across the **east** and reaching 2000mm/79in in the **west**. Frequent **fogs** cling to the **southeast coast** from June into August. The intensity of winter's cold and summer's heat can vary noticeably from year to year across Hokkaido. At the other end of the archipelago, Japan's sub-tropical islands stay fairly dry in winter, although the weather can turn surprisingly chilly and overcast even as far southwest as **Okinawa**. Summer is hot, humid and sunny, punctuated with typhoons that may strike these isolated islands at full ferocity. Japan was pummelled by ten typhoons in 2005, beating the old record of six set in 1995. The deadliest of the barrage, Typhoon Tokage, killed 79 people.

Kazakhstan

This is the behemoth of Central Asian countries, a sprawling land that spans 40° in longitude. That's one-tenth of the way around the globe, fully twice the breadth of India. But Kazakhstan features less weather variety than you might expect across such a vast area, mainly as a result of its landlocked location. Only the **Caspian** and **Aral Seas** provide any marine influence, and it's minimal at best. **Aktau**, on the east Caspian coast, averages less than 150mm/6in of rain a year, and winds top 54kph/34mph on more than eighty days a year. Decades of irrigation have drained most of the Aral, devastating the region's economy and ecology. The **northern steppes**, brushed by bands of summer showers and thunderstorms, average roughly 500mm/20in in annual moisture. Wet spells may settle into the mountainous southern fringe, including **Almaty**, from March into May, followed by early-summer

Alma-Ata

43°14'N, 76°56'E
Elev: 847m (2778ft)

Tucked into the foothills of the Kirghiz range, Alma-Ata runs a bit cooler than the nearby steppes in summer, although days may turn hot and nights chilly in this arid region. Winter brings even more spectacular temperature swings, with frequent light snow and fog. Spring days can be damp and thundery; late summer is drier and more temperate.

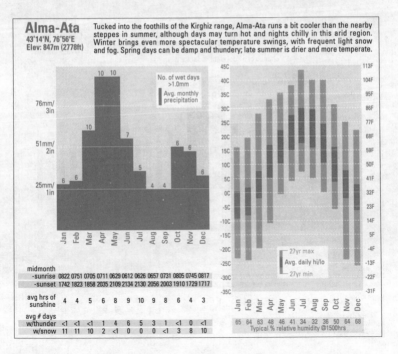

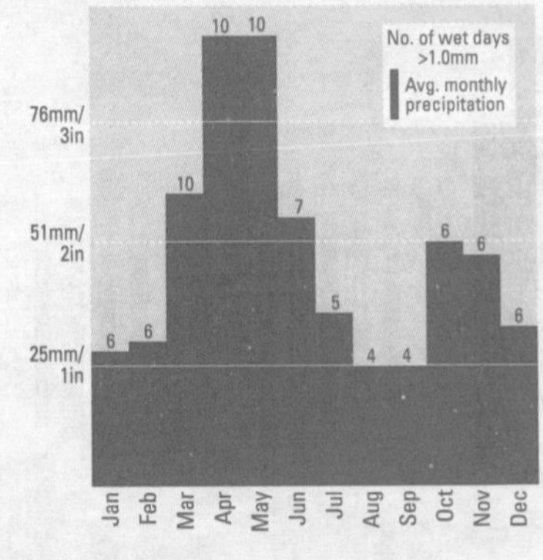

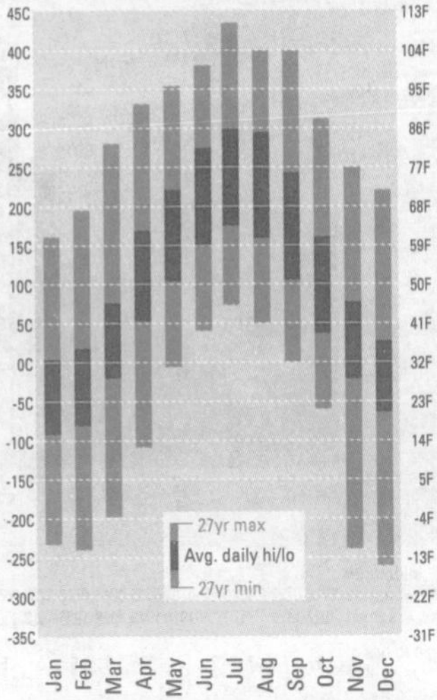

	Jan	Feb	Mar	Apr	May	Jun	Jul	Aug	Sep	Oct	Nov	Dec
midmonth -sunrise	0822	0751	0705	0711	0629	0612	0626	0657	0731	0805	0745	0817
-sunset	1742	1823	1858	2035	2109	2134	2130	2056	2003	1910	1729	1717
avg hrs of sunshine	4	4	5	6	8	9	10	9	8	6	4	3
avg # days w/thunder	<1	<1	<1	1	4	6	5	3	1	<1	0	<1
w/snow	11	11	10	2	<1	0	0	0	<1	3	8	10
Typical % relative humidity @1500hrs	65	64	63	48	46	41	34	32	36	50	64	68

thunderstorms. Temperatures across Kazakhstan combine the dramatic seasonal shifts of Russia with plenty of wind and the large daily range of an arid climate. Winters are somewhat cloudy and quite cold: most spots dip below −10°C/14°F regularly. Relief comes to the north-facing mountain slopes with the downslope **föhn** winds, which can howl at over 160kph/100mph through narrow gaps such as the **Dzhungarskiy** northeast of Almaty. Summers are pleasantly warm at altitude and hot on the plains and plateau.

Malaysia

Brunei | Singapore

Seasonality is barely a concern when planning travel to this moist equatorial country. Temperatures in **Malaysia** hardly deviate through the year, with highs near sea level usually in the range of 30–32°C/86–90°F, and lows around 22–25°C/72–77°F. The range drops about 6°C/11°F for every 1000m/3300ft in elevation. Rainfall is well distributed through the year, with a slight increase during the **monsoon transition periods** (April–May and October–November). By and large, you can expect to get a shower or storm

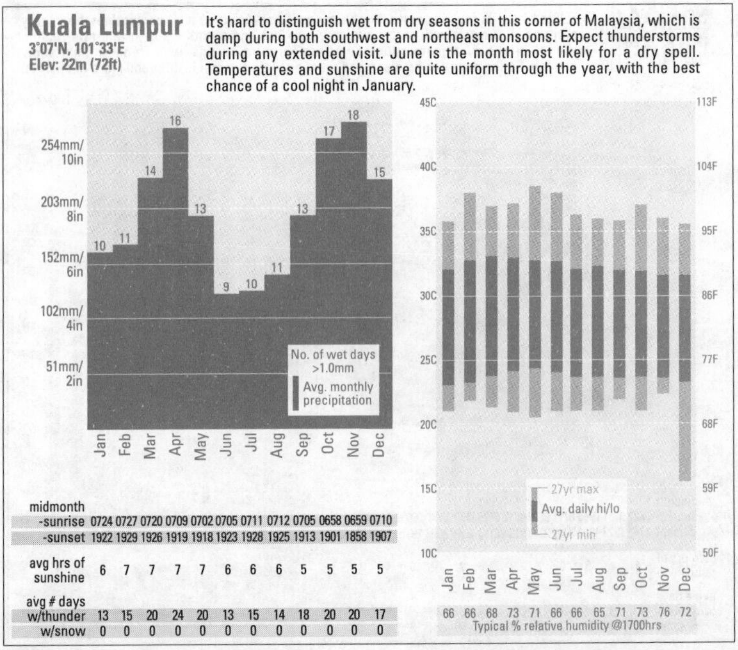

Kuching

1°29'N, 110°20'E
Elev: 27m (89ft)

There's more seasonal temperature variation in Kuching than you'd expect so close to the equator. It's mainly as a result of the torrential monsoon rains that bring afternoon highs down a few notches on most days from November–February. Mid-year is hotter and thundery, but only half as wet (yet still soggy).

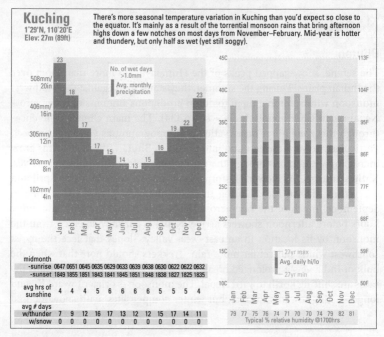

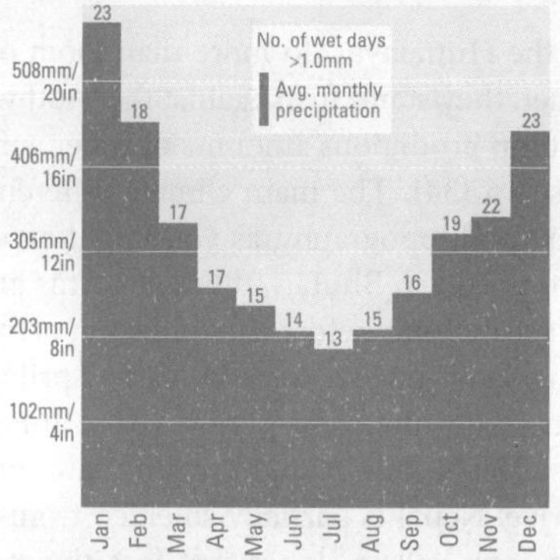

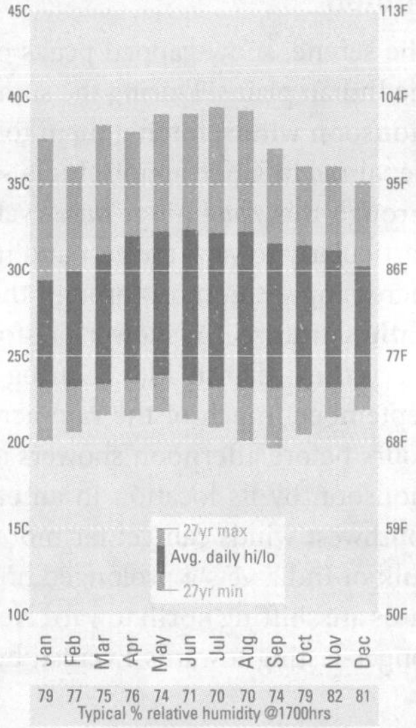

	Jan	Feb	Mar	Apr	May	Jun	Jul	Aug	Sep	Oct	Nov	Dec
midmonth -sunrise	0647	0651	0645	0635	0629	0633	0639	0638	0630	0622	0622	0632
-sunset	1849	1855	1851	1843	1841	1845	1851	1848	1838	1827	1825	1835
avg hrs of sunshine	4	4	4	5	6	6	6	6	5	5	5	4
avg # days w/thunder	7	9	12	16	17	13	12	12	15	17	14	11
w/snow	0	0	0	0	0	0	0	0	0	0	0	0

every other day. The exceptions are those spots that face the winter monsoon winds head on, such as the northeast coast of the **Malayan Peninsula** and parts of north and east **Borneo**. Here, the rains can be torrential (over 600mm/24in per month in spots) from late autumn into early spring. In **Singapore**, the rains become almost twice as heavy during this period as they are from May to September. The tiny nation of **Brunei**, tucked within northwest Borneo, reflects the Malaysian climatic norm.

Maldives

Perhaps the most telling aspect of the Maldives' climate isn't local but global. The slow worldwide rise in sea level threatens to inundate much of this island nation within the next century. There's little high ground here: the tallest atolls barely top 2m/6ft. Since the Maldives straddle the equator, wet seasons vary, but they tend toward the summer half of the year. Temperatures hardly vary at all, staying sultry even for a tropical locale. Only the northernmost Maldives are at risk from tropical cyclones.

Nepal

Bhutan

The serene, snow-capped peaks of the **Himalayas** do more than loom over the Indian plains. During the summer, they stand firm against the southwest **monsoon winds**, forcing them to dump prodigious amounts of water across Nepal (as in Cherrapunji, India – see p.334). The main climate differences through this zone hinge on elevation and topography, as you might expect. Particularly toward the east and into northern **Bhutan**, thunderstorms grow increasingly common through the spring long before the monsoon arrives. **Kathmandu** sees a shower or storm about once a week by early April and every third day in May. During the core monsoon period (from June to September) much of the rain across Nepal falls at night, with a few sunny hours before afternoon showers arrive. Nepal is partially shielded from the monsoon by its location in an east–west valley; slopes that face the moist southwest winds can get far more rain. Trekkers should note that when the bulk of India gets a prolonged break in the monsoon, it typically means the rains are shifting north toward Nepal. The monsoon is briefer toward **K2** and longer-lasting towards **Everest**, but winter compensates, with snow focused

Kathmandu

27°42'N, 85°22'E
Elev: 1337m (4385ft)

West of nearby Darjeeling and at a lower elevation, Kathmandu gets a more pronounced warm-up on its typical bright winter days, with nights chilly but freezes rare. Showers, thunder, and humidity increase steadily through the mild spring. The June–September monsoon isn't quite as wet here as than on higher ground nearby.

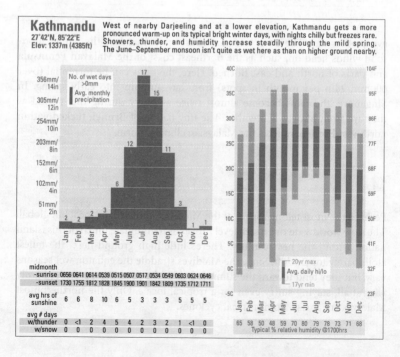

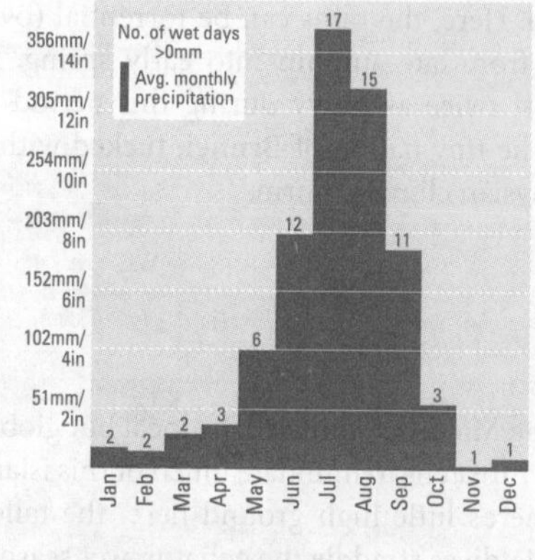

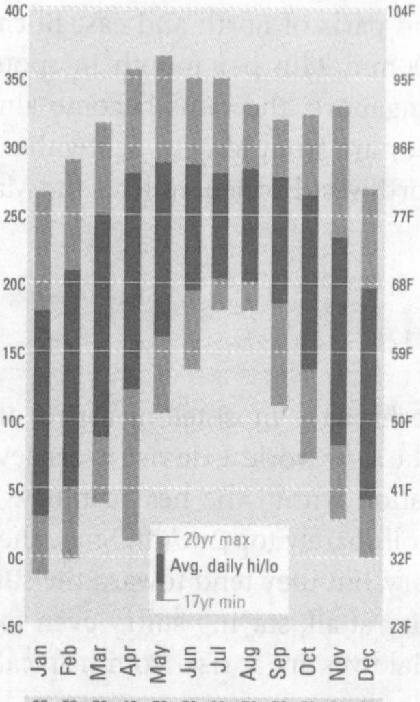

	Jan	Feb	Mar	Apr	May	Jun	Jul	Aug	Sep	Oct	Nov	Dec
midmonth -sunrise	0656	0641	0614	0539	0515	0507	0517	0534	0549	0603	0624	0646
-sunset	1730	1755	1812	1828	1845	1900	1901	1842	1809	1735	1712	1711
avg hrs of sunshine	6	6	8	10	6	5	3	3	3	5	5	5
avg # days w/thunder	0	<1	2	4	5	4	2	3	2	1	<1	0
w/snow	0	0	0	0	0	0	0	0	0	0	0	0

Jan	Feb	Mar	Apr	May	Jun	Jul	Aug	Sep	Oct	Nov	Dec
65	58	50	48	59	70	80	79	78	73	71	68

Typical % relative humidity @1700hrs

towards the K2 region, which sits far enough north to intercept mid-latitude systems. Below the wind-whipped peaks of the highest Himalayas, winter is typically sunny and crisp, with morning frost and occasional fog.

Philippines

Even the warm **South China** and **Philippines Sea** aren't quite enough to keep this chain of tropical islands from the effects of the Asian monsoon. While **Mindanao** and the **southern Philippines** stay warm to hot year round, **Luzon** (including **Manila**) experiences a few degrees of cooling, especially in the evening, from November to March. Winter nights in Luzon occasionally drop below 16°C/60°F even at sea level. This is one of the most frequent stopover points for northwest Pacific **typhoons**, which helps make the July-to-September period the wettest for much of the Philippines. As with Malaysia, north- and east-facing slopes tend to be at their dampest during the winter. Most other areas enjoy relatively dry winters, though showers or storms may arrive once or twice a week.

Manila

14°31'N, 121°00'E
Elev: 15m (49ft)

In this monsoonal regime, it only rains on a handful of days from November–April, when temperatures are still tropical but occasionally moderated by a stray Asian front. Spring is the breeziest but hottest season. Tropical cyclones can be quite destructive; the core of the season runs from August–October.

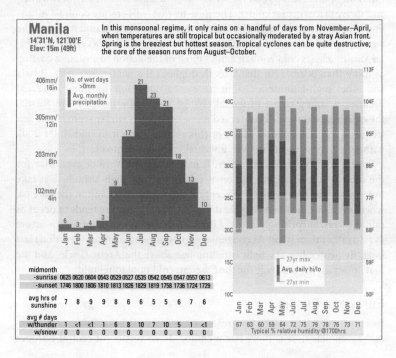

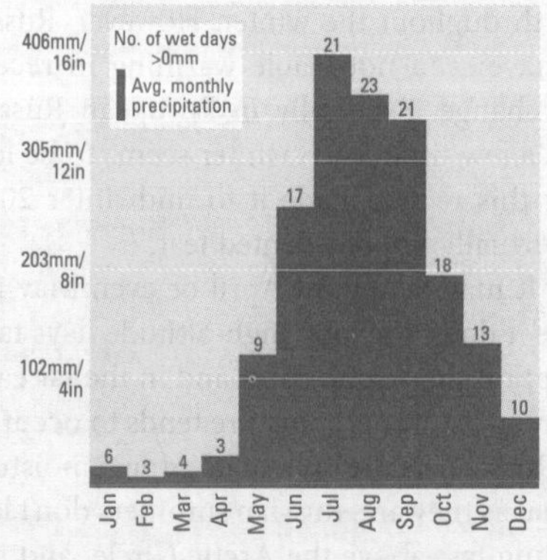

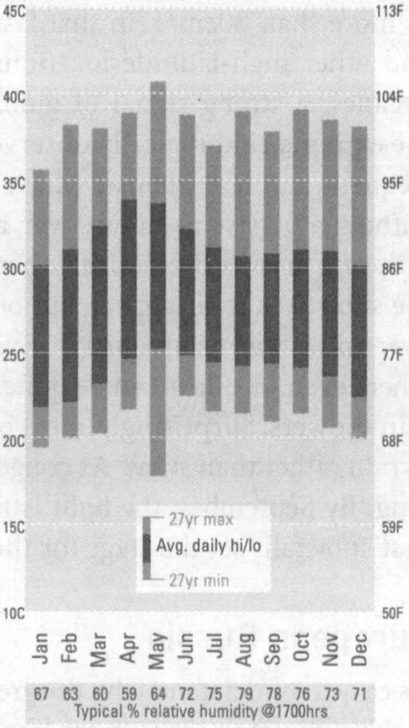

	Jan	Feb	Mar	Apr	May	Jun	Jul	Aug	Sep	Oct	Nov	Dec
midmonth -sunrise	0625	0620	0604	0543	0529	0527	0535	0542	0545	0547	0557	0613
-sunset	1746	1800	1806	1810	1813	1826	1829	1819	1758	1736	1724	1729
avg hrs of sunshine	7	8	9	9	8	6	6	5	5	6	7	6
avg # days w/thunder	1	<1	<1	1	6	8	10	7	10	5	1	<1
w/snow	0	0	0	0	0	0	0	0	0	0	0	0

Russia (including European Russia)

Armenia | Azerbaijan | Georgia

Although its reputation precedes it, **Russia**'s climate is extreme only after a particular fashion. There's no sugar-coating the fact that Russian winters are extremely cold. Yet much of the country enjoys warm mid-summer days that average above 20°C/68°F, and there are even a few bona fide hot spells. Thunderstorms are seldom intense outside of the mountains and the southwestern steppes. Periods of rain and snow, even if lengthy, tend to be on the light side across much of the country. The only places that see more than 1000mm/39in of moisture in a typical year are parts of the **Pacific coast** and some **mountain stations**.

The fact that Russia straddles Europe and Asia has meteorological, as well as political, implications. **European Russia**, home to much of the country's populace, is close enough to the **Atlantic** for modified maritime air to sweep in regularly. East of the **Urals**, the sparsely settled **Asian** portion of Russia is dominated by a mammoth pool of cold air that settles in each autumn and stays until spring. It's the lower elevations that feel the cold most sharply; hillsides can average as much as 20°C/38°F warmer than valleys. Even so, virtually all of **Siberia** dips below –40°C/–40°F at least once in a typical winter, and many places get far colder. The west-to-east temperature contrast means that Muscovites seeking relative warmth are just as likely to find it toward the west as toward the south. No other country can match Russia's extensive winter **snowpack**: it's cold enough across most of the nation to sustain a coating of more than 30cm/12in that lasts throughout the winter. However, Russia and other high-latitude locations have seen noticeable warming in recent decades, a strong signal of global change. Especially in European Russia, the extreme cold typically observed a few times each winter seems to be less frequent and less intense. Much of this region made it to midwinter 2007 without a persistent snow cover, a virtually unprecedented feat.

Spring arrives dramatically here. It may take until April or even May for the snowpack to disappear, but once it does, the long high-altitude days take over and it warms up quickly. Except on the **Arctic coast** and in the far east, where even July days can be quite dreary, summer moisture tends to occur as rain showers. Surprisingly, most of Russia gets the bulk of its annual moisture as rain rather than snow. As congenial as they are, Russian summers don't last long. By September, the light is fading fast above the Arctic Circle, and the first snowfalls set the stage for the Siberian cold pool to rebuild.

European Russia

It's easy to understand why the greatest cities of Russia lie on its western fringes. The climate here is more like a chilled-down version of central Europe's than the bitter beast that rules Siberia. The coast of the **Barents Sea**, though it's north of the Arctic Circle, is tempered by the warm **Gulf Stream**. Winters on the coast are peppered with snow showers and run a shade less cold than in **Moscow**, while summers tend to be dank and drizzly; spring brings the best shot at crisp, clear weather. The forested heart of European Russia, including **St Petersburg** and Moscow, gets rain or snow about every other day on average, with the frequency dropping slightly toward the south. Summer showers may be complemented by a thunderstorm every few days, especially toward the **Ukraine**. These rains usually pass more quickly than the winter snows, which are seldom heavy, but often persistent. When the summer rains fail, drought can be widespread and smoke from forest fires may hang above the region. In the steppes of the south, warm, dry winds known as *sukhovei* can parch crops. Temperatures in European Russia seldom descend to Siberian levels, but you'll be lucky to see a single day climb above 6°C/43°F between December and March. Spring and autumn are rather brief affairs here, with predictable timing. Frosts arrive in October, the snowpack is in place by late November, and the spring thaw kicks in during April.

The climate is a bit warmer and far more varied in and near the **Caucasus Mountains**, extending from the southern tip of Russia into **Georgia** and **Azerbaijan**. Screaming winds that sail through the **Transcaucasus valley** – typically reaching hurricane force at least once per year around **K'ut'aisi** – produce dramatic winter warm-ups, especially toward the western end. The Caucasus themselves are quite cool in the summer; in the winter, sunny days between snowstorms help make the mountains more appealing than the cloudy lowlands just to the north. Spring brings frequent rain and thunder to the western Caucasus and the plateau at the heart of **Armenia**. Drier, warmer weather then settles into the high country for the summer. The western Transcaucasus, sheltered from Siberian air, is almost sub-tropical. Average highs top 30°C/86°F in July, and winter readings below –10°C/14°F are rare. Rainfall along the moist **Black Sea coast** is sometimes heavy but can also vary sharply from month to month. At the eastern end of the Transcaucasus, the lower **Kura Valley** and the **Baku region** of Azerbaijan enjoy plenty of bright, crisp winter weather and dusty summer heat. The same applies to Armenia, although extremely cold winter air may pool for days within the basins that pockmark the plateau.

Siberia

This is the world capital of continental climates. With the **Arctic** frozen most of the year, there's no large, open body of water nearby to keep temperatures in check. This works to **Siberia**'s advantage in the summer, when the air can warm to shirtsleeve levels and beyond. The city of **Yakutsk**, which has sunk below –60°C/–76°F in winter, has been known to soar to 38°C/100°F in sum-

Moscow

55°58'N, 37°25'E
Elev: 190m (623ft)

Moscow winters are similar to those in northern Minnesota or southern Quebec, except for lighter snow and less dramatic temperature swings. The air can remain near freezing for days in cloudy, dank spells, but it only rarely dips to –25°C/–13°F. Late spring and summer are sunny and mild, with frequent thunderstorms and a few hot spells.

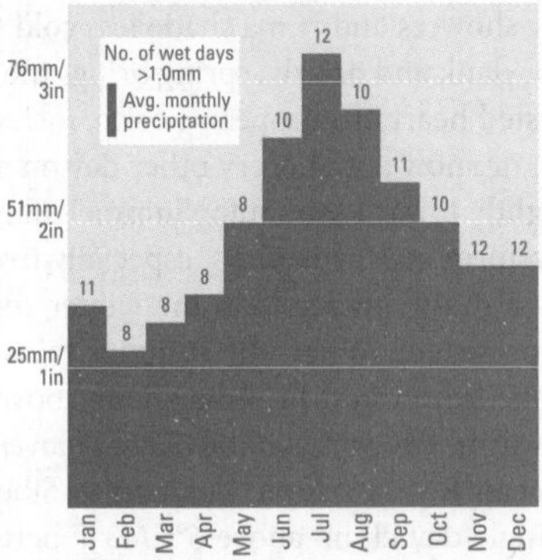

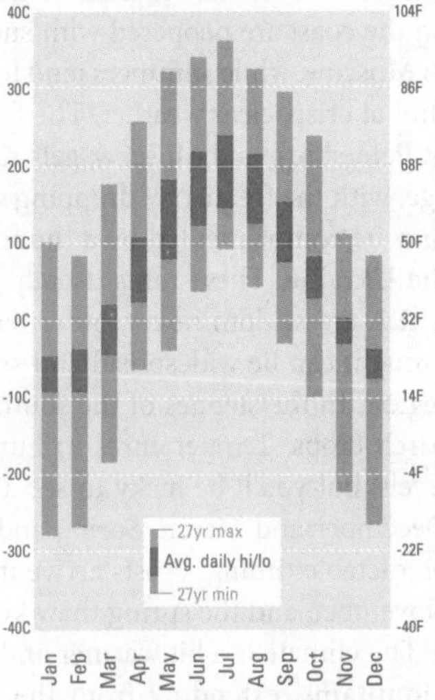

	Jan	Feb	Mar	Apr	May	Jun	Jul	Aug	Sep	Oct	Nov	Dec
midmonth -sunrise	0852	0756	0647	0626	0518	0444	0505	0601	0701	0801	0806	0855
-sunset	1628	1734	1833	2036	2136	2218	2207	2108	1949	1831	1623	1555
avg hrs of sunshine	1	3	4	6	9	9	9	8	5	2	1	1
avg # days w/thunder	<1	<1	1	1	7	9	9	5	1	<1	<1	1
w/snow	26	21	17	10	1	<1	0	0	1	9	19	26
Typical % relative humidity @1500hrs	81	76	68	57	51	57	60	63	67	73	82	84

St Petersburg

59°55'N, 30°15'E
Elev: 20m (66ft)

Short heat waves may strike St Petersburg from June–August, with typical summer readings comparable to London's. In winter, rounds of bitter cold alternate with near-freezing stretches. The Neva River normally ices over in December and thaws in early April.

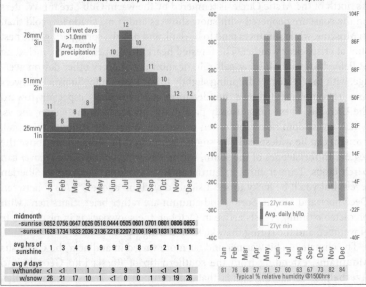

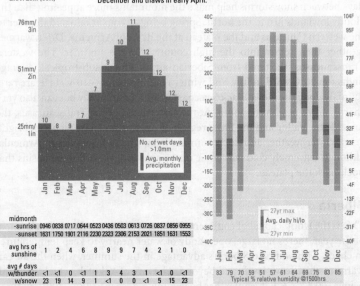

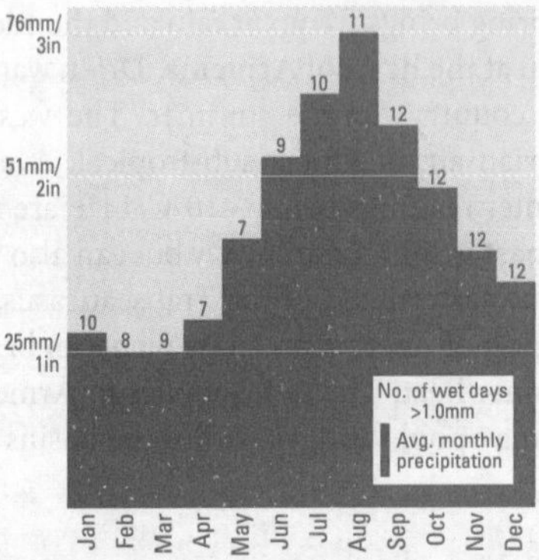

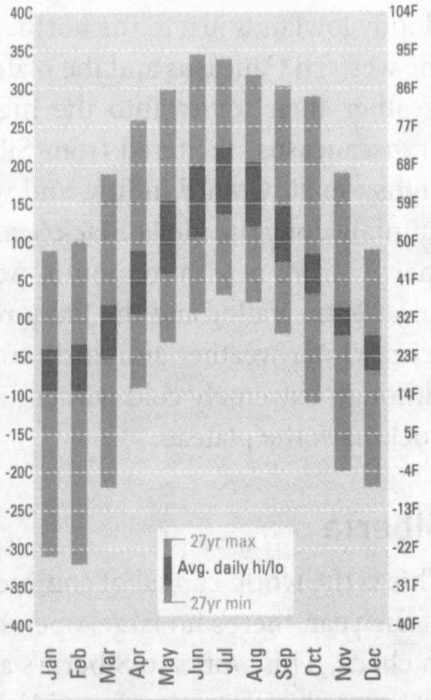

	Jan	Feb	Mar	Apr	May	Jun	Jul	Aug	Sep	Oct	Nov	Dec
midmonth -sunrise	0946	0838	0717	0644	0523	0436	0503	0613	0726	0837	0856	0955
-sunset	1631	1750	1901	2116	2230	2323	2306	2153	2021	1851	1631	1553
avg hrs of sunshine	1	2	4	6	8	9	9	7	4	2	1	0
avg # days w/thunder	<1	<1	0	<1	1	3	4	3	1	<1	0	<1
w/snow	23	19	14	9	1	<1	0	0	<1	5	15	23
Typical % relative humidity @1500hrs	83	79	70	59	51	57	61	64	69	75	83	85

Verkhoyansk

67°33'N, 133°23'E
Elev: 137m (449ft)

Few places can match Verkhoyansk for mild summers combined with extreme winter cold. The winds are calm when it's coldest, averaging under 5kph/3mph from November–March. Snow is frequent but very light. Summers are very sunny; most days are comfortable and dry, but it can be either hot or chilly at times.

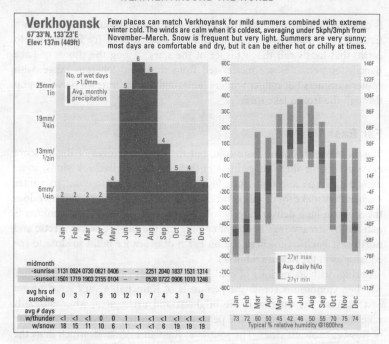

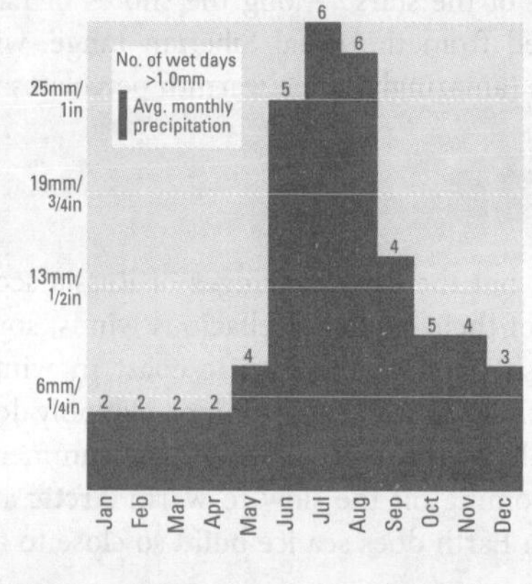

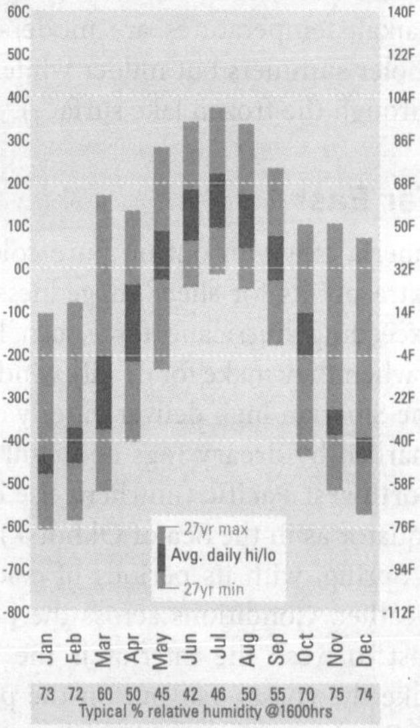

	Jan	Feb	Mar	Apr	May	Jun	Jul	Aug	Sep	Oct	Nov	Dec
midmonth -sunrise	1131	0924	0730	0621	0406	–	–	2251	2040	1837	1531	1314
-sunset	1501	1719	1903	2155	0104	–	–	0528	0722	0906	1010	1248
avg hrs of sunshine	0	3	7	9	10	12	11	7	4	3	1	0
avg # days w/thunder	<1	<1	<1	0	0	1	1	<1	<1	<1	<1	<1
w/snow	18	15	11	10	6	1	<1	<1	6	19	19	19

mer. Even near the **Arctic coast**, where clouds and chilly onshore winds are almost constant in summer, strong south winds can bring startling warmth. Yet frosts are possible across northern Siberia even in July. Most days are mild rather than hot, with showers becoming less frequent as the summer progresses. During autumn and spring, **western Siberia** sees wild gyrations in temperature as Atlantic and Siberian air masses twirl around low-pressure centres crossing the forested taiga. Temperatures are relentlessly brutal during mid-winter in the heart of the Asian cold pool, which is often centred roughly from **Lake Baikal** to **Verhoyansk**. Yet during the worst of it, the sky

How winter came to Russia's aid

Like a judo expert manipulating another's strength to his advantage, Russia learned through years of experience how to use the power of wintertime in its own defence. Would-be conquerors who extended their stay past the autumn equinox did so at their own risk, as **Napoleon** discovered in 1812. That September, some 400,000 of his troops invaded northwest Russia. Less than 40,000 of them straggled back to France after Russian forces, disease and early winter weather had laid waste to the army. History repeated itself in 1941 and 1942, when successive autumn campaigns from **Hilter**'s German forces failed miserably across European Russia. In 1942, the prolonged siege of **Stalingrad** (now Volvograd) played itself out through waves of bitter cold and snow.

is clear and the air remarkably still. You may hear the sound of your breath freezing, or a faint rustle, produced by tiny particles of **ice fog** settling to earth, that locals call the "whisper of the stars". Along the shores of Lake Baikal, temperatures are moderated from the usual Siberian range, with cooler summers but milder winters (amazingly, some warmth percolates up through the frozen lake surface).

Far East

Siberia may win out on pure cold, but the **eastern fringes** of Russia score extra points for sheer ruggedness of their climate. Hellacious winds, some exceeding hurricane force, can lash parts of the **Pacific coast** in winter – when they make for deadly wind chill – as well as in summer. Not only does the Siberian high deliver bitterly cold air through the winter, but summer is marked by **dreary fogs** and chill coming off the slow-to-warm **Arctic** and **northwest Pacific** (nowhere else on Earth does sea ice build so close to the equator as in the **Sea of Okhotsk**).

Spring, with its periods of cold, clear calm, provides the most tolerable weather. Conditions across the peninsula of **Kamchatka** vary greatly from east to west: the interior is the brightest and driest area, while blizzards rake the coasts. Far out on the peninsula's south tip, **Petropavlovsk** is one of Russia's snowiest cities. The southeast Russian mainland, which hugs the Sea of Japan, gets a monsoon-driven climate more akin to northeast China's. **Vladivostok** is actually colder than Moscow by winter, but distinctly humid during the heart of its mild, rainy summer.

Tajikistan

Kyrgyzstan

Almost half of **Tajikistan** sits more than 3000m/9900ft above sea level. What's distinctive about the desolate highlands of Tajikistan and **Kyrgyzstan** is how dry they are. Pinched off from the Indian monsoon by the Himalayas and Tibetan plateau, these peaks get most of their winter snow as weak systems arrive from the west. Across the **Pamirs**, the snow line can rise to 5000m/16,400ft during the summer. Still, snowpacks of more than 20cm/79in are not uncommon in favoured spots during the winter, and temperatures at elevation can plummet to dangerous levels. Summers are mild in the highlands and warmer below. A hellish, localized east wind called the *harmsil* has been known to raise readings in lower Tajikistan as high as 47°C/117°F.

Thailand

Myanmar

Cradled by mountains to its east and west, **Thailand** manages to avoid the worst extremes a monsoon climate can dish out. Although it's decidedly moist here, sunshine is widespread, especially in winter, which is almost rain-free across the country. Even during summer, a given day has roughly even odds of staying dry, outside of the mountainous regions where showers and thunderstorms gather regularly. The higher elevations and latitude of northern Thailand and **Myanmar** allow for some cool spells from December into February, with lows typically dropping below 16°C/60°F and sometimes into the 5–10°C/41–50°F range, even cooler at elevation. **Bangkok** and the nearby delta experience hot afternoons year-round, with evenings a shade less humid in the winter. March and April bring torrid pre-monsoon heat not unlike India's. Further south, the **Isthmus of Kra** and points beyond are similar to Malaysia, with little temperature variation and rains on the east coast in winter and the west coast in summer. May through to October is the rainy season across the bulk of Thailand, with the heaviest weather focused in the late summer. Thanks to its sheltered location, Thailand only rarely experiences direct typhoon landfalls, but the remnants of systems crossing Vietnam and Laos may bring a spell of wind and rain.

Bangkok
13°44'N, 100°34'E
Elev: 20m (66ft)

Though hot and humid, this monsoonal climate is a bit gentler than India's. Average highs stay above 30°C/86°F year round. Lows are marginally cooler during the sunny, dry winter, and a good sea breeze eases the spring heat. Rainfall peaks in May, with a slight lull before a bigger peak in September.

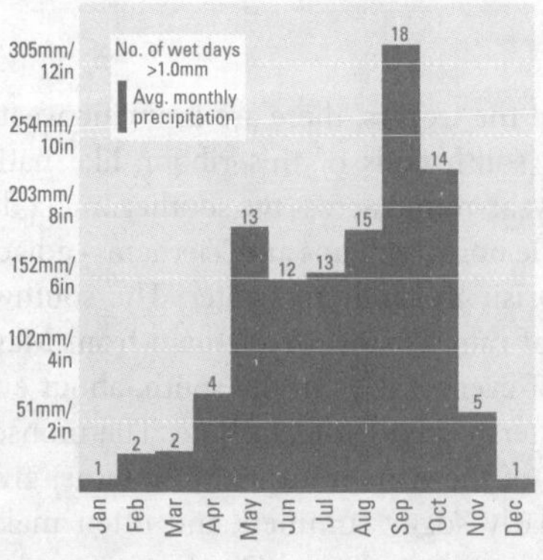

	Jan	Feb	Mar	Apr	May	Jun	Jul	Aug	Sep	Oct	Nov	Dec
midmonth -sunrise	0645	0641	0626	0605	0551	0550	0558	0605	0607	0608	0617	0633
-sunset	1809	1823	1828	1831	1837	1846	1849	1840	1819	1759	1747	1752
avg hrs of sunshine	9	9	9	9	7	6	6	5	5	6	8	8
avg # days w/thunder	<1	1	2	6	15	7	10	10	16	12	3	<1
w/snow	0	0	0	0	0	0	0	0	0	0	0	0

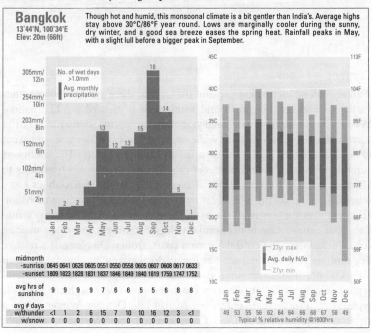

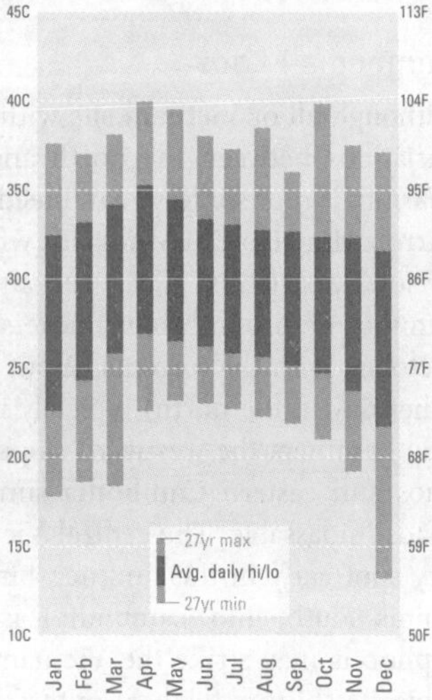

Jan	Feb	Mar	Apr	May	Jun	Jul	Aug	Sep	Oct	Nov	Dec
49	53	55	56	62	64	64	66	68	67	58	49

Typical % relative humidity @1600hrs

Uzbekistan

Turkmenistan

Dust may be the most vivid weather feature across most of **Uzbekistan** and its southern neighbour, **Turkmenistan**. Each spring, great clouds of dust whipped up from the **Kara Kum desert** cascade across the land. They often follow in the wake of blustery thunderstorms that drop little if any rain. The dustiest regions are in central Turkmenistan, with sixty or more gritty days a year, but a pale haze can spread well to the east and last for days. Points along northern mountain slopes, such as **Tashkent**, are most prone to light winter rain or snow. They're also the most likely to be warmed by southerly **föhn winds** that scour out the typically chilly air filtering down from Russia. Even the heart of the desert gets a few days of light snow and a frequent high overcast in winter. Although it's hardly sultry, there's a surprising amount of moisture in the air during the virtually rainless summer; the relative humidities are low simply because it's so hot, with some days reaching 45°C/113°F. Autumn arrives quickly, before which September offers a pleasing blend of warm temperatures, sunshine and little rainfall. The mountains of far east Uzbekistan have an array of micro-climates packed into a small area. For instance, the lower **Fergana Valley** is prone to howling east winds in winter and a steady west breeze in summer (as revealed by east-leaning trees).

Vietnam

Cambodia | Laos

Although all of **Vietnam** sits within the tropics, there are dramatic weather variations between the north and south ends of this ribbon-like nation. Warmth is more or less guaranteed year-round across the southernmost delta surrounding **Ho Chi Minh** city, while northern **Laos** and Vietnam – adjacent to southeast China – can get surprisingly chilly in winter. The southwest summer monsoon brings buckets of rain to most of Vietnam from May to October, with the torrents (2 out of every 3 days in the south, about every other day in the north) typically interspersed with sunshine. The monsoon also hammers the westward slopes of the **Annamitique Mountains**, giving Laos and eastern **Cambodia** similarly soggy summers and often making roads impassable. The central Vietnam coast, around **Da Nang**, is a relative dry spot early in the summer, but it gets drenched in autumn (along with points south, into Cambodia) as the monsoon begins to retreat. Several **typhoons** may strike the Vietnam coast in a given year, with the landfalls typically shifting further south from August into October. Winter brings a string of gorgeous days to Laos, Cambodia and south Vietnam, with light

Ho Chi Minh City
10°49'N, 106°40'E
Elev: 19m (62ft)

As in much of Southeast Asia, the second half of the year is the wettest. The monsoon arrives in late spring, with showers and storms roughly every other day thereafter. Typhoons – a serious threat here – ratchet up the rain and flood potential in September and October. It's usually dry and sunny, though still muggy, from December–March.

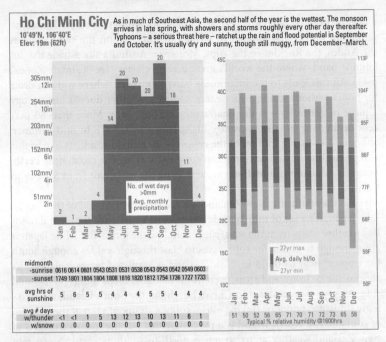

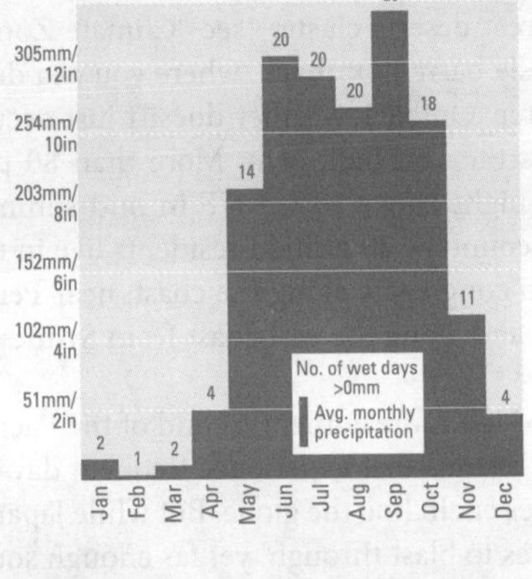

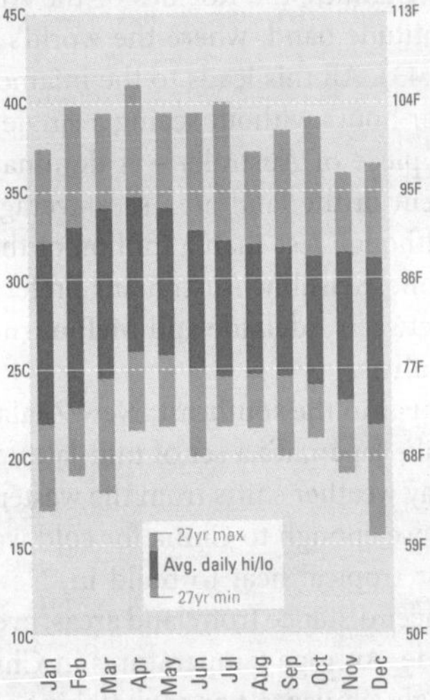

51 50 52 56 65 71 70 71 72 73 65 58
Typical % relative humidity @1600hrs

midmonth	Jan	Feb	Mar	Apr	May	Jun	Jul	Aug	Sep	Oct	Nov	Dec
-sunrise	0616	0614	0601	0543	0531	0531	0538	0543	0543	0542	0549	0603
-sunset	1749	1801	1804	1804	1808	1816	1820	1812	1754	1736	1727	1733
avg hrs of sunshine	5	6	5	5	4	4	4	5	5	4	4	4
avg # days w/thunder	<1	<1	1	5	13	12	13	10	13	11	6	1
w/snow	0	0	0	0	0	0	0	0	0	0	0	0

rains only once or twice a month. Temperatures are usually close to ideal in Laos (although there are wide daily swings, and it can turn either chilly or hot), while conditions get increasingly muggy into spring across the south peninsula. The central Vietnam coast is at its driest from February to June; further north, the *crachin* ("spitting") rains bring 3- to 5-day spells of cool drizzle every week or two around **Hanoi**, where temperatures can dip below 7°C/46°F.

Australasia/South Pacific

Water – or the lack of it – has everything to do with the weather across this vast realm where the **Pacific** and **Indian Oceans** meet. The island continent of **Australia** and its much smaller neighbours are surrounded by mammoth stretches of sea. If you took a globe and shifted it so New Zealand were at the top, all of the world's other big land areas (except Antarctica) would lie across the bottom half.

Given all the water lapping at its shore, it's ironic that Australia is the driest of the world's six settled continents. Most of the Australian coastline gets a healthy dose of rain each year, but inland it's a different story. There's no mountain range high enough to wring out moisture on the same scale as the Andes, the Rockies or the Alps. Moreover, Australia sits astride the 30° latitude band, where the world's great deserts cluster (see "Climate Zones", p.43). All this leads to the infamously barren **Outback**, where you can drive for hours without seeing a single tree. Outback weather doesn't just occupy a piece of Australia – it dominates the vast bulk of it. More than 80 percent of the land area sees average highs above 33°C/91°F in mid-summer, although less than a million of the country's 20 million residents live in this zone. Sensibly, Australians prefer to congregate along the coast: near **Perth**, between **Adelaide** and **Melbourne**, and along the east coast from **Sydney** to **Cairns**.

Far to the southeast, **New Zealand** is a more authentic child of the Pacific. Like Japan, it's a set of mid-latitude islands that experience frequent day-to-day weather shifts from the westerlies encircling the globe. But while Japan is close enough to China for cold waves to blast through, yet far enough south for tropical heat to build in, New Zealand is insulated by its isolation. Its sheer distance from land areas, even from Australia, keeps things on the mild side. Average temperatures in **Christchurch**, on the **South Island**, are strikingly similar to London's (taking into account a six-month flip of the seasons, of course). The **North Island** is a touch warmer, but the main variations in New Zealand weather aren't so much north-to-south as east-to-west: rain and snow are far heavier on the western slopes than they are on the east.

From New Zealand to the equator, the **southwest Pacific** is peppered with hundreds of islands that range from volcanic peaks to flat-as-a-pancake atolls. The weather across this zone is close to the quintessential tropical ideal, albeit a bit on the moist side. **Tropical cyclones** are a threat, although they're somewhat less common than across the northwest Pacific. Most of the rains are in the afternoon downpour category. Temperatures rarely dip below 18°C/64°F or rise above 32°C/90°F.

Australia

Alice Springs | Brisbane | Cairns | Darwin | Melbourne | Perth | Sydney

You can't look at a map of Australian rainfall and temperature without being struck by the effect of the **Great Dividing Range**. This low-slung set of peaks shelters the population belt of Australia's southeast coast from the dry, blast-furnace heat found through the vast interior. The **Outback**'s well-deserved reputation for scorching summer heat is balanced by its invigorating winter, with sunny days ideal for bushwalking and nights that can chill below freezing almost anywhere. Great stretches of the Outback qualify as desert, but their low rainfall averages can be misleading, as long stretches of dry weather are punctuated by wayfaring tropical lows that drift across the country and produce rare bouts of substantial rain that can flood roads and strand people for days. **Alice Springs**, in the heart of the Outback, got 205mm/8.1in (close to its annual average) on a single day in March 1988.

On top of the Dividing Range is the nation's coolest climate outside Tasmania, with sublime summer weather and just enough snow to support a few ski resorts. Follow the range south across the **Bass Strait** to **Tasmania**, and the weather takes on a maritime cast, with coolness year round and rain or snow on more than half the days of the year. Australia's north coast is part of the great sweep of the **Indian monsoon**. Buckets of rain can fall, just as in southeast Asia, but virtually all of the rain arrives between December and April. After each summer's deluge comes a winter that's warm and humid but virtually free of rain from May to October.

The country's typical rainfall pattern is tweaked by two big influences: **tropical cyclones** and **El Niño**. Cyclones can sometimes linger for days and drop incredible amounts of rain in the north – more than 500mm/20in in a single day. El Niño tends to raise the odds for **drought** across the north and east, while La Niña makes **flooding** more likely. Drought has become more persistent across southern Australia in recent years, and climate-change projections show this trend continuing.

Victoria and New South Wales

Like a microcosm of the nation, this corner of Australia has a moist coastal climate on one side and a dry Outback on the other, with mountains in between. It's easy to see why **Sydney** and **Melbourne** were the first great cities to emerge down under. Along with their superb harbours, both have frequently magnificent weather. Winters are reliably cool but never frigid: it's seldom close to freezing in central Sydney (although snow fell in 1836), and near Melbourne, frost only rarely extends from the hills into town. In the Great Dividing Range, **Canberra** (at an elevation of 570m/1870ft) has typically frosty winter mornings and snow about once every other year, and above 1500m/4900ft the ground can stay white for several months each winter. As with much of Australia, the most dramatic extremes across the southeast tend to occur in summer rather than winter. That's when the sharpest contrast exists between a then-baking continent and an always-cool **Southern Ocean**. From spring through early autumn, thunderstorms often rumble eastward from the Dividing Range to the New South Wales coast from Sydney to **Brisbane**. They can bring torrential rain, frequent lightning, and occasionally large hail (one of the world's most expensive hailstorms pounded Sydney in 1999; see box, p.76). Although it's cloudier and cooler than Sydney in the winter, Melbourne gets only half as much rain in a typical year, and it's more prone to blasts of summer heat escaping the interior. However, Sydney can simmer as well. In the great heat wave of January 1939, both cities hit all-time highs exceeding 45°C/113°F, while great stretches of the interior were ravaged by bush fires. This perennial threat can send a smoky pall to the coasts, destroying lives and property inland. Walkers should listen for warnings and avoid the bush during extremely hot and dry spells. Cool changes (cold fronts) sweep in from the south every few days during the summer, reducing the coastal temperature by as much as 20°C/36°F in an hour's time.

Queensland

The split personality of the southeastern states extends to **Queensland**, but with a tropical tilt. Here, with the Dividing Range pushed even closer to the water, the contrast between the soggy coast and the parched interior is astonishing. In the wet season of 2000, **Bellenden Ker** (just south of **Cairns**) set a national rainfall record with 12,461mm/491in. Up and down the touristy stretch from **Brisbane** to **Port Douglas**, the mountains can wring rain out of southeasterly trade winds any time of year. The November-to-April "Wet" is by far the most likely time to produce travel-dampening deluges in which the rain doesn't stop for days on end, while August and September are the driest months. **Townsville** is a relative dry spot: the curve of its coastline reduces the impact of the trades, so only about 1000mm/39in of rain falls there each year. **Tropical cyclones** are a serious threat along the entire Queensland coast during the Wet; you're sure to hear if any big ones approach. None made landfall through the 1980s and 1990s, but Larry – the worst cyclone to hit Australia in decades – brought sustained winds of more than 185kph/115mph to the Innsfail area, just south of Cairns, in March 2006. Despite the severe damage wrought by Larry, there were no direct fatalities, a sign of increased awareness and pumped-up evacuation plans for the growing numbers of residents and tourists along the Queensland coast. Apart from cyclones, **flooding** can close roads any time during the Wet, especially in the far north. Most of Queensland, though, lies to the west of the mountains, from the ultra-tropical **Cape York Peninsula** to the parched **Great Artesian Basin**. The latter may experience widespread flooding without rain toward the end of a Wet, as rivers run off from the soaked mountains into the basin and toward **Lake Eyre** (see opposite). Otherwise, western Queensland is much like the rest of the Outback: mostly dry year round, with summer heat and winter relief.

South Australia

Acre for acre, this is the driest part of an already dry country. The bulk of **South Australia** gets less than 300mm/12in of annual rainfall. As in the rest of the Outback, ample water supplies are a must for travellers venturing north from the coastal hills and wineries into the desert. While it's not quite as hot on average here as in northwest Australia, most summer days soar above 30°C/86°F from **Adelaide** north, and readings above 35°C/95°F often envelop the whole state (**Oodnadatta** has topped 50°C/122°F). Winters are typically dry and cool across the desert; along the coast, they can be as cloudy and damp as in Melbourne, although the rain is usually a bit lighter. Perhaps the greatest weather-related spectacle is the moistening of **Lake Eyre**, in the state's northeast corner. Over 15 percent of Australia drains into this huge, usually dry salt pan. Every few years, often during **La Niña**, the lake partially fills. Waterfowl flock to the area, and brine shrimp and other sea creatures materialize.

Northern Territory

There's a vast difference between the **monsoon regime** of **Darwin** and the classic desert climate of Alice Springs. Unlike the eastern states, this contrast is triggered not by a mountain range but purely by **latitude**. The northernmost quarter of the state – the **Top End**, which includes **Kakadu National Park** – is perpetually humid. High temperatures average 30°C/86°F or better year round, and the lows are only a few degrees cooler in winter than in summer. Most of the seasonal variation shows up in rainfall. The Top End shifts from insufferable heat during the monsoon build-up of October and November to near-daily thunderstorms in the summer, then to a bone-dry period from June to August when rain is virtually unheard of. Small, but intense, **tropical cyclones** can flay the coastline anytime during the Wet. Cyclone Tracy destroyed 70 percent of the buildings in Darwin on Christmas Day 1974, and Monica – the strongest cyclone on record to hit the Northern Territory – raked the state's north coast in April 2006. A gradual transition zone south of Darwin leads to the hot, dry regime that covers the state's southern half. Mid-summer nights at **Uluru (Ayers Rock)** typically hover above 21°C/70°F, and many afternoons top 38°C/100°F. The remnants of a cyclone or other low-pressure centre can bring a rare prolonged spell of rain to the Outback from summer into autumn. Fair-weather seekers will do best at Uluru from April to September, when each month averages only 1 or 2 wet days and daytime highs are typically close to room temperature.

Western Australia

If you want the most extreme weather Australia can dish out, this is a good place to start. Along the beautiful, but empty, northwest coast near the **Hamersley Range**, a single **tropical cyclone** can drift inland and halt a dry spell with a flood worthy of Noah. The town of **Onslow** has seen annual rainfall as low as 15mm/0.6 inches and as high as 1085mm/43in. Most of the nation's strongest tropical cyclones from 1950 to 2000 struck the Western Australia coast. With population and media outlets so scarce, travellers on the northwest coast during the Wet should watch for signs of impending cyclones (see p.83) and take heed. To the east, the **Great Sandy Desert** is Australia's hottest region, with an average high in January above 38°C/100°F; the town of **Marble Bar** hit this mark for 160 consecutive days during 1923–24. The southern half of the desert is a bit more tolerable. Cool changes sweep well inland from the west every week or so during the summer, and the immediate coast from **Carnarvon** south gets almost daily relief through medicinal sea breezes, such as the "Fremantle doctor" that graces Perth and the "Esperance doctor" that cools the **Nullarbor coast** to the southeast. When the doctor's absent, towns on the Nullarbor shore can jump from typical summer highs of around 26°C/79°F to readings above 45°C/113°F. The southwest strip from **Perth** to **Cape Leeuwin** is surprisingly Mediterranean, with toasty, dry summers and mild, soggy winters. It's damper here in July than anywhere else along the nation's south coast.

Tasmania

Peering at the rest of the continent from the poleward side of 40°S, **Tasmania** is a distinct region of its own in terms of weather as well as geography. Summer is the time to visit if you want conventionally nice weather, though even in January you should pack a jacket and raincoat. The mountains and west coast can get drenched any time of year. **Hobart** and other points to the east get lighter amounts, rarely pouring in the Sydney style – but even here, it's wet nearly every other day in summer and almost 2 out of 3 days in winter. There's nothing to block Tasmania's mountains from chilly upper-level winds, and **snow** can fall any time of year on short notice, so hikers should be well prepared. In winter, temperatures at sea-level Hobart never dip much below freezing, although snow falls across the elevated suburbs once or twice each year. Cool summer days are the norm in Hobart, though on very rare occasions a burst of heat pushes across the Bass Strait from Melbourne. After the air climbs and descends the island's peaks, it can send thermometers across southern Tasmania above 30°C/86°F on a few days each summer. In general, the air across Tasmania is uncommonly fresh, as the prevailing westerlies arrive along a continent-free path.

Alice Springs

23°49'S, 133°53'E
Elev: 546m (1791ft)

Though it's not Australia's hottest spot, Alice Springs is the toastiest place many visitors will encounter in Oz. Summer days average only about 5°C/8°F cooler than those in Phoenix; the dry air ensures some relief at night. Day-to-night temperature extremes are sharpest in the winter, with mild afternoons following cold nights. Rare rainy spells can bring travel-disrupting floods.

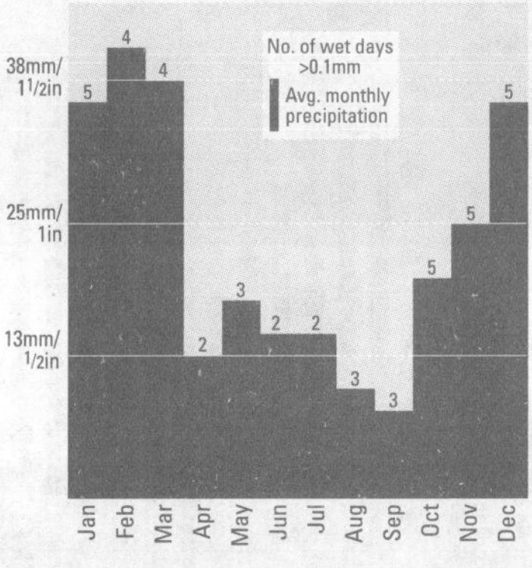

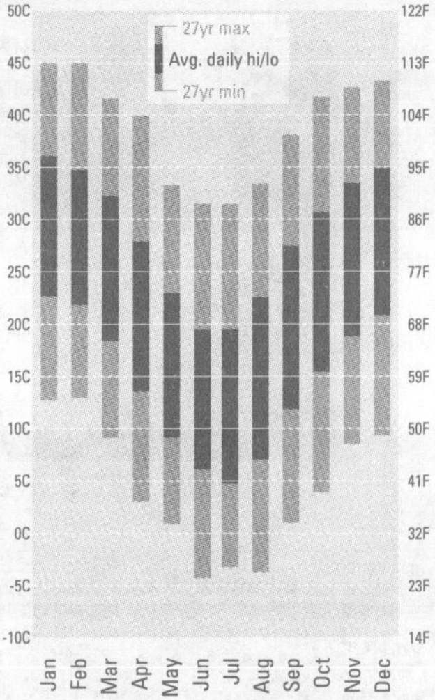

	Jan	Feb	Mar	Apr	May	Jun	Jul	Aug	Sep	Oct	Nov	Dec
midmonth -sunrise	0600	0622	0636	0648	0701	0715	0717	0701	0632	0602	0541	0542
-sunset	1927	1915	1851	1821	1800	1755	1804	1817	1828	1839	1857	1917
avg hrs of sunshine	10	10	10	9	9	9	9	10	10	10	10	10
avg # days w/thunder	3	2	1	1	<1	<1	<1	<1	1	3	4	4
w/snow	0	0	0	0	0	0	0	0	0	0	0	0

Jan	Feb	Mar	Apr	May	Jun	Jul	Aug	Sep	Oct	Nov	Dec
23	20	24	25	30	36	30	19	19	19	18	25

Typical % relative humidity @1700hrs

Brisbane

27°23'S, 153°07'E
Elev: 4m (13ft)

It may be the Miami of Australia, but Brisbane is distinctly cooler than its American counterpart. A July night offers a 50/50 chance of dropping below 10°C/50°F. Summers are warm but seldom blazing and far less humid than Cairns. Sunshine is prevalent year-round, with thunderstorms (and tropical-cyclone risk) focused during the Wet.

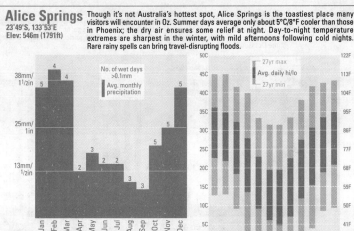

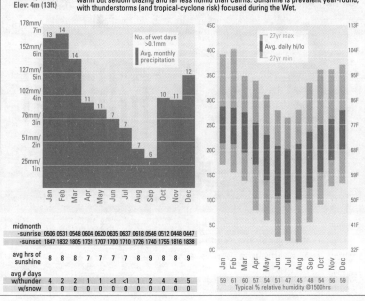

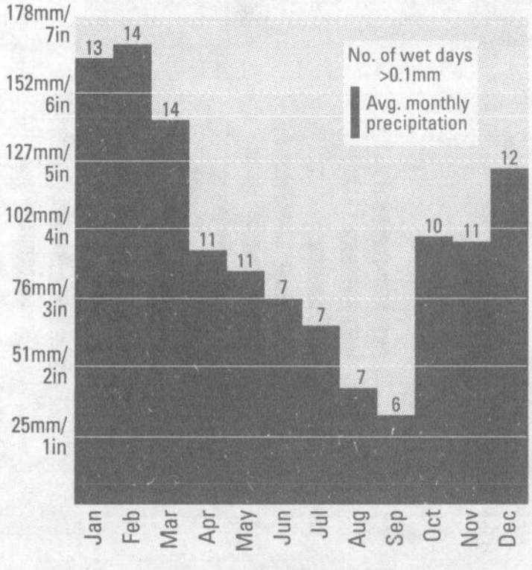

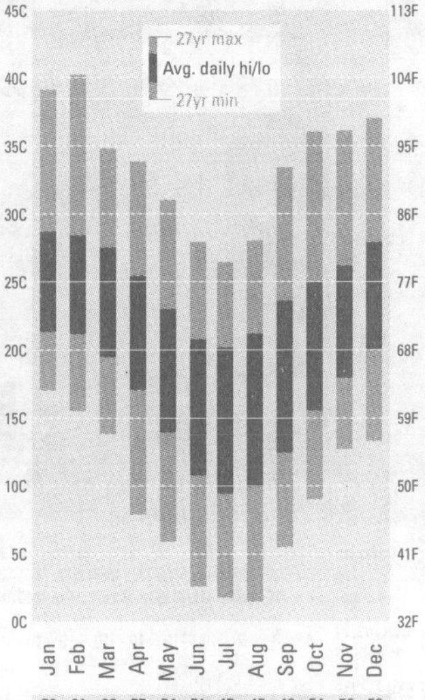

	Jan	Feb	Mar	Apr	May	Jun	Jul	Aug	Sep	Oct	Nov	Dec
midmonth -sunrise	0506	0531	0548	0604	0620	0635	0637	0618	0546	0512	0448	0447
-sunset	1847	1832	1805	1731	1707	1700	1710	1726	1740	1755	1816	1838
avg hrs of sunshine	8	8	8	7	7	7	7	8	9	8	8	9
avg # days w/thunder	4	2	2	1	1	<1	<1	1	2	4	4	5
w/snow	0	0	0	0	0	0	0	0	0	0	0	0

Jan	Feb	Mar	Apr	May	Jun	Jul	Aug	Sep	Oct	Nov	Dec
59	61	60	57	54	51	47	45	48	54	56	59

Typical % relative humidity @1500hrs

Cairns

16°53'S, 145°45'E
Elev: 3m (10ft)

Cairns' tropical climate ranges from somewhat muggy during the dry season (May–October) to full-blown steamy during the Wet, when flooding may hinder travel and tropical cyclones are a threat. Most summer days manage at least some sun. Winter brings only occasional showers, with the sunniest period from September into October.

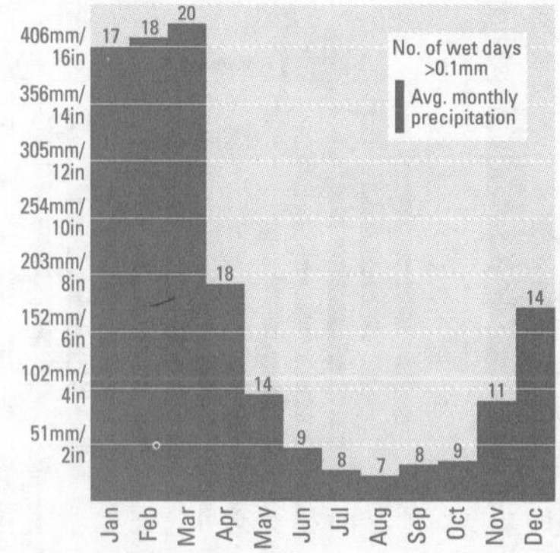

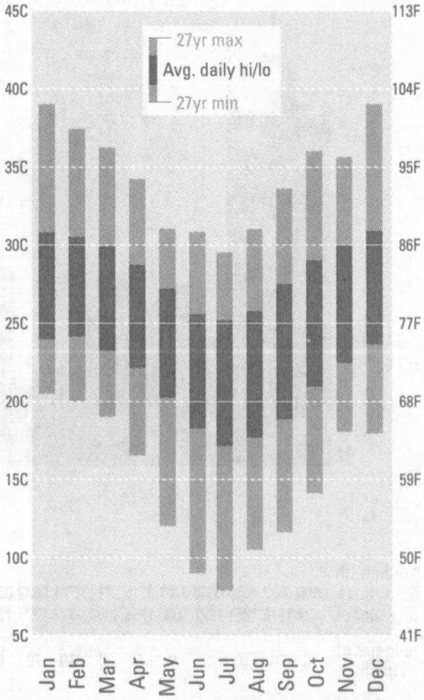

	Jan	Feb	Mar	Apr	May	Jun	Jul	Aug	Sep	Oct	Nov	Dec
midmonth -sunrise	0555	0612	0620	0625	0633	0644	0647	0636	0613	0549	0535	0538
-sunset	1857	1850	1832	1809	1753	1751	1759	1808	1812	1817	1829	1846
avg hrs of sunshine	7	6	6	6	6	8	7	8	8	9	8	8
avg # days w/thunder	4	4	2	<1	<1	<1	0	<1	0	1	2	3
w/snow	0	0	0	0	0	0	0	0	0	0	0	0
Typical % relative humidity @1500hrs	66	69	67	66	65	61	58	56	54	57	60	61

Darwin

12°26'S, 130°53'E
Elev: 31m (102ft)

For wet/dry contrast, Darwin is hard to beat. Rains are virtually absent in mid-winter but torrential during the wet season, with spectacular thunderstorms (many build offshore at night). It's hot year round: highs hover around 30–35°C/86–95°F, with cooler nights in winter. Heat and humidity peak in the spring; late autumn is less oppressive.

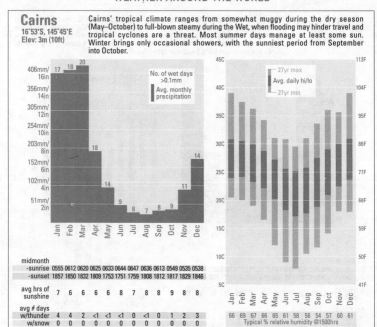

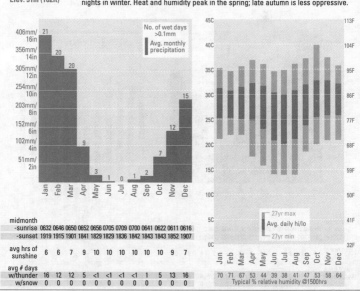

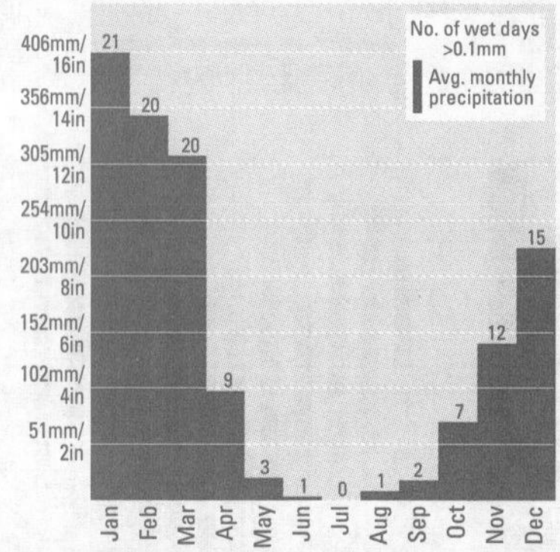

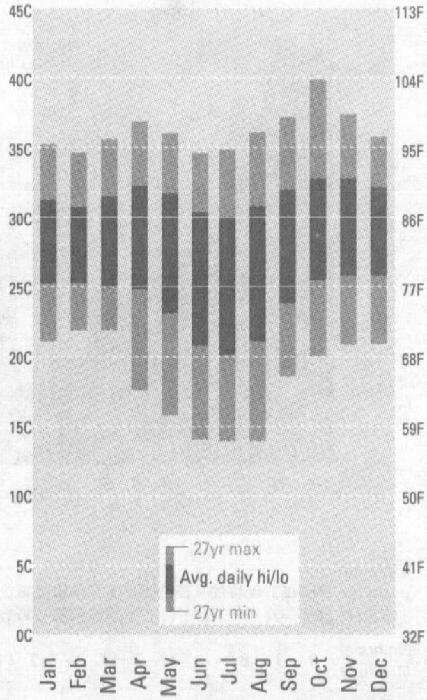

	Jan	Feb	Mar	Apr	May	Jun	Jul	Aug	Sep	Oct	Nov	Dec
midmonth -sunrise	0632	0646	0650	0652	0656	0705	0709	0700	0641	0622	0611	0616
-sunset	1919	1915	1901	1841	1829	1829	1836	1842	1843	1843	1852	1907
avg hrs of sunshine	6	6	7	9	10	10	10	10	10	10	9	7
avg # days w/thunder	16	12	12	5	<1	<1	<1	<1	1	5	13	16
w/snow	0	0	0	0	0	0	0	0	0	0	0	0
Typical % relative humidity @1500hrs	70	71	67	53	44	39	38	41	47	53	58	64

Melbourne

37°41'S, 144°51'E
Elev: 132m (433ft)

Locals insist that Melbourne can have three seasons in one day. Variations occur easily in this city flanked by the parched Outback and the chilly Southern Ocean. Cool changes take the edge off hot summer stretches that can top 32°C/90°F for days. Winters tend to be chilly and drizzly, with more wet days but lighter rain than Sydney.

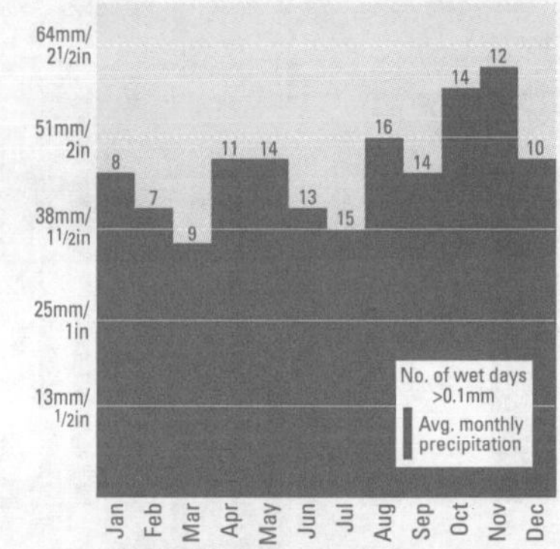

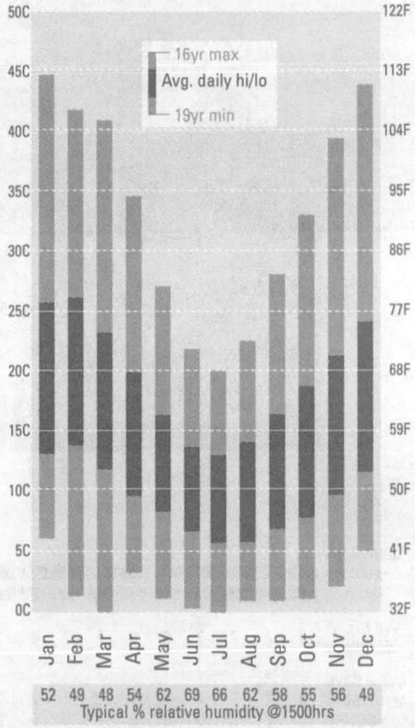

	Jan	Feb	Mar	Apr	May	Jun	Jul	Aug	Sep	Oct	Nov	Dec
midmonth -sunrise	0515	0550	0618	0746	0813	0834	0833	0806	0722	0636	0501	0453
-sunset	1944	1919	1840	1855	1820	1808	1820	1845	1911	1937	1910	1938
avg hrs of sunshine	9	8	7	6	4	4	4	5	6	7	7	8
avg # days w/thunder	2	1	1	1	<1	<1	<1	1	1	2	2	2
w/snow	0	0	0	0	0	0	<1	0	0	0	0	0

Perth

31°56'S, 115°58'E
Elev: 20m (66ft)

Perth's sunny summers have a split personality. Shots of intense heat from the Outback are tempered by the Fremantle doctor, a sea breeze that arrives like clockwork at mid-day. As in Los Angeles, readings may be far hotter just inland. Most winter days are wet, but mild temperatures and the sub-tropical latitude brighten things up a bit.

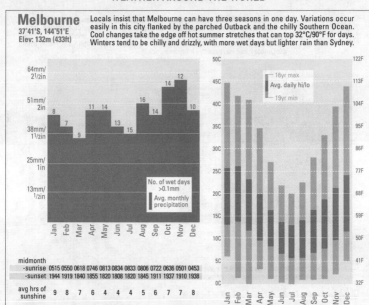

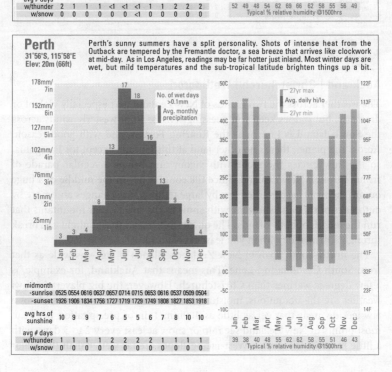

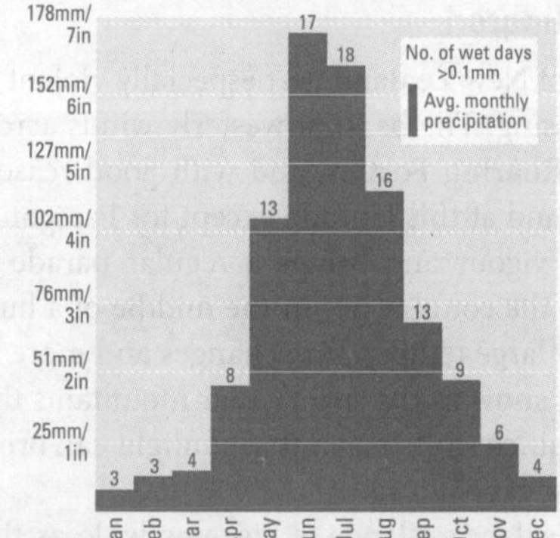

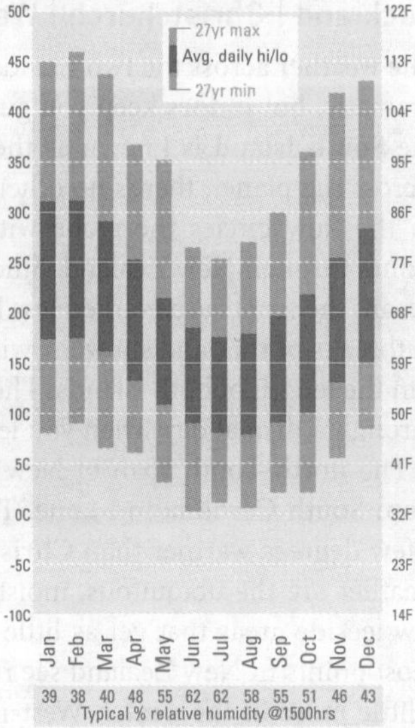

	Jan	Feb	Mar	Apr	May	Jun	Jul	Aug	Sep	Oct	Nov	Dec
midmonth -sunrise	0525	0554	0616	0637	0657	0714	0715	0653	0616	0537	0509	0504
-sunset	1926	1906	1834	1756	1727	1719	1729	1749	1808	1827	1853	1918
avg hrs of sunshine	10	9	9	7	6	5	5	6	7	8	10	10
avg # days w/thunder	1	1	1	1	2	2	2	2	1	1	1	1
w/snow	0	0	0	0	0	0	0	0	0	0	0	0

Sydney
33°56'S, 151°10'E
Elev: 6m (20ft)

For a nation dominated by wet and dry seasons, Sydney's moisture is well distributed. Rain falls every two or three days throughout the year, seldom lasting long. From spring into autumn, thunderstorms may bring squally winds and big hail. Winters are coolish and summers mild for the latitude, although west winds from the Outback can increase the heat dramatically.

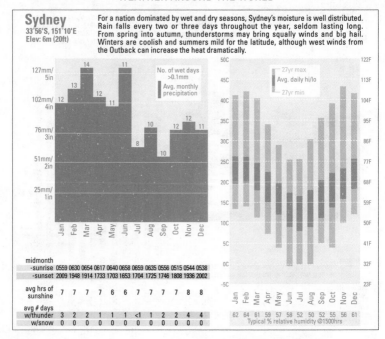

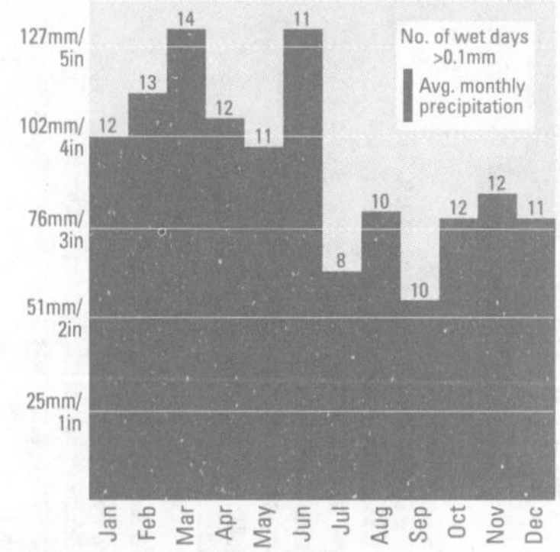

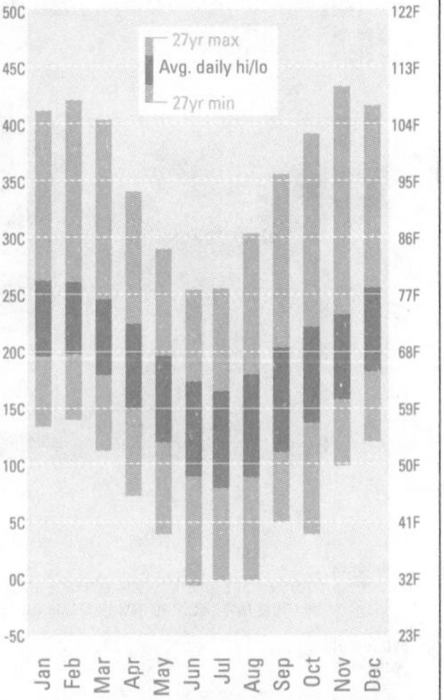

	Jan	Feb	Mar	Apr	May	Jun	Jul	Aug	Sep	Oct	Nov	Dec
midmonth -sunrise	0559	0630	0654	0617	0640	0658	0659	0635	0556	0515	0544	0538
-sunset	2009	1948	1914	1733	1703	1653	1704	1725	1746	1808	1936	2002
avg hrs of sunshine	7	7	7	7	6	6	7	7	7	7	8	8
avg # days w/thunder	3	2	2	1	1	1	<1	1	2	2	4	4
w/snow	0	0	0	0	0	0	0	0	0	0	0	0

Jan	Feb	Mar	Apr	May	Jun	Jul	Aug	Sep	Oct	Nov	Dec
62	64	61	59	57	58	52	50	52	55	56	61

Typical % relative humidity @1500hrs

New Zealand

Auckland | Christchurch | Wellington

The weather across the two islands of **New Zealand** isn't especially violent or dramatic, but it does keep you guessing. The band of westerly winds across the **South Island** is known as the **Roaring Forties**, and with good reason. Across the planet, there's no other land at this latitude except for Patagonia, so the flow circles the globe with vigour and brings a regular parade of **fronts** through New Zealand. Since the country lies in the middle of a huge ocean, the fronts are marked less by large temperature changes and more by outbreaks of rain and showers, with **snow** in the spectacular mountains that run the length of both islands. The quick shifts mean that sunlight can break through a dismal day when you least expect it to.

The north–south span of New Zealand's islands is big – as wide as that from South Carolina to Maine. This means that **Auckland**, for example, is a few degrees warmer than **Christchurch**. However, the big players in Kiwi weather are the ubiquitous, moisture-intercepting **mountains**. Outside of a few leeside areas that get as little as 400mm/16in of precipitation per year, most points in New Zealand see rain or snow at least every 2 to 3 days, with a little more in the winter. West-facing slopes are far wetter than the major cities, especially on the South Island. **Hokitika** – on the coast just west of Arthur's Pass National Park – averages over 2800cm/110in of precipitation a year, with each month getting a healthy dose. Glaciers and snowfields dot the rugged peaks of the **Southern Alps**, where total annual precipitation (including snowmelt) exceeds 8000mm/315in. Snow also coats the lower elevations of the eastern South Island a few times each winter. Thunder is a rare event across this maritime land.

Wind is far more of a factor in New Zealand than in Australia. Gales channelled through the **Cook Strait** keep the people of **Wellington** hanging onto their hats. To the east of **the Alps**, strong westerly flow is sometimes forced over the mountains to create a warm, dry *föhn*-style wind called the **norwester**. It's a welcome visitor when it breaks up a prolonged spell of sub-freezing cold over snow-covered lowlands. In summer, the same type of downslope wind can trigger unsettling heat on rare occasions. The most spectacular case occurred as the remnants of a tropical cyclone threw energy across the Alps on February 7, 1973. Christchurch and other towns soared to the vicinity of 42°C/108°F, well above the nation's previous record high. Lying closer to the sub-tropics, the **North Island** is less prone to such gyrations. At sea level the temperature seldom goes below 0°C/32°F or above 30°C/86°F, although frosts are frequent in many areas and the volcanic peaks can get colder still. The eastern lowlands can hit 35°C/95°F with a good downslope

Auckland

37°01'S, 174°48'E
Elev: 6m (20ft)

New Zealand's northernmost major city, Auckland is close enough to the tropics for summer to hold a touch of humidity. True heat is rare, though, with the typical summer day remaining below 25°C/77°F. Winters are much like San Francisco's, with dependably cool temperatures but more frequent rain – almost two out of every three days.

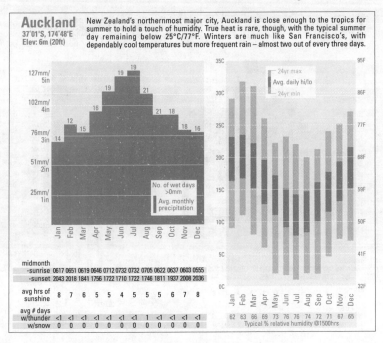

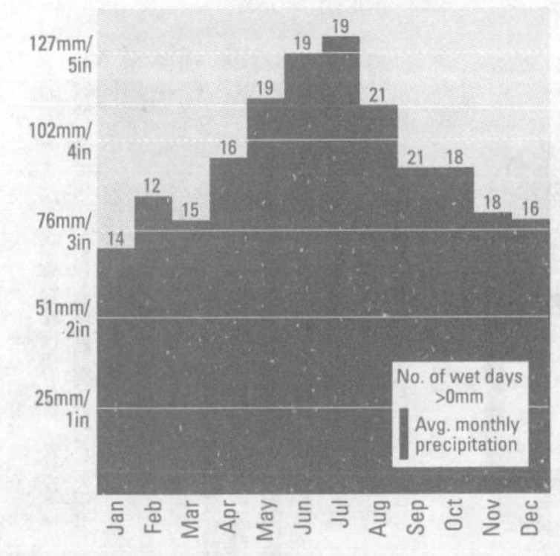

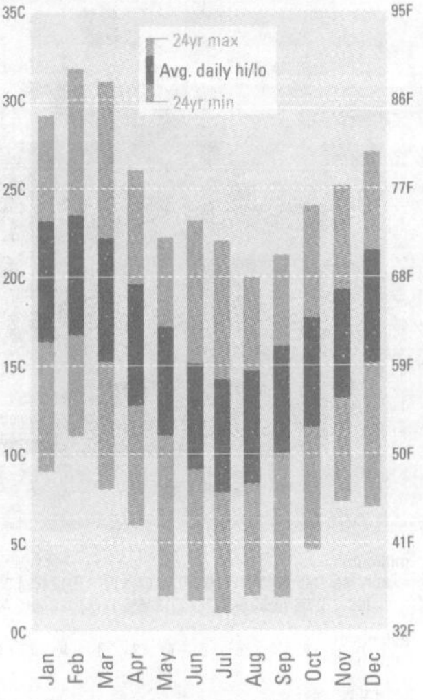

	Jan	Feb	Mar	Apr	May	Jun	Jul	Aug	Sep	Oct	Nov	Dec
midmonth -sunrise	0617	0651	0619	0646	0712	0732	0732	0705	0622	0637	0603	0555
-sunset	2043	2018	1841	1756	1722	1710	1722	1746	1811	1937	2008	2036
avg hrs of sunshine	8	7	6	5	5	4	5	5	5	6	7	8
avg # days w/thunder	<1	<1	<1	<1	<1	<1	<1	1	<1	<1	<1	<1
w/snow	0	0	0	0	0	0	0	0	0	0	0	0

	Jan	Feb	Mar	Apr	May	Jun	Jul	Aug	Sep	Oct	Nov	Dec
Typical % relative humidity @1500hrs	62	63	66	69	73	76	76	74	72	71	67	65

Christchurch

43°29'S, 172°33'E
Elev: 34m (112ft)

Despite a few light snows, hard freezes are rare in a Christchurch winter; many nights are frosty, though. Air quality in winter may be reduced by persistent inversions. Summer temperatures vary, with a few toasty days and some nights that drop below 10°C/50°F. The fairly light rain is well distributed through the year.

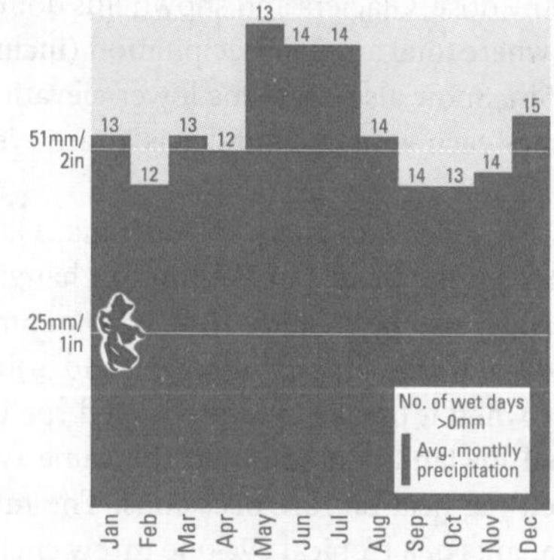

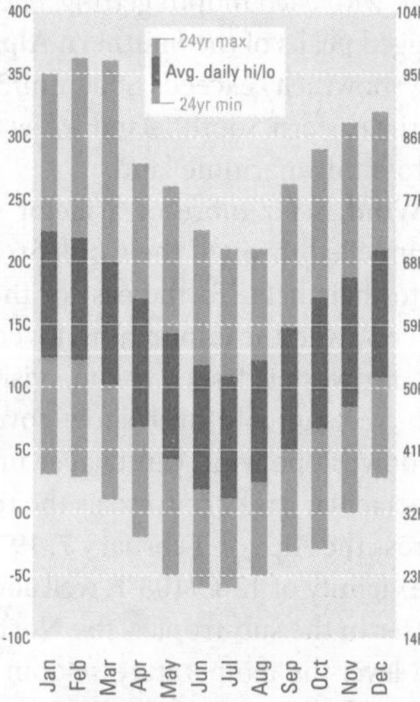

	Jan	Feb	Mar	Apr	May	Jun	Jul	Aug	Sep	Oct	Nov	Dec
Typical % relative humidity @1500hrs	55	59	61	63	66	69	69	67	62	59	60	58

midmonth	Jan	Feb	Mar	Apr	May	Jun	Jul	Aug	Sep	Oct	Nov	Dec
-sunrise	0507	0549	0625	0802	0836	0901	0859	0825	0733	0639	0456	0443
-sunset	2010	1938	1852	1857	1816	1759	1813	1844	1918	1953	1933	2007
avg hrs of sunshine	7	7	5	5	4	4	4	5	5	7	7	7
avg # days w/thunder	<1	<1	<1	<1	0	<1	<1	0	<1	<1	<1	1
w/snow	0	0	0	0	<1	1	<1	1	<1	0	0	0

Wellington

41°17'S, 174°46'E
Elev: 128m (420ft)

It's with good reason this town is called Windy Wellington. Gusts of over 72kph/45mph buffet the city for more than 130 days each year. Summers feel much cooler here than in Christchurch or Auckland, with the ocean breeze taking the edge off incipient warm spells. Although winter is rather grey, chilly and windy, frosts are rare.

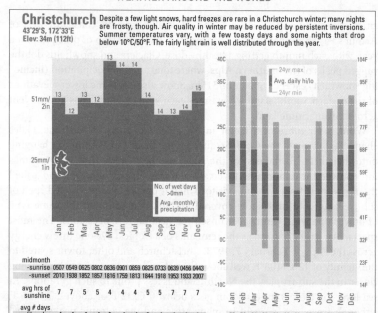

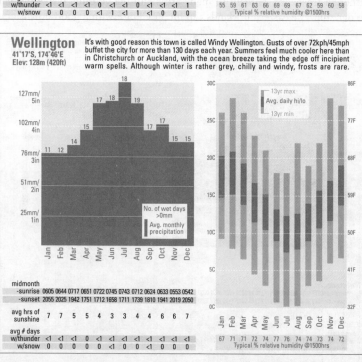

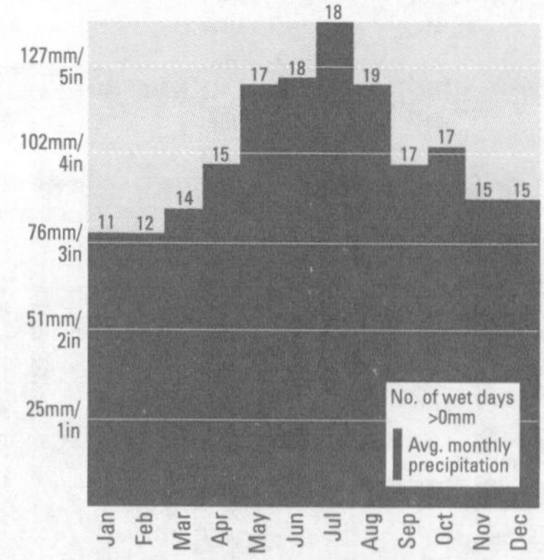

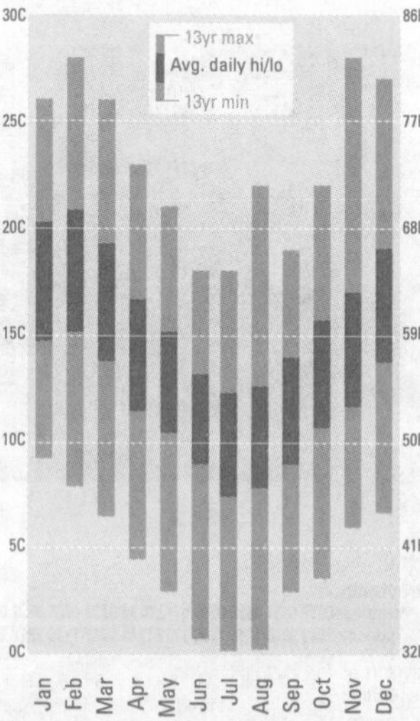

	Jan	Feb	Mar	Apr	May	Jun	Jul	Aug	Sep	Oct	Nov	Dec
Typical % relative humidity @1500hrs	67	71	71	72	74	77	76	74	74	73	74	72

midmonth	Jan	Feb	Mar	Apr	May	Jun	Jul	Aug	Sep	Oct	Nov	Dec
-sunrise	0605	0644	0717	0651	0722	0745	0743	0712	0624	0633	0553	0542
-sunset	2055	2025	1942	1751	1712	1658	1711	1739	1810	1941	2019	2050
avg hrs of sunshine	7	7	5	5	4	3	3	4	4	6	6	7
avg # days w/thunder	<1	<1	<1	0	<1	<1	<1	<1	<1	<1	<1	<1
w/snow	0	0	0	0	0	<1	0	0	<1	0	0	0

wind in summer. Both the North and South islands have a fairly dependable sea breeze that helps keep inland heat from plaguing the coastlines.

South Pacific

Papeete | Suva

Richard Rodgers and Oscar Hammerstein picked the right venue for a lush, warm musical when they composed *South Pacific*. This enormous expanse of island-dotted ocean simply screams "tropics" – or perhaps it gently whispers that refrain.

The only real weather worry in these parts are **tropical cyclones (hurricanes)**. In general, they're most likely to threaten the west part of the region, near the **Solomon Islands, Tuvalu** and **Vanuatu**. However, El **Niño** tends to shift the activity so that islands as far east as **Fiji, Samoa** and even **Tahiti** are at higher than usual risk of a cyclone. **La Niña**, in contrast, tends to focus the activity even more toward the west than usual. Most cyclones move to the west or southwest, then recurve to the east after they've moved south of 20°S, away from the main belt of South Pacific islands. (See "El Niño and La Niña", p.114).

Papeete
17°33'S, 149°37'W
Elev: 2m (7ft)

The volcanic peaks of Aorai and Orohena help shield Papeete from Tahiti's heaviest rain. Hot, humid summers bring frequent downpours and the risk of a tropical cyclone, mainly during El Niño. Islands and atolls to the north are considerably drier. Showers are much more sparse in the winter, when trade winds shift from northeast to southeast.

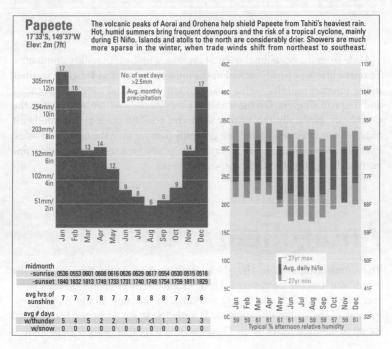

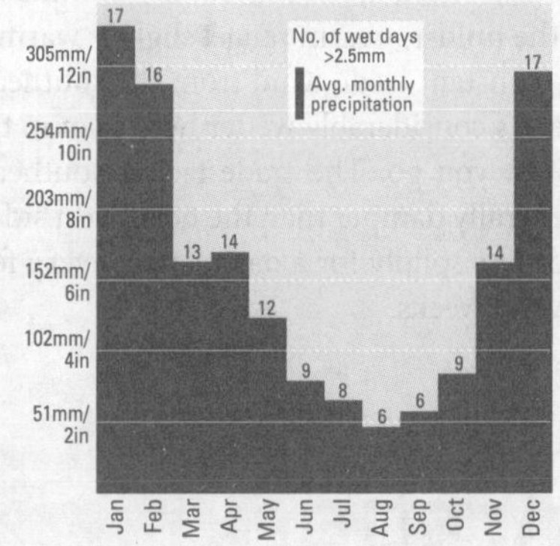

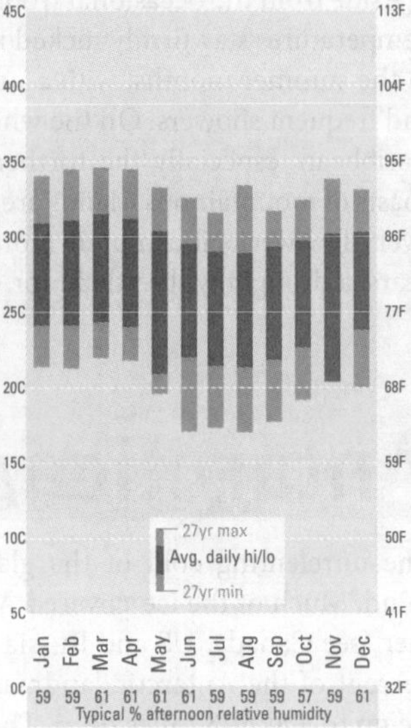

	Jan	Feb	Mar	Apr	May	Jun	Jul	Aug	Sep	Oct	Nov	Dec
midmonth -sunrise	0536	0553	0601	0608	0616	0626	0629	0617	0554	0530	0515	0518
-sunset	1840	1832	1813	1749	1733	1731	1740	1749	1754	1759	1811	1829
avg hrs of sunshine	7	7	7	8	7	7	8	8	8	7	7	6
avg # days w/thunder	5	4	5	2	2	1	1	<1	1	1	2	3
w/snow	0	0	0	0	0	0	0	0	0	0	0	0

Suva
18°08'S, 178°26'E
Elev: 6m (20ft)

Tropical sun notwithstanding, Suva averages over 1000mm/70 inches of rain from December–May. The wet season may bring a tropical cyclone, especially during El Niño. In winter, showers are still frequent but less intense and thundery, and nights can dip below 20°C/68°F. Rains are lighter on the northwest coasts of Viti Levu and Vanua Levu.

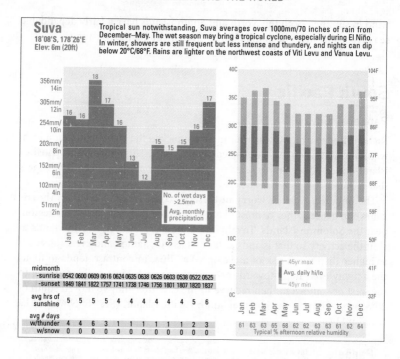

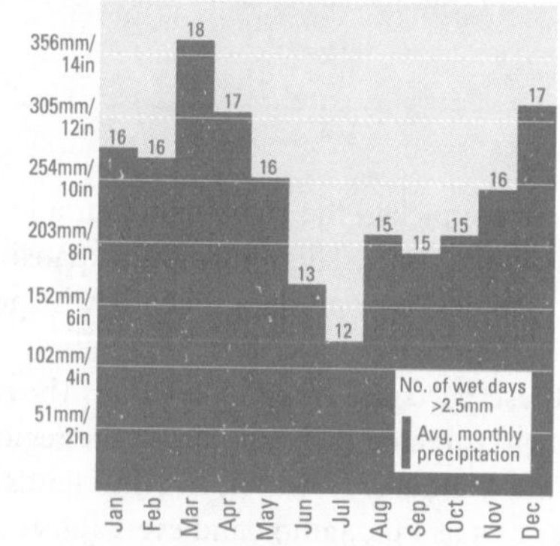

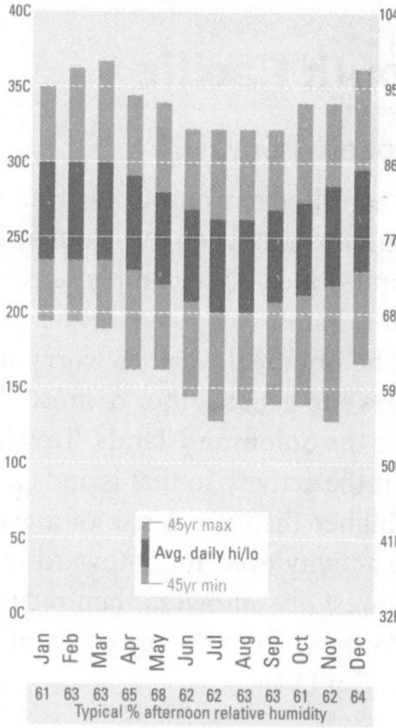

midmonth	Jan	Feb	Mar	Apr	May	Jun	Jul	Aug	Sep	Oct	Nov	Dec
-sunrise	0542	0600	0609	0616	0624	0635	0638	0626	0603	0538	0522	0525
-sunset	1849	1841	1822	1757	1741	1738	1746	1756	1801	1807	1820	1837
avg hrs of sunshine	5	5	5	5	4	4	4	4	4	4	5	6
avg # days w/thunder	4	4	6	3	1	1	1	1	1	1	2	3
w/snow	0	0	0	0	0	0	0	0	0	0	0	0

Aside from the occasional tropical cyclone, a moist tranquillity dominates. Temperatures stay firmly tucked in the mild-to-warm range, slightly warmer in the summer months, with a near-constant trade wind from the southeast and frequent showers. On the whole, it's considerably wetter here than in the Caribbean, especially the further west you go. The trade-facing southeast coasts of mountainous islands are generally damper than the northwest sides. Every few weeks it can pour more or less solidly for a day or two; every few years, a drought will settle in for several weeks.

Antarctica/Arctic

The unrelenting cold of the glacial continent of **Antarctica** boggles the mind. Much of the ice-covered **Arctic Ocean** melts or breaks up each summer (see Canada, US and Russia sections). By contrast, all but roughly two percent of the Antarctic landmass is sealed year-round beneath a mass of ice up to 4.8km/3 miles deep. There's so much ice that, were it all to melt, it would raise the planet's sea level by more than 70m/230ft. Furthermore, a seasonal ice pack roughly doubles the breadth of Antarctica's ice each winter and spring.

Nearly all of this cold continent lies within the Antarctic Circle, but it's by no means a symmetric land. Most tourists only make it to the **Antarctic Peninsula** – a narrow, volcanic tendril that extends toward South America as if reaching for Patagonia. In mid-summer, the peninsula's temperatures seldom get far above freezing point or much below it, making spells of either snow or rain possible. Cruise ships passing through the **Drake Passage** and around the peninsular islands are often forced to adapt their itineraries to changeable weather, including fogs and coastal storms. The same upper-level vortex that helps create the famed ozone hole also steers low-pressure centres around Antarctica's perimeter, where ice meets sea. Particularly in spring and autumn, streams of cold air pour off the elevated plateau and flow down the continent's margins toward offshore lows. The winds that result can easily exceed hurricane force, occasionally topping 320kph/200mph in spots where they flow through mountainous channels. Coastal storms can produce "herbies" (hurricane-strength blizzards) and "whirlies" (tornado-like whirlwinds that spin up in the unstable air caught beneath an outflowing cold pool). The relentless outflow also helps drive continental ice seaward, thus compensating for the inland snow that's light but that never melts.

If Antarctica has a metropolis, it's the **McMurdo** and **Scott research bases,** nestled on an island off the rugged Scott Coast. Roughly a thousand scientists and other visitors assemble here each summer, and a few dozen hang on through winter. Near 78°S, this is the closest coastal settlement to the South Pole. Summer days at McMurdo and Scott often push above freezing, and the winter cold – while admittedly fierce – pales next to that of the interior.

Most of Antarctica lies in the Eastern Hemisphere, forming a grand plateau visited by hardly anyone but research teams. Even with 24-hour sunlight, mid-summer temperatures across the plateau never even come close to the freezing point, and when the sun disappears this is the coldest place on Earth (Vostok notched the world's lowest low at –88°C/–127°F). Winter temperatures bob up and down within a bone-chilling range, inching above –30°C/–22°F at best, and dipping below –73°C/–100°F at worst. Rather than bottoming out in July, the average highs and lows change little from April to September – a "coreless" winter unique to the Antarctic. Clear, calm spells produce the most bitter cold. It takes only a little wind (it's breezy but seldom gale-force here) to disrupt the tissue-thin inversion and bring down air that may be 25°C/40°F warmer. Ice crystals often materialize in the air itself rather than falling from above. Although covered by ice and snow, Antarctica is technically a desert: some spots haven't seen a measurable rainfall or snowfall for millennia.

Amundsen-Scott

90°00'S, 0°00'E
Elev: 2835m (9299ft)

The South Pole is no place for the thin-skinned. In this land of perpetual cold, a reading above –18°C/0°F is big news. When the sun disappears in March for six months, average temperatures drop like a rock. The worst cold occurs when it's clear and calm. On the plus side, winds and snowfall are usually light.

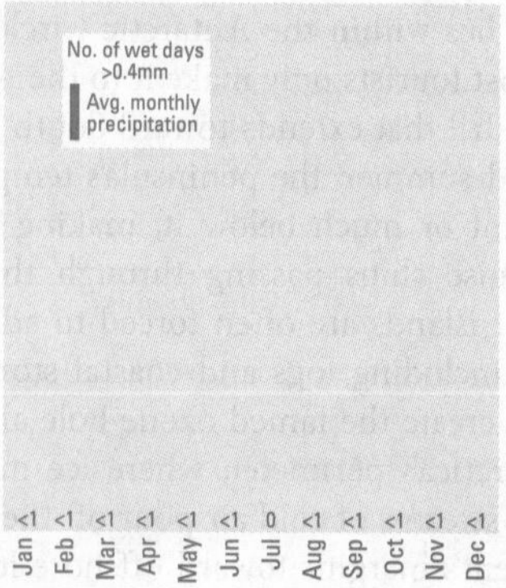

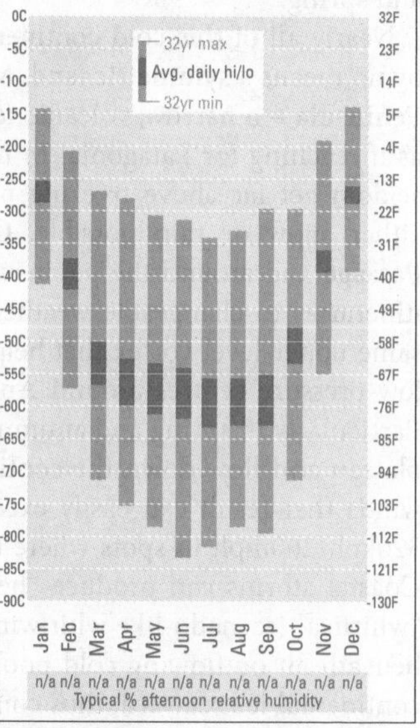

	Jan	Feb	Mar	Apr	May	Jun	Jul	Aug	Sep	Oct	Nov	Dec
midmonth -sunrise	–	–	–	–	–	–	–	–	–	–	–	–
-sunset	–	–	–	–	–	–	–	–	–	–	–	–
avg hrs of sunshine	0	0	–	–	–	–	–	–	–	0	0	0
avg # days w/thunder	<1	<1	<1	<1	<1	0	0	0	<1	<1	<1	<1
w/snow	22	19	14	12	19	19	19	19	12	18	18	21

McMurdo Sound

77°52'S, 166°58'E
Elev: 8m (26ft)

The unrelenting cold of Antarctica shows some mercy in this coastal settlement. Under the 24-hour mid-summer sun, temperatures oscillate around the freezing mark. From April–September, the frigidity is virtually constant, and winds can be fierce. Light snows are frequent year round.

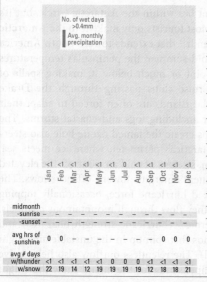

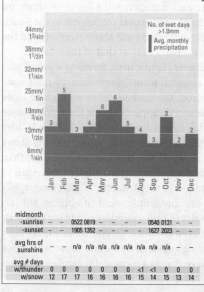

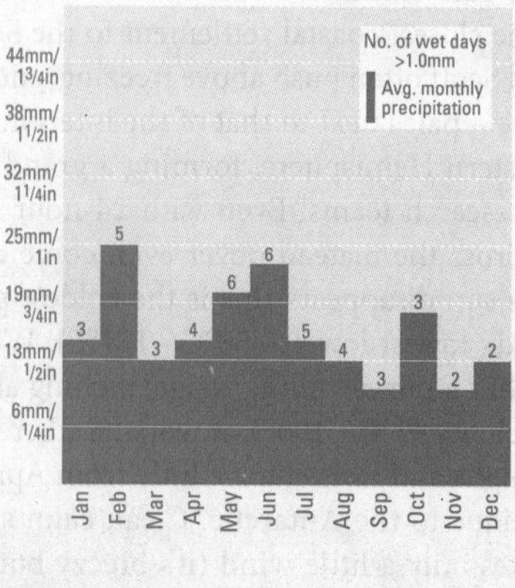

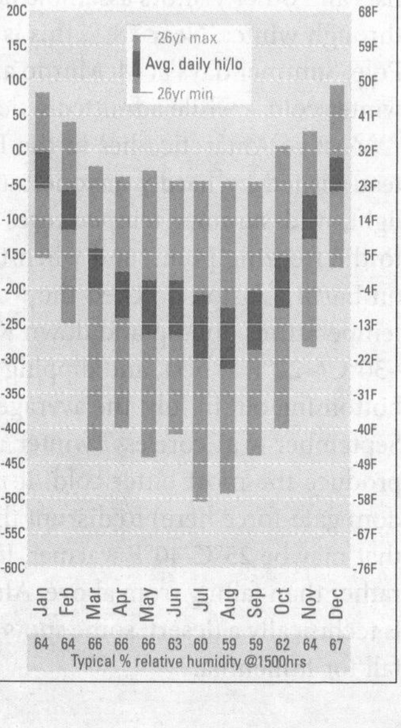

	Jan	Feb	Mar	Apr	May	Jun	Jul	Aug	Sep	Oct	Nov	Dec
midmonth -sunrise	–	–	0522	0819	–	–	–	–	0540	0131	–	–
-sunset	–	–	1905	1352	–	–	–	–	1627	2023	–	–
avg hrs of sunshine	–	–	n/a	n/a	n/a	n/a	n/a	n/a	n/a	n/a	–	–
avg # days w/thunder	0	0	0	0	0	0	0	<1	<1	0	0	0
w/snow	12	17	17	16	16	16	16	15	14	15	13	14

Chapter six
Weather on other planets

Weather on other planets

Earth's atmosphere certainly has its wild side. But elsewhere in the solar system, you'll find places that get hotter, colder and windier than our home turf ever does. Each of the other seven planets of the solar system (eight, if you count the recently demoted Pluto … now technically a humble dwarf planet) has a mix of sunlight, chemistry and orbital patterns that leads to its own distinctive set of weather conditions.

Inside the inner planets

The first rule of planetary weather is an obvious one – planets are generally cooler the farther they are from the Sun. But temperature also depends strongly on the thickness of each planet's atmosphere. When that insulating blanket is thinner than ours, temperatures can vary greatly between daylight and darkness. **Mercury** offers the most vivid example. Due to its small size, Mercury has a relatively puny gravitational field that can't hang on to much of an atmosphere. The best it can manage is a few molecules burned off the surface by the intense sunlight (Mercury is only about 40 percent as far from the Sun as Earth is). Absent an atmosphere, Mercury also lacks clouds, winds, and most everything else that we think of as weather. The exception is some very big swings in temperature, from around 400°C/750°F in the sunshine to –178°C/–290°F at night. The contrast is accentuated by Mercury's extremely long days and nights. Because the planet spins so slowly on its axis but flies so quickly around the Sun, each point on the planet gets sunny and hot for an entire year (which lasts 88 Earth days), then goes dark and cold for the next year.

If Mercury hasn't much of an atmosphere, **Venus** may have too much of one. The air pressure on the Venusian surface is almost 100 times that on Earth. Surface winds are tepid, typically blowing at no more than about 16kph/10mph, but these gentle breezes pack substantial force due to the atmosphere's heft. Of course, you wouldn't have much chance to experience any of this if you were suddenly transported to Venus, since you'd be burnt to a crisp almost instantly. Temperatures on Venus typically run close to a blistering 460°C/860°F, and they only drop to about 445°C/833°F at their lowest.

This is the result of a runaway greenhouse effect, one that famed cosmologist Carl Sagan first explored in the 1960s.

It's believed that Venus may have had vast oceans early in its history. At some point – perhaps as far back as four billion years ago – the Sun's slowly increasing firepower would have begun to evaporate an increasing amount of Venusian water. Since water itself is a greenhouse gas, presumably this led to warmer temperatures, which evaporated even more water, and so forth.

Today, the oceans as well as the water vapour are gone, broken into lighter molecules of hydrogen and oxygen that escaped in the scorching heat. What's left is mainly carbon dioxide, which makes up 97 percent of Venus's thick, hazy atmosphere. This greenhouse gas keeps the planet toasty regardless of the time of day or the time of year. (In case you're wondering, global warming isn't expected to produce this stark an outcome on Earth. We don't have enough fossil fuel – and we're too far from the Sun – to fully replicate the Venusian atmosphere.)

Despite the incredible heat on Venus, a thick deck of clouds made of sulphuric acid sits at heights of about 48–64km/30–40 miles, which helps to make the surface surprisingly dim, perhaps only about as bright as twilight on Earth. Within the clouds, there's a zone where the air pressure and temperature aren't too far from earthbound norms. That might make this region the most hospitable part of Venus for humans to explore someday, although winds there can whip along at speeds up to 360kph/225mph.

Compared to Venus, **Mars** is a cold orb indeed. Its temperatures can drop to a very chilly –140°C/–220°F, and at best they reach around room temperature. As with Venus, the vast majority of the Martian atmosphere (about 95 percent) is made up of carbon dioxide. But Mars is much further from the Sun than Venus, and its atmosphere is very thin (only about 1 percent as dense as ours), so there isn't anything close to a runaway greenhouse effect. Like Earth, Mars has year-round ice caps at both poles; these are primarily made of carbon dioxide, though there's also evidence of frozen water in both ice caps.

Early in its history, when the still-young planet bristled with gas-spewing volcanoes, Mars might have been warm enough to support oceans. But the planet has less than half the gravitational force of Earth, so it's believed that many of the components of that early atmosphere escaped to space as volcanic activity died down and the planet cooled. Unlike Venus, some of Mars' strongest winds are near the surface, where they can kick up gigantic dust storms (the source of the planet's characteristic tan colour). Colossal dust devils can grow as big as Earth's biggest tornadoes – as wide as 2km/1.2 miles and as tall as 10km/6 miles – dwarfing anything seen across our own deserts. Mars also experiences mists over its polar regions during winter, and clouds of water vapour or carbon dioxide occasionally form.

Gas giants and icy moons

Despite their frigidity, the four planets past Mars (Jupiter, Saturn, Neptune,and Uranus) manage to produce some interesting weather. One of their moons even has an atmosphere surprisingly similar to Earth's. All four planets have been dubbed gas giants, because the bulk of their interiors are made up of gases rather than solids (though Neptune and Saturn also hold large amounts of ice). The makeup of these planets' atmospheres resembles the Sun more than earthly air: hydrogen predominates, followed by helium. Small amounts of methane, ammonia, sulphides and other ingredients help produce their wide variety of clouds and colours as seen from Earth.

Jupiter, the solar system's largest planet by far, features an ever-evolving array of clouds, as well as vivid lightning (observed by NASA) and spectacular aurorae. The clouds, which consist of ammonia and other compounds as well as water, change within an overall pattern long noted by astronomers: alternating bright and dark bands at varying latitudes (see photo). The bands reflect a more intricate version of a closer-to-home pattern. On Earth, air tends to rise and form clouds along the Equator and at roughly 60°N and S, while sinking air and dry conditions are common around 30°N and S and at the poles (see diagram, p.24). These bands are a product of incoming sunlight and Earth's rotation. Jupiter spins far more

NASA Jet Propulsion Laboratory (NASA-JPL)

Jupiter's Great Red Spot, snapped by Voyager 1 in 1979, when the spacecraft was 9.2 million kms/5.7 million miles from the planet.

rapidly – almost two and a half times faster than Earth – and it's encircled by more than a dozen alternating light and dark bands. NASA's **Cassini** spacecraft has spotted tiny, bright storms – too small to be visible from Earth – speckling the darker bands.

Since there's no solid surface to help brake Jupiter's circulation, its winds can scream at more than 600kph/380mph, often spinning around long-lasting cyclones and anticyclones. The most famous is the **Great Red Spot** (see photo opposite), which is bigger than Earth itself. It's the Methuselah of weather systems: first observed in the mid-1600s, the spot is still going strong. Though it's often referred to as a storm, the Great Red Spot is actually an anticyclone, which usually brings fair weather on Earth. The spot gets a continual supply of spin from its location, nestled between two of Jupiter's circulation bands and their alternating winds. Spacecraft have tracked other, smaller circulations across Jupiter, but none with the persistence and size of the Great Red Spot.

Saturn may have spectacular rings, but its visible weather features are more muted than Jupiter's. Its latitudinal bands are less vivid, and except near the poles, its winds typically blow from a constant direction (west), albeit at astounding speeds – sometimes topping 1600kph/1000mph. Like Jupiter, Saturn experiences water-based thunderstorms and lightning. Saturn's large-scale storm systems are generally shorter-lived than Jupiter's. A recurring feature called the **Great White Spot** sets in about every 30 years, roughly the amount of time Saturn takes to complete an orbit – but it only lasts for a few weeks at a time. The next appearance of the Great White Spot is expected in 2020.

Of the four outer planets, **Uranus** has the most humdrum weather. The atmospheric pots on both Jupiter and Saturn get stirred by the immense heat generated by their respective cores (both emit far more heat than they receive from the Sun). Uranus is much less dense, though, and with relatively little heat coming into its atmosphere from either its core or the Sun, there's isn't much action to observe. Perhaps the most distinctive thing about weather on Uranus is the incredibly slow transitions between light and dark, thanks to the planet's odd rotation. Unlike the other planets, which feature only slight season-producing tilts, Uranus lies nearly on its side, spinning around an axis just eight degrees off its orbital plane. Much like our own North and South Pole, this means that each year has just one sunrise and one sunset, in the case of Uranus, each separated by nearly 42 Earth years (since a Uranian year lasts about 84 of ours).

Neptune, in contrast, is a surprisingly lively place weatherwise. It boasts the strongest winds observed anywhere in the solar system, ripping along at speeds estimated to be as high as 2500kph/1600mph. The winds feed into massive, but surprisingly transient storms. In 1989 NASA's Voyager 2 discovered a cyclone the size of the Indian Ocean, quickly dubbed the Great

Dark Spot. Five years later, the **Hubble Space Telescope** found that the Great Dark Spot was gone, with a similar storm now evident in the opposite hemisphere.

Pluto hasn't got much of an atmosphere at all – its air is almost a million times less dense than Earth's. The biggest changes in Plutonian air are driven by the planet's markedly elliptical orbit. During its 248-year-long trip around the Sun, Pluto gets within 4.5 billion km/2.8 million miles of the Sun at its closest – closer even than Neptune – but it's more than 7.3 billion km/4.5 billion miles at its farthest. Some of the planet's icy coating evaporates as it nears the Sun and then freezes again during the more distant periods.

More than 100 moons have been found spinning around the four gas giants. Nearly all of these lack an atmosphere and thus are featureless weatherwise, though many are either composed largely of ice or coated with it (such as Europa, which orbits Jupiter and which may harbour liquid water below its ice-covered surface). Neptune's most famous moon, **Triton**, manages an ultra-thin atmosphere, fed by volcanoes that emit nitrogen and methane. A few other moons harbour a smattering of trace gases. But the only moon with a bona fide atmosphere is Saturn's **Titan**, which is actually larger than Mercury. Titan boasts a surface air pressure even higher than Earth's, and like our own atmosphere, that air consists mainly of nitrogen. Smaller amounts of methane and other hydrocarbons react in sunlight to produce an orange, smog-like haze that shrouds the surface in perpetual dusk. Occasionally methane and ethane drop from the Titanian sky in the form of rainstorms. Right now Titan is a dependably cold place – temperatures average around −180°C/−292°F – but some astronomers have speculated that it could serve as a potential platform for life in a few billion years, as the Sun expands and warms its atmosphere.

For more on planetary weather and how to observe it via telescope, see *The Rough Guide to the Universe*.

Chapter seven Resources

Resources

Books

Since weather is indelibly sewn into countless fictional narratives, we've focused the reading list below on the more tractable realm of non-fiction. (For a superb essay on weather in literature, see the *Encyclopedia of Climate and Weather* listed overleaf). Many of these books are available in paperback; a few are out of print but worth seeking out. Those listed below that are aimed at specialists should nonetheless prove rewarding for a patient layperson.

General Meteorology

Historical

Aristotle, Meteorologica. There are plenty of translations of this classic. Any of them will provide a glimpse into this Greek genius's style of interpreting weather.

H. Howard Frisinjer, The History of Meteorology: to 1800 (American Meteorological Society, 1983). One of the best summaries of how our understanding of weather evolved from ancient myth into the age of empiricism.

Mark Monmonier, Air Apparent: How Meteorologists Learned to Map, Predict, and Dramatize Weather (University of Chicago, 1999). Fact-filled walk through the history of weather mapping and forecasting and how the two developed hand in hand.

Reference/Anthologies

C. Donald Ahrens, Meteorology Today, 8th ed (Brooks Cole, 2006). One of the best of the textbooks used in undergraduate meteorology classes. Readable, colourful and full of facts and ideas.

Chris Burt, Extreme Weather: A Guide and Record Book (W.W. Norton, 2004). Stuffed with statistics on all manner of wild weather: the hottest, coldest, wettest, driest, etc. Focused on the US, but includes many world records as well. And it's also beautifully illustrated.

Storm Dunlop, Oxford Dictionary of Weather (Oxford University Press, 2001). Handy, authoritative and inexpensive guide to more than 1800 terms, with useful cross-referencing.

Glossary of Meteorology, 2nd ed (American Meteorological Society, 2000). The ultimate guide to more than 12,000 weather words.

Roger Pielke, Sr, and Roger Pielke, Jr (eds), Storms, Vol. I and II (Routledge, 1999). Not for the casual reader, but a great source of informed summaries from world experts on all sorts of violent weather and its social impacts.

Gavin Pretor-Phinney, The Cloudspotter's Guide (Sceptre, 2006). Rather than serving as a pictorial guide to clouds, this fact-packed volume by the founder of the Cloud Appreciation Society gives a wealth of information on each cloud type.

Stephen Schneider, ed, Encyclopedia of Climate and Weather (Oxford University Press, 1996). Stellar two-volume set, both comprehensive and succinct, spanning topics from weather and religion to atmospheric scattering. Clean, clear writing and sharp black-and-white graphics.

Phenomena

Tornadoes

Wallace Akin, The Forgotten Storm: The Great Tri-State Tornado of 1925 (Lyons Press, 2002). America's deadliest tornado comes to vivid life in Akin's narrative, a blend of survivor accounts and background on the ravaged area from Missouri to Indiana.

Howard Bluestein, Tornado Alley: Monster Storms of the Great Plains (Oxford University Press, 1999). Engaging reflections, clear explanations and striking photos from a pioneer storm chaser and researcher.

Marlene Bradford, Scanning the Skies (University of Oklahoma Press, 2001). The saga of how scientists in Tornado Alley figured out how to detect and warn for twisters. Rare historical tidbits set in a straightforward narrative.

Keay Davidson, Twister: The Science of Tornadoes and the Making of an Adventure Movie (Pocket Books, 1996). Despite the title, Davidson quickly leaves the movie behind and proceeds with an entertaining, informative look at the history of tornadoes and the drive to understand them.

Thomas Grazulis, The Tornado: Nature's Ultimate Windstorm (University of Oklahoma Press, 2001). Authoritative one-stop primer, with plenty of well-researched detail in a lay-friendly style.

Hurricanes

Willie Drye, Storm of the Century: The Labor Day Hurricane of 1935 (National Geographic, 2002). Meticulous, gripping account of how negligence, coincidence, and atmospheric fury led to the deaths of hundreds of veterans trapped on the Florida Keys in the strongest hurricane ever to hit US shores.

Pete Davies, Inside the Hurricane: Face to Face with Nature's Deadliest Storms (Henry Holt, 2000). Finely woven portrait of hurricanes, their impacts and the people in Miami who study and predict them. Davies skilfully captures the mix of adrenaline, curiosity and healthy respect that drives hurricane specialists.

Kerry Emanuel, Divine Wind: The History and Science of Hurricanes (Oxford University Press, 2005). A world-class expert on tropical cyclones looks at hurricanes from scientific, aesthetic and historical angles. If you want only one book on hurricanes, this beautifully written and elegantly designed volume is hard to beat.

Jed Horne, Breach of Faith: Hurricane Katrina and the Near Death of a Great American City (Random House, 2006). One of the most moving and authentic of the many you-are-there accounts of Katrina, written by a Pulitzer-Prize-winning journalist and long-time New Orleans resident.

Sebastian Junger, The Perfect Storm (William Norton, 1997). The hugely popular story of a never-named hurricane and the ill-fated fishing boat that crossed its path. Junger employs the meteorological and human mysteries swirling around the disaster to maximum effect.

Erik Larson, Isaac's Storm (Crown, 1999). Best-selling saga of the 1900 Galveston hurricane, with a moving portrayal of the brothers Isaac and Joseph Cline and their intersections with calamity.

Bob Sheets and Jack Williams, Hurricane Watch: Forecasting the Deadliest Storms on Earth (Vintage, 2001). Lively, fact-packed look at tropical cyclones – along with first-person stories from Hurricane Andrew – from a veteran forecaster and a leading weather writer.

Ivor von Heerden, The Storm: What Went Wrong and Why During Hurricane Katrina (Viking, 2006). Incisive analysis of the engineering and human failures that produced the New Orleans debacle, from a local expert who gained fame for his dire predictions (which soon proved accurate) issued as Katrina stormed ashore.

Snow

Judd Caplovich, Blizzard! The Great Storm of '88 (VeRo Publishing Company, 1987). The frigid scoop on America's most infamous snowstorm.

Weather and cinema

Storms and seasons have been part of human drama since ancient cultures began crafting dance, music and theatre to appease the gods. Shakespeare leaned on weather as both a plot device and metaphor, from the oceanic fury in *The Tempest* to the climactic storm in *King Lear*. However, stage dramas were inherently limited in their ability to portray realistic weather. With the advent of film came an entirely new way of adding literal atmosphere to a story.

Some of the most evocative weather sequences of all time were filmed during and just after the silent era. In **D.W. Griffith**'s prototypical melodrama *Way Down East* (1920), Lillian Gish finds herself wandering through a blizzard, finally collapsing onto an ice floe and floating toward a waterfall in the film's climax. A few years later Gish endured *The Wind* (1928), which blows sand through door frames and into windows in seemingly every gritty frame. (Gish's character also encounters one of the silver screen's first-ever tornadoes.) The vivid bursts of snow that envelop Klondike goldseekers in **Charlie Chaplin**'s *The Gold Rush* (1925) fling themselves into the action in a similar way. Elegant shots of raindrops, rivulets, and torrents punctuate *Rain* (1932), starring Joan Crawford and Walter Huston. Buster Keaton, meanwhile, encountered one of the period's most impressive storm sequences in *Steamboat Bill, Jr.* (1928). The movie's climax was to have featured a flood, but the idea was abandoned after the US Mississippi Valley was ravaged by the epic flooding of 1927.

By the late 1930s, bigger budgets and more sophisticated filming techniques raised the weather bar dramatically. **John Ford**'s *The Hurricane* (1937) was the *Twister* of its era, and one of the first major films to focus on a single, momentous weather event. The film was an early showcase for sarong-clad Dorothy Lamour, who also starred in the semi-knockoff *Typhoon* (1940). Presaging the disaster flicks of the 1970s, *The Hurricane* is packed with drama, including a baby's birth at the height of the storm. The film gets bonus points among weather nuts for showing an actual anemometer display and for the line, "I've never seen the barometer so low!" Equally impressive is *The Rains Came* (1939), which depicts the ravaging of an Indian town by an earthquake and flood, compounded by monsoon rains, fire and plague. Thanks to its astoundingly vivid visuals this movie won the first Oscar for special effects.

Judy Garland comes face to face with the cinema's most famous tornado in *The Wizard of Oz* (1939). Created with a hodgepodge of materials that included a muslin sock, smoke, glass, cotton, dust and a burlap bag – plus horizontal rods at top and bottom to produce the back-and-forth snaking motion – this eerily realistic twister has terrified generations of American children. The background footage was recycled for a subsequent MGM musical, *Cabin In The Sky* (1943).

The science-fiction craze of the 1950s and 1960s mostly bypassed weather, one exception being the deliciously titled serial *Zombies Of The Stratosphere* (1952): it includes a young Leonard Nimoy as part of a gang of Martians bent on exploding Earth from its orbit so that Mars can move in and benefit from a milder climate. The weather-control theme –a sign of Cold War tension as well as ongoing experiments in weather modification – cropped up in a few other films of the time, including *In Like Flint* (1965), a James Bond spoof with James Coburn battling a group of mad scientists out to control Earth's weather.

It wasn't until the 1970s that the inherent drama of weather provided irresistible fodder for a decade of wildly popular yet critically panned disaster flicks. *Avalanche* (1978) and *Hurricane* (1979) are both prime examples, the latter being a remake of the 1937 classic with special effects that seem little improved after four decades. Disaster-film mogul **Irwin Allen**'s first made-for-TV entry in the genre was *Flood!* (1976).

Movies set in the agricultural heartland of the US can't avoid the weather. You can practically taste the misery of the 1930s Dust Bowl in the stark, John Steinbeck classic *The Grapes of Wrath* (1940), directed by Ford. In 1984 a tornado-packing thunderstorm opened *Country*, instantly conveying the struggle of farm life, while the film ends in the literal and figurative bitterness of midwinter.

After years of cameo appearances, perhaps it was inevitable that tornadoes would someday be the star of a film. That day came with *Twister* (1996), which launched a new era of weather-disaster films bolstered by digital effects. Next came *Hard Rain* (1998), which merges the disaster and heist genres by following thieves through flood-inundated streets. Another weather blockbuster was *The Perfect Storm* (2000), based on Sebastian Junger's hit book about a fishing boat caught in a 1991 Atlantic maelstrom.

Who would have expected the cinema to latch onto TV weathercasters themselves as protagonists? First came the bemused **Steve Martin** in *L.A. Story* (1990), while in the funny yet evocative *Groundhog Day* (1993), **Bill Murray** portrays a small-market weathercaster in Pennsylvania who finds himself reliving the same day again and again. This mini-genre turned slightly darker with *The Weather Man* (2005) and *Lucky Numbers* (2000). Then there's the cult-hit sex farce from Japan, *A Weather Woman* (1995). After the lead character flashes her underwear on camera, ribald complications ensue. The follow-up, *A Weather Woman Returns* (2000), covers similar ground.

On a broader scale, some mysteries and horror films use weather to set the tone for an entire picture. Suspense master **Alfred Hitchcock** opened *The Lady Vanishes* (1938) with a trainload of people stuck overnight in a Balkans snowstorm, while heavy snowfalls serve a similar function in the *The Shining* (1980) and *Misery* (1990), both adapted from **Stephen King** novels. *The Fog* (1978) is itself the villain, as ghosts within the fog terrorize a coastal town. In quiet contrast, wintry settings are used to frame suburban dysfunction in *The Ice Storm* (1997) and to symbolize a woman's bitterness in *Smilla's Sense Of Snow* (1997).

Global climate change might seem a ripe topic for epic cinema. However, the distributed and diffuse nature of the global-warming threat makes it hard to encompass in a two-hour story. *Soylent Green* (1973) was the first major film to make mention of the greenhouse effect, while in *The Arrival* (1996), aliens intent on occupying Earth pollute the atmosphere to goose the greenhouse effect and melt polar ice. Two other, lower-profile films that employed global change were *The March* (1990) and *Prey* (1997).

The first true attempt to address climate change in a Hollywood blockbuster, *Waterworld* (1995), was one of the costliest films ever made. The melting of icecaps isn't so much the topic of *Waterworld* as it is the backdrop for a more conventional story of good guys facing off against bad. A much bigger hit in cinemas was *The Day After Tomorrow* (2004). Dennis Quaid plays a US paleoclimatologist who discovers that global warming has triggered a shutdown of North Atlantic currents. Within days, the Northern Hemisphere is plunged into an ice-age-scale deep freeze, and Hall is tromping through snowfields up the East Coast to rescue his son from the icebound New York Public Library. In reality, a shutdown of North Atlantic currents would take decades to unfold. Still, many activists hailed the film as a sign (albeit flawed) of growing public interest in climate change. Former US vice president **Al Gore** became an unlikely movie star with *An Inconvenient Truth* (2006), which put Gore's road show on global warming into cinematic form. Playing to packed houses for weeks, it rode the wave of two big trends – interest in global warming as well as a surge of hugely popular US documentaries in recent years.

Ruth Kirk, Snow (1977; reissued by University of Washington, 1998). Thoughtful musings on snow and its effects on people, animals and climate around the world.

David Laskin, The Children's Blizzard (Harper, 2004). Vivid account of the frigid fury of a 1888 blizzard that killed hundreds across the US Great Plains, including many schoolchildren trapped while walking home.

Bernard Mergen, Snow in America (Smithsonian Institution Press, 1997). The kaleidoscope of snow and how we experience it, turned in surprising directions on US culture and history.

Wind and storms

Jan DeBlieu, Wind: How the Flow of Air Has Shaped Life, Myth, & the Land (Houghton Mifflin, 1998). In the best nature-writing tradition, an introspective and informative look at air in motion through the ages.

Gretel Ehrlich, A Match to the Heart: One Woman's Story of Being Struck by Lightning (Penguin, 1995). Idiosyncratic, wonderfully crafted tale of Ehrlich's medical and spiritual journey back from a lightning strike's painful aftermath.

Scott Huler, Defining the Wind: The Beaufort Scale, and How a 19th-Century Admiral Turned Science into Poetry (Crown, 2004). A look at how the world's best-known wind scale came to be. Written in the spirit of Dana Sobel's Longitude and sprinkled with the author's self-proclaimed entrancement with the scale.

Eric Pinder, Tying Down the Wind: Adventures in the Worst Weather on Earth (Putnam, 2000). Elegant prose on a potpourri of weather topics, rooted in the author's experience as an observer on wind-scoured Mount Washington, New Hampshire.

Jeffrey Rosenfeld, Eye of the Storm: Inside the World's Deadliest Hurricanes, Tornadoes, and Blizzards (Perseus, 2000). Engaging survey of science and human experience at the heart of some of our stormiest weather.

Flood and drought

John M. Barry, Rising Tide: The Great Mississippi Flood of 1927 & How it Changed America (Simon & Schuster, 1997). A heart-rending look at America's biggest flood disaster and its lasting effect on US policy, culture and people, including the dispossessed African-Americans of the delta.

Timothy Egan, The Worst Hard Time: The Untold Story of Those Who Survived the Great American Dust Bowl (Houghton Mifflin, 2005). More than 70 years after the US Great Plains were nearly destroyed by heat and drought, Egan brings this almost-forgotten era back to visceral life with telling first-person accounts.

Michael Glantz (ed), Drought Follows the Plow (Cambridge University Press, 1994). Compact profiles of how culture and climate intersect to shape – or not shape – the course of drought across central Asia, Australia, Brazil and several African nations.

El Niño and La Niña

Cesar Caviedes, El Niño in History (University of Florida Press, 2001). Wide-ranging survey from a veteran researcher on historical events linked to a greater or lesser extent to El Niño, including some unexpected ones (Napoleon's march on Russia, for instance).

Mike Davis, Late Victorian Holocausts: El Niño Famines and the Making of the Third World (Verso, 2001). As with his book on Los Angeles, *Ecology of Fear*, Davis uses an eclectic set of research to support a provocative thesis – in this case, that intense droughts of the late 1800s in India, China and Brazil, likely triggered by El Niños, worked in tandem with imperialist designs to starve tens of millions.

Michael Glantz, Currents of Change: El Niño's Impact on Climate and Society, 2nd ed (Cambridge University Press, 2000). Compact and readable, this paperback is a solid one-stop reference on El Niño for scholars and laypeople alike.

J. Madeleine Nash, El Niño: Unlocking the Secrets of the Master Weather-Maker (Warner, 2002). Vivid portraits of El Niño researchers in action and the climate phenomenon's effect on people and the planet.

S. George Philander, Our Affair with El Niño: How We Transformed an Enchanting Peruvian Current into a Global Climate Hazard (Princeton University Press, 2004). Engaging, offbeat look at El Niño from an eminent researcher who explains the often-perplexing phenomenon from the vantage points of music, art and poetry.

Richard Hamblyn, The Invention of Clouds (Farrar, Strauss & Giroux, 2001). The story of Luke Howard – the man who named cirrus, nimbus, cumulus and stratus – and the science of Howard's day. Eloquently written and cleverly designed.

Raymond L. Lee, Jr, and Alistair B. Fraser, The Rainbow Bridge: Rainbows in Art, Myth and Science (Pennsylvania State University Press, 2001). Lee and Fraser weave their way through the intermingled culture and science of rainbows. Fact-filled narrative, stunning photos, and artwork from across cultures.

Optics and the Sky

Robert Greenler, Rainbows, Halos, and Glories (Elton-Wolf, 1999). Detailed, thoughtful, and – appropriately – illuminating, with first-person reflections and anecdotes throughout. Striking colour photos.

Forecasting

George Freier, Weather Proverbs: How 600 Proverbs, Sayings and Poems Accurately Explain Our Weather (Fisher Books, 1992). Crisp explanations of the meteorology behind hundreds of weather sayings.

Stewart A. Kingsbury, Mildred E. Kingsbury and Wolfgang Mieder, Weather Wisdom (Peter Lang, 1996). Just-the-facts assemblage of 4435 weather sayings, indexed by keyword or significant date.

Frederik Nebeker, Calculating the Weather: Meteorology in the 20th Century (Academic Press, 1995). How weather forecasting shifted from an art into a computational science. Lots of background on the key players in the field, with the technical detail leavened by engaging prose.

Alan Watts, Instant Weather Forecasting (Sheridan House, 2001). Find the photo that best matches the current sky, then consult the adjacent table to learn the conditions to come. A slender perennial in print since 1968. The only drawback is that the ultra-specific outlooks become less useful outside the UK and western Europe.

Climate Change

Mark Bowen, Thin Ice: Unlocking the Secrets of Climate in the World's Highest Mountains (2005). A masterful, in-depth look at tropical glaciers and those who study them. Himself a climber and a doctorate-level physicist, Bowen digs deeper than most into the science.

Gale Christensen, Greenhouse: The 200-Year Story of Global Warming (Walker & Company, 1999). Marvellously literate look at the growth of world industry and the parallel realization of its effect on our atmosphere.

John Cox, Climate Crash (2005). Wonderfully written report on the science of abrupt climate change and its roots in the various paleoclimatic clues offered by glaciers, fossils and the like. Cox conveys the excitement of the research chase, the difficulty of the work, and the mixed signals it provides about our climatic future.

Kirstin Dow and Thomas Downing, The Atlas of Climate Change: Mapping the World's Greatest Challenge (Earthscan Publications, 2006). Colourful, enticing paperpack with dozens of world maps that each depict a key variable of climate change, from fossil-fuel use to possible impacts on agriculture and water supplies.

Tim Flannery, The Weather Makers: How

Man Is Changing the Climate and What It Means for Life on Earth (2005). The Australian zoologist, museum director and renowned writer makes good use of his gift for pithy explanations and apt metaphor as he walks readers through climate history, global-warming science, political and technological options, and what individuals can do.

James Rodger Fleming, Historical Perspectives on Climate Change (Oxford University Press, 1998). From the Enlightenment to the late twentieth century, a survey of theories and perceptions around climate and its evolution. Both scholarly and lively.

Ross Gelbspan, Boiling Point: How Politicians, Big Oil and Coal, Journalists, and Activists Have Fueled the Climate Crisis – and What We Can Do to Avert Disaster (Basic Books, 2004). Passionate, informed polemics from an award-winning investigative journalist. Gelbspan expands the list of climate villains from the sceptics and corporate interests profiled in *The Heat Is On* (Addison-Wesley, 1997) to include the media and "compromised activists."

Al Gore, An Inconvenient Truth (2006). Handsome paperback version of the documentary by the same name, with stunning graphics and elaboration on many of Gore's key points.

John Houghton, Global Warming: The Complete Briefing–Third Edition (2004). Readers who crave scientific meat but don't have the time or inclination to pore over IPCC reports should be delighted with this primer, first published in 1997. Houghton led two of the IPCC's major assessments (see Websites) and speaks with both authority and humanity.

Elizabeth Kolbert, Field Notes from a Catastrophe: Man, Nature, and Climate Change (2006). Among the most popular of the 2006 vintage of climate-change books. Kolbert's dry, crisp prose reveals how climate change is already affecting the world today and how it's altered past civilizations.

James Lovelock, The Revenge of Gaia (2006). The man who told us Earth is a self-sustaining system now worries that we're pushing that system beyond its capacity to heal. In this slender and sometimes gloomy book, Lovelock argues we may need nuclear power to avoid what he sees as the truly catastrophic implications of "global heating."

Mark Lynas, High Tide (2004). One of the best first-person accounts of global warming. Lynas hops from Britain to Alaska to China to Peru and elsewhere, describing how climate change is affecting people and landscapes. Both an engaging travelogue and an approachable introduction to climate-change science.

Mark Maslin, A Very Short Introduction to Global Warming (2004). Maslin zips through the high points of global-change science and policy in this handy, well-written extended essay.

Spencer Weart, The Discovery of Global Warming (2004). Superb chronicle of climate-change research over the past century. Weart's sparkling prose is both concise and precise.

Charles Wohlforth, The Whale and the Supercomputer: On the Northern Front of Climate Change (2004). The best book to date on what climate change is doing to the people, creatures and landscape of Alaska – and, by extension, high latitudes around the world.

Weather around the world

Colin Buckle, Weather and Climate in Africa (Addison Wesley Longman, 1996). Two books in one: an introduction to meteorology and a storehouse of detail on African weather. Dense with facts and figures but always readable.

Bob Crowder, The Wonders of the Weather (AGPS Press, 1995). Handsome, beautifully written guidebook to weather in general and Australia's in particular.

Tom Fort, Under the Weather: The Story of our National Obsession (Century, 2006). Idiosyncratic, kaleidoscopic view of the British preoccupation with weather.

Alexander Frater, Chasing the Monsoon (Henry Holt, 1990). Delectable portraits of people and weather along a journey northward on the crest of India's summer monsoon.

Jon Krakauer, Into Thin Air (Villard, 1997). Not so much a weather book – there's little meteorological discussion per se – as a gripping account of how it feels to encounter vicious snow, cold and wind on Mt Everest.

H. E. Landsberg (ed in chief), World Survey of Climatology (Elsevier, 1969–84; out of print). Enormous fifteen-volume set that covers the entire globe's weather and climate, region by region. Still unrivalled in its depth and quality.

David Laskin, Braving the Elements: The Stormy History of American Weather (Doubleday, 1997). Entertaining, eclectic survey of how explorers, preachers, scientists and other Americans came more or less to terms with their unique climate.

Susan Solomon, The Coldest March (Yale University Press, 2001). Antarctica's brutal climate, compellingly illuminated through the 1912 journey of Robert Falcon Scott and the extra-harsh weather that led to his death.

Robin Stirling, The Weather of Britain (Giles de la Mare, 1997). Handsomely produced paperback with a thorough blend of narrative and statistics on all manner of British weather.

Jack Williams, USA Today: The Weather Book (Vintage Books, 1997). A colourful treasure trove of information on US weather. Tons of graphics and profiles of experts.

Websites

The ever-shifting nature of weather, and the vast volumes of meteorological data out there, make the atmosphere a natural for the Web. As with all websites, evolution is the norm. Most of these sites are unlikely to change addresses, but if you can't find a given site, try using Google to track it down.

Organizations

While the membership requirements vary from group to group, most of the organizations below have a category for people who lack formal training in meteorology but who want to learn more.

American Meteorological Society
www.ametsoc.org
The largest US group devoted to the atmosphere, with over 11,000 members. Publishes an online newsletter and a print journal each month.

Australian Meteorological and Oceanographic Society
www.amos.org.au
Sponsors awards, conferences and regional chapters across the nation.

National Weather Association
www.nwas.org
A US group oriented toward working forecasters, with a large contingent of weather enthusiasts as well. Back issues of monthly printed newsletters are available online.

Royal Meteorological Society
www.royal-met-soc.org.uk
The pioneering RMS was founded as the British Meteorological Society in 1850. Publishes the monthly magazine *Weather*.

World Meteorological Organization
www.wmo.ch
An agency of the United Nations, the WMO serves as a conduit for vital international collaboration and co-operation, including world weather observing and analysis of climate change. (See p.391 for member links.)

Media

Each of the sites below offers local, regional and global forecasts in one form or another. The differences among them are mainly in format, emphasis and the extent of background material. Many sites allow you to specify your city or postal code and get weather data customized to your location on return visits to the site.

Australian Weather News and Links
www.australianweathernews.com
Superb one-stop-shopping site for news, forecasts, climate data and much else on Aussie weather, all packaged in a rock-solid interface.

BBC
www.bbc.co.uk/weather
Elegantly designed, yet fast-loading, interface – covers the weather at hand and on the way, including a raft of features and explanations of TV-weather symbols.

CNN
www.cnn.com/WEATHER
Includes extensive coverage of world weather events from the vast CNN network.

USA Today Weather Forecasts and News
www.usatoday.com/weather/wfront.htm
This newspaper's devotion to weather extends to its excellent website, with thorough lay-oriented discussions of many weather elements.

The Weather Channel
www.weather.com
Operated by the 24-hour US-based TV network, with deeper background on many stories covered only briefly on the air. Heavy on lifestyle-oriented features, including sports, travel and home/garden.

The Weather Notebook
www.mountwashington.org/notebook
Years of archives from this now-defunct two-minute radio feature, which covered the science and experience of weather, can be sampled in print and audio form. Produced by the Mount Washington Observatory.

Weatherwise
www.weatherwise.org
The leading US weather magazine offers many of its accessibly written features online at no charge; others are available at nominal cost.

Weather sites – commercial

Several private weather vendors continue to offer a range of Web services at no cost to the user, recouping their expenses through advertising and other means.

AccuWeather
www.accuweather.com
This giant weather firm's site is big on personalized features. Home pages are pegged to interests (agriculture, golf, health) and forecasts keyed to US zip codes. True aficionados can pony up for access to AccuWeather Professional, a fee-based service with expanded data and commentary from weather experts.

Intellicast
www.intellicast.com
Amid the banner ads on this site, you'll find a good deal of meaty weather data, including extensive radar and satellite links.

Unisys Weather
weather.unisys.com
Formerly based at a university, this site offers a sterling line-up of output from the leading US and European computer models, along with the usual forecasts and current conditions.

Weather Underground
www.wunderground.com
Another spinoff from a university-founded site, offering expert commentary and easy-to-access forecasts, plus current and historical data, for cities around the world.

Weather sites – government

Bureau of Meteorology (Australia)
www.bom.gov.au
Easy-to-navigate home page from Australia's official weather agency, offering forecasts and current conditions, attractive climatological maps and a variety of educational products.

Environment Canada/Canadian Weather
weatheroffice.ec.gc.ca/canada_e.html (English interface)
Produced by the Meteorological Service of Canada, this site uses point-and-click navigation to display current and expected weather conditions across Canada and beyond.

European Centre for Medium-range Weather Prediction
www.ecmwf.int
A heavy hitter in numerical weather prediction, especially in the realm of extended outlooks. Interactive forecast access and good background on computer modelling and climate, all presented in a straightforward interface.

GOES Hot Stuff
rsd.gsfc.nasa.gov/goes/text/hotstuff.html
Entrancing gallery of images collected by US satellites from above North and Central America. Arranged in reverse chronological order, so you can hunt for an image of the storm that knocked tiles off your roof exactly four years and two months ago.

The Met Office (UK)
www.metoffice.com
The British government's weather agency offers a streamlined site with a clean, bright interface and a wide array of features.

National Oceanic and Atmospheric Administration
www.noaa.gov
The agency that operates much of the US research on weather. Plenty of news updates, background material and an extensive photo archive.

National Weather Service/NWS Internet Weather Source (US)
weather.noaa.gov
The most convenient way to access the wealth of data and forecasts available through the NWS, the US government's weather agency. The low-graphics interface comes in handy if your Internet connection is on the slow side.

Weather sites – university and non-profit

NCAR/RAP Real-Time Weather Data
www.rap.ucar.edu/weather
One of the best sites for current US data. Satellite photos, surface and upper-air data, radar imagery and computer-model output are all here, updated promptly in an easy-to-navigate format.

The UK Weather Information Site
www.weather.org.uk
Nothing fancy – a solid, basic set of links to sources on British weather and climate.

UM Weather
cirrus.sprl.umich.edu/wxnet
From the University of Michigan, a top-notch spectrum of weather resources, with a focus on current conditions, satellite and radar data, and computer-generated forecasts. Breadth and depth are both unusually strong. Includes weather-cam links to over 700 sites across North America.

World Weather Information Service
worldweather.wmo.int
Operated by the World Meteorological Organization, this site offers data and forecasts from most of the world's national weather services (see "Weather agencies around the world", p.391).

Weather sites – educational and reference

Bad Meteorology FAQ
www.ems.psu.edu/~fraser/Bad/BadFAQ
The bottom line on why air doesn't "hold" water vapour and other common, but technically inaccurate, explanations.

Quotes about climate
www.bom.gov.au/bmrc/clfor/cfstaff/nnn/nnn_climate_quotes.htm
Courtesy of Australian climatologist Neville Nicholls, an intriguing collection of what people have said about climate (and weather) through history.

University Corporation for Atmospheric Research (UCAR)
www.ucar.edu
This unique US-based consortium, which operates the National Center for Atmospheric Research, includes 70 universities. Lots of background on recent findings from the lab, the field and computer models.

WW2010: The Weather World 2010 Project
ww2010.atmos.uiuc.edu/(Gh)/home.rxml
Impeccably produced site walks you through the full potpourri of weather and climate with an extensive set of interactive modules. Great for browsing as well as fact-finding. Graphics are colourful but not browser-overwhelming.

Climate-related sites

Because many governments now reserve some climate data to sell to clients at hefty fees, only a sub-set of the world's storehouse of climate information can be found online. Even sites that do provide data often attach stringent caveats to its use. Still, a lot can be unearthed.

NCDC CLIMVIS (Climate Visualization)
lwf.ncdc.noaa.gov/oa/climate/onlineprod/drought/xmgr.html
On-the-fly graphics allow you to plot weather history for any one of dozens of US locations. Also includes a global climatological database with similar graphics.

NOAA Climate Diagnostics Center/US Interactive Climate Pages
www.cdc.noaa.gov/Usclimate/
Specify a variable (temperature, precipitation, etc) and you can produce a map showing its departure from normal across the US, going back more than 100 years.

NOAA National Climatic Data Center (NCDC)/Climate Extremes
lwf.ncdc.noaa.gov/oa/climate/severeweather/extremes.html
This is the place to find hard data on where it's been unusually hot, cold, wet or dry. Full reports each month on US and global temperatures and precipitation and how they rank against long-term averages.

Weatherbase
www.weatherbase.com
For a quick read on average conditions at that obscure destination you're about to visit – or for London, Hong Kong or Rio – this no-frills site delivers the goods quickly. Includes data on more than 15,000 towns and cities.

Special interest Websites

You can take your search even further by trekking through the many specialist sites on the Web. Below are listed just some of those on offer.

Tornadoes and severe storms

Australian Severe Weather
australiasevereweather.com
With this site's stunning photos of awesome Aussie weather, it's easy to get distracted from the wealth of detail provided by folks who observe and document severe storms throughout the country.

The Tornado Project
www.tornadoproject.com
One of the most authoritative sites on tornado history, climatology and behaviour, especially as it pertains to the US.

TORRO
www.torro.org.uk
The group that pioneered the study of Britain's often-weak but surprisingly frequent tornadoes. Unbeatable source of detail on severe weather across Britain and Europe.

US NOAA Storm Prediction Center
www.spc.noaa.gov
The agency that issues US tornado watches and warnings provides loads of tornado statistics and background on forecasting, including experimental outlooks that appear here before they go public.

Hurricanes

Atlantic Tropical Weather Center
www.atwc.org
A vast array of links, with multiple choices for almost every conceivable hurricane-related data type from sea-surface temperatures to computer-model output.

FAQ: Hurricanes, Typhoons, and Tropical Cyclones
www.aoml.noaa.gov/hrd/tcfaq/tcfaqHED.html
A gold mine of accessibly written background covering virtually all major aspects of hurricanes, including the science behind their formation, behaviour and demise.

NOAA Atlantic Oceanographic and Meteorological Laboratory
www.aoml.noaa.gov
Learn about "hurricane hunter" flights for research and the latest findings on how hurricanes are being better monitored and more accurately predicted.

NOAA National Hurricane Center
www.nhc.noaa.gov
The official source of US hurricane watches and warnings. Lots of climatology and other background detail for hurricanes across the Atlantic and northeast Pacific.

El Niño and La Niña

ENSO Information (International Research Institute for Climate and Society)
iri.columbia.edu/climate/ENSO
Excellent source of data and background on El Niño and La Niña prediction, including the latest IRI forecasts.

ENSO Wrap-Up (Australian Bureau of Meteorology)
www.bom.gov.au/climate/enso/
Includes a discussion of BOM's take on current ENSO conditions, as well as educational links and background on El Niño and La Niña impacts on Australia.

NOAA El Niño/La Niña Pages
www.elnino.noaa.gov
www.elnino.noaa.gov/lanina.html
Great jumping-off points for learning about these sibling phenomena. Includes graphics, explanations, current and projected status, and plenty of internal and external links.

What Is El Niño/What Is La Niña?
www.pmel.noaa.gov/tao/elnino/el-nino-story.html
www.pmel.noaa.gov/tao/elnino/la-nina-story.html
Another strong pair of NOAA sites, with emphasis on data from the network of monitoring buoys stationed in the Pacific Ocean that keeps an eye on Niño and Niña progress.

Lightning

NASA MSFC/A Lightning Primer
thunder.msfc.nasa.gov
From NASA's Marshall Space Flight Center (which measures lightning from above), a pleasing survey of lightning knowledge past and present, including links to hot research topics and the latest data.

National Lightning Safety Institute
www.lightningsafety.com
A group that takes lightning seriously, this consulting firm provides a wide range of statistics, explanations, safety tips, medical facts and other eye-opening stuff.

Climate change

CSIRO Atmospheric Research
www.cmar.csiro.au
From Australia's largest research agency (the Commonwealth Scientific and Industrial Research Organization), a broad palette of information on air pollution, climate change and other pressing environmental issues pertaining to weather and climate.

The Discovery of Global Warming (American Institute of Physics)
aip.org/history/climate
A tremendous source of background on the history of climate-change science. Includes all of the information from Weart's top-notch book *The Discovery of Global Warming*, plus much more. The mass of information can be accessed in several useful ways.

Environment Canada/Climate Change
www.ec.gc.ca/climate/home-e.html (English interface)
A look at how global change might affect Canada and the world.

Hadley Centre for Climate Prediction and Research
www.met-office.gov.uk/research/hadleycentre
Part of the UK Met. Office, this world-renowned modelling centre has crafted excellent summaries of how the atmosphere is simulated inside computers and what the results tell us.

Intergovernmental Panel on Climate Change
www.ipcc.ch
Full text from many of the IPCC's global-consensus reports on climate change is available here. Not always easy reading, but the most thorough and influential summaries to date on where our atmosphere is going.

NASA's Global Change Master Directory/Learning Center
gcmd.gsfc.nasa.gov/Resources/Learning/index.html
A handy entreé into NASA's varied and sprawling set of online resources on global change. Includes a guide to global-change data and a lay-oriented FAQ.

US National Assessment
www.nacc.usgcrp.org
Reports from the congressionally mandated project to examine regional US impacts of climate change. The US Climate Change Science Program (www.climatescience.gov) is now carrying out a set of targeted assessments on a wide range of climate-change topics.

Blogs

Some of the world's most knowledgeable and articulate weather fans now share their thoughts regularly and invite feedback from readers via blogs. Below is a sampling from this fast-growing medium. A few weather blogs allow you to get up-to-the-minute postings via RSS feeds.

AccuWeather Community Weather Blog
http://wwwa.accuweather.com/adcbin/public/community_blog.asp
Observations on a range of wild weather from Jesse Ferrell, a Web-weather pioneer.

Jeff Masters (Weather Underground)
http://www.wunderground.com/blog/JeffMasters
Masters, a PhD meteorologist and former hurricane hunter, follows the progress of each year's season, pulling together useful clues from a variety of sources.

NZ Weather Blog
nzweather.wordpress.com
News stories pertaining to Kiwi weather and touching on climate change and other relevant topics.

RealClimate
www.realclimate.org
Run by an international team of top scientists, RealClimate addresses burning questions on global change in a style aimed at motivated laypeople, with plenty of cross-references.

The Weather Channel Blog
www.weather.com/blog/weather
Senior meteorologist Stu Ostrow and other experts from TWC hold forth on topics that range from global warming to weather-related pop-songs.

The Weather Guys (USA Today)
blogs.usatoday.com/weather
A colourful range of weather-related tidbits, from the lighthearted to the serious, hosted by editors Bob Swanson and Doyle Rice.

Weather agencies around the world

Even where private meteorological firms are prevalent, national governments are typically charged with gathering information on weather across the land and informing people of its progress, especially when anything hazardous is brewing. (See "Forecasts and how to read them", p.120, for more on the public- and private-sector roles in weather forecasting.) Below are links to member agencies of the **World Meteorological Organization** that provide weather data on the Web. For the latest member list and links, check the WMO site, www.wmo.ch/web-en/member.html. Of course, in order to get forecasts and other information from these pages, you'll need to know the local language – or you'll want to use a Web-based, on-the-fly translator. A few sites, such as Canada's, are wholly or largely bilingual; many others offer some of their basic information in several languages. A convenient alternative is the WMO's one-stop shop, the World Weather Information Service, worldweather.wmo.int. It offers selected conditions and forecasts from nearly all of these national sites in a single format.

Algeria,
Office National de la Météorologie
www.meteo.dz

Antigua and Barbuda,
Meteorological Services
www.antiguamet.gov.ag

Argentina, **Servicio Meteorológico Nacional**
www.meteofa.mil.ar

Armenia, **Department of Hydrometeorology of the Republic of Armenia (Armhydromet)**
www.meteo.am

Australia, **Bureau of Meteorology**
www.bom.gov.au

Austria, **Central Institute for Meteorology and Geodynamics**
www.zamg.ac.at

Azerbaijan, **National Hydrometeorological Department**
www.azhydromet.com

Bahamas, **Department of Meteorology**
www.bahamasweather.org.bg

Bahrain, **Meteorological Directorate**
www.bahrainweather.com

Barbados **Meteorological Service**
http://www.agriculture.gov.bb/default.asp?V_DOC_ID=1150

Belarus, **Hydrometeorological Centre**
www.pogoda.by

Belgium, **The Royal Meteorological Institute of Belgium**
www.meteo.be

Belize, **National Meteorological Service**
www.hydromet.gov.bz

Benin, **Service Météorologique National du Bénin**
www.meteo-benin.net

Bolivia, **Servicio Nacional de Meteorologia e Hidrologia**
www.senamhi.gov.bo

Bosnia and Herzegovina, **Federal Meteorological Institute**
www.fmzbih.co.ba

Botswana, **Department of Meteorological Services**
www.weather.info.bw

Brazil, **Instituto Nacional de Meteolorogia**
www.inmet.gov.br

Brunei **Meteorological Service**
www.bruneiweather.com.bn

Bulgaria, **National Institute of Meteorology and Hydrology**
www.meteo.bg

Cameroon, **Meteo Cameroon**
www.meteo-cameroon.net

Canada, **Meteorological Service of Canada**
www.msc-smc.ec.gc.ca

Chile, **Direccion Meteorologica de Chile**
www.meteochile.cl

China, **The China Meteorological Administration (English site)**
211.147.16.25/ywwz

Colombia, **Instituto de Hidrologîa, Meteorologica y Estudios Ambientales**
www.ideam.gov.co

Congo, **Direction de la Meteorologie Nationale**
www.meteo-congo-brazza.net

Costa Rica, **Instituto Meteorológico Nacional**
www.imn.ac.cr

Croatia,
Meteorological and Hydrological Service
meteo.hr

Cuba, **Instituto de Meteorología**
www.met.inf.cu

Czech Republic, **Czech Hydrometeorological Institute**
www.chmi.cz

Democratic Republic of the Congo, **National Agency of Meteorology**
www.meteo-congo-kinshasa.net

Denmark,
The Danish Meteorological Institute
www.dmi.dk

Dominica **Meteorological Services**
www.meteo.dm

Dominican Republic,
Oficina Nacional de Meteorologia
www.onamet.gov.do

Ecuador, **Instituto Nacional de Meteorología e Hidrología**
www.inamhi.gov.ec

Egypt, **Egyptian Meteorological Authority**
www.nwp.gov.eg

El Salvador, **Servicio Meteorológico**
www.snet.gob.sv/meteorologia

Eritrea Meteorological Services
www.punchdown.org/rvb/meteo/MetUpdates.html

Estonia, Meteoroloogia ja Hüdroloogia Instituut
www.emhi.ee

Ethiopia, National Meteorological Agency
www.ethiomet.gov.et

Fiji, Fiji Meteorological Service
www.met.gov.fj

Finland, The Finnish National Weather Service/The Finnish Meteorological Institute
www.fmi.fi

France, Météo France
www.meteo.fr

Gabon, Météo Gabon
www.meteo-gabon.net

Georgia, State Department of Hydrometeorology
www.hydromet.ge

Germany, Deutscher Wetterdienst
www.dwd.de

Ghana
Meteorological Services Department
www.meteo.gov.gh

Greece, Hellenic National Meteorological Service
www.hnms.gr

Guinea, Le Service Météorologique National
www.meteo-guinee-conakry.net

Guinea-Bisseau, Meteorologia de Guine Bissau
www.meteo-guinee-bissau.net

Haiti Weather Net
www.haitiweather.net

Honduras, Servicio Meteorologico Nacional
www.smn.gob.hn

Hungary,
Hungarian Meteorological Service
www.met.hu

Iceland,
The Icelandic Meteorological Office
www.vedur.is

India, India Meteorological Department
www.imd.ernet.in

Indonesia, Indonesian Meteorological and Geophysical Agency
www.bmg.go.id

Iran, Islamic Republic of Iran Meteorological Organization
www.irimet.net

Ireland, Met Éireann
www.met.ie

Israel, Israel Meteorological Service
www.ims.gov.il

Italy, Servizio Meteorologico
www.meteoam.it

Jamaica Meteorological Service
www.metservice.gov.jm

Japan Meteorological Agency
www.jma.go.jp/jma/indexe.html

Jordan Meteorological Department
www.jometeo.gov.jo

Kazakhstan
www.meteo.kz

Kenya Meteorological Department
www.meteo.go.ke

Korea Meteorological Administration
www.weather.go.kr

Kuwait International Airport
www.kuwait-airport.com.kw/Index_e.htm

Kyrgyz Republic, Main Hydrometeorological Administration
www.meteo.ktnet.kg

Latvia, Latvian Environment, Geology and Meteorology Agency
www.meteo.lv

Lesotho Meteorological Services
www.lesmet.org.ls

Libya, Libyan Meteorological Department
www.lnmc.org.ly

Lithuania, Lithuanian Hydrometeorological Service
www.meteo.lt

Luxembourg, Administration de l'Aéroport – Météo
www.aeroport.public.lu/fr/meteo

Macedonia Hydrometeorological Service
www.meteo.gov.mk

Madagascar Sur Ranet
www.meteo-madagascar.net

Malawi Meteorological Services
www.metmalawi.com

Malaysia,
Malaysian Meteorological Service
www.kjc.gov.my

Maldives Department of Meteorology
www.meteorology.gov.mv

Mali, Météo Nationale du Mali
www.meteo-mali.net

Malta, Malta International Airport Meteorological Office
www.maltairport.com/weather

Mauritius Meteorological Services
metservice.intnet.mu

Mexico, Servicio Meteorológico Nacional
smn.cna.gob.mx/SMN.html

Moldovia,
Serviciului Hidrometeorologic de Stat
www.meteo.md

Mongolia Institute of Meteorology and Hydrology
www.env.pmis.gov.mn/Namhem

Montenegro Republic
Hydrometeorological Service
www.pps.gov.sc/meteo

Morocco, Direction de la Météorologie Nationale
www.marocmeteo.ma

Mozambique,
Instituto Nacional de Meteorologia
www.inam.gov.mz

Myanmar Department of Meteorology & Hydrology
www.dmh.gov.mm

Namibia Meteorological Service
www.meteona.com

Nepal, Department of Hydrology and Meteorology
www.dhm.gov.np

Netherlands, Royal Netherlands Meteorological Institute
www.knmi.nl

New Zealand, Meteorological Service of New Zealand, Ltd
www.metservice.co.nz

Nicaragua,
Dirección General de Meteorologia
www.ineter.gob.ni

Niger, Direction de la Meteorologie Nationale du Niger
www.meteo-niger.net

Niue Island Meteorological Service
informet.net/niuemet

Norway,
Norwegian Meteorological Institute
www.dnmi.no

Oman, Meteorological Department
www.met.gov.om

Pakistan Meteorological Department
www.pakmet.com.pk

Panamá Hidrometeorologia
www.hidromet.com.pa

Paraguay, Dirección de Meteorología e Hidrología
www.pakmet.com.pk

Peru, Servicio Nacional de Meteorología e Hidrología
www.senamhi.gob.pe

Philippines, PAGASA (Philippine Atmospheric, Geophysical & Astronomical Services Adm.)
www.pagasa.dost.gov.ph

Poland, Institute of Meteorology and Water Management
www.imgw.pl

Portugal,
Instituto de Meteorologia
www.meteo.pt

Romania, National Meteorological Administration
www.inmh.ro

Russia, Russian Federal Service for Hydrometeorology and Environment Monitoring
www.mecom.ru

St. Lucia Met. Service
www.slumet.gov.lc

Samoa Meteorology Division
www.meteorology.gov.ws

Saudi Arabia Presidency of Meteorology & Environment
www.mepa.org.sa

Senegal Direction de la Météorologie Nationale
www.meteo-senegal.net

Serbia, Hydrometeorological Service
www.hidmet.sr.gov.yu

Seychelles,
Seychelles Meteorological Service
www.seychelles.net/meteo

Singapore,
Meteorological Services Division
www.gov.sg/metsin

Slovakia,
Slovak Hydrometeorological Institute
www.shmu.sk

Slovenia, Environmental Agency
www.rzs-hm.si

South Africa, South African Weather Service
www.weathersa.co.za

Spain,
Instituto Nacional De Meteorologia
www.inm.es

Sri Lanka Department of Meteorology
www.meteo.slt.lk

Suriname,
Meteorologische Dienst Suriname
www.meteosur.sr

Swaziland Meteorological Service
www.swazimet.gov.sz

Sweden, Swedish Meteorological and Hydrological Institute
www.smhi.se

Switzerland, Federal Office of Meteorology and Climatology
www.meteoswiss.ch

Tanzania Meteorological Agency
www.meteo.gov.tz

Thailand Meteorological Department
www.tmd.go.th

Togo,
Service de la Meteorologie Nationale
www.meteo-togo.net

Tonga Meteorological Services
www.mca.gov.to/met

Trinidad and Tobago
Meteorological Service
www.metoffice.gov.tt

Tunisia
www.meteo.tn

Turkey,
Turkish State Meteorological Service
www.meteor.gov.tr

Uganda
Department of Meteorology
www.meteo-uganda.net

Ukraine Hydrometeorological Center
www.meteo.com.ua

United Arab Emirates, Meteorological Department
www.uaemet.gov.ae

United Kingdom, The Met Office
www.metoffice.gov.uk

United States, National Weather Service
www.weather.gov

Uruguay,
Dirección Nacional de Meteorología
www.meteorologia.com.uy

Venezuela Air Force Weather Service
www.meteorologia.mil.ve

Vietnam Hydrometeorological Service
www.nchmf.gov.vn

Zambia Meteorological Department
www.meteo-zambia.net

Zimbabwe Meteorological Services
www.weather.co.zw

Weather and health

For all the comfort we get from heating, air conditioning, humidifiers and dehumidifiers, the weather outside still shapes how we feel each day. Even the names of some illnesses betray their real or presumed links to the atmosphere. The term malaria, for instance, was coined from "bad air".

Centuries before Western science came along, scholars and laypeople alike surmised that the air had much to do with our health. In one of the earliest surviving medical guides – the *Nei Ching* (produced in the 2500s BC) – Chinese emperor **Huang Ti** noted that diseases often developed in sync with the seasons: diarrhoea in the summer, colds and fevers as the weather turned colder. Almost 3000 years ago, **Hindu physicians** laid down the Ayurvedic principle of balancing heat, cold and moisture as it manifests both inside and outside of us. **Hippocrates**, the titan of Greek medicine, envisioned the link between internal and external states in terms of inner humours (phlegm, blood, bile) that could be shifted unhealthily by an imbalance among the outer elements: fire (dry and hot), water (cold and wet), air (hot and wet) and earth (cold and dry).

The importance of **weather in medicine** grew with the zest for observation that followed the scientific renaissance. Physicians were the central observers in several key weather networks across the US and Europe in the late 1700s and early 1800s; they took note of weather features and hunted for links between weather and disease. Later in the nineteenth century, doctors turned away from the sky and toward the microscope, while meteorology came into its own as a distinct discipline. A reunion of sorts is now taking place: the threat of climate change is bringing public health experts and atmospheric scientists together to map out how global patterns of illness might shift in the years to come.

Temperature and its effects

Warm-blooded creatures that we are, humans are constantly pulling on sweaters or stripping off shirts to prevent themselves from getting too cold or too hot. It's a valid concern. If body temperature wanders more than a few degrees from the standard 37°C/98.6°F (well, not quite; see box, opposite), then the complications can be severe. The sweat that pours out of your pores on a hot day – as much as a litre/one quart every half-hour – is the chief way your body works to cool itself. Should the evaporation from sweat fail to relieve the body's overheating – whether due to over-exertion, high humidity or some other factor – then the lassitude of heat exhaustion can follow. Potentially fatal **heat stroke** results when one's core body temperature exceeds 41°C/106°F and the sweating mechanism slows or halts. The "Bedouin solution" of loose-fitting white garb is an ingenious weapon against heat: the white reflects sunlight, while the loose fabric allows air to circulate and hastens evaporation of sweat from the skin.

Exposure to sub-freezing cold, especially in a strong wind, is a recipe for **frostbite**, which literally freezes tissue near the skin. Frostbite can also damage tiny vessels and turn blood into sludge, which is why frostbitten tissue needs to be warmed very slowly and gently. Even when the temperature stays above freezing, cold can take its toll. Feet that are swathed in damp, chilly socks for days on end can suffer a painful restriction of blood flow and potential damage associated with **trench foot**. It was the bane of thousands of foxhole-stationed troops in World War I, and is still a threat for hard-core adventure travellers. If one's whole body is wet, then surprisingly mild temperatures can produce the adverse mental and physical effects of **hypothermia**.

Heat and **wind-chill indices** (see p.402) can help one better assess the effect of humidity on a hot day, or the impact of wind on a chilly one. They can also help society prepare for the most dangerous spells of severe heat and cold. Mortality rises only gradually as the temperature drops, mainly due to the indirect effect of the flu and other ailments as people spend more time together indoors. However, a sustained heat wave can cause a spike of up to 100 percent in death rates, especially when the average temperature spends more than a couple of days above a threshold value that seems to vary from place to place. The threshold effect is muted in regions such as the US South, where people are accustomed to serious heat, and their homes and offices – even when modest – are typically well-ventilated. The mortality spikes are sharpest in northern cities, where heat-absorbing buildings are packed closely together and home air conditioning is less common. Over 500 Chicagoans died in a 1995 heat wave, many of them elderly who kept windows closed, to foil crime, and air conditioners (if they had any) off, to save money. Excess heat also kills people in many other indirect ways, such as accidents, alcohol and even homicide.

A new take on body heat

For those of us familiar with English units of measurement, few numbers are as instantly recognizable as 98.6. For over a century, it's been famous as the average temperature of the human body in Fahrenheit. As it turns out, we've been misled – and all because of an error in switching from English to metric units and back again. Back in the late 1800s, German scientist Carl Reinhold August Wunderlich analysed data from thousands of Europeans. In converting to the Celsius units that are standard in the world of science, the average from Wunderlich's data set was rounded to an even 37.0°C – which, in turn, converts back to 98.6°F. More recent studies have found that the real average was, and is, closer to 36.8°C/98.2°F. Perhaps more importantly, body temperature varies not only from person to person and by the body part where the temperature is taken, but also by the time of day and even the time of the month (for menstruating women, that is).

Weather safety tips

Tornado

- Take shelter underground if possible. Get under a table or cover yourself with a blanket or cushions.
- If no below-ground shelter is available, go to a windowless, interior room (such as a bathroom or closet) in the lowest level of a well-constructed building.
- Avoid mobile homes, vehicles and large rooms with free-span ceilings.

Hurricane/typhoon/tropical cyclone

- If you live in a high-risk area, strengthen your home with window shutters and roof clips, and keep items such as batteries and bottled water on hand.
- Heed your local warnings and evacuation notices. Don't delay leaving if you are told to evacuate. Avoid driving through flooded roads.
- Stay sheltered through the duration of the hurricane. The clear calm in the hurricane's eye is usually followed by the return of severe conditions within minutes.

Lightning

- If thunder follows a lightning flash by less than 30 seconds, the storm is likely close enough for a bolt to strike you. Seek shelter, preferably inside a closed building or a vehicle with all its windows up.
- If caught in the open, crouch as low as possible, but keep as little of your body touching the ground as possible. Avoid high ground, open spaces and trees.
- When indoors during a lightning storm, avoid using a corded telephone, and stay away from doors and windows.

Heat waves are one of the biggest threats in scenarios of future global warming. On average, overnight lows are expected to rise more than daytime highs, and dangerous bouts of heat may spread poleward to places that have been spared up to now. One study showed that heat-related deaths could rise more than sevenfold globally by the year 2050 if temperatures climb as computer models predict they will. (See p.102 for more on heat and cold waves.)

Pollution and allergens

Sometimes the air itself makes us sick. The horrors of prolonged pollution episodes that once killed great swaths of people (see box, p.69) have passed, yet air quality remains a pressing issue, especially in urban areas. The noxious compounds produced by cars and factories can irritate our eyes, nose and throat; they also put asthmatics at risk and lower our aerobic efficiency. Recent studies have shed light on tiny particulates that were once out of instrumental sight and thus out of mind. We now know that these particulates – some less than 0.1mm/0.004in wide – can lodge deep in the lungs, with

Flash flood

- Get to the highest ground possible immediately, even if it means abandoning your home or car.
- Never cross a flooded road; as little as 500mm/20in of water – below the top of most tyres – can float a car or truck and send it into harm's way.
- Beware of canyons if there is any risk of heavy rain nearby, especially in deserts or semi-arid regions. It need not rain where you are for a deadly flash flood to flow your way.

Extreme heat

- Use a fan to ventilate spaces that lack air conditioning.
- Drink plenty of water and other non-diuretic, non-alcoholic fluids.
- If someone shows signs of heat exhaustion (profuse sweating, fatigue, cold skin), get the person out of the heat, elevate the lower body and apply cool compresses.
- If heat stroke develops (hot, dry skin and rapid pulse, with a body temperature above 41°C/106°F), get medical help immediately. Apply a cold bath or cold sponges. Do not provide anything to drink.

Extreme cold

- Dress in layers (for added insulation).
- Cover head, face and hands as much as possible.
- If signs of hypothermia develop (uncontrolled shivering, clumsiness, disorientation), get the person out of the cold, elevate the lower body, and give her or him something warm to drink.

the bloodstream their only point of exit. They've been blamed for increasing heart-attack rates by up to 50 percent on polluted days. Fortunately, most cities are now adept at monitoring **air quality** and keeping the public apprised of it, not only when smog is at hand but also when light winds and an inversion (relatively cool, stable air near the surface) are expected to trap pollution over the next day or so. Of course, natural events such as volcanoes or large-scale, lightning-triggered forest fires can also degrade air quality.

As **pollen** and **mould spores** drift through the air, they can torment millions of allergy sufferers, as well as the growing ranks of asthmatics; fatal bouts of **asthma** in the US rose by some 40 percent during the 1990s. Allergy forecasts are now a staple of Internet weather sites and many TV weathercasts. In general, pollen outbreaks are worst when it's dry and windy; breezes hasten the spread of pollen, while the low humidity helps induce flowers to release pollen and reduces the opportunity for moisture to bind with the pollen and bring it down to earth. Many mould spores like to travel on a dry wind as well, even though mould itself tends to thrive when it's moist. The best weather for allergy relief, aside from a killing frost, is a solid rain – although the gust fronts that often precede thunderstorms have triggered major rounds of asthma attacks in London and Melbourne, among other places.

Infectious disease and climate change

The links between **climate** and **infection** were once as hard to ferret out as the infectious agents themselves. It's possible that the so-called medieval climatic optimum – the several centuries of mild weather between about 950 and 1400 AD that helped the Norse to settle Greenland – also helped rats and fleas to thrive across Europe, paving the way for the **Black Death** that killed a third of Europeans between 1347 and 1352.

Another warming of the planet is now under way, this one apparently triggered by human activity, and microbes may be responding. For instance, **malaria** is moving into previously untouched parts of Africa's central highlands, possibly due to a scarcity of cool, mosquito-killing temperatures. Increased settlement in the highlands may also be a factor in the upward curve. Still, for whatever reasons, malaria now infects an estimated half a billion people each year, four times more than in 1990. **Dengue** – another potentially fatal, mosquito-borne virus – is also on the increase, with 50 million cases each year. One study estimates that a 1°C/1.8°F rise in average temperature could raise the risk of dengue by up to 47 percent.

Shifts in rainfall also make a difference. Some 25 million Africans are plagued with **onchocerciasis**, a painful skin and eye condition caused by the growth of tiny worms. These worms reach humans through the bite of a blackfly; in years when **heavy rains** keep the rivers full and the flies breeding, the risk of onchocerciasis can spike severalfold. Some health effects can even be put at the door of everyone's favourite climatic scapegoat, **El Niño**. For example, waters warmed by El Niño off the coast of South America may have triggered a 1991 cholera epidemic in and near Peru by fostering the growth of bacteria that infected seafood.

Wind, altitude and sun

Although they're too fleeting for science to get a good handle on their health effects, the furious **winds** that scream down mountain ranges – such as France's *mistral* and California's *Santa Ana* – have a reputation for making people edgy and cranky. One possible culprit is the sharp pressure changes associated with these ill winds. A rapid **drop in air pressure** can lead to a build-up and expansion of fluid within sensitive tissues and joints, possibly triggering the pain that people can sense – like a weather change – "in their bones".

Changes in **air mass** can also shift the balance of ions in the atmosphere. The arrival of a thunderstorm or a high wind can add to the count of positive ions, while fair weather tends to produce more negative ions. For decades, the makers of negative ion generators have claimed their devices can make an indoor space more healthy – but research has failed to show that such machines can produce enough of the short-lived ions to have an effect. If you're looking for major impact from electricity in the atmospheric, try **lightning**. A strike lasting a fraction of a second can send 300,000 volts through your body. Even when a bolt doesn't kill, it can cause severe burns, and most survivors experience some amount of neurological or psychological damage.

Virtually everyone feels the effects of **altitude** when they head to the mountains. At the altitude of Denver, Colorado (1610m/5280ft) there's about 15 percent less oxygen than at sea level. In La Paz, Bolivia (4010m/13,170ft), the reduction is more than 40 percent. One result is a bona fide difference between mountain natives and flatlanders. Studies have shown that people who head for the hills as adults never completely adapt – even after living there for decades, their bodies work less efficiently than those of their native-born neighbours. On a quick visit to the mountains, especially above 2400m/7900ft, be prepared for a couple of days of headaches, fatigue and perhaps a nosebleed, and be sure to drink plenty of fluids (but watch out for alcohol – its punch is increased with height).

Sunshine may be the most underrated risk at altitude, and it's no slouch at sea level, either – although life would be dull indeed without it. As sunlight wanes in the winter, **seasonal affective disorder** (or SAD) brings hard-to-shake depression to millions of people in mid-latitudes. Specially tailored banks of indoor lighting seem to help. The problem with sunshine itself is **ultraviolet radiation**, which is only partially filtered by the stratosphere's ozone layer. As our skin absorbs UV, it synthesizes vitamin D, but along with this benefit comes the risk of **sunburn**. Basically, the higher the sun in the sky and the higher your elevation, the greater the influx of UV rays – thus, the common-sense rule of avoiding direct sun and/or using suncreen during the midday hours. Bright ground reflects sunlight and enhances its effects, so a ski run in January may send more UV your way than a forested trail in July.

As more people spend more time outdoors, **skin cancer** rates are rising in the US, Australia and elsewhere. The slow, slight erosion of the ozone layer above mid-latitude lands (much less severe than the seasonal polar ozone hole), may be goosing this trend by allowing greater amounts of biologically active UV **(UVB)** to reach the Earth. While you're protecting yourself with sunscreen for your skin, put sunglasses over your eyes: UVB can hasten macular degeneration and other eye problems.

Conversions

Temperature

To convert from degrees F to C, subtract 32, then multiply by 5 and divide by 9.

To convert from degrees C to F, multiply by 9, divide by 5 and add 32.

To convert from degrees C to K, add 273.15.

Wind speed

1mph = .869 knots = 1.61kph
1kph = 0.54 knots = 0.62mph

1 knot = 1.15mph = 1.85kph

Atmospheric pressure

1 millibar = 33.8 inches of mercury

1 inch of mercury = 29.5 millibars

Temperature chart

-40°C = -40°F
-30°C = -22°F
-20°C = -4°F
-10°C = 14°F
0°C = 32°F (freezing)
10°C = 50°F
20°C = 68°F
30°C = 86°F
40°C = 104°F

Heat and cold indices

Scientists have worked for decades to incorporate the effects of moisture and wind on heat and cold, respectively. A new **wind-chill index** was released in 2001 by the US and Canadian governments, updating a formula that had stood since the 1970s. The new values show somewhat less of a wind-chill effect, largely because the former index assumed that the wind at measurement height 10m/33ft equated to the wind at face level. Future versions of the wind-chill table may incorporate other effects.

Heat indices are most popular in the US. **The International Society of Biometeorology** is working on a new scale that will incorporate the effects of both heat and cold in a single index.

Wind-chill index, imperial version: www.nws.noaa.gov/om/windchill

Wind-chill index, metric version: www.msc.ec.gc.ca/windchill/charts_tables_e.cfm

US heat index: www.nws.noaa.gov/om/heat

Intensity scales

Hurricanes

Tropical cyclones whose winds reach 119 kph/74 mph are also known as **hurricanes, severe cyclonic storms, severe tropical cyclones** or **typhoons,** depending on what part of the world you're in (see p.83). Meteorologists rate their intensity by directly measuring winds through instrument packages dropped by airplane and/or by inferring wind speed through the motion and appearance of clouds on satellite. Those whose winds reach 178 kph/111 mph in the northeast Pacific and Atlantic are dubbed **major** or **intense hurricanes**. If a typhoon in the northwest Pacific packs winds of at least 242 kph/150 mph, it becomes a **supertyphoon**. On the Saffir-Simpson scale, only two hurricanes have struck the US while at Category 5 strength: the Florida Keys' Labor Day hurricane (1935) and Hurricane Camille (1969).

Saffir-Simpson Hurricane Scale (US)

Category	Sustained wind speed	Typical storm surge
1	**119–153 kph/74–95 mph**	**122–152cm/4–5ft**
Damage primarily to unanchored mobile homes, shrubbery, trees, signs; some coastal road flooding and minor pier damage		
2	**154–177 kph/96–110 mph**	**183–244cm/6–8ft**
Some roofing material, door, and window damage of buildings. Some trees blown down. Considerable damage to mobile homes, poorly constructed signs and piers. Small craft in unprotected anchorages break moorings.		
3	**178–209 kph/111–130 mph**	**274–366cm/9–12ft**
Some structural damage to small homes and sheds. Foliage blown off trees; large trees blown down. Mobile homes and poorly constructed signs destroyed. Coastal flooding destroys smaller structures; larger ones damaged by floating debris.		
4	**210–249 kph/131–155 mph**	**396–549cm/13–18ft**
Roofs fail on some small homes. Shrubs, trees and all signs down. Mobile homes destroyed. Extensive door and window damage. Major damage to lower floors of structures near shore.		
5	**>249 kph/>155 mph**	**>549cm/>18ft**
Complete roof failure on many buildings. Some buildings destroyed. Severe and extensive window and door damage. Major damage to lower floors of all structures located less than 4.6m/15ft above sea level and within 460m/1500ft of the shoreline.		

Australian Cyclone Severity Scale

Note that this scale is pegged to highest wind gust rather than highest sustained wind speed. Typically, the strongest gusts in a tropical cyclone are about 15% higher than the peak sustained wind speed.

Category	Highest wind gust
1	**<125kph/<78mph**

Negligible house damage. Damage to some crops, trees and caravans. Craft may drag moorings.

2	**125–169kph/78–105mph**

Minor house damage. Significant damage to signs, trees and caravans. Heavy damage to some crops. Risk of power failure. Small craft may break moorings.

3	**170–224kph/106–139mph**

Some roof and structural damage. Some caravans destroyed. Power failure likely.

4	**225–279kph/140–173mph**

Significant roofing loss and structural damage. Many caravans destroyed and blown away. Dangerous airborne debris. Widespread power failure.

5	**>279kph/>173mph**

Extremely dangerous with widespread destruction.

Tornadoes

The US and Britain each have their own scales for measuring tornado intensity. Both are based not on observations while the storm is in progress, but rather on damage surveys. If a rural US tornado causes no damage, its rating is F0 no matter how strong its winds are. More recently, wind measurements from Doppler radar have been used as an independent measure of tornado intensity, and improved storm surveys have shown that damage can be incurred with weaker winds than once believed possible. As a result, American scientists and engineers have developed an Enhanced F scale (EF scale), to be used in classifying US tornadoes from 1 February 2007 onward. Older tornadoes will keep their original F ratings.

In the US, only a quarter of US tornadoes were rated F3 or better on the original Fujita scale and only one in a thousand was an F5 (it's believed that ratio won't change much under the new system). No tornadoes were ever rated F6 on the original scale, and the new scale eliminates that category.

Enhanced Fujita-Pearson Scale (US)

The damage descriptions below apply to the EF scale and vary somewhat from those used in the original F scale (whose wind speeds are given for comparison). The estimated wind speeds in the EF scale correspond to the highest gusts lasting at least three seconds, which correlate well to certain types of damage but differ slightly from a standard wind measurement. For more details on the new scale, see www.spc.noaa.gov/efscale.

Rating	Name	Estimated wind speed for EF rating (and for original F rating)
EF0	light	105–137 kph/65–85mph (F0: 64–116kph/40–72mph)

Some damage to chimneys, branches, signs; shallow-rooted trees pushed over.

EF1	moderate	138–179kph/86–110mph (F1: 117–180kph/73–112 mph)

Roofs torn off frame homes; mobile homes demolished; moving autos pushed off the roads; windows and glass doors broken.

EF2	considerable	180–217kph/111–135mph (F2: 181–252kph/113–157mph)

Roofs torn off well-constructed houses; cars lifted off ground; foundations of frame homes shifted; large trees snapped or uprooted.

EF3	severe	218–266kph/136–165mph (F3: 254–331kph/158–206mph)

Some walls torn off well-constructed houses; trains overturned; most trees in forest uprooted; heavy cars lifted and thrown; structures with weak foundations blown some distance.

EF4	devastating	267–320kph/166–200mph (F4: 333–419kph/207–260 mph)

Well-constructed houses levelled; structures with weak foundations blown off some distance; cars thrown; trees debarked.

EF5	incredible	Greater than 320kph/200mph (F5: 420–512kph/261–318 mph)

Strong frame houses lifted off foundations and swept away; brick houses completely wiped off foundations with strongest winds; automobile-sized missiles fly through the air for distances of more than 100m/330ft; cars thrown and large missiles generated.

TORRO (Tornado and Storm Research Organisation) Intensity Scale (Britain)

Rating	Estimated wind speed
T0	63–87kph/39–54mph

Loose light litter raised from ground-level in spirals. Tents, marquees seriously disturbed; most exposed tiles, slates on roofs dislodged. Twigs snapped; trail visible through crops.

T1	89–116kph/55–72mph

Deckchairs, small plants, heavy litter becomes airborne; minor damage to sheds. More serious dislodging of tiles, slates, chimney pots. Wooden fences flattened. Slight damage to hedges and trees.

T2	118–148kph/73–92mph

Heavy mobile homes displaced, light caravans blown over, garden sheds destroyed, garage roofs torn away, much damage to tiled roofs and chimney stacks. General damage to trees, some big branches twisted or snapped off, small trees uprooted.

T3	150–184kph/93–114mph

Mobile homes overturned/badly damaged; light caravans destroyed; garages and weak outbuildings destroyed; house roof timbers considerably exposed. Some of the bigger trees snapped or uprooted.

T4	186–219kph/115–136mph

Motor cars levitated. Mobile homes airborne/destroyed; sheds airborne for considerable distances; entire roofs removed from some houses; roof timbers of stronger brick or stone houses completely exposed; gable ends torn away. Numerous trees uprooted or snapped.

T5	220–258kph/137–160mph

Heavy motor vehicles levitated; more serious building damage that for T4, yet house walls usually remaining; the oldest, weakest buildings may collapse completely.

T6 **260–299kph/161–186mph**
Strongly built houses lose entire roofs and perhaps also a wall; more of the less-strong buildings collapse.

T7 **301–341kph/187–212mph**
Wooden-frame houses wholly demolished; some walls of stone or brick houses beaten down or collapse; steel-framed warehouse-type constructions may buckle slightly. Locomotives thrown over. Noticeable debarking of trees by flying debris.

T8 **343–386kph/213–240mph**
Motor cars hurled great distances. Wooden-framed houses and their contents dispersed over long distances; stone or brick houses irreparably damaged; steel-framed buildings buckled.

T9 **388–433kph/241–269mph**
Many steel-framed buildings badly damaged; locomotives or trains hurled some distances. Complete debarking of any standing tree-trunks.

T10 **435–481kph/270–299mph**
Entire frame houses and similar buildings lifted bodily from foundations and carried some distances. Steel-reinforced concrete buildings may be severely damaged.

Wind speed

The Beaufort Scale

The granddaddy of wind-related scales, this one was devised 200 years ago by British rear admiral Sir Francis Beaufort as a purely descriptive device, linked to the behaviour of ships in the wind. Only in 1926 were wind speeds officially attached to Beaufort's scale. It's still used in UK shipping forecasts (broadcast on BBC Radio 4) and in other outlooks.

Force	Wind speed	Name
0	**0kph/0mph**	**calm**
smoke rises vertically		
1	**1–5kph/1–3mph**	**light**
air direction of wind shown by smoke drift, but not by wind vanes		
2	**6–11kph/4–7mph**	**light breeze**
wind felt on face; leaves rustle; ordinary vane moved by wind		
3	**12–19kph/8–12mph**	**gentle breeze**
leaves and small twigs in constant motion; wind extends light flag		
4	**20–28kph/13–18mph**	**mod. breeze**
raises dust and loose paper; small branches are moved		
5	**29–38kph/19–24mph**	**fresh breeze**
small trees in leaf begin to sway; crested wavelets form on inland waters		
6	**39–49kph/25–31mph**	**strong breeze**
large branches in motion; whistling heard in telegraph wires; umbrellas used with difficulty		
7	**50–61kph/32–38mph**	**near gale**
whole trees in motion; inconvenience felt when walking against wind		

8	**62–74kph/39–46mph**	**gale**
breaks twigs off trees; generally impedes progress		
9	**75–88kph/47–54mph**	**strong gale**
slight structural damage occurs [chimney pots and slates removed]		
10	**89–103kph/55–63mph**	**storm**
seldom experienced inland; trees uprooted; considerable structural damage occurs		
11	**104–117kph/64–72mph**	**violent storm**
very rarely experienced; accompanied by widespread damage		
12	**≥117kph/>72mph**	**hurricane**

Glossary

For a full list of cloud types, see the box at the end of the glossary and the colour section.

absolute zero: The lowest temperature considered attainable in the known universe: 0°K/–273°C/–459°F.

adiabatic ascent: An air parcel rising with its relative humidity remaining below 100 percent, so that no moisture is condensed and no heat thus added.

advection: Fancy term for the wind moving weather elements around (temperature, moisture and the like).

aerosol: Dust, soot, salt or other airborne particle.

air mass: Large air parcel with relatively uniform qualities of temperature and moisture, separated from other air masses by fronts.

albedo (solar): Percentage of sunlight reflected from a surface.

altocumulus: See box p.415.

altostratus: See box p.415.

anemometer: Instrument for measuring wind speed.

anticyclone: A high-pressure centre, with descending air at its core, usually accompanied by fair weather. Winds flow outward from an anticyclone, turning clockwise in the Northern Hemisphere and counterclockwise in the Southern Hemisphere.

anvil: Sheet of cirrus cloud atop a thunderstorm.

aphelion: The point at which a planet is farthest from the Sun (see p.21).

aurora: Electrical activity and the associated light show stimulated in Earth's upper atmosphere by plasma arriving from the Sun, usually most visible toward the poles and during winter. The Northern Hemisphere sees the aurora borealis; down under, it's the aurora australis.

ball lightning: Mysterious form of lightning that takes shape as a glowing sphere which lasts from a few seconds to a few minutes. It's been observed inside houses and even aeroplanes. Ball lightning's existence is now accepted by most experts, but its cause remains unknown.

barometer: Instrument used to measure atmospheric pressure.

blizzard: Commonly used to describe any windy, frigid snowstorm. In the precise US definition, the wind must be sustained or gusting often to at least 56kph/35mph, with falling or blowing snow limiting visibility to 0.4km/0.25 miles for at least three hours.

blocking pattern: Large-scale pattern in which the jet stream locks into position and highs and lows stay fixed, sometimes for weeks. Often produces repetitive weather.

bora: A strong downslope wind that may accompany the passage of a cold front in and near mountains. (For local names, see p.233.)

boundary layer: Roughly the lowest 1km/0.6 mile of the atmosphere, where temperature shifts are greatest and the influences of land and sea are most pronounced.

carbon dioxide (CO_2): The most prevalent of the greenhouse gases added to the atmosphere by fossil-fuel burning. Also produced in large amounts by plant respiration.

ceiling: The height of the lowest level at which the sky is mostly or fully cloud-covered.

climate: Aggregate of the weather conditions prevailing over time at a given point.

climate change: Evolution of climate, as opposed to short-term weather changes. Often refers to changes triggered by human activity, although it can also denote natural processes.

cirrus: See box p.415.

cirrocumulus: See box p.415.

cirrostratus: See box p.415.

cold front: Narrow zone denoting where a cooler air mass is moving beneath a warmer one.

computer model: In weather and climate, sophisticated software that replicates the atmosphere and its behaviour.

condensation: The process of water vapour being deposited as liquid on a particle or surface when an air parcel reaches **saturation** or **supersaturation**. Condensation adds heat to the air.

convection: Parcels of air that rise due to buoyancy (relatively warm air positioned below relatively cool air). Often refers to the showers and thunderstorms that may result.

Coriolis force: The apparent shift in wind direction to the right as air flows toward either pole. A side effect of Earth's rotation.

crepuscular rays: Alternating light and dark bands of diverging sunlight that are produced when the sun is shadowed, especially near sunrise and sunset.

cumulonimbus: See box p.415.

cumulus: See box p.415.

cut-off low: Centre of low pressure divorced from the jet stream and thus stationary for days.

cyclone: General term for a low-pressure centre. Once a common US synonym for **tornado**, it also refers to those **tropical cyclones** that reach hurricane or typhoon strength across the Indian and far southwest Pacific Oceans.

depression: Synonym for **cyclone**.

dew: Moisture condensed on or near the ground, typically after a clear, calm night.

dew point: Temperature to which an air parcel must drop, at constant pressure, for the **relative humidity** to reach 100 percent, or **saturation**. Relative humidity changes with temperature for a closed parcel of air, but the dew point does not.

Doppler effect: Technique used in weather radar to infer the speed of winds by measuring the change in frequency of microwaves after they reflect off rain, snow or ice within a cloud.

downburst: Core of strong wind that descends from a shower or thunderstorm due to cooling from evaporation. See also **microburst**.

drizzle: Raindrops smaller than 0.5mm/0.02 inch.

dropsonde: Package of weather sensors dropped from above, usually parachuted from an airplane.

drought: Extended period with precipitation far enough below normal to affect ecosystems and/or societies.

dry line: Narrow zone that separates air masses with differing moisture content but little temperature contrast.

dust devil: Small **vortex** that forms over a sun-heated surface, especially in arid or semi-arid regions.

easterly wave: Westward-moving zone of low pressure in the tropics, typically with showers and thunderstorms. Can evolve into a tropical cyclone in certain locations and seasons.

El Niño: Periodic warming (typically lasting 1–2 years and occurring roughly every 2–7 years) at and just below the surface of the eastern tropical Pacific Ocean. Sometimes refers to the global weather shifts that result.

El Niño/Southern Oscillation (ENSO): The entire phenomenon that includes **El Niño**, **La Niña**, and the related cycles of weather and climate in and near the tropical Pacific.

ensemble prediction: The use of multiple computer-model simulations to find a range of possible weather or climate outcomes.

equinox (vernal and autumnal): The dates in late March and September on which spring and autumn begin, when Earth's tilt is neither toward or away from the sun. The exact dates vary somewhat from year to year due to our imperfect calendar system.

evaporation: The process of water entering the atmosphere as a vapour after escaping from a liquid (such as the ocean). Evaporation removes heat from the air.

extratropical cyclone: Low-pressure centre that forms outside the tropics.

eye: The core of a mature hurricane, typhoon or tropical cyclone, featuring nearly calm winds, few clouds and warm, descending air. The eye is surrounded by an eyewall that bears the storm's most intense winds.

flash flood: Intense, localized flood that develops quickly and can cause severe damage.

fog: A cloud at ground level that obscures visibility to less than 1km/0.6 miles.

föhn: A strong downslope wind that brings warming and drying in and near mountains: **föhn wall** of thick cloud often forms on the upwind side, where it may be raining or snowing heavily. (For local names, see p.233.)

front: Narrow zone separating air masses of different qualities. Fronts may be **warm**, **cold** or **occluded**.

frost: Ice crystals deposited on or near the ground, typically after a clear, calm night that dips below freezing.

funnel cloud: Narrow, rotating cloud that descends from a thunderstorm and may evolve into a **tornado**.

gale: Wind that exceeds 63kph/39mph.

geostationary satellite: Satellite that orbits Earth at the same rate Earth rotates, thus keeping it above a fixed domain. See **polar-orbiting satellite**.

glaze: Ice produced by rain, mist, drizzle or fog droplets that fall onto a below-freezing surface.

global warming: Rise in the average air temperature at Earth's surface. Most often refers to warming due to human activity as well as to the associated effects on climate and society.

greenhouse effect: Process by which greenhouse gases such as carbon dioxide and water vapour allow sunlight to reach Earth's surface but absorb longer-wave radiation as it leaves Earth, thus raising the average temperature. Greenhouse gases from human activity are believed to be producing an enhanced greenhouse effect beyond the one in place for millennia.

gust: Pulse of wind, usually lasting a few seconds, that may be 20 percent or more stronger than the sustained wind speed.

hail: Ice falling from convective clouds (US); rain or melted snow that freezes before it hits the ground (UK). See **sleet**.

halo: Complete or partial ring surrounding the moon or sun, usually produced in a layer of high, thin cloud.

harmattan: Dry, often-dusty northeast wind that dominates the African Sahel in winter.

haze: Reduced visibility at relative humidities below 100 percent due to aerosols or small water droplets, such as those forming on salt in marine air.

heat burst: Rare, brief, intense warming (sometimes above 35°C/95°F) produced by **downbursts** from a decaying thunderstorm, especially at night.

heat island: The overall climatic warming produced in an urban area as the pavement and buildings absorb and retain heat.

heat lightning: Colloquial term for distant lightning, often tinted red, orange or yellow as its light passes through the atmosphere on the horizon.

hoar frost: Opaque mixture of **frost**, frozen **dew** and/or **rime**.

hook echo: Radar image produced by rain wrapping around an intense thunderstorm updraught in which a tornado may develop. (A hook echo can be present without a tornado, and vice versa.)

humidity: See **relative humidity**.

hurricane: A **tropical cyclone** in the Atlantic, Caribbean or northwest Pacific Ocean with sustained winds of at least 119kph/74mph. Hurricanes are called **typhoons** in the northwest Pacific and **tropical cyclones** in the South Pacific and Indian oceans.

hygrometer: Instrument used to measure atmospheric relative humidity.

ice age: A period (typically many thousands of years long) when glaciers cover a greatly expanded part of Earth. When capitalized, Ice Age refers to the most recent event, which began roughly two million years ago and ended about 10,000 years ago.

ice fog: Ice crystals that form in very cold low-level air, obscuring visibility.

ice pellets: See **sleet**.

Indian summer: Colloquial term for a period of mild, dry weather in autumn that follows one or more significant frosts.

infrared: The wavelength of light just above

the visible range. See **ultraviolet**.

instability (atmospheric): Condition in which a lifted parcel of air will be warmer than its surroundings, thus allowing it to rise further (unless some other process intervenes).

intertropical convergence zone (ITCZ): Semipermanent band of converging trade winds and associated showers and thunderstorms that encircles the globe near the equator and shifts north and south with the sun's seasonal path.

inversion: Condition in which air warms overlays cooler air, thus making the lower layer extremely stable. Often a factor in big-city pollution.

ionosphere: The region within the upper atmosphere where intense solar radiation at low wavelengths strips electrons from atoms, producing ions. The ionosphere is collocated with the **thermosphere** and upper **mesosphere**.

iridescence: Rainbow-like patches of colour observed near the sun in high- and mid-level clouds.

isobar: Line that connects points of equal air pressure, typically after adjusting for altitude.

jet stream: One of several bands of strong upper-level winds that meander and encircle Earth at polar and subtropical latitudes. Within the jet stream are jet streaks, which have markedly stronger winds and can influence low-level storm development.

lake-effect snow: Often-intense bands of snow that form near large bodies of water, such as the US Great Lakes, due to instability and variations in friction over land and water.

land breeze: A sea breeze in reverse, typically blowing offshore late at night as the land cools more than the water.

landspout: Weak tornado that forms through a process similar to that of a **waterspout**.

La Niña: Periodic cooling (typically lasting 1–2 years and occurring roughly every 2–7 years) at and just below the surface of the eastern tropical Pacific Ocean. Sometimes refers to the global weather shifts that result.

lapse rate: Change in air temperature (normally a cooling) with height. For well-mixed air, the lapse rate is about 10°C/1000m or 5.5°F/1000ft. The overall global average is somewhat lower, around 6°C/1000m or 3.3°F/1000ft, due to inversions and other areas of relatively stable air.

latent heat (of condensation/freezing): Energy released when a molecule of water vapour condenses or a water droplet freezes.

lidar: Acronym for "light detection and ranging"; an instrument analogous to radar, but using lasers instead of microwaves to sense elements smaller than those detected by radar.

low-level jet: Localized, temporary core of intense winds, typically around 1.6km/1 mile above ground level, that may develop at certain seasons or in particular weather situations.

mammatus: See box p.415.

meridional flow: Winds blowing along a north–south axis, often bringing warmer- or colder-than-normal temperatures.

mesocyclone: Area of rotating winds that develops within many severe thunderstorms and from which a tornado may form.

mesoscale: "Middle-scale"; refers to weather systems, such as thunderstorms, that are larger than an urban area but smaller than an average-sized European country or US state.

mesoscale convective system (MCS): Large, persistent and organized complex of thunderstorms.

mesosphere: Atmospheric layer between the stratosphere and thermosphere, about 50–80km/30–50 miles in height, featuring temperatures that drop with altitude.

meteorology: The study of weather and, more generally, the atmosphere as a whole.

methane: A major greenhouse gas produced in large amounts by industrial sources as well as by termites, cows and rice paddies. Less prevalent than carbon dioxide, but more powerful per molecule in its greenhouse impact.

microburst: An intense **downburst** less than 4km/2.5 miles in diameter.

microclimate: The aggregate of weather observed in a local area, especially when landforms result in strong variations of climate across short distances.

mid-latitudes: The zone between the tropics and poles in each hemisphere; extends roughly from 30° to 60° N and S.

millibar: Unit of atmospheric pressure equal to 1 hectopascal or roughly 0.03 inches of mercury.

mist: Small airborne water droplets that reduce visibility to a lesser extent than fog.

model: See computer model.

monsoon: Annual cycle of wet and dry periods and associated air flow, driven mainly by large-scale differences in heating between land and ocean. Its phases are sometimes called the "winter monsoon" or "summer monsoon".

multicell storm: Thunderstorm composed of short-lived cells unlikely to produce prolonged severe weather.

nephoscope: Instrument used to measure cloud motion.

nimbostratus: See box p.415.

North Atlantic Oscillation: Irregular cycle of atmospheric pressure that spans the North Atlantic and influences weather across northeastern North America and Europe.

nowcast: Forecast that covers a very short time period (on the order of an hour), based largely on an extension of current conditions.

numerical weather prediction: Formal name for the use of mathematical and physical relationships, encoded in **computer models**, to generate weather forecasts.

occluded front: Narrow zone separating two air masses where a **cold front** has overtaken a **warm front**. The cooler of the two air masses may lie on either side of the occluded front.

orographic: Pertaining to higher terrain, as in mountain-induced wind or rain.

overshooting top: A small core of cumulus cloud visible above a thunderstorm **anvil** at the top of an intense updraught.

ozone (O_3): A three-atom molecule of oxygen; also a greenhouse gas. The ozone layer in the stratosphere shields Earth from harmful ultraviolet sunlight, while ozone generated near ground level is a pollutant.

ozone hole: Depletion in stratospheric ozone over and near Antarctica and, less dramatically, over and near the Arctic. The hole forms and disappears over a few weeks each spring, as sunlight returns and triggers ozone destruction.

Pacific Decadal Oscillation: Slowly varying cycle in near-surface water temperatures across the central North Pacific that influences climate in and near the region.

perihelion: The point at which a planet is closest to the Sun (see p.21).

permafrost: Ground at high latitudes that remains frozen year-round, sometimes with layers of unfrozen ground atop, below or nearby.

polar front: Semipermanent boundary that separates polar from mid-latitude air. Typically overlaid by the polar **jet stream**.

polar-orbiting satellite: Satellite that moves along a north–south axis, thus passing over both the North and South poles while Earth rotates beneath. See **geostationary satellite**.

precipitable water: Amount of water vapour and cloud droplets through the depth of the atmosphere above a point. Usually expressed as the depth that would accumulate if it all fell as liquid.

precipitation: Rain, ice, snow or other moisture falling to the ground.

pressure: Weight of the air above a given point. The pressure gradient is the difference in pressure between two points.

prevailing westerlies: General west-to-east flow that persists in both the northern and southern **mid-latitudes**.

profiler: Upward-pointing type of Doppler radar that measures wind speed and direction at various heights.

psychrometer: Instrument used to infer

atmospheric humidity by measuring wet-bulb temperature.

radiometer (atmospheric): Upward-pointing instrument that measures humidity and temperature by detecting the amount of radiation absorbed at various heights.

radiosonde: Balloon-borne instrument package that measures weather elements as it ascends.

relative humidity: Amount of water vapour in a parcel of air compared to the largest amount that could exist in that parcel (assuming the pressure and temperature did not change).

return flow: Poleward circulation of air that typically brings a warm-up after a cold air mass has passed by.

ridge: Elongated area of high pressure.

rime: Ice deposited from **supercooled droplets** within **fog** directly onto a surface.

Roaring Forties: Strong westerly winds that circle the globe across the expanse of open ocean between 40°S and 50°S.

saturation (of air with respect to water): The state of air at 100 percent **relative humidity**, when no more water vapour can be present without either **condensation** or **supersaturation**.

sea breeze: Wind that blows onshore in the afternoon as the land heats up more than the adjacent sea.

sleet: Mixed rain and snow (UK); rain or melted snow that freezes before it hits the ground (US). The latter are referred to as **hail** or **ice pellets** in the UK.

smog: Layer of ground-based pollution. Originally a conflation of "smoke" and "fog", the term is now used as a synonym for pollution even when little moisture is present.

solar constant: Amount of energy intercepted by an imaginary surface on the outer edge of Earth facing the Sun; roughly 1370 watts per square metre.

solar cycle: Regular peaks and intervening lulls in sunspots and other measures of solar activity, as generated by the reversal of the Sun's magnetic polarity about every eleven years.

solstice (winter and summer): The dates in late December and June on which winter and summer begin, respectively (the opposite applies in the Southern Hemisphere) when Earth's tilt is directly away from or toward the sun. The exact dates vary somewhat from year to year due to our imperfect calendar system.

sounding: Profile of the atmosphere at various heights, obtained through a radiosonde or other means.

Southern Oscillation: See **El Niño**.

squall: Prolonged, intense wind gust; technically, a sudden jump in wind speed of at least 29kph/18mph that produces a speed of at least 40kph/25mph and that lasts at least one minute, typically much less than an hour. Often associated with thunderstorms arranged in squall lines or with **tropical cyclones**.

stability: Condition in which a lifted parcel of air will be cooler than its surroundings, thus inducing it to sink back to its original level (unless some other process intervenes).

standard atmosphere: One of several vertical profiles of the atmosphere created as an average of the conditions observed across the world over time.

storm surge: Brief, sudden and often-devastating rise in ocean levels of up to 6m/20ft or more, produced by the winds of a **tropical** or **extratropical cyclone**. The low pressure within a hurricane or typhoon usually makes only a slight contribution to the storm surge.

stratocumulus: See box p.415.

stratus: See box p.415.

sublimation (of water vapour): Conversion of ice directly to water vapour without an intervening liquid stage. In dry climates, snow can sublimate and disappear without producing slush.

supercell: Long-lived thunderstorm with an organized structure and the potential to produce a tornado or other severe weather.

supercooled droplets: Water that cannot freeze at temperatures below 0°C/32°F due to the absence of an aerosol particle or other surface.

supersaturation: Air at a relative humidity above 100 percent due to a lack of surfaces for condensation. Since cloud particles are so prevalent, supersaturation is believed to be very rare and brief.

teleconnection: Relation between climatic patterns or weather events that occur at widely separated points, as with the tendency of an **El Niño** in the Pacific to be associated with **drought** across parts of Africa.

thermosphere: Thin, outermost layer of the atmosphere above the mesosphere, from roughly 80 to 650km or 50–400 miles. Temperatures increase with height in the thermosphere.

tornado: Violently rotating column of air extending to the ground from a severe thunderstorm.

tropical cyclone: General term for a closed centre of low pressure with tropical characteristics. Tropical cyclones whose winds exceed 119kph/74mph are named **hurricanes** in the Atlantic, Caribbean and northeast Pacific Ocean; **typhoons** in the northwest Pacific; and **cyclones** in the far southwest Pacific and Indian Ocean.

tropical depression: A centre of low pressure in the tropics whose winds are below 63kph/39mph.

tropical storm: An organized tropical system with winds between 63kph/39mph and 119kph/74mph – stronger than a **tropical depression** but weaker than a **hurricane**, **typhoon**, or **cyclone**.

troposphere: Lowest layer of the atmosphere, where most weather is generated. Its height is controlled by weather features and seasonal shifts; the average is close to 10km/6 miles over the poles and 16km/10 miles in the tropics.

trough: Elongated area of low pressure.

typhoon: See **hurricane** and **cyclone**.

ultraviolet: Wavelength of light just below the visible range; responsible for sunburn and other health effects.

Universal Time: Equal to Greenwich Mean Time; the global standard used in meteorology to synchronize observations. 1200 UT is equivalent to 7:00 AM Eastern Standard Time in New York; 0000 UT on the 13th of the month is the same as 7:00 PM EST on the 12th.

updraught: A rising parcel of air, especially within a thunderstorm.

virga: Precipitation that evaporates before reaching the ground. See also box opposite.

vortex: A region of closed circulation; often used as a synonym for **cyclone**.

vorticity: A measure of circulation around a low, calculated as a function of the wind's curvature and strength.

wall cloud: A lowered portion of a thunderstorm cloud base that may precede tornado formation.

warm front: Narrow zone denoting where a warmer air mass is flowing over a cooler one.

waterspout: Rapidly rotating column of air descending from a shower or thunderstorm to an ocean, lake or other body of water. Waterspouts and **landspouts** form through a similar mechanism and are usually weaker than the most intense **tornadoes**.

weather: The behaviour of the atmosphere at a point in time and space, especially as measured by instruments and experienced by humans. See **climate**.

wind chill: The effect of wind in intensifying the perception and actual effects of cold air.

zonal flow: Wind flowing along a west-to-east axis.

Cloud types

The ever-changing nature of clouds makes them ideal springboards for the imagination. That same quality, though, had scientists tussling for centuries on how to classify these evanescent creatures. Each one is "a small catastrophe, a world of vapour that dies before our eyes", according to writer Richard Hamblyn. It wasn't until 1802 that **Luke Howard**, an English lay scientist, came up with the basis for the durable nomenclature still used today. It adopts four Latin terms: **stratus** (stretched or extended), **cumulus** (heaped), **cirrus** (curly, fibrous) and **nimbus** (a generic term for cloud). Stratus clouds, which tend to cover broad areas, are produced when moist air is gradually lifted over large areas. The bubbles of cumulus result from stronger rising motion focused in small pockets. Cirrus and its ice crystals form at heights well above the freezing level, while nimbus clouds – whatever their height – are those that produce some form of precipitation.

Howard originally coined seven cloud types using these roots. Over the course of the nineteenth century, Howard's original cumulostratus became **cumulonimbus**, and three other types were added. The set of ten shown below was finalized by international agreement in 1896. Over a dozen other terms have been incorporated in this system, most of them subtypes of the ten biggies. Exceptions include **virga** – rain that evaporates as it falls to form a streaky cloud – and **mammatus**, the breast-shaped appendages that often hang from the tops of violent storms. These cloud types, as well as many of those shown below with their approximate base-level heights, are illustrated in the two colour sections of this book.

(6–16km/4–10 miles)

cirrocumulus
Patches of small, high clouds, usually ice crystals; often resembling the scales of fish.

cirrostratus
Sheets of ice crystals; typically thin enough for the sun or moon to shine through.

cirrus
Feathery wisps of ice crystals; also called mares' tails.

(3–10km/2–6 miles)

altocumulus
Patches of small clouds, usually arranged in rows or arrays.

altostratus
Thick sheets of crystals and/or droplets, often heralding a period of rain or snow.

(less than 4km/2.5 miles)

cumulonimbus
A thunderstorm cloud extending up to 16km/10 miles or higher, featuring cumulus-style features below and a cirrus anvil on top.

cumulus
Puffy, fair-weather clouds; known as **towering cumulus** if they build upward into middle levels.

nimbostratus
Stratus that produce rain, snow or other precipitation.

stratocumulus
Sheets of cloud with bubbles of cumulus-type development; especially common in summer over cool oceans.

stratus
Low, thick, grey clouds; known as **fog** at ground level.

Index

weather chart references are highlighted in colour